NORTH AMERICAN TUNNELING 2002

PROCEEDINGS OF THE NORTH AMERICAN TUNNELING CONFERENCE 2002,
18–22 MAY 2002, SEATTLE, WASHINGTON, USA

North American Tunneling 2002

Edited by

Levent Ozdemir

Colorado School of Mines, Golden, Colorado, USA

CRC Press is an imprint of the
Taylor & Francis Group, an **informa** business
A BALKEMA BOOK

Published by:
CRC Press/Balkema
P.O. Box 447, 2300 AK Leiden, The Netherlands
e-mail: Pub.NL@taylorandfrancis.com
www.crcpress.com – www.taylorandfrancis.com

ISBN 13: 978-90-5809-376-9 (hbk)

Visit the Taylor & Francis Web site at
http://www.taylorandfrancis.com

and the CRC Press Web site at
http://www.crcpress.com

North American Tunneling 2002, Ozdemir (ed.)
© 2002 Swets & Zeitlinger, Lisse, ISBN 90 5809 376 X

Table of contents

Session 2, Track 2 – Underground spaces and policies for infrastructure network

Session 2, Track 3 – Tunnel lining techniques

Session 3, Track 1 – Contracting practices

Session 3b, Track 2 – Policy and decisions about underground space

Session 3, Track 3 – Ground treatment and grouting techniques

Session 1, Track 1

Geotechnical baseline reports

North American Tunneling 2002, Ozdemir (ed.)
© 2002 Swets & Zeitlinger, Lisse, ISBN 90 5809 376 X

Rock mass conditions as baseline values for TBM performance evaluation

Gerald L. Dollinger, Ph. D.
Geologic Consultant – Underground Excavation, Bellevue, WA

John H. Raymer
Tunneling Geologist – Jordan, Jones and Goulding, Norcross, GA

ABSTRACT: Rock-mass conditions are a critical TBM performance parameter. In hard-rock tunnels, the size, spacing, and orientation of rock fractures can be more important than the rock strength. Yet, information about the fractures is seldom included in geotechnical pre-bid documents in a form that is useful for TBM boreability evaluations. The geotechnical investigations for the Chattahoochee Tunnel in Cobb County, Georgia included an extensive analysis of fractures based on the NTNU system. Fracturing factors (k_s) for the rock mass were developed from measurements of 2400 joints in the rock core. Baseline k_s values were included in the Contract Documents and were used as part of the Engineer's boreability estimate. The baseline k_s values increased the estimated penetration rate two to three times over what it would have been based on rock-strength alone.

1 INTRODUCTION

Accurate estimates of tunnel boring machine (TBM) performance are an important part of both a design engineer's and contractor's analyses of hard-rock tunneling projects. Without such estimates, it is impossible for owners or contractors to make realistic evaluations of the total time and cost required to complete a project. This results in taking high risks and often leads to costly overruns followed by lengthy and bitter disputes between owner and contractor.

The orientation and spacing of fractures are the most important parameters in estimating TBM performance in tunnels where the rock is very strong. In these situations, TBMs are limited by the thrust capability of the machine, rather than by the available cutter-head torque. Unfortunately, the potential effect of fractures on TBM performance is seldom evaluated in pre-bid geotechnical investigations in the United States.

The University of Trondheim, Norway (NTNU) has developed an empirical method for describing fractures and quantifying their effect on TBM performance (Bruland, 1998). In this method, the size, spacing, and orientation of the fractures are used to determine an index value called the "fracturing factor," abbreviated k_s. Low k_s values indicate wide fracture spacings and unfavorable fracture orientations, resulting in little benefit to TBM performance; high k_s values, on-the-other-hand, indicate close fracture spacings and favorable fracture orientations, resulting in higher TBM performance. In this context, it is important to note that TBM performance refers only to instantaneous penetration rate, which is the cutter penetration that occurs at a given average cutter load. It is also important to note that very high k_s values indicate a highly fractured rock and probably difficult ground. In this situation, the TBM will probably be torque limited and the overall advance rate will be limited by the time required to install rock support.

Fracturing factors (k_s) were estimated for the Chattahoochee Tunnel in Cobb County, Georgia, as part of the pre-bid geotechnical investigation. These factors were included as baseline values in the pre-bid documents, and were used by the Engineer for making performance assessments for use by the Owner. This report describes how the k_s values were determined from the fractures observed in the cores and how the k_s affected the Engineer's assessment of TBM performance.

2 PROJECT DESCRIPTION

The Chattahoochee Tunnel is a deep sewer tunnel located in Cobb County (metro Atlanta), Georgia. It is 15,125 meters long, has an excavated diameter of 5.5 meters, and ranges in depth from 28 to 108 meters below ground surface. The tunnel is being excavated up slope from south to north in two consecutive drives. The tunnel is owned by the Cobb County Water System. The geotechnical work on which this paper is based was performed by the tunnel designer Jordan, Jones, & Goulding, Inc. The tunnel is currently under construction by Gilbert/Healy, LP and is scheduled for completion in 2004.

The geotechnical investigation included geological field mapping, 50 core borings, geotechnical analysis of the core, and laboratory testing of rock properties. The results were reduced, interpreted with respect to tunneling, and presented in the contract documents as the geotechnical baseline for the project. An Engineer's estimate of TBM performance was performed by Dr. Gerald Dollinger using the reported baseline values. Although not part of the contract documents, this report was made available to the Contractor as additional information.

2.1 *Rock conditions*

The ground along the Chattahoochee Tunnel consists of medium grade metamorphic rock overlain by 5 to 50 meters of progressively weathered rock and residual soil. The bedrock consists of about 65 percent quartz-feldspar gneiss and 20 percent quartz-mica schist, with the remainder consisting of amphibolite, granite, quartzite, and other rock types. Most of these rocks are strongly foliated, making them anisotropic. The compositional layering of the various rock types typically parallels the foliation.

The rock along the Chattahoochee Tunnel is strong, with an average unconfined compressive strength (UCS) of approximately 220 MPa normal to foliation, which is the strong direction. In the gneiss, amphibolite, and quartzite, the rock is approximately two times stronger normal to foliation than parallel to it. In the schist, the rock is generally greater than four times stronger normal to foliation than parallel to it.

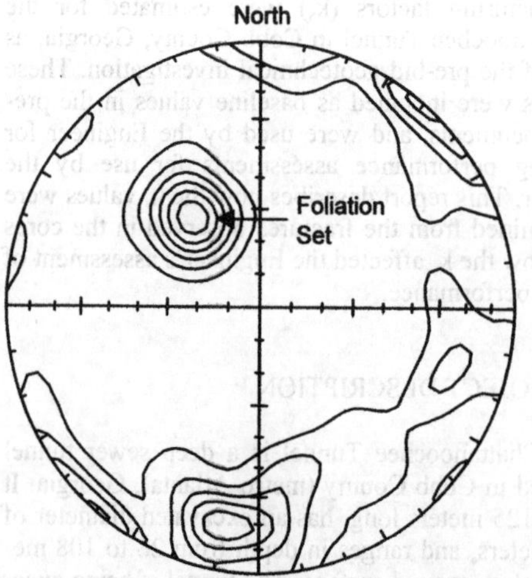

Figure 1. Stereonet of fractures based on field mapping. Contours indicate relative abundance.

2.2 *Fracture patterns*

The fracture pattern for the tunnel was established using geologic field mapping. Figure 1 shows that the fractures are concentrated in several sets. The most common fracture set forms along the foliation because it is the principal plane of weakness in the rock. The foliation set has a fairly consistent orientation along the tunnel alignment, with an average strike of N.52°E., and an average dip of 35 degrees SE. At a smaller scale, however, the foliation is observed to undulate, with variations in dip and strike typically in the range of ±20 degrees.

3 FRACTURING FACTOR ANALYSIS

Fracturing factor (k_s) is an index value used to quantify the benefit that fractures provide to TBM performance. Fracturing factors are estimated from the size, spacing, and orientation of the rock fractures relative to the tunnel axis. The fracturing factor concept was developed by NTNU and is described in Bruland (1998). The NTNU approach is an empirical method based around post-excavation tunnel mapping. This approach was successfully adapted for use with rock cores during the pre-bid geotechnical investigation for the Chattahoochee Tunnel.

3.1 *Basic concepts*

Fracturing factors (k_s) are calculated for each fracture set. A fracture set is a group of fractures that are approximately parallel to one another. The contoured peaks on Figure 1 indicate that at least three major fracture sets occur regularly along the Chattahoochee Tunnel. The k_s value of each fracture set is determined from Figure 2 based on the size, spacing, and orientation of the fractures in the set. Based on this graph, the k_s value for a fracture set can range from a base value of 0.36 for massive rock (St Class 0), to greater than 4 for highly fractured rock (St Class IV). The total fracturing factor for the rock mass ($k_{s\text{-tot}}$) is equal to the sum of the k_s values for each set of fractures in the rock, minus the base value of 0.36 for each set beyond the first (i.e. the sum of all k_s values, minus $0.36 \times (n\text{-}1)$, where n is the total number of fracture sets analyzed).

The fracture size is defined relative to the size of the tunnel. Fractures that completely cross the tunnel are classified as "Sp" fractures. Smaller fractures are classified as "St" fractures. While this distinction is practical for post-excavation tunnel mapping, it is impractical for core analysis because the cores are too small to distinguish fracture size at tunnel scale. In the Chattahoochee Tunnel investigation, all fractures were conservatively classified as St fractures, because smaller fractures impart less benefit to TBM performance.

Fracture spacing is defined in the NTNU system as classes designated by Roman numerals. Each class has a single spacing value, as opposed to covering a range. For the Chattahoochee Tunnel, a spacing range was defined around each class so that all fracture spacings would fall clearly into one class or another. The NTNU classes and how they were adapted for the Chattahoochee Tunnel are listed in Table 1.

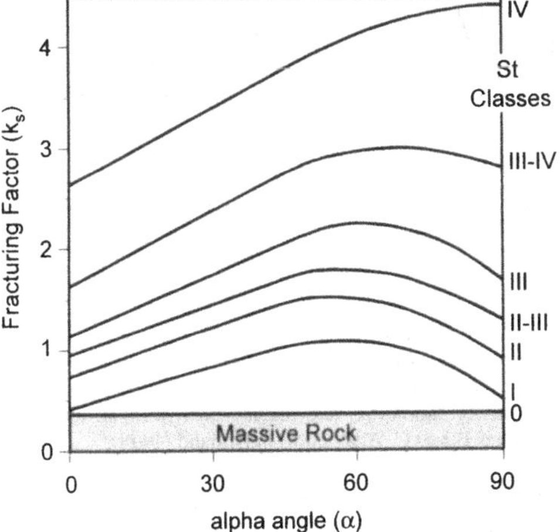

Figure 2. Diagram for estimating fracturing factor from alpha angle and fracture spacing (St class). Adapted from Bruland (1998, Fig. 2, 3).

Table 1. Fracture class definitions used for the Chattahoochee Tunnel. Adapted from Bruland (1998)

Fracture Class	NTNU (cm)	Chattahoochee Tunnel Spacing (cm)
0	massive	greater than 200
0-I	160	200 to 100
I-	80	100 to 50
I	40	50 to 25
II	20	25 to 12.5
III	10	12.5 to 6.3
III-IV	5	6.3 or less

Fracture orientation is defined relative to the tunnel axis in terms of the "alpha angle." The alpha angle is the minimum angle between the fracture plane and the tunnel axis and is calculated as:

$$\alpha = \arcsin(\sin \alpha_f - \sin(\alpha_t - \alpha_s)) \quad (1)$$

where α_s is the strike of the fracture surface, α_f is the dip of the fracture surface, and α_t is the strike of the tunnel axis.

As shown by Figure 2, k_s increases as fracture spacing decreases and the alpha angle approaches about 60 degrees, depending on the St Class. Widely spaced fractures oriented parallel to the tunnel axis ($\alpha = 0°$) provide the least benefit to TBM performance, and closely spaced fractures oriented around 60 degrees or higher to the tunnel axis provide the most benefit.

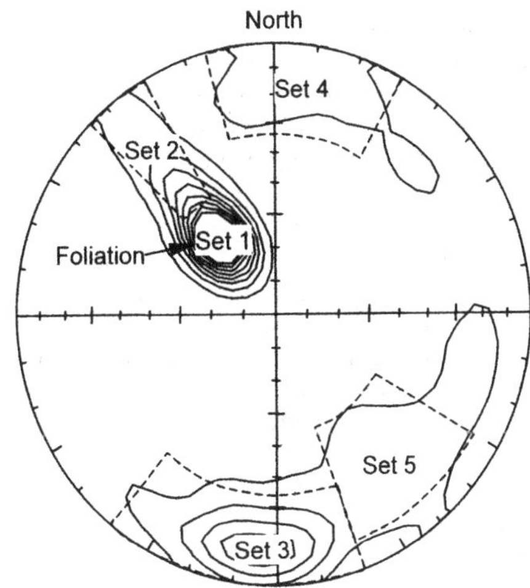

Figure 3. Contoured stereoplot of fractures measured in cores. Dashed areas indicate fracture sets specifically analyzed for k_s.

3.2 *Chattahoochee Tunnel analysis*

The k_s values for the Chattahoochee Tunnel were based on detailed orientation measurements of over 2400 fractures from the rock cores along the alignment. These fracture orientations were first plotted and statistically contoured on a stereonet (Figure 3). Based on this stereonet plot, five principal fracture sets were identified and separate k_s values were estimated for each set. These k_s values were then summed using the method as described above to obtain $k_{s\text{-tot}}$ for the tunnel.

The orientation of each fracture plane was estimated in real-world coordinates as strike and dip. The estimate of dip was made by measuring the angle between the fracture plane and the axis of the rock core. Since the cores were assumed to be vertical, the dip was equal to 90 degrees minus that angle. The strike of the fracture plane was estimated by measuring the rotation angle of the fracture. This angle is the clockwise angle around the core axis between the down-dip direction of the fracture plane and the down-dip direction of the foliation plane. Since the average strike of foliation was known from the field mapping as N.52°E., the estimated strike of each fracture plane was equal to N.52°E. plus the rotation angle.

Once the fracture orientations were estimated, each fracture was weighted by the secant of its dip angle. This was done to account for the bias of vertical boreholes against steeply dipping fractures. The close correspondence of Figure 3 to Figure 1 shows that using the foliation for orienting the cores was valid from a statistical perspective.

Fracture spacing was evaluated separately for each fracture set. The spacing was first measured as the distance parallel to the core axis between two adjacent fractures in a set. The core distances between adjacent fractures in a set were then multiplied by the cosine of the average dip angle of the two adjacent fractures to obtain the true, perpendicular spacing between the fractures. These true spacings were then accumulated in the appropriate fracture classes as defined in Table 1. The result of this analysis is shown in Table 2 as the percentage of rock in each NTNU spacing class. This procedure was repeated for each of the five fracture sets.

Table 2. Percentage of rock in each fracture class per set

St Class	Set 1	Set 2	Set 3	Set 4	Set 5
O	62.5%	89.8%	62.4%	74.1%	83.1%
O-I	8.1%	6.5%	13.0%	12.4%	9.1%
I-	5.8%	1.8%	12.2%	8.2%	4.0%
I	10.6%	1.1%	7.0%	2.7%	2.1%
II	8.3%	0.5%	3.7%	1.6%	0.6%
III	3.0%	0.1%	0.9%	0.6%	0.4%
III-IV	1.7%	0.1%	0.9%	0.4%	0.8%

The k_s values for each St class in each fracture set were obtained using Figure 2. The k_s value for the set was taken as the average of the k_s values for each St class, weighted by the percentage of rock in each St class, as listed in Table 2. The $k_{s\text{-tot}}$ for the tunnel was then calculated as the sum of the k_s values for each set, as described previously.

4 EFFECT ON PERFORMANCE ESTIMATES

The effect of increasing the fracturing factor on TBM penetration rates in hard rock tunnels is illustrated in Figure 4. This beneficial effect is especially marked in stronger rock that is lightly to moderately fractured. Under these conditions, the penetration rate is limited mainly by the amount of thrust that can be applied to the cutters. The presence of fractures improves TBM performance because the fractures are planes of weakness in the rock. These planes of weakness assist the crushing and chipping of the rock that occurs beneath and adjacent to the cutters during excavation.

The actual improvement expected from changes in the fracturing factor is shown in Figure 4 for three different rock strengths. This figure shows that relatively small increases in k_s will result in very large increases in instantaneous penetration rate. The percentage increase is especially large between k_s 0.36 and 1.4, where the curves are steep.

Fracturing factors between 0.36 and 1.4 represent the transition from massive rock into highly fractured and possibly blocky rock. In massive rock (k_s 0.36), there are not enough fractures to assist in

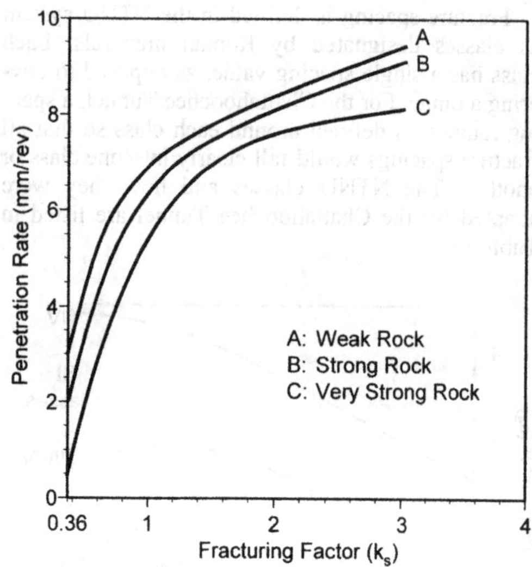

Figure 4. Effect of fracturing factor on penetration rate. Shale, limestone, and low grade metamorphic rocks are typical of rocks between lines A and B. Medium to high-grade metamorphic rocks, granodiorite, and diabase are typical of rocks between lines B and C. Adapted from Bruland (1998).

TBM performance. In highly fractured or blocky rock (e.g. k_s 1.4 and higher), the many fractures can create difficult ground conditions where overall TBM advance may be limited by the time needed to install extensive support.

Closer inspection of the curves shows that increases in k_s has a greater beneficial effect in stronger rock than in weaker rock. Increasing the k_s value from 0.36 to 0.50, for example, results in a 350 percent penetration rate increase for the very strong rock, while the corresponding increase for the weak rock is only about 150 percent. Rock with a k_s of 0.5 is not likely to be blocky or difficult.

Similar dramatic changes in the estimated instantaneous penetration rates for the Chattahoochee Tunnel were obtained when the fracturing factors from core analysis were considered. Because the strength of the Chattahoochee Tunnel rock is strong to very strong, the k_s values derived from the core analysis resulted in substantial increases in the expected instantaneous penetration rates over the instantaneous penetration rates that had originally been estimated for massive rock conditions ($k_{s\text{-tot}}$ = 0.36). This increase in penetration rate was on the order of two to three times for much of the tunnel.

5 CONCLUSIONS

Fractures in the rock have a major effect on TBM performance in hard rock tunnels. This effect is most pronounced when the rock is very strong. In weak rock where the machine is torque limited, fractures provide less benefit and may reduce the overall TBM advance rate due to the time required to install support.

Because of the effect of fractures on TBM performance, pre-bid geotechnical documents should contain baseline descriptions of the rock-mass conditions that are relevant to estimating TBM performance. One method by which this can be done is to estimate NTNU fracturing factors (k_s) for the rock mass. These factors are a function of the size, spacing, and orientation of the fractures in the rock mass relative to the tunnel axis, and can be used with various predictor models for estimating TBM penetration rates. The NTNU method was adapted for the Chattahoochee Tunnel and used to develop baseline fracturing factors for the project. This is believed to be the first pre-bid application of this method in the United States.

The method used for the Chattahoochee Tunnel followed the NTNU approach as closely as possible. This method was labor intensive and involved individual orientation measurements of thousands of fractures. The dip of each fracture was determined by measuring the angle between the fracture plane and the core axis. The strike of each fracture was obtained by orienting the core using the trend of the regional foliation. While orienting the core based on foliation worked well for the Chattahoochee Tunnel project, it was wholly dependent on the consistent orientation of foliation along the alignment. Having a reference within the rock to orient the core is rare and cannot be expected for most projects. To adapt the Chattahoochee Tunnel method for general usage, another method of core orientation would be required, such as drilling oriented core.

The authors are currently developing simpler methods for estimating fracturing factor values from rock core. These methods are expected to be less labor intensive and do not require oriented core, which is very costly. These new methods should make estimating baseline fracturing factors much more practical, putting them well within the scope of most tunneling projects.

REFERENCE

Bruland, Amund 1998. *Hard rock tunnel boring: Advance rate and cutter wear.* Trondheim: Norges teknisk-naturvitenskapelige universitet.

North American Tunneling 2002, Ozdemir (ed.)
© 2002 Swets & Zeitlinger, Lisse, ISBN 90 5809 376 X

Geotechnical characterization for a 34-mile long hard rock tunnel in San Diego, California

M.E. Schmoll & D.J. Young
URS Corporation, Oakland, California

D.L. Schug
URS Corporation, San Diego, California

ABSTRACT: The San Diego County Water Authority is completing feasibility-level planning studies for a new water conveyance system to transport Colorado River water from the Imperial Valley to San Diego. The geologic and geotechnical conditions of two alternative pipeline routes have been evaluated by URS Corporation consisting of large diameter open-cut pipelines and tunnels. One route would lift water 3,800 feet in elevation over the Peninsular Ranges while the other route includes a 34-mile long tunnel through the mountains, resulting in a lift of only 1,500 feet.

The 34-mile long tunnel would pass through the hard crystalline granitic, and metamorphic rocks of Mesozoic age collectively known as the Peninsular Range Batholith. Key geotechnical design issues for the tunnel, which has up to 4,900 feet of cover, include the potential for high pressure, high volume groundwater inflows and the potential surface dewatering related impacts, high ambient temperatures at tunnel depth due to the natural geothermal gradient and high in situ stress requiring special ground support. Major challenges for the project included the prediction of the rock mass conditions at tunnel depth utilizing core borings that were typically less than 300 feet in depth and the development of a groundwater model to estimate the length of pre-excavation grouting to mitigate impacts to the groundwater resources and for constructability.

1 PROJECT DESCRIPTION

The San Diego County Water Authority (Authority) is studying the feasibility of constructing a water conveyance facility between Imperial Valley and San Vicente Reservoir in San Diego County. The water currently delivered by the Authority is imported to southern California through the State Water Project facilities and the Colorado River Aqueduct to Riverside County and is then transported via large diameter pipelines operated by the Metropolitan Water District of Southern California and the Authority to San Diego County. More than 90 percent of San Diego County's water needs are delivered through the aqueducts. The Authority is studying options to diversify its water supply to ensure reliability and to meet the region's future needs.

Two alternative alignments (designated the 5A and 5C alignments) are being considered as shown on Figure 1. These two alignments were screened out of several alternatives alignments identified in a previous conceptual study completed in 1996. The 1996 study did not include any subsurface investigations.

Both alignments begin at Drop No. 1 on the All-American Canal and end at the San Vicente Reservoir near the eastern margin of the San Diego metropolitan area. The first approximate 46 miles of both alignments would consist of a new gravity flow concrete-lined canal parallel to the existing All-American Canal, which parallels the U.S./Mexico border. An additional 12 miles of concrete-lined canal parallel to the north-south trending Westside Main Canal would be required for the 5C alignment.

Figure 1. Project Location Map.

From a new forebay pump station adjacent to either the Westside Main Canal for the 5A alignment or the All-American Canal for the 5C alignment, the conveyance system would enter a pressurized pipeline/tunnel to the terminus point at San Vicente Reservoir. The pressurized section of the 5A alignment would involve predominantly tunnel construction, whereas the pressurized section 5C alignment would use predominantly open-trenched construction. Both alignments share two common tunnel segments totaling about 7 miles in length between a portal just upstream of El Capitan Reservoir to San Vicente Reservoir (see Figure 1). The total lengths of the 5A and 5C alignments are approximately 129 and 139 miles, respectively including canals. Only the pressurized portions of the alignments were evaluated as part of the geotechnical evaluations by URS Corporation. The results of these evaluations were then used by the Authority's consultant, Black & Veatch to update the conceptual cost estimates developed during the 1996 study. Water transfer volumes of 300,000 acre feet/year (AF/yr), 400,000 AF/yr and 500,000 AF/yr were evaluated.

The key geologic and land use criteria used to lay out the alignments in the 1996 study include the following:

1 Avoid crossing active faults in tunnel
2 Avoid areas of known geothermal activity in tunnel
3 Avoid Indian Reservations and Federal Wilderness areas
4 Minimize impacts to the groundwater resources
5 Minimize surface disruptions in Forest Service and BLM lands
6 Minimize impacts to environmentally sensitive areas

The pressurized section of the 5A "tunnel" alignment starts at about elevation –30 feet at a new forebay at the Westside Main Canal and extends westward in a cut-and-cover pipeline approximately 30 miles to the eastern front of the Peninsular Ranges in the Anza Borrego Desert. Two pumping stations would be required between the forebay and the tunnel portal to lift the water a total of 1,500 feet. The pipeline would then enter a tunnel at about elevation 970 feet and extend for approximately 34 miles before portaling out into a short cut-and-cover segment across a canyon just upstream of El Capitan Reservoir. The pipeline then enters a tunnel again for about 3.5 miles to a 0.5 mile cut-and-cover segment across the San Diego River Valley where it would again enter a tunnel again for about 3.6 miles to the end point at San Vicente Reservoir at an elevation of about 700 feet.

1.1 *Alignment consideration*

During the 1996 study the 34-mile 5A tunnel alignment included four deep construction shafts spaced at about 6 to 7 mile intervals. These shafts were included to provide multiple points of access to the deep tunnel to keep individual tunnel drives less than about 7 miles in length. However, the cost estimate for the 1996 study showed that these deep shafts (up to 3,600 feet deep) were about one-third of the total tunnel construction cost and took up to 3 years of time to construct. Preliminary evaluations at the start of the current study showed that at least two of the construction shafts could be eliminated by doing longer tunnel drives. Only two shafts were considered for the current study; a 3,080-foot deep shaft and a 1,600-foot deep shaft at the approximate one-third points along the tunnel alignment. Alternatives to the shafts were also evaluated including the construction of decline tunnels from the ground surface to tunnel depth. However, the length of the declines were several miles long since the cover remains consistently high north and south of the alignment and the cost of these decline tunnels were considered to be more than the cost of the shafts.

Once the tunnel cost and schedule estimates for the project were underway, it became evident that a tunnel mined from the east portal would reach the 3,080-foot deep shaft before the shaft could be excavated to the tunnel depth. Similarly, a tunnel drive starting at the west portal would also nearly reach the 1,600-foot deep shaft before this shaft could reach tunnel depth. Another issue that became apparent as the cost estimates were developed was related to the cost of pumping tunnel groundwater inflows out of the deep shafts. The power requirements and size of the piping to lift the groundwater (estimated to be up to 18,000 gpm) out of the shafts were viewed as a major obstacle. Based on these issues, inquiries were made to TBM manufactures and reviews were made of similar tunnel case histories as to the maximum feasible length of a tunnel drive from a single portal access point. Elimination of the shafts would result in two long drives for the 34-mile tunnel; one heading east from the portal near El Capitan Reservoir and one heading west from the Anza Borrego portal. There were no apparent reasons why tunnel drives of this length could not be completed as long as the ventilation requirements could be maintained, the tunnel muck could be efficiently removed from the heading and groundwater inflows could be removed from the tunnel. Based on these considerations, it was decided to eliminate the two remaining construction shafts and plan the construction of the long 5A tunnel from the two portals. To accommodate the muck removal and ventilation requirements, the size of the tunnel excavation was increased to 15-foot diameter to allow for a conveyor belt muck removal system and to accommodate the large piping for removal of groundwater. Revisions to the tunnel grade were also made so that both tunnel segments are mined upgrade to facilitate removal of the groundwater from the tunnel heading. A permanent vent shaft would also be required at the middle high point of the tunnel to provide passive venting during filling and draining of the tunnel. This vent shaft could be constructed with raised-bore methods.

1.2 *Regional geology and seismicity*

The 5A alignment extends across two major physiographic and geologic provinces: 1) the Peninsular Ranges of southern and Baja California, and 2) the Salton Trough within the Imperial Valley. The Peninsular Ranges are comprised of igneous and metamorphic rock of Mesozoic age collectively known as the Peninsular Range Batholith (PRB). The Salton Trough is a deep sedimentary basin containing Tertiary and Quaternary age sedimentary deposits. The 34-mile long 5A tunnel is entirely within the rocks of the PRB.

The rock units making up the PRB comprise a westward tilted block, which is bounded along its eastern edge by the active Elsinore fault. The east portal of the 5A tunnel is located west of the Elsinore fault. There are no major active faults cutting the PRB to the west of the Elsinore fault, therefore the PRB block has been tectonically stable as the terrain is not extensively faulted, especially compared to the Salton Trough (which contains the active San Jacinto and San Andreas Fault systems).

The igneous rocks within the PRB consist of several plutons, or large rock bodies with similar minerology and cooling history. Within the 5A alignment, the rocks comprising these plutons are largely of granitic composition with the predominant minerals consisting of quartz and feldspar, with lesser biotite, hornblende and pyroxene. From east to west, mapped granitic plutons along the 5A tunnel include the: La Posta Pluton, Cibbett Flat Pluton, Pine Valley Monzogranite, and the Alpine Tonalite as shown on Figure 2. Several gabbro plutons would also be encountered in the tunnel. The gabbro is devoid of quartz and consists chiefly of plagioclase feldspar, pyroxene and olivine. Gabbro plutons along the SA tunnel include Guatay Peak, Poser Mountain, and Viejas Mountain.

Portions of the 5A tunnel will pass through metamorphic rocks of the Julian Schist. These rocks predate the PRB, and represent ancient sedimentary rocks disrupted and metamorphosed by emplacement of the plutonic rocks. The Julian Schist contains an assemblage of metamorphic rocks ranging from mostly quartzite and gneiss in the western PRB, with marble and schist making up a large percentage in the eastern PRB. The more extensive exposures of the Julian Schist are assumed to have near vertical boundaries and extend to the tunnel depth based on the vertical contacts observed at the ground surface; smaller, less extensive surface outcrops of the Julian Schist may not extend to the depth of the tunnel.

The portion of the tunnel between the Cibbett Flat and Pine Valley Plutons would pass through a belt of granitic rocks with metamorphic texture described as the Cuyamaca-Laguna Mountain Shear Zone (CLMSZ) (Walawender, 2000). The rocks within the CLMSZ are interpreted to have experienced ductile deformation, rather than brittle crustal faulting. Therefore, the rock mass within the CLMSZ is not anticipated to be highly fractured or sheared as would be expected within a brittle fault zone.

The 5A tunnel will intersect several regional linear landforms recognizable on very high altitude photographs. These regional lineament features are thought to be expressions of rock foliation, fracture zones and inactive faults that likely extend to tunnel depth. The major regional lineaments crossing the tunnel are likely to represent highly fractured and sheared rock and potential sources of groundwater inflow to the tunnel.

2 GEOTECHNICAL INVESTIGATIONS

Geotechnical investigations completed to provided a preliminary characterization of the ground conditions along the 5A tunnel included the following:

1 Review of previous investigations and relevant tunnel construction case histories
2 Review of remote sensing data and aerial photographs
3 Geologic mapping
4 Rock core borings
5 Hydraulic conductivity (packer) testing
6 Down-hole geophysical surveys, and
7 Laboratory testing

A total of 12 HQ-wireline core borings were completed along the 5A tunnel alignment to investigate the rock mass conditions and to obtain samples for laboratory testing. The inclination of the borings ranged from 60 to 90 degrees (from horizontal). The borings were completed at portal locations and along the tunnel alignment in an effort to obtain representative rock samples of the major plutons. The depths of the borings were typically 200 to 300 feet, or at least 100 feet into the unweathered rock. However, one boring was drilled to a depth of 1,000 feet into the CLMSZ and one boring was drilled to a depth of 1,750 feet in a relatively low cover area, reaching tunnel depth. All borings were completed as open-hole piezometers and were monitored over a period of about 6 months before being abandoned by backfilling them with cement grout.

Hydraulic conductivity (packer) testing was completed in all core borings at approximate 20-foot intervals for the entire length of the boring below the weathered rock. The packer testing was completed using either a single packer as the boring was advanced, or more typically using a double straddle packer at the completion of drilling. A pressure

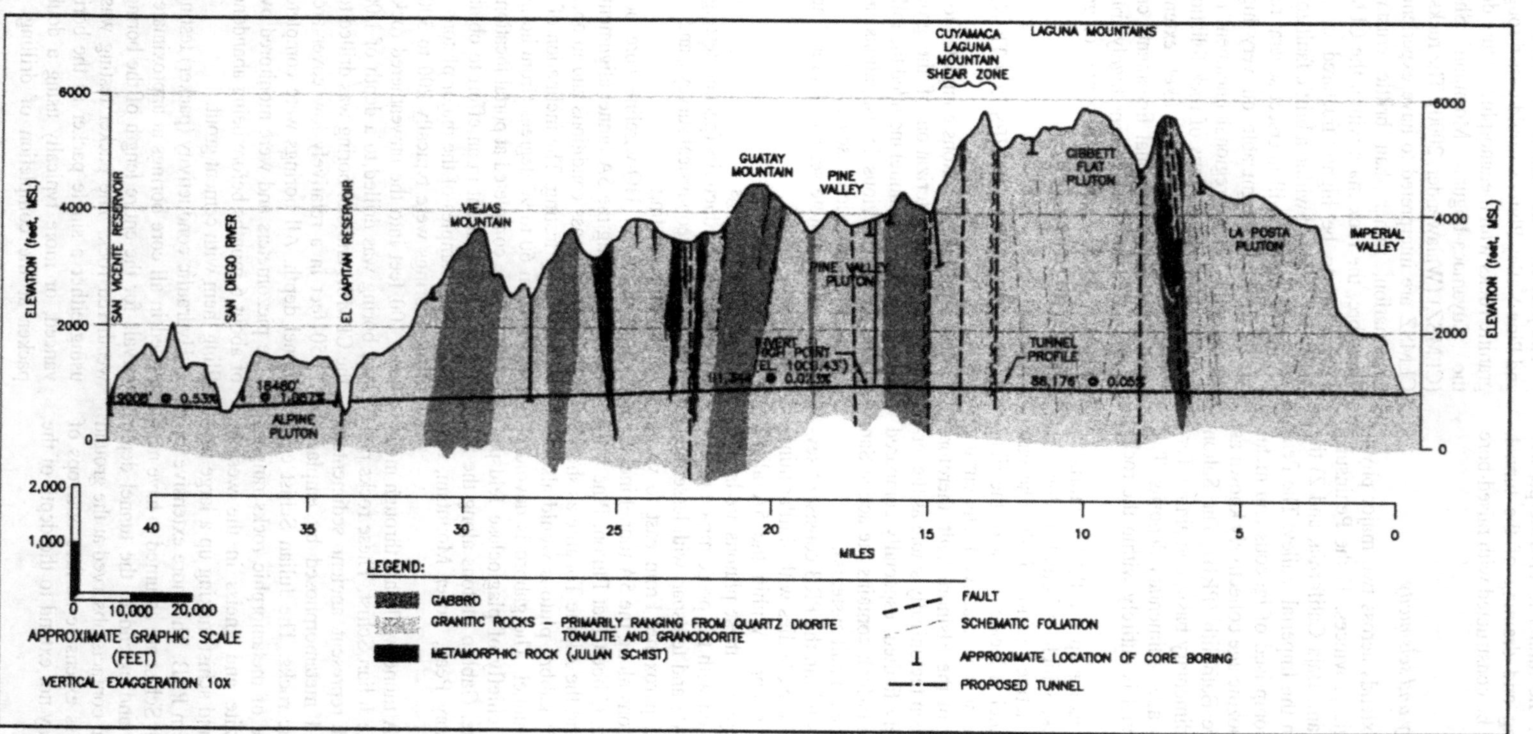

Figure 2. Geologic profile along 5A tunnel.

transducer was used monitor water pressure within the packer test zone to eliminate the need to calculate friction losses and to insure the packers were sealed. A total of 214 packer tests were completed in the 12 core borings. Hydraulic jacking tests were also performed in the low tunnel cover areas near the portals to evaluate in situ stress conditions and for making estimates of the required length of steel lining.

Down-hole geophysical surveys were conducted in all the core borings. The surveys included continuous optical televiewer logging to characterize the in situ frequency and orientation of rock discontinuities such as rock foliation, fractures and shear zones. Temperature logging was also completed in the two deep core borings to measure the local geothermal gradient.

Laboratory testing on the recovered rock core samples included point load testing at approximate 10-foot intervals, unconfined compression tests and microscopic petrographic (thin section) examinations. Representative samples from each of the major rock types were also sent to the Earth Mechanics Institute at the Colorado School of Mines for TBM performance testing, which included Brazillian Tensile strength, Punch Penetration Response, Cerchar Abrasivity Index, unconfined compression and petrographic analysis.

3 GROUND CHARACTERIZATION

3.1 *Approach to characterization*

The goal of the ground characterization program was to subdivide the range of anticipated conditions along the project alignment into segments with uniform characteristics related to penetration rate, tunnel stability, and groundwater inflow. Criteria for subdividing the tunnel included the plutons identified by surface mapping or subdivision by rock type or minerology. Rock strength data showed less scatter when grouped by rock type than by pluton, so the alignment was divided into 20 segments (18 tunnel segments) based on rock type as shown on Table 1. The next step in the characterization process involved summarizing rock, discontinuities and groundwater properties and design values for evaluating tunneling requirements in each of the 18 tunnel segments. Because groundwater inflow, rock quality and rock stress conditions are related to cover, this data was summarized for each segment in Table 1.

Igneous rocks were named according to the International Union of Geological Sciences (IUGS) classification scheme. If significant metamorphic texture was present, a metamorphic rock name was used, following metamorphic rock name convention published by Compton (1985).

Rock and discontinuity properties were assigned to each segment including: rock type, lineaments crossed, rock strength (point load, UCS, Brazilian Tensile, Punch Penetration), Cerchar Abrasivity, percent hard minerals, fracture frequency, degree of foliation and the orientation of predominate joint sets relative to the tunnel drive.

3.2 *Rock mass properties*

3.2.1 *Rock types*

Based on visual and laboratory analysis of core samples from the borings and results of the geologic mapping, six predominant igneous rock types and one metamorphic rock type were identified along the tunnel alignment. The rock types are:

1 granodiorite,
2 mafic tonalite,
3 tonalite,
4 monzogranite,
5 gabbro,
6 diorite; and
7 the metamorphic rock type is gneiss.

All of the crystalline rocks are strong to very strong and contain relatively high percentages of quartz with the exception of the gabbro and diorite, which contained little or no quartz. Quartz content for the rocks ranged from about 15% for the mafic tonalite to 33% for the monzogranite.

3.2.2 *Surface weathering*

One of the key challenges for the project included the prediction of the rock mass conditions at tunnel depth utilizing core borings that were typically less than 300 feet in depth. The depth of rock mass weathering ranged from a little as 15 feet to as much as 175 feet with an average depth of 40 to 60 feet.

3.2.3 *Rock strength*

Below the weathered zone the rock generally appeared uniform in strength and degree of weathering except within shear zones, The unconfined compressive strength of the rocks ranged from a low of about 10,000 psi to over 44,000 psi. Figure 3 shows the range and average unconfined compressive strength of rocks tested.

In order to evaluate rock strength verses depth and whether potential stress relief during coring could have disturbed the samples, the stress-strain curves obtained during unconfined compressive strength testing were reviewed and the results of the point load tests were plotted for the 1,000 and 1,750-foot deep borings. Figure 4 shows the point load index verses depth for the 1,750 boring. Based on this plot and a review of the unconfined compression tests completed at various depths in this boring, there does not appear to be any noticeable stress relief of the core nor a change in rock strength with depth.

Table 1. 5A Tunnel Corridor Summar Table.

Segment	Length (mi.)	No. of Lineaments	Rock Name	Appx. Tunnel Elevation (ft.)	Ground Cover Elevation (ft.)			Average Groundwater Elevation in Borehole (ft.)	Average Tunnel Depth Below Groundwater (ft.)
					Max	Min	Avg.		
1A	2.3	5 minor	Granodiorite	1000	2900	1000	2000	1980	980
1B	4.3	15 minor	Granodiorite	1000	4550	2900	4000	3980	2980
1C	1.0	5 minor	Granodiorite	1000	5700	4550	5100	5080	4080
2	0.8	2 major 1 minor	Gneiss	950	5700	4750	5300	5280	4330
3	6.1	4 major 14 minor	Mafic Tonalite	900	5800	4750	5400	5350	4450
4	0.9	3 minor	Tonalite	900	4900	4000	4500	4490	3590
5	0.2	1 minor 1 major	Mixed Monzogranite and Mafic Tonalite (1/2 of each)	900	4100	4100	4100	4090	3190
6	1.1	4 minor 1 major	Gabbro	900	4300	4000	4200	4170	3270
7	2.6	4 minor 1 major	Tonalite	900	4000	3700	3900	3880	2980
8	1.3	1 minor	Mixed Gabbro, Tonalite, and Gneiss (1/3 of each)	900	4500	3900	4200	4130	3230
9	1.1	none	Gabbro	880	4500	4000	4300	4230	3350
10	4.8	8 minor 2 major	Mixed Gneiss and Granodiorite (1/2 of each)	850	3900	3200	3500	3450	2600
11	0.9	none	Gabbro	850	3600	3200	3500	3460	2610
12	1.0	1 minor	Mixed Granodiorite, Tonalite, and Diorite (1/3 of each)	830	3100	2500	2800	2760	1930
13	2.0	4 minor	Gabbro	830	3600	2500	3200	3160	2330
14A	1.8	2 minor	Tonalite	800	3400	1900	2800	2760	1960
14B	1.8	1 major	Tonalite	800	1900	1100	1500	1460	660
15	0.1		OPEN CUT PIPELINE	800	1100	1100	1100	1060	260
16	3.5	10 minor	Tonalite	700	1500	1700	1300	1250	550
17	0.5		OPEN CUT PIPELINE	600	700	600	600	560	-40
18	0.6	1 minor	Tonalite	620	1400	600	1000	950	330
19	0.5	1 minor	Gneiss	650	1400	1200	1300	1250	600
20	2.5	10 minor	Tonalite	700	2000	1000	1500	1450	750

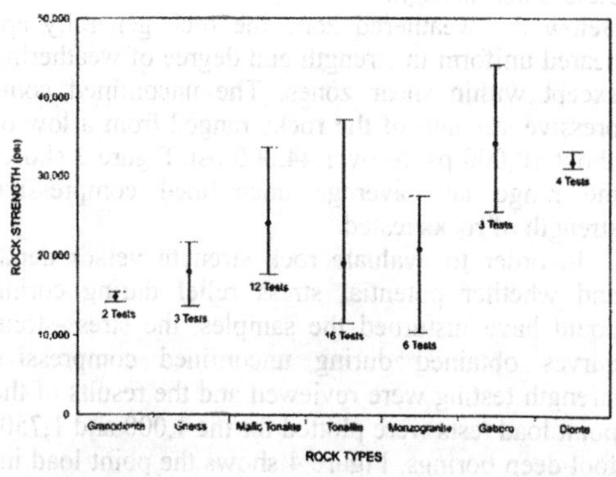

Figure 3. Rock Strength by Rock Type.

3.2.4 *Rock mass discontinuities and permeability*
Although all of the borings were drilled at least 100 feet into the unweathered rock, the rock mass was often moderately fractured to the bottom of the shallow borings, even when located away from known lineaments and shear zones. The 1,000-foot deep boring was drilled into the Cuyamaca-Laguna Mountain Shear Zone and thus the degree of rock fracturing remained relatively constant for the entire length of this boring. However, the 1,750-foot deep boring was drilled away from known shear zones and is thought to be more representative of the rock mass conditions that will be encountered along the majority of the tunnel alignment. To evaluate the degree of fracturing verses depth, the fracture frequency (number of recorded naturally occurring fractures per foot of core) were plotted for the deep boring as shown on Figure 5. Even though unweathered rock was encountered at a depth of 45 feet in this boring, the rock remained relatively fractured to a depth of about 200 to 250 feet. Below this depth the rock was relatively unfractured except for within a minor shear encountered at 1,450 feet.

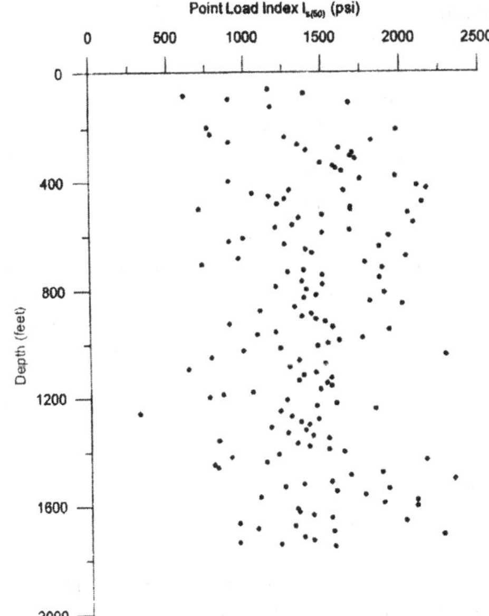

Figure 4. Point Load Index vs. Depth for 1,750-foot Boring.

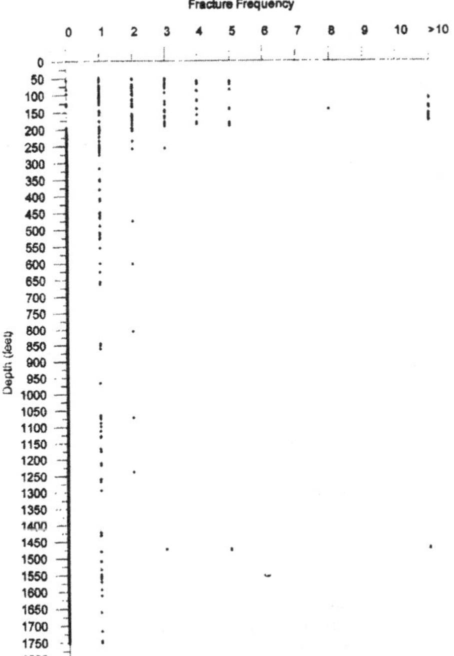

Figure 5. Fracture Frequency vs. Depth for 1,650-foot Boring.

Another indication that the rock mass is generally more fractured in the upper 200 to 300 feet than would be expected at a tunnel depth of over 1,000 feet was obtained from packer testing. Groundwater at tunnel depth exists only in the open fractures of the essentially impermeable hard igneous and metamorphic rocks. In order for groundwater to flow in the hard rock mass, the fractures in which water is present must be open and interconnected. The degree of fracturing (fracture frequency) and openness of the fractures are generally expected to be higher near the ground surface and to decrease with depth.

As a result, most groundwater and groundwater flow is expected to be perched in the more highly fractured near-surface rock mass.

Results of the packer testing in the borings generally confirm these observations. A histogram showing the distribution of the 214 permeability values measured in all of the borings combined is shown on Figure 6. This plot shows that about 63 percent of the rock mass had a permeability of 1×10^{-7} cm/sec or less. A similar plot for the 1,750-foot deep boring shown on Figure 7 showed that 87 percent of the rock mass had a permeability of 1×10^{-7} cm/sec or less. The permeability distribution and the degree of rock fracturing from the 1,750-foot deep boring is thought to be more representative of the overall rock mass excluding lineaments and major shear zones, which were not encountered in the preliminary drilling program. The number and magnitude of faults or shear zones shown on Table 1 were assumed to be represented by the lineaments mapped from the satellite and aerial photographs. The minor lineaments were assumed to be 50 feet in width and the major lineaments were assumed to be 300 feet in width based on a comparison of similar lineaments encountered in previously constructed tunnels in granitic rock terrain (MK, 1992 and WC, 1997).

3.2.5 *Hot water, high rock temperatures*

Hot springs and wells that encounter hot water are present throughout southern California. Hot water and associated high ground temperatures present a significant hazard to tunneling at the depths contemplated. The Phase I study identified tunnel alignments using avoidance of hot springs and active

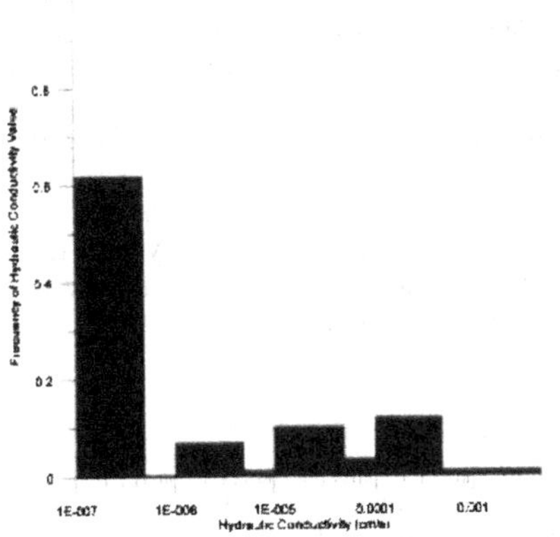

Figure 6. Packer Test Results From All Core Borings.

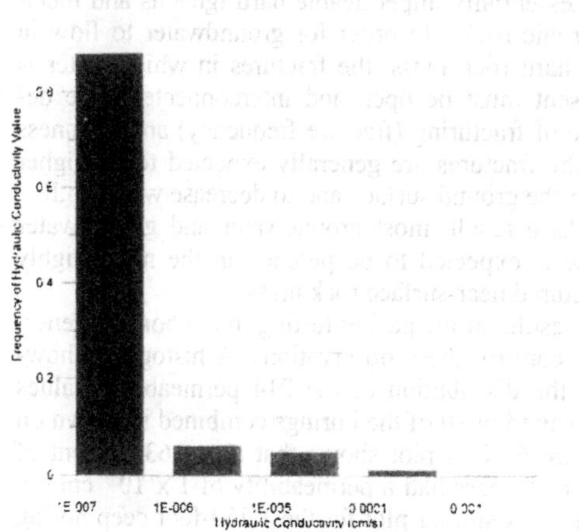

Figure 7. Packer Test Results From 1,750-foot Boring.

faults as a routing criterion. The avoidance of active fault zones in the tunnel sections is significant because within the PRB, there is a general correlation between hot springs and thermal wells with active faults and zones of seismicity (Gastil and Bertine, 1986). As a result, the 5A corridor does not cross near known mapped thermal springs or thermal wells and crosses the active Elsinore-Laguna Salada fault zone in pipeline.

Water temperatures were measured over the length of two of the deeper borings, C5A-9A (total depth 1,000 feet) and C5A-13 (total depth 1,750 feet). The temperature measurements were taken several days after completion of drilling to allow the hole to stabilize. The measured water temperatures are considered to be representative of the rock temperature also. Temperatures in C5A-9A ranged from 16.6° to 20.5 °C (62° to 70° F) at depths ranging from 94 feet to 921 feet. In C5A-13 temperatures ranged from 19°C to 26.8 °C (66° to 80° F) at depths of 100 feet to 1740 feet. In both cases temperature increased uniformly with depth. These temperature ranges represent relatively low gradients of 0.0047 °C (0.0085° F) per foot, or approximately 1 °C per 210 feet (1° F per 118 feet). Assuming a surface temperature of 14 °C (57° F) at the highest elevation area along the 5A alignment where the tunnel cover is about 4,900 feet would suggest a maximum temperature at tunnel depth of approximately 37 °C (99° F). This is a relatively high temperature for workers to be exposed to since the humidity in the tunnel will be very high.

3.2.6 *Radon gas*

Significant levels of radon gas were encountered in groundwater samples with measured levels of 9,180 and 989 pCi/L from Borings C5A-9A and C5A-13, respectively. Maximum levels of groundwater radon concentration recorded in the samples were over 2 times the safe drinking water level developed by the National Research Council committee. The safe drinking water level was calculated on the basis of water-to-air transfer because the inhalation pathway dominates other means of ingestion or absorption contributing to overall cancer risk from radon exposure (Hopke et al., 2000). Ventilation airflow rates within the tunnel during construction would need to be several times larger than that of residential indoor air to eliminate the potential for a radon gas hazard. Given the standard means of ventilation typically used in tunneling projects, a protective exposure level for workers should be obtained; however, air monitoring for radon gas levels will likely be required during tunnel construction as well as worker radiation monitoring.

3.3 *Rock mass classification*

The rock mass was evaluated in terms of the RMR system (Bieniawski, 1989), and the Q-system (Barton et al., 1974) and basic classification values were assigned to each core run. Following more detailed evaluation of in-situ conditions at tunnel depth, the Q-system ratings were modified to reflect in-situ stress conditions. The Q-system was used to divide the range of anticipated conditions into the following categories: massive, moderately blocky or schistose rock, blocky and seamy rock, and fault zones. The distributions of Q values from the boreholes were used to estimate the probability of occurrence of each ground category. A description of each ground category is provided in Table 2. The propensity for overbreak and rock stress problems are predicted based on relevant project experience and the Q-system rating.

Table 2. Summary of Ground Categories - 5A Tunnel.

Ground Category	Description	Rock Mass Quality (Q)
Massive, Moderately Blocky, Schistose	Strong, fresh rock with or without foliation, joint spacing 1 to 2 feet. Joint surfaces may be locally slightly attered. Occasional thin shears or seams (less than 6 inches wide) may be intercepted. Spalling expected in some areas	>4
Blocky and Seamy	Strong, fresh rock with joint spacing 0.5 to 2 feet. Joint surfaces may be weathered or altered with some slickensides or opening. Occasional shears or seams (less than 10 feet wide) may be intercepted, requiring supplemental support.	0.1–4
Wide Faults and Shears	Zones greater than 10 feet wide, consisting of rock ranging from highly fractured to sheared or crushed, with variable degrees of decompostion. Rock mass immediately adjacent to shear or fault zones anticipated to be blocky to seamy for tens of feet.	<0.1

3.4 *Anticipated tunneling conditions*

Anticipated rock conditions were evaluated based on results of rock mass classification, using boring log and cover depth information, as well as reviewing geologic mapping, air photo studies, petrographic examination, laboratory testing and groundwater inflow evaluations.

Based on the rock mass classification discussed above, and allowing for encountering faults associated with air photo lineaments, the tunnels are anticipated to encounter 82% massive to moderately jointed or schistose rock, 11% blocky and seamy rock and 7% fault or shear zones. Tunneling in massive to moderately jointed or schistose rock under the proposed 1,100 to 4,900 feet of maximum cover raises the potential for encountering spalling conditions (Proctor and White, 1968). An evaluation of the potential for spalling is included below. Blocky and seamy rock (Q less than 4) will be associated with frequent rockfall in the TBM area as discussed by Barton (Barton, 1990). An irregular tunnel perimeter will result. Faults and shears may be charged with water and exhibit low stand-up time resulting in face instability and overbreak above the cutterhead. Ravelling and squeezing conditions may be associated with fault zones as well. Probe drilling and formation grouting will result in an improvement in ground behavior prior to excavation.

3.5 *Groundwater characterization*

Groundwater is an important resource in eastern San Diego County. Except for the western one-third of the 5A alignment, imported water is not available and water for agricultural and domestic use comes entirely from surface storage and groundwater wells. The surface and groundwater resources are also an important natural resource on the U.S. Forest Service, BLM and Anza Borrego State Park lands overlying the tunnel. Nearly the entire 5A tunnel will be constructed below the groundwater table. Estimates of groundwater inflows to the tunnel excavations are needed to: (1) evaluate safety and working conditions during tunnel construction; (2) evaluate portal or shaft discharge quantities and the need for tunnel grouting or water-tight linings; and (3) evaluate the interaction between the tunnel and the groundwater regime.

Groundwater modeling was conducted to investigate these issues: the peak inflow of groundwater to the tunnel heading, the peak cumulative inflow to the tunnels during construction, the steady-state cumulative inflow under long term conditions, and the impact of the tunnel inflows on the overlying groundwater elevations. Several different approaches were used to calculate the rate of groundwater discharge into the tunnels.

In order to make estimates of groundwater inflow into the tunnel during and after construction and to evaluate the potential impact to the groundwater level, a numerical model was developed by Dr. David Huntley at San Diego State University. Using the information obtained from the field investigation (including results of packer tests, rock core fracture frequency; downhole optical logs and groundwater level measurements in the piezometers), a conceptual geologic model was developed and used in the numerical model. Several scenarios were assumed and the results compared to tunnel groundwater inflows in previously constructed tunnels in similar geologic environments including the Metropolitan Water District's San Jacinto (MWD, 1940) and the San Bernardino tunnel (DWR, 1967) in southern California and the Manapouri tailrace tunnels in New Zealand (WC, 1997). The numerical model was then refined using the results and interpretations of the field data combined with observed case history inflows to make estimated inflows for the proposed tunnel.

Some of the key elements and criteria used to estimate the groundwater inflows in the model were that:

1 the groundwater inflows would be mitigated to a point that the groundwater table above the tunnel alignment would not be lowered more than 1-foot on average, during and following construction, and
2 during tunneling two probe holes would be advanced in front of the TBM and all significant fault and shear zones that produce water would be grouted to a permeability of 3×10^{-5} cm/sec.

The model showed that small volume groundwater inflows into the tunnel segments with low cover (less than 1,000 feet) had a significant impact on the groundwater table. For example a groundwater inflow in excess of 0.08 gallons per minute per foot (gpm/ft) of tunnel would result in a lowering of the groundwater level. In contrast, groundwater inflow into a tunnel with 3,000 feet of cover would need to exceed 20 gpm/ft of tunnel before having an impact of the groundwater level due to the much larger storage and recharge area above the tunnel. Because of this sensitivity, the two tunnel segments between the west portal of the 34-mile tunnel and San Vicente Reservoir (Segments 16 and 18 through 20 from Table 1) were planned to use precast, gasketed segments, since it was not considered feasible to control groundwater inflows with grouting alone to maintain the criteria of 1-foot maximum drawdown of the groundwater level. However, the use of gasketed segments was not considered feasible for the deeper 34-mile tunnel segment since the gaskets and strength of the segments cannot withstand the high hydrostatic heads. Groundwater inflows into this tunnel would need to be controlled by preexcavation grouting; however, realistic grouting closure criteria can be established based on a packer test permeability of 3×10^{-5} cm/sec.

Refinements to the model were made including a reduction of the total available hydraulic head at tunnel depth based on review of the initial head flow pressures measured in the San Jacinto tunnel and the Manapouri tailrace tunnels, which were generally about ½ of the total available head.

To make a comparison of the numerical model inflow quantities, a semi-empirical groundwater inflow method developed by Heuer (1995) was used, using the results of the packer test permeability distribution for the 1,750-foot deep boring (Figure 7). This method essentially confirmed predicted inflows except under very high head situations (over 3,500 feet), where the project estimates exceeded estimates based on Heuer by a factor of three (1995).

The groundwater inflow quantities estimated by the model after grouting of the lineaments and an additional 3 to 4 percent of the tunnel length to prevent impacts to the groundwater resources resulted in a maximum cumulative inflow of 18,000 gpm for eastern segment and 15,000 gpm for the western segment. The predicted amount of grouting using ordinary portland cement resulted in a grouted length of 5,900 feet for the eastern segment and 5,100 feet for the western segment, or about 8 to 10 percent of the total tunnel length. Use of high-early and microfine cements were also evaluated.

3.6 *Anticipated spalling conditions*

Spalling ground can occur when in-situ stress concentrations around the tunnel perimeter exceed the strength of the rock mass. Goodman (1989) suggests that where the major in-situ stress exceeds 25 percent of the rock unconfined compressive strength (or compressive strength divided by in-situ stress is less than 4), cracking and spalling that results in overbreak can occur. Implicit in the Strength Reduction Factor of Barton et al. (1974) is a threshold of 5, below which mild to heavy spalling or rock burst is said to occur. While Goodman and Barton are relatively consistent on prediction of spalling, other case histories suggest different relationships between rock strength and in-situ stress.

The potential for spalling ground was evaluated by comparing the average rock unconfined compressive strength in each tunnel segment to the estimated overburden pressure.

One possible explanation for the seemingly wide variation in the occurrence of spalling is the effect of lateral in-situ stress. When the in-situ lateral stress is the same as the in-situ vertical stress, the stress concentration around a circular opening in an elastic medium is twice the in-situ stress (Hoek and Brown, 1982). However, if the lateral stress differs from the vertical stress, the maximum stress concentrations in the tunnel perimeter increase sharply. Table 3 summarizes the maximum stress concentration in the tunnel perimeter as a function of the ratio of vertical to horizontal stress.

Table 3. Maximum Stress Concentration in Tunnel Perimeter Northern Alignments Geological Study.

K[1]	Maximum Stress Concentration in Circular Tunnel Perimeter[2]
0	3
0.5	2.5
1	2
1.5	3.5
2	5

Notes: [1] K = Lateral in Situ Stress / Vertical in Situ Stress

[2] Stress concentration values represent factors applied to vertical in-situ stress, which is normally estimated as the product of the depth of cover and average rock unit weight. These values apply to elastic, isotropic conditions around a circular tunnel and are developed using either the Kirsch equations or the principle of superposition based on 2-dimensional boundary element modeling of stresses around a circular opening as presented in Hoek & Brown, 1982.

A second explanation that is likely to complement the first is anisotropic rock strength and the effects of partially healed discontinuities on rock mass strength. At the first Manapouri Tailrace tunnel the spalling behavior was reportedly observed when the strike of rock mass discontinuities and/or foliation aligned sub-parallel with the tunnel excavation (WC, 1997) which demonstrates a relationship between anisotropic rock mass strength and spalling. Selection of rock samples for laboratory testing of intact rock strength is by necessity biased toward higher intact rock strength. For comparison, consider the rock strength test cases summarized in Table 4. Based on this summary the strength of nearly intact rock can be 34–38 % lower than intact rock.

Table 4. Effects of Geologic Structure on UCS Results[1] Northern Alignments Geological Study.

Case	Rock Type	Boring	Depth (feet bgs)	UCS[2] (psi) Structurally Controlled Failure[3]	UCS[2] (psi) Intact Rock Strength
1	Monzogranite	C5A-9A	910.7 – 914.2	11,000 12,800	17,900
2	Gneiss	C5A-12	201.0 – 202.7	17,100 9,400	21,500

Notes: [1] Data from Table 2A of GDR.

[2] Only test results from CSM are shown.

[3] Structurally controlled failures of Case 1 resulted in a 34% decrease in strength over the intact rock strength. For Case 2, decrease in strength was 38%.

In-situ stress conditions are difficult to predict without in-situ testing; however, Hoek and Brown (1982) provide a chart for estimating the ratio of vertical to lateral stresses (K_o) that is based on stress measurements obtained from the United States and abroad. Considering depths ranging from 3,000 to

5,000 feet, their work indicates that K_o typically ranges between 0.5 and 1.5 with and average of 1.0. This would generally support the use of the vertical overburden stress for evaluating spalling ground conditions.

Cover over the proposed tunnel ranges from negligible at the portals to a maximum of about 4,900 feet within Segment 3. Assuming an average total unit weight of 175 pcf, this results in a maximum estimated overburden pressure of about 5,950 psi. Assuming that the 25 percent rule is an appropriate indicator of spalling conditions, a tunnel excavation would be susceptible to spalling with rock unconfined compressive strengths less than 24,000 psi. Over 60% of the samples tested had strength lower than this value. Based on this assumption, it was determined that there is a potential for spalling ground conditions in segments 1B, 1C, 2, 3, 4, 7 and 8. Consistent with the Barton relative stress classification, these areas typically exhibit a ratio of rock compressive strength to in-situ stress ranging between 2.5 and 5. Where spalling conditions are expected along these segments, overburden heights above the tunnel range between 2,800 and 4,900 feet. This condition is present over 15+ miles of the tunnel alignment; hence, it is considered critical with respect to primary support and lining requirements and hydraulic roughness.

3.7 *Estimated TBM performance*

Instantaneous penetration rates were estimated for each segment by Dr. Levent Ozdemir of the Earth Mechanics Institute, of the Colorado School of Mines (CSM). Penetration rate estimates were based on a summary of rock and geologic properties for each tunnel segment. A final comparison was made to check the CSM results with historical instantaneous penetration rate data obtained from completed tunnel projects in California and abroad.

Penetration rates estimated by the CSM Method are a function of tunnel geometry, rock mass characteristics and machine parameters. Tunnel geometry parameters are the tunnel diameter and the tunnel axis orientation. Rock mass characteristics include the rock strength and hardness properties, degree of fracturing and orientation, and foliation. Machine parameters involve the TBM diameter, cutter diameter and spacing, net thrust per cutter, cutterhead RPM, number of cutters and installed cutterhead power.

The assumed tunnel diameter for this project is 15 feet. This diameter has been selected to accommodate plant and equipment in the tunnel necessary for the long tunnel drives envisioned. This excavation size ensures the necessary hydraulic requirements are achieved even if thick linings are deemed necessary. The tunnel axis orientation with respect to joint orientation can have an impact on penetration rate. Tunnel drive orientation along the project varies. In general, the tunnel drive direction for the 5A corridor is east-west. The orientation measured by azimuth, ranges between 248 degrees (segment 1) to 305 degrees (segments 17, 18, 19 and 20).

The degree and orientation of jointing was characterized using the Norwegian Institute of Technology (NTH, 1994) system as presented in Table 5. No distinction between joints and fissures was attempted because joint persistence data was not available from core logs.

Table 5. Norwegian Institute of Technology Joint Classification.

NTH Joint Class	Distance Between Joints (cm/inches)
0	-
0-I	160/63
I	80/31
I-II	40/16
II	20/8
III	10/4
IV	5/2

The joint orientation with respect to the tunnel axis orientation is expressed as the alpha angle (α):

$$\alpha = \arcsin(\sin \alpha f * \sin(\alpha t - \alpha s))$$

where: α s = strike angle of each joint set
α t = dip angle of each joint set
α f = tunnel axis orientation.

The alpha angle is the minimum angle between the tunnel axis and a normal to the discontinuity plane. This angle is independent of the tunnel drive direction. Joint orientation for each segment was identified by plotting borehole and surface mapping data on stereoplots. Major joint sets were determined along with their strike and dip angles. Where multiple joint sets were identified, an average alpha angle was computed for the tunnel segment.

The CSM model employs a variety of laboratory tests to estimate rock drillability (non-fractured rock) or the "basic" TBM penetration rate. These laboratory tests include unconfined compressive strength, tensile strength, punch penetration and point load index. All of these tests were conducted on select rock core samples obtained during the field investigation. Estimated "basic" rates are then adjusted to account for joint effects using the NTH joint classification and foliation.

Two machine specifications were developed for the CSM model: a "standard" TBM and a "high power" TBM. Both assumed a 15-foot (4.54 m) TBM diameter with 33 cutters spaced at 3.2 inches (81 mm). Assumed cutter diameter for the standard and high power TBM's was 17 inches (432 mm) and 19 inches (483 mm), respectively. The standard TBM has an operating thrust per cutter of 55 kips

(245 kn) with a cutter head rotational speed of 12.5 rpm. The high power TBM employs an operating thrust per cutter of 70 kips with a rotational speed of 13.3 rpm. Installed power for the standard and high power TBM's was assumed to be 2532 and 2954 hp, respectively. Certain advances in cutter and TBM technology between the time of the study and envisioned construction timeframe were anticipated for the high-powered TBM.

The instantaneous penetration rate, typically expressed in inches per cutterhead revolution, is dependent upon the "basic" penetration rate, joint effects and equivalent thrust per cutter. The net penetration rate, expressed in feet per hour, is a function of instantaneous penetration rate and the cutterhead RPM:

$$I = i_o * RPM * 60 \text{ (feet/hour)}$$

where: i_o = instantaneous penetration rate (inches/revolution).

To identify a range in penetration rates for each segment and TBM, CSM developed models for the worst-case, best-case and most likely scenarios. Maximum and minimum rates for the standard TBM ranged between 2.9 feet/hour for gabbro and 19.9 feet/hour for granodiorite, gneiss and tonalite. Maximum and minimum rates for the high power TBM ranged between 4.4 feet/hour for gabbro and 19.9 feet/hour for granodiorite, gneiss and tonalite. Weighted averages for the overall project penetration rate were estimated to range between 9.9 and 16.9 feet/hour for the standard TBM and between 14.9 and 18.5 feet/hour for the high power TBM. Overall average penetration rates for the most likely scenario were 13.1 feet/hour (0.21 feet/rev) for the standard TBM and 17.4 feet/hour (0.26 inches/rev) for the high power TBM.

3.7.1 *Comparison between CSM method and case histories*

The estimated TBM penetration rates developed using the CSM Method were compared to actual rates obtained at relatively recent projects in California. This comparison provides a way to confirm that recommended penetration rates are consistent with actual TBM performance at similar projects and that reasonable construction costs and schedules can be developed.

One project considered to be relatively similar and in near proximity to the 5A tunnel is the Cowles Mountain Tunnel located in San Diego. The Cowles Mountain Tunnel involved an 11.25-foot diameter tunnel constructed in very strong granitic rocks including granodiorite, monzogranite and tonalite. An average unconfined compressive strength of about 40 ksi was reported for these rock types as compared to the average of about 22 ksi for the rock types obtained for the 5A tunnel Typical joint spacing observed during the construction of the Cowles Mountain Tunnel was approximately 1 to 2 feet, which compares favorably with those assumed for the 5A tunnel. Cutter diameter was 17 inches with an average thrust per cutter of about 55 kips. Reported instantaneous penetration rates ranged between 0.08 to 0.16 inches/rev with and average of about 0.12 inches/rev. The average penetration rate is considerably lower than the average rate estimated for the 5A tunnel (0.21 in/rev - standard TBM); however, the average rock compressive strength at Cowles Mountain was nearly double.

Numerous TBM tunnel projects located in the Sierra Nevada Mountains can be compared and contrasted with the 5A tunnel project. These projects include Grizzly Power, Kerchoff 2, Calaveras-North Forks and Sandbar. Most of these tunnels were constructed in granodiorites and/or gabbro with average unconfined compressive strengths on the order of 20 ksi, which is comparable to the 5A tunnel average strength. In contrast to the 5A tunnel, joint spacing is typically about 5 to 10 feet in the Sierra Nevada's; hence, the NTH joint classification is 0 to I. Assuming similar rock strength and hardness, the low degree of fracturing would tend to result in lower penetration rates when compared to a typical joint class of I to II for the 5A tunnel. Reported instantaneous penetration rates for these projects ranged between 0.16 to 0.18 inches/rev. Thrust per cutter was typically 45 to 50 kips for these projects; hence, it can be assumed that a 55 kip/cutter (5A tunnel) thrust would result in higher penetration rates that approach 0.20 inches/rev.

In summary, CSM's average estimated TBM penetration rates (standard TBM) compare favorably with those obtained at completed projects when rock mass and machine parameter variables are considered. Based on an overall review of the estimated rates, it was recommended that the average rates developed for the standard TBM be used for determining construction costs and schedules. Although some new TBMs are being fitted with front-loading 19-inch cutters, their performance is relatively unproven compared to 17-inch cutter machines. The improvements in performance offered by a high-powered TBM equipped with 19 inch cutters should be considered in evaluating project cost contingencies.

4 CONCLUSIONS

Construction of the 34-mile long 5A tunnel is geotechnically feasible on the basis of the limited geologic and hydrogeologic data collected during the study. The tunnel will encounter hard, massive, abrasive granitic and metamorphic rocks with zones of highly fractured and sheared ground that will likely produce high volume, high pressure groundwater inflows.

As currently envisioned, this tunnel would be constructed from two separate headings each 17 miles in length. Tunnel drives of this length from a single heading, combined with the high cover (up to 4,900 feet with an average of nearly 3,000 feet) have not been constructed to date. Key geologic issues that impact the cost and construction schedule of this long tunnel include:

1 potentially high rock temperatures due to the earth's natural geothermal gradient
2 high in-situ stresses requiring additional ground support
3 several significant lineaments (fault zones) which must be crossed requiring pre-excavation grouting and support, and
4 potentially high pressure/high volume groundwater inflows requiring pre-excavation grouting to reduce flows to a manageable level.

Future design-level studies will be needed to better characterize the rock and groundwater conditions at the depth of the tunnel including ground temperatures at tunnel depth, in situ stress conditions, degree of rock fracturing, rock mass permeability and rock strength.

5 ACKNOWLEDGEMENTS

The authors wish to thank the San Diego County Water Authority, and Richard Pyle, Project Manager for their cooperation and assistance during the course of this study. Thanks are also due to the individuals who provided technical support during the project including P.E. Sperry, G. Korbin, T. Brekke, M. Walawender, D. Huntley and L. Ozdemir.

REFERENCES

Barton, N., Lien, R. and Lunde, J., 1974, Engineering classification of rock masses for the design of tunnel support: rock mechanics, Vol. 6, No.4, 1974, pp. 189-236.

Barton, N., 2000, *TBM Tunneling in Jointed and Faulted Rock*, Balkema, Rotterdam.

Bieniawski, Z.T., 1989, Engineering rock mass classification, John Wiley and sons, New York.

Black & Veatch, 1996, Feasibility Level Engineering for Facilities to Transfer Water from the Imperial Irrigation District, prepared for San Diego County Water Authority.

Compton, R.R., 1985, Geology in the field, John Wiley and sons, New York.

Gastil, G., Bertine, K., 1986, Correlation between Seismicity and the Distribution of Thermal and Carbonate Water in Southern and Baja California, United States and Mexico. Geology, v. 14, p. 287-290.

Goodman, R.E., 1989, Introduction to Rock Mechanics, Second Edition, John Wiley and Sons, New York.

Heuer, R.E., 1995, Estimating rock tunnel water inflow, in Williamson, G.E. and I.M. Gowring, eds., Proceeding of the Rapid Excavation and Tunneling Conference, San Francisco, June.

Hoek, E. and Brown, E.T., 1982, "Underground Excavations in Rock", Institute of Mining and Metallurgy, London. 2nd Edition.

Hopke, P.K. et al. 2000. Health Risks Due to Radon in Drinking Water. *Environmental Science and Technology*. Vol. 34, No. 6.

Morrison Knudsen Corporation, 1992, Geologic field mapping sheets for Cowles Mountain Tunnel, Contract 412, Prepared for San Diego County Water Authority, 49 p.

NTH, 1994, Hard Rock Tunnel Boring, Project Report I-94, University of Trondheim, The Norwegian Institute of Technology.

Proctor, R.V., White, T.L., 1968, *Rock Tunneling with Steal Supports*, with an introduction to Tunnel Geology by Karl Terzaghi, Commercial Shearing, Inc., Youngstown Ohio, revised.

Walawender, M., 2000, The Peninsular Ranges-A geological guide to San Diego's back country, Kendall-Hunt Publishing.

Woodward-Clyde, 1997, Second Manapouri Tailrace Tunnel, Geotechnical Baseline Report, Vol. 4 of 6, prepared for Electricity Corporation of New Zealand.

North American Tunneling 2002, Ozdemir (ed.)
© 2002 Swets & Zeitlinger, Lisse, ISBN 90 5809 376 X

Compensation for boulder obstructions

Steven W. Hunt, P.E.
Montgomery Watson Harza, Milwaukee, Wisconsin

ABSTRACT: Compensation for boulder obstruction removal is discussed with consideration of boulder conditions anticipated, baselining of boulder obstruction quantities and methods of removing boulder obstructions. Boulder obstruction removal costs and experiences on eight projects are presented.

1 INTRODUCTION

A boulder obstruction occurs when a boulder is encountered at the heading that stops reasonable forward progress of a tunnel. Generally this means that one of the following conditions has occurred:

- The obstructing boulder is too large to be routinely fractured and/or ingested through the tunnel face and tunnel mucking system.
- The obstructing boulder requires removal by supplementary means such as drilling and splitting or blasting through the heading or from an excavation made from outside of the tunnel.

Removal of a boulder obstruction requires extra effort and expense that is generally not included in the base tunneling rate and productivity. If it is included in the base tunneling rate and productivity, more time and elaborate equipment was probably provided to handle boulders. Either way, boulder obstruction removal adds cost to tunneling and requires compensation directly or indirectly.

Compensation can be directly linked to the extent of boulder obstruction removal work and have an associated pay item in the contract, or it can be incidental and made part of the tunneling unit rate. In both cases, bidders must estimate the equipment, effort and cost for boulder obstruction removal. In order to make this estimate, information on the quantity and character of boulder conditions is essential. Proper subsurface exploration and accurate geotechnical baseline report preparation are important. An approach for contractual boulder compensation should be selected after subsurface conditions are understood, boulder risk has been assessed and tunnel size and anticipated mining methods are designed.

2 BOULDER BASELINING

2.1 *Desk Study and Subsurface Exploration*

Before selecting a boulder obstruction compensation method, the geologic character of boulder deposits and probability of encountering boulders should be assessed. Hunt & Angulo, 1999 suggested six boulder occurrence characteristics that should be investigated by a desk study and subsurface investigation:

- Frequency
- Distribution
- Sizes
- Shapes
- Rock composition
- Matrix soil composition

A contractor when selecting excavation methods and tunnel boring machine features generally needs reliable information on all of these characteristics. In addition, a contractor should know the approximate matrix soil permeability and groundwater heads. The thoroughness of investigating and reporting boulder occurrence characteristics should be based anticipated risks and consequences of boulder occurrence for the tunnel size, depth and length, anticipated ground conditions, surface constraints and likely methods of tunneling.

2.2 *Subsurface Exploration*

Proper subsurface exploration is the most important component required to baseline boulder conditions. It is the source of site-specific data on geologic and boulders conditions. Hunt and Angulo, 1999 concluded that problems from unanticipated boulder obstructions often result from an inadequate desk study and subsurface investigation and listed the following as primary reasons why many subsurface investigations and geotechnical reports fail to adequately predict boulder occurrence:

1. Lack of geologic assessment in planning, evaluating and reporting of subsurface exploration data resulting in a poor understanding of the geologic setting, character and uncertainties.
2. Lack of a phased exploration program that progressively reduces key uncertainties that remain from the previous phase.
3. Lack of focused drilling instructions and monitoring of drilling by a geotechnical engineer or geologist (or one with proper training and experience).
4. Inadequate documentation and reporting of drilling observations and historical information.
5. Over reliance on drilling refusal or extremely high Standard Penetration blow counts as the primary indicators of cobbles and boulders.
6. Insufficient exploration budget (including boring spacing) and lack of redundancy in exploration methods causing over-reliance on only conventional drilling methods that are often chosen for speed and cost of drilling rather than sensitivity to cobbles and boulders.

Where a desk study concludes that boulder obstructions may have a significant impact on tunneling, special, non-typical exploration and testing methods should be considered for the subsurface investigation program. These special methods might include:

- Multiple methods of drilling and probing (for example both tri-cone rotary wash and hollow-stem auger drilling.
- Continuous or near-continuous sampling.
- Use of rotosonic drilling and coring.
- Use of large (500-1000 mm) diameter probe holes.
- Directional drilled horizontal probe holes.
- Geophysical methods.

Explanations of exploration methods for tunneling and their pros and cons are discussed by a number of authors, including: Brierley, Howard, & Romley, 1991; Cording et al, 1989; Davis & Oothoudt, 1997; De Pasquale & Pinelli, 1998; Essex, 1993; Frank & Daniels, 2000; Hindle, 1995; Hunt & Angulo, 1999; Neyer, 1985; Schmidt, 1974; and Smirnoff, & Lundin, 1985.

2.3 *Boulder Baselining*

2.3.1 *Baselining Issues*

Boulder baselining involves interpretation of desk study information and subsurface exploration results to provide a contractual representation of expected boulder conditions and to quantify characteristics affecting compensation for boulder obstruction removal. Baselining should be performed in accordance with the guidelines presented in *Geotechnical Baseline Reports for Underground Construction* (Technical Committee, 1997) and other commentaries on baselining, e.g. Essex & Klein, 2000.

The boulder characteristics that should be quantified depend on the boulder obstruction compensation method that is selected. Characteristics such as rock composition, shape, and distribution within the soil matrix, and location relative to tunnel heading are important and should be addressed, but are seldom quantified for payment. More commonly, quantification consists of one or more of the following estimates:

- Number of boulders or boulder obstructions anticipated (may be further restricted by minimize size or volume).
- Volume of boulders or boulder obstructions anticipated (may be restricted by minimize volume).
- Crew hours (or TBM downtime) required to access and fracture boulder obstructions (might also include time required to repair damaged or worn cutters).

If size is a criterion, the meaning of boulder size must be clarified. It may mean the approximate diameter, width, length or breadth of a boulder. Determination of boulder size during tunneling is difficult and is usually only an estimate. If intact boulders are freed from the soil and available for measurement, a large caliper could be used to measure size. More commonly, a ruler is laid across the boulder and a size is recorded. Either way, an important question is what dimension to measure. The size will be different if it is taken as the largest distance versus the length in the long axis.

Measuring size is even more difficult for embedded, partially exposed boulders. Figure 1 shows a nest of three boulders and several cobbles exposed in a shaft wall at a tunnel eye. Measurements can only be made of exposed boulder surfaces. Even less boulder surface area is typically exposed in a tunnel heading. A 300 to 400 mm exposure could be the small end of an embedded 1500 mm oblong boulder. After blasting and resumption of mining, the actual boulder dimensions would be nearly impossible to determine. The inaccuracy in determining boulder size should be considered when baselining boulder quantities.

Figure 1. Boulder Nest in Silty Clay Till Matrix.

2.3.2 *Boulder Baselining in Milwaukee*

One approach to boulder baselining is to correlate approximate typical ranges in boulder volume as a percentage of excavated tunnel volume to common, distinct geologic units within a region. Another approach is to develop a semi-empirical correlation of interpreted boulder frequency in borings to boulder volume as a percent of excavated tunnel volume.

The author has used both methods to baseline boulder obstructions in the Milwaukee area. Boulder encounter data for five Milwaukee and one Chicago area projects is given in Table 1 of Hunt & Angulo, 1999. Since that paper was published, data from three additional Milwaukee area projects was processed and is summarized in Table 1. These tables indicate that estimated average boulder volumes ranged from 0.01 to 1.82 percent and averaged 0.97 percent of the excavated tunnel volumes. Within some 60 m long segments, boulder volumes were as large as 11 percent of the excavated tunnel volume.

A correlation between total estimated lengths of boulders encountered during drilling through soil units penetrated by tunnels to the estimated boulder volume encountered as a percentage of excavated tunnel volume is shown in Figure 2. This relationship required careful interpretation of boring logs to estimate lengths of boulders encountered within the strata being tunnel through. It also required that boulder volumes be estimated using the rough boulder size information documented in the field. Boulder volumes were estimated as described in Hunt & Angulo, 1999.

Similar correlations could be developed for other localities where data exists on boulder volumes encountered during tunneling. Where no data from previous tunneling exists, boulder quantity estimating will be more difficult and should rely on combinations of boulder indications from borings and the special exploration methods previously mentioned.

Table 1. Boulder Encounter Data – Additional Projects.

Case No.	6	7	8
Project Name	Oklahoma Ave. Relief Sewer	Oak Creek Southwest	Miller, 37th and State MIS
Tunnel Length	725 m (2380ft)	1524 m (5000ft)	383 m (1255ft)
Excavated Diameter	1.52 m (5.0ft)	1.40 m (4.6ft)	2.64 m (8.7 ft)
No. of Borings	12	12	9
Avg. Boring Spacing	68.6 m (225 ft)	132.6 m (435ft)	44.2 m (145 ft)
Boulder Length / Boring Length in Till/Outwash	6.9%	1.8%	8.9%
Est'd No. of Boulders Hit in Borings	15	11	29
% Borings That Hit a Boulder	58%	58%	100%
Reported No. of Boulders in Tunnel	151	156	346
Reported No. of Boulder Obstructions	71	156	60
Avg. Boulders per 30m of Tunnel	6.2	3.1	27
Max. Boulders per 100m Tunnel	59	-	154
Estimated Boulder Volume, m^3 (yd^3)	12.3 (16.1)	36.8 (48.1)	33.7 (44.2)
Avg. % Boulders by Volume in Tunnel	0.43%	1.57%	1.62%
Max. % Boulders by Volume, 60m (200 ft) Tunnel Segment +/-	2.3%	-	2.9%

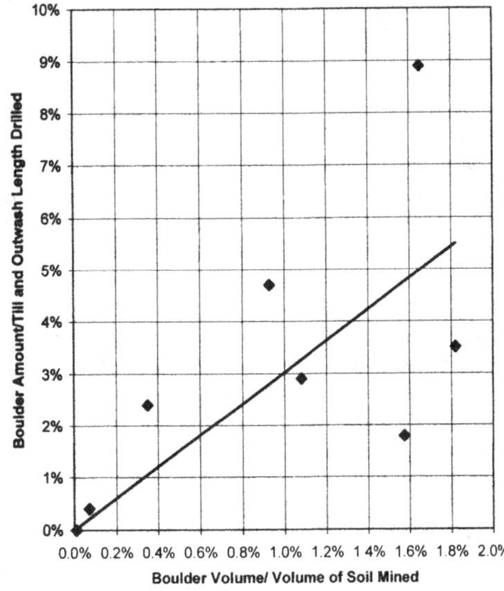

Figure 2. Milwaukee Correlation for Boulder Baselining.

3 REMOVAL OF BOULDER OBSTRUCTIONS

3.1 *Possible Methods*

Cobbles and boulders might be handled during tunneling by one or more of the following methods:

- Pushing aside by TBM.
- Intact removal through TBM.
- Piece removal through TBM after fracturing.
- Removal from outside of TBM.

These methods are briefly discussed below.

3.2 *Pushing Aside*

In some circumstances, boulders might be pushed aside before being fractured or ingested. To be pushed aside, requires mining with a partially closed face TBM through soil that is sufficiently soft or loose to allow boulder penetration into the soil in response to the forces being applied to the heading. Partial contact with a boulder at the perimeter of the TBM increases chances of it being pushed aside. Boulders encountered within weak or loose soil that are not pushed aside and are too big to be ingested are likely to roll around in front of the cutters (possibly causing damage) or to become lodged in a cutting wheel door.

3.3 *Intact Ingestion*

Boulders can be ingested intact or mostly intact when the openings at the heading are large enough for boulders to pass through. The boulder "size" that may pass through the heading openings depends on the size and shape of both the boulder and openings. Smooth, spherical boulders will generally pass through cutting wheel openings more readily than oblong or angular boulders. Oblong and angular boulders are more likely to become stuck in an opening or roll around in front of the heading until fractured.

Most rotary cutting wheel TBMs are able to ingest boulders with sizes smaller than 15 to 30 percent of the excavated diameter. Digger shield and mechanical open-face shield type TBMs can ingest boulders up to approximately half of the excavated diameter, but mucking of large boulders generally requires splitting, which might be more efficiently done before the boulder is excavated. Backhoe type digger shields are relatively rugged and can generally handle large boulders. Rotating arm type mechanical digger shields are less rugged and may be excessively damaged by nests of hard cobbles and boulders. Staheli et al, 1999 present a case history where an Akkerman mechanical open-face shield sustained extensive damage during mining through high compressive strength cobbles and boulders in California.

Once a boulder passes through the heading, it must also pass through the mucking system. If screw augers are used and diameters are too small or if conveyor openings through the plenum bulkhead are too small, a boulder can become lodged in the plenum or mucking device. Generally, heading openings are sized to be consistent with the ingestible size limits within the plenum or bars are placed across cutting wheel openings to reduce rock sizes that may pass through.

3.4 *Fracturing and Ingestion*

Boulders can be fractured and ingested by one of three methods:

- Cutter fracturing at the heading.
- Mechanical splitting through openings.
- Crushing within the plenum.

3.4.1 *Cutter Fracturing*

The ability of cutters to fracture boulders depends on cutter type (and degree of wear), rock composition and strength, boulder shape, orientation, and soil matrix strength, (Dowden & Robinson, 2001). Cutter types might include:

- Drag bits.
- Pick bits.
- Roller (strawberry) cutters.
- Single or multi-disc cutters.

Drag and pick bits may be capable of fracturing weak rock boulders and scattered moderately hard boulders, but are generally ineffective at fracturing nested boulders and hard rock boulders. Castro, Webb and Nonnweiler, 2001 and Staheli et al, 1999 present a case history where extensive carbide tipped pick bit damage occurred during mining through high compressive strength cobbles and boulders in California.

Roller cutters are rugged and capable of chipping away boulder pieces, however, the chipping action is generally slow within moderate to high strength rock. Roller cutter inefficiencies may reduce mining production too much to be a viable boulder fracturing method in most situations.

Disc cutters (Figure 3) can be very effective at fracturing boulders provided that the soil matrix is sufficiently strong to hold the boulder long enough for fracturing. Navin et al, 1995 provide a discussion of required soil matrix strength by comparison of embedded boulder bearing capacity to applied cutter force. If the matrix is too loose or soft, boulders will tend to roll around and cause excess cutter damage.

The size, configuration and design of disc cutters and seals are also important factors. The effectiveness of single and multi-disc cutters of various designs are discussed by Ozdemir, 1995. Dowden & Robinson also discuss the effectiveness of various disc cutter designs and the use of back-loading disc cutters.

3.4.2 *Mechanical Splitting*

If heading access is available from inside the tunnel and the heading is within low permeability zones of cohesive soil or granular soils that have been dewatered or stabilized by grouting, boulder obstructions

Figure 3. Slurry shield TBM with drag bits and disc cutters.

may be mechanically split through the cutting wheel doors. Mechanical splitting is generally accomplished by one of three methods:

- Blasting with explosives inserted into a drilled hole or placed on the boulder surface.
- Insertion of a hydraulic powered rock splitter into a drilled hole.
- Insertion of rapidly expanding mortar into a drilled hole.

Where the heading is within the water bearing granular deposits that cannot be effectively dewatered or grouted, an option for heading access is air pressurization of the plenum. Some earth-pressure-balance or mix-shield type TBMs with diameters of approximately 2 m can be furnished with sealed plenum having a man-lock portal for access to the heading to repair cutters or drill and split boulders. The compressed air provides stability and groundwater control to soil exposed when cutting head doors are open.

3.4.3 *Crushing Within Plenum*

Slurry and mix-shield TBMs generally include a rock crusher within the plenum to break boulders that have passed through the cutting head opening into a small enough size to be pumped with the slurry. Generally, cutting head openings and rock crushers are sized and configured to handle boulders up to from 20 to 35 percent of the excavated diameter. Rock crushers types include roller, cone and jaw types. Dowden & Robinson, 2001 provide more discussion on boulder crushing capabilities of slurry shield TBMs.

3.5 *Removal From Outside TBM*

When a boulder obstruction is encountered with a microtunneling TBM that is not equipped with rock cutters and the boulder is too large for crushers, the obstruction has to be removed from outside of the TBM. The following options might be considered:

- Pushing into a drilled hole.
- Drilling and blasting.
- Removal from a temporary access shaft or tunnel.

3.5.1 *Pushing Into a Drilled Hole*

Boulder pushing into a drilled hole (Figure 4) has been used in Milwaukee with limited success. The method requires surface access for drilling equipment over or close to the boulder obstruction and that no significant subsurface utilities or structures are present. Furthermore, the drilling equipment must be able to advance a sufficiently large and stable hole (generally 0.8 to 1.2 m) to a depth of one to two meters below the TBM. After a hole has been completed, jacking is resumed in an attempt to push the boulder or boulders into the hole. If this is successful, the temporary drilled hole must be properly backfilled. In addition, remaining void space around

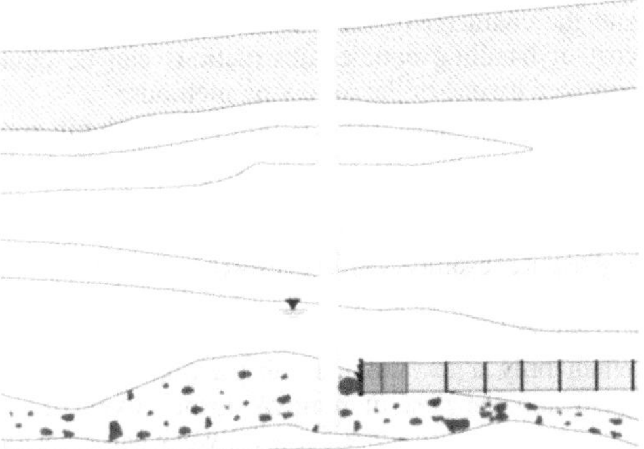

Figure 4. TBM pushing of boulder into a drilled hole.

the boulder within the hole below the tunnel must be filled with grout.

3.5.2 *Surface Drilling and Blasting*

If blasting is allowed, drilling into a boulder from the ground surface and blasting it may be feasible. Successful boulder fracturing will depend on rock composition and luck at drilling a hole and placing a charge at a strategic location relative to the boulder position and size. If successful, this method would be cost-effective, but the risk of failure is high.

3.5.3 *Removal From a Temporary Access Shaft*

Removal of a boulder obstruction from a temporary access shaft may be feasible if surface access is available and a one to two meter diameter cased hole or lined access shaft can be rapidly advanced over or close to the boulder. If the boulder is within a granular aquifer, dewatering or grouting will be required to maintain stability. Once accessed, the boulder may be split and removed in pieces. The temporary shaft must then be abandoned and the surface restored.

3.5.4 *Removal From an Access Tunnel*

A boulder obstruction might also be removed by hand-mining an access tunnel from a nearby work-shaft, if one exists. Temporary ground support might be provided by timber sets, a steel casing, liner plates or other means. If the boulder is within a granular aquifer, dewatering or grouting will likely be required to maintain stability. After boulder removal, tunneling can resume.

4 COMPENSATION FOR BOULDER OBSTRUCTIONS

4.1 *Compensation Options*

Tunneling through cobbles and boulders generally requires additional effort, equipment expense, and time depending on the tunneling method being used and the character of the cobbles and boulders. The cost of handling boulder obstructions can be compensated by one of the following methods:

- Incidental – part of other unit rates or lump sum items.
- Pay item.
- Contract modification.
- Dispute resolution (a last resort).

4.1.1 *Incidental*

When only a few boulder obstructions are anticipated and the cost of removal is not likely to be high, boulder obstruction removal work can be made incidental. Even if boulder obstruction removal is made incidental, boulder baselining should still be completed to more clearly establish boulder conditions and extent that boulder obstructions are expected.

If boulder obstruction baselining is ignored, the low bidder may assume that no boulder obstructions are anticipated and provide no allowance for removal work. Worse yet, a microtunneling TBM with no face access or rock disc cutters might be selected. Then if boulder obstructions are encountered during tunneling and significant removal costs develop, the contractor will likely file a differing site condition or defective specification claim in an attempt to recover extra costs and schedule delays.

4.1.2 *By Contract Modification*

Payment for boulder obstruction removal by contract modification may or may not be a cost-effective method. If boulder obstructions are anticipated and the contractor has suitable equipment to quickly and efficiently remove boulder obstructions, then this method may be a good choice. It may be the only practical vehicle for compensation when boulder obstructions are not anticipated or when scope of the boulder obstruction removal work cannot be sufficiently determined prior to bidding. Unless an allowance is made to cover contract modifications, the cost for boulder removal work will not be in the project budget. If getting approval and funding for unanticipated work is likely to be a significant problem then the boulder obstruction removal should be handled as either an anticipated and defined incidental item or as a pay item.

In many cases, the price to be paid for crew time and expenses under a contract modification condition is likely to be much greater than a bid boulder obstruction removal pay item, particularly if the scope of removal work is reasonably well known. If only a few obstructions are anticipated, the extra cost of rates that are negotiated as part of contract modification may not be large. However, if many boulder obstructions are anticipated, the cost of removal should be significantly less if unit rates have been bid.

4.2 *Pay Item Compensation*

4.2.1 *Compensation by Number Encountered*

A simple method of boulder obstruction compensation is payment at a unit rate per obstruction that is bid. This method would provide the same payment for removal of small boulder obstructions that might be split with one blast taking 15 minutes of time as a big boulder obstruction requiring several blasts and over an hour of time. It would also pay the same whether or not the boulder is within a wet sand matrix that must be dewatered or grouted for stability as within a stable cemented granular soil or stiff cohesive soil. Uncertainties such as these will likely lead to contingencies in the bid price.

4.2.2 *Compensation by Size*

Payment based on boulder size may be a more equitable method but requires size measurement. Size measurement is often difficult as previously discussed. The measurement method should be carefully defined to avoid disputes.

One use of payment by size is to avoid payment for boulders anticipated to be small enough to pass through the cutting wheel doors and mucking system. This size might be defined as 20 percent of the excavated tunnel diameter or smaller.

Another option is categorize boulders into two or three size ranges and to pay at bid rates for quantities of each size encountered. This approach may be more equitable and reduce contingencies needed to cover the potential cost of removing very large boulders.

4.2.3 *Compensation by Volume*

Measurement and payment for boulders by volume removed has been used on many projects. Horn & Ciancia, 1989, describe how boulders of ½ yd^3 or more were compensated at a rate of US $1,800 per yd^3 during hand mining within a 3.2 m diameter open-face shield on the Red Hook project in New York. Casey & Ruggiero, 1981 provide additional details on the boulder removal method for this project. These two references indicate that 80 percent of the boulders had to be hydraulically split to fit through the conveyor mucking system. The references do not clarify how boulder volumes were determined.

The volume method of compensation is not considered viable with most tunneling methods, particularly those where boulders are fractured with cutters or split through cutting wheel doors. It may be a viable method when tunneling with a shield or digger shield where whole boulders or fragments may be readily removed from the tunnel and sorted for measurement. However, determining boulder volume accurately is not easy. Boulders have irregular shapes that allow only an approximate boulder volume to be calculated based on dimensions. Perhaps the only accurate method of measurement would be to suspend boulders into a tank of water and compute their volume base on increased height of the water.

4.2.4 *Compensation by Weight*

Boulder obstruction removal compensation by weight of boulders removed has been used on some projects and should be easier to measure than volume. The payment item could be limited to boulders greater than a selected weight or size. Bids could also be taken for ranges of weights to allow contractors to bid more for removal of large boulders.

With this method, the contractor must segregate and stockpile boulders and qualifying boulder fragments. However, separating, stockpiling and weighing qualifying boulders is additional work that may not be practical for many projects, particularly where most boulders will be fractured rather than excavated whole or in large pieces.

4.2.5 *Compensation by Crew Hours*

Payment for boulder obstruction removal by TBM crew hours should be an equitable method of compensation, provided that an obstruction event and qualifying removal work are clearly defined. Mason, Berry & Hatem, 1999 provide a thorough discussion of issues associated with payment for obstruction removal by crew hours. Key aspects of the method include:

- Listing of potential obstructions. The contract should indicate the various anticipated *potential* obstructions and anticipated obstruction locations. However, in most cases, obstruction removal should be specified as covered work regardless of the location where the obstruction is encountered to avoid claims that a boulder obstruction was encountered at a different segment of alignment than encountered in the borings. Where some obstruction locations could cost substantially more in removal cost (such as below a freeway, building or watercourse), a separate pay item should be considered for these locations.
- Contractual definitions. Obstructions to tunneling progress might be defined with a time limit, for example: "man-made or man-placed objects, materials, structures or boulders, which unavoidably and completely stop the progress of the excavation or subsurface for more than 1 hour, despite the contractor's diligent efforts." In addition, the contract definitions should clarify the removal work included to help field staff properly measure crew hours for payment.
- Notification and determination. The contractor should be required to provide immediate notification to the resident engineer when an obstruction event has occurred. The owner/engineer should be obligated to promptly investigate the event and then provide notice of any disagreement that a qualifying obstruction was encountered within a brief period such as three days.
- Means and methods. Mason, Berry & Hatem, 1999 recommend that means and methods of boulder removal not be specified, but left to the contractor. However, consideration should be given to describing some potential methods anticipated based on unique aspects of the project.
- Measurement and payment. In order to provide budget to pay for boulder obstruction removal by crew hours, estimates are needed on how many boulder obstructions are likely to be encountered. Crew hours for obstruction removal should be estimated, baselined and included as a bid item quantity. In order to prevent unbalanced bidding, Mason, Berry & Hatem, 1999 recommend that a minimum price be stipulated, e.g., minimum number of crew hours or minimum unit rate per crew hour.
- Quantity Variations. The contract should indicate that crew hour unit prices will apply to quantities within a range such as ± 25 percent.
- Schedule. Mason, Berry & Hatem, 1999 recommend not allowing obstruction removal time to impact the schedule unless the time exceeds 125 percent of the baselined removal time.

5 PROJECT EXPERIENCES

5.1 *Boulder Obstruction Removal Costs*

Boulder obstruction removal costs varied widely for the eight projects discussed in Hunt & Angulo, 1999 and this paper. Average boulder obstruction removal prices (translated into 2001 dollars using a 3 percent inflation rate) for these projects are listed in Table 2.

Table 2. Boulder Obstruction Removal Costs in Milwaukee Area.

Case	Boulder Obstruction Compensation Method	No. of Obstructions Encountered	Average Cost Per Obstruction, 2001 U.S. $
1	T&M as DSC, blast from TBM	232	$372
2	T&M as DSC, blast from TBM	262	$1,620
3	Bid, exterior removal shaft	7	$5,450
3*	Bid, exterior removal shaft *	2*	$21,800*
4	Ingested with muck	5	$0
5	T&M as DSC, blast from TBM	112	$1,500 +/-
6	Bid, blasting from TBM	25	$2,500
6	T&M as mod, blast from TBM	46	$630
6*	Bid, blasting from TBM)*	25*	$1,500*
7	Bid, blasting from TBM	156	$90
8	Bid, blasting from TBM	60	$3,480
8*	Bid, disc cutter fracture, ingest*	55*	$1*

* Price of second lowest bidder based on quantities listed in bid.

Prices ranged from $0 to $21,800 per boulder obstruction. Discussions of boulder obstruction removal aspects of these projects are given below.

5.2 *Project Discussions*

5.2.1 *Case 1 – Interplant Solids Pipeline*
Some scattered boulders were anticipated on this project, but quantities were not baselined and an obstruction removal pay item was not provided. The contractor encountered 232 boulder obstructions that required 82.5 hours of extra crew time to remove by blasting through cutting wheel openings. A differing site condition claim was filed that was settled for an amount of $57,123. This corresponds to an average boulder obstruction removal cost of only $372 (in 2001 dollars – see Table 2). The low rate was due to the efficiency of boulder blasting (an average of 21 minutes per boulder) and that some boulder removal costs had been built into the unit rate for tunneling.

5.2.2 *Case 2 – South Pennsylvania Avenue*
Boulder obstructions were anticipated, but were specified to be covered under the differing side condition clause for extra work:

"Work directly related to tunnel excavation through cobbles and boulders that require size reduction (by drilling, blasting splitting, etc.) or special manipulation in order for the tunnel boring machine to proceed, will be considered a Differing Site Condition. Upon review and approval of the Engineer, payment for such work will be made in accordance with Article 'Payment for Modifications' of the General Conditions."

Probable boulders were reported in the Geotechnical Report, but quantities were not baselined. The contract budget included no allowance for boulder obstruction removal costs.

The contractor elected to use an open-face, rotary cutting wheel TBM that was advanced ahead of 1067 mm (42-inch) ID jacked pipe. Although small, the TBM was large enough to efficiency blast the 232 boulders encountered. These boulders were mostly embedded in a matrix of low permeability silty clay till so face instability was not a problem during blasting. Mining delay times for blasting generally ranged from 15 to 60 minutes per boulder and averaged 32 minutes. Average boulder obstruction removal costs were $1,620 as listed in Table 2.

The specifications did not provide any significant restrictions on TBM selection for the project. Although the low bidder selected an appropriate open-face rotary cutting wheel machine, the second lowest bidder (by only about $50) planned to use an Isecki Unclemole, a microtunneling TBM with no face access and capability of handling only cobbles and very small boulders. Had this bidder won the project, the contract modification cost for boulder obstruction removal would likely have been 5 to 10 times higher. Either a different TBM would have had to have been mobilized after encountering the numerous obstructions or hundreds of exterior boulder removal shafts would have been required.

The lessons learned on this project are that boulder quantities should have been baselined and TBMs without capability to handle numerous large boulders should have been prohibited.

5.2.3 *Case 3 – Ramsey Avenue Relief Sewer*
Boulder obstructions were baselined in the Geotechnical Report and a pay item was included for two boulder obstructions. No minimum price was specified, but a minimum removal scope was specified in the contract:

"An obstruction occurrence is when boulder(s) or other obstructions are encountered of a size and concentration that can not be reasonably excavated by the tunnel boring machine. When an occurrence develops, the Contractor shall determine appropriate methods of boulder or obstruction removal. These methods may involve drilling of a shaft adjacent to the boulder or obstruction to allow man access for removal or to provide an opening into which the obstruction may be pushed. The allowance for boulder or obstruction removal shall include <u>at least</u>: boulder removal [equipment] mobilization, 8 crew hours of excavating and boulder removal, equipment time (assuming 65 vertical foot excavation), material, equipment and man hours for backfilling, and associated tunnel or jacking delay time."

The contractor elected to jack 762 mm (30-inch) ID concrete pipe behind an Isecki Unclemole, a microtunneling TBM with no face access and capability of handling cobbles but not boulders. No boulder obstructions were encountered during the first 80 percent of the tunneling. However, with 6.4 m remaining in the second to last drive, the TBM encountered refusal to advance against an approximately 600 mm diameter, high strength igneous (gabbro) boulder situated within a water-bearing granular soil matrix (outwash).

Rather than construct an exterior boulder removal shaft as had been planned, the contractor elected to hand mine a 1219 mm (48-inch) ID access tunnel from the receiving shaft that was located about 6 m away. Due to haste to remove the boulder, dewatering wells were not installed within the granular soil layer, which had approximately 15 m of head based on a nearby piezometer. Instability developed in the tunnel heading and six flowing sand blow-ins occurred over a period of two months. Ultimately void backfill grouting and permeation grouting with chemical grout were required to stabilize the ground and remove the boulder. By this time, the friction forces on the TBM and pipe had become excessive due to soil set-up. The TBM and jacked pipe could not be moved. The TBM had to be pulled from the pipe after being encompassed by a temporary 1219 mm steel casing that was jacked around it.

Even though boulder obstructions were anticipated and a bid item was provided, the contractor filed a differing site condition claim for approximately $600,000 in time and expense. The claim was denied, but in order to avoid litigation, payment for this boulder obstruction was increased from the bid amount of $5,000 to a settlement amount of approximately $100,000.

This case shows that cost of removing boulder obstructions from an exterior shaft or tunnel may be much larger than the low bidder's unit price. A minimum unit price for obstruction removal by shaft or tunnel may have improved this contract.

5.2.4 *Case 4 – CT-7 Collector System*
Scattered boulders were anticipated on this project, but quantities were not baselined and an obstruction removal pay item was not provided. Five boulders were encountered but none obstructed mining progress. All of the boulders were encountered within low permeability silty clay till which allowed mining in non-earth pressure balance mode. The 3.54 m (11.6 ft) diameter Lovat TBM had cutter wheel openings of approximately 700 mm (2.3 feet) which allowed all of the boulders to pass through the cutting wheel openings and the mucking system.

Based on an anticipated low occurrence of mostly small boulders, the size of tunnel involved, and that face access for splitting was an option if large boulders were encountered, the lack of a boulder obstruction removal provision was appropriate for this project.

5.2.5 *Elgin Northeast Interceptor*
The contract documents for this project did not characterize boulder obstruction risk and did not have a payment provision. The contractor anticipated that a few boulders would be encountered in silty clay till and outwash deposits and selected an open-face, rotary cutting wheel TBM to provide face access. The TBM was advanced ahead of 1219 mm (48-inch) ID jacked pipe. Dewatering wells were provided to control water at the open-face TBM heading.

Instead of a few boulders and moderate inflows to dewatering wells, the contractor encountered 112 boulder obstructions and inflows over 40 l/s (640 gpm). The geotechnical report had not properly anticipated the large boulder quantities or high groundwater inflows. The borings had only been drilled to 0 to 2 m below tunnel invert. The tunnel zone aquifer was found to extend over 10 m below the tunnel and had a much higher transmissivity than expected.

These poorly characterized and non-baselined subsurface conditions resulted in significant extra costs averaging approximately $1,500 per boulder obstruction. The contractor filed a differing site condition claim. The claim and an appeal were denied. A lawsuit in federal court resulted.

The lesson learned from this case is that poorly characterized and non-baselined boulder obstruction conditions will likely lead to contractor problems, delays, extra costs and a dispute. The project owner would have experienced less overall cost by providing a proper geotechnical baseline report and by including a payment provision for boulder obstruction removal.

5.2.6 *Case 6 – Oklahoma Ave. Relief Sewer*
Boulder obstructions were baselined in the Geotechnical Report and a pay item was included for 25 boulder obstructions that were defined as follows:

"A boulder obstruction occurs when a large boulder is encountered at the heading of a tunnel that stops forward progress because it is too large to be broken and/or ingested through the TBM cutting wheel or tunnel mucking system and requires removal by supplementary means such as drilling and splitting through the heading or excavating from outside of the TBM. Boulders that have a size of 30 percent or more larger than the excavated diameter of the TBM are considered too large to be broken and/or ingested through a TBM. Boulders that have a size less than 30 percent or more larger than the excavated diameter of the TBM are considered small enough to be broken and/or ingested through a TBM without supplementary means."

The contractor anticipated that boulders would be encountered in silty clay till and outwash deposits and selected an open-face, rotary cutting wheel TBM to allow face access for blasting. The TBM was advanced ahead of 1219 mm (48-inch) ID jacked pipe. Dewatering wells were provided when mining through outwash deposits to control water at the open-face TBM heading.

The 25 baselined boulder obstructions proved to be insufficient. A total of 71 boulder obstructions were encountered. Most of the extra boulders occurred within three drives where the tunnel was unexpectedly within 5 m of bedrock. The borings in this area had extended only about 2 to 3 m below tunnel invert and had just missed the surface of a bedrock ridge. Because of tunneling in much closer tunnel proximity to bedrock, much higher boulder concentrations were present within the overlying till sheet than had been estimated.

Although boulder quantities exceeded the baselined amount, the extra boulders did not cause a contractual problem. The tunneling system with face access was appropriate for the ground conditions. In fact, since the boulder obstruction quantity exceeded 15 percent of the 25 obstructions included in the bid, the additional obstructions were compensated based on the TBM crew time that was required to blast boulders. The contractor was so efficient at blasting boulders, that the average cost for the additional 46 boulder obstructions was only $630 each. The amount in the bid for the anticipated 25 obstructions was $2,500 each (Table 2).

Two lessons can be learned from this project. First, borings along the tunnel alignment should probably extend at least 4 to 6 m (15 to 20 feet) deeper than tunnel invert and not just two to three tunnel diameters. Second, properly written provisions for boulder obstruction payment based on crew time may have been a more equitable payment method than unit rate payment per obstruction encountered.

5.2.7 *Case 7 – Oak Creek Southwest Interceptor*

The final designer on this project anticipated boulder obstructions and provided three payment items and two quantities:

- 400 boulder obstructions ranging from 300 mm to 600 mm in size (12 to 24 inches).
- 70 boulder obstructions ranging from 600 mm to 1200 mm in size (24 to 48 inches).
- Payment for boulder obstructions over 1200 mm as a contract modification based on time and expense required for removal.

The method that was used to baseline boulder quantities is not known, however, it was not very accurate. Field records indicate that totals of 54 and 102 small and medium size boulders were encountered, respectively, for a total of 156 boulders or 314 less than baselined. In addition, the contractor alleged that one of the boulders counted in the medium size range actually exceeded 1200 mm in size and should be compensated based on time and material rather than at the bid unit price.

The contractor had only bid $80 and $100 dollars each, respectively, for removing small and medium sized boulders. These low amounts were due to plans to excavate the tunnel by hand mining within a digger shield rather than as pipe jacking behind an open-face rotary cutting wheel TBM or other shield. After short drives, the contractor determined that hand mining was too difficult and slow. An open-face, rotary cutting wheel TBM was then mobilized. It was advanced ahead of 1067 mm (42-inch) ID jacked pipe. The final sewer pipe (600 mm or 24-inch) was then installed as carrier pipe and the resulting annulus was grouted.

The lessons learned are that boulder obstruction baseline quantities are difficult to predict and even harder to predict by size. Payment per obstruction based on multiple sizes did not result in a big difference in bid price and could have been handled as one size. Payment for boulders larger than 1200 mm based on time and expense may have been a reasonable provision, but determination of size in the field was an issue. The initially exposed portion of the disputed boulder suggested a size less than 1200 mm. After blasting and removal, the true size could not be accurately measured. If all boulder obstructions had been compensated as time and expense, then the size would not have been an issue.

5.2.8 *Case 8 – Miller 37th & State Relief Sewer*

Boulder obstructions were baselined in the Geotechnical Report and a pay item was included for 55 boulder obstructions that were defined as follows:

"A boulder obstruction occurs when a boulder is encountered at the heading of a tunnel that stops or significantly inhibits forward progress because it is too large to be broken and/or ingested through the TBM cutting wheel or tunnel mucking system and requires removal by supplementary means such as drilling and splitting through the heading or from an excavation made from outside of the tunnel. Boulders that have a size of 20 percent or more larger than the excavated tunnel diameter will be considered boulder obstructions. Boulders that have a size less than 20 percent of the excavated tunnel diameter will not be considered boulder obstructions."

In addition to baselining boulder quantities, the contract documents prohibited dewatering because of an overlying gasoline contaminated aquifer that was being remediated (Hunt, Bate & Persaud, 2001). The contract documents also specified use of a pressurized face tunnel boring machine and that disc cutters be considered for boulder fracturing unless a compressed air plenum was used to allow face access for splitting.

The contractor proposed use of a non-pressurized, open-face, rotary wheel TBM with flood doors. They proposed to perform permeation grouting if needed to stabilize the heading when blasting boulders or if needed to prevent detrimental groundwater drawdowns and contaminant migration. The TBM was equipped with drag cutters (Figure 5). The contractor's proposed method was accepted.

Approximately 346 boulders were encountered during mining. Of these, 60 or 5 more than baselined were considered boulder obstructions. Figure 6 shows typical muck with numerous cobbles, small boulders and blasted boulder fragments.

Compensation for the 60 boulder obstructions was made at a unit price of $3,480 per obstruction, as bid. No compensation was made for the estimated 286 smaller boulders (300 mm to 528 mm or 12-inch to 21-inch) that were less than 20 percent of the excavated diameter.

The contractor experienced significant damage to the drag bit and gage cutters during tunneling. Tun-

Figure 5. Modified 2.64 m OD Tabor TBM with flood doors.

Figure 6. Muck with numerous cobbles and boulders.

nel drives had to be stopped twice to allow hand-mining through the cutter wheel doors to access the cutting wheel for drag bit and gage cutter replacement. Fortunately, cutter repairs occurred at locations where good stand-up time and low groundwater heads allowed successful hand mining.

In spite of the delays and costs associated with cutter repairs and the difficult, slow progress through boulder fields, no differing site condition claims were made. The geotechnical baseline report had appropriately characterized the bouldery nature of the ground and the number of boulder obstructions included as a pay item (55) was within 15 percent of the actual number of obstructions encountered (60).

The second lowest bidder on this project had proposed to mine with a Lovat earth pressure balance TBM equipped with back loading disc cutters. As a result, no boulder obstructions were anticipated and this contractor submitted a unit price of $1 per boulder obstruction (Case 8* in Table 2).

The lessons learned on this project include the following:

- Boulder obstructions can be handled by different methods, including methods not conforming to specifications. The contractor was willing to gamble that extra costs of cutter repairs and grouting to stabilize the heading during boulder blasting or cutterhead repairs would be more cost effective than procuring a more expensive TBM that would have been better suited for the bouldery ground conditions.
- Proper boulder baselining and payment provision for 55 boulder obstructions in the contract resulted in a suitable boulder removal budget for the contractor to handle the difficult bouldery ground conditions. If boulders conditions had not been baselined and no unit rate compensation provided, differing site condition claims would have been unavoidable.

6 CONCLUSIONS

The following conclusions were made from this boulder obstruction removal compensation study:

1. Proper subsurface exploration and boulder condition baselining is very important so that tunneling equipment that is suitable for the subsurface and boulder conditions can be provided.
2. Boulder quantities can be reasonably estimated if boulder occurrence records from similar geologic units are correlated with indications of boulders from properly logged borings.
3. Compensation for boulder obstruction removal by number of obstructions encountered is an equitable method where a thorough subsurface investigation and boulder baselining have been completed.
4. Compensation for boulder obstruction removal by quantities encountered at various sizes would be a more equitable method where a large range in boulder sizes is anticipated.
5. Compensation for boulder obstruction removal by volume or weight is not practical for most tunneling situations, but may be a reasonable method when mining with an open-face shield in stable ground.
6. Compensation for boulder obstruction removal by crew time is probably the most equitable method of those considered. It resulted in the least cost per obstruction removed on several Milwaukee projects. This method allows contractors to include less contingency for boulder removal and to bid reasonable hourly crew rates. The method does, however, require more care in preparing specifications and payment item definitions to clarify scope of work covered and qualifying boulder obstruction events.

REFERENCES

Brierley G.S., Howard, A.L. & Romley R.E. 1991. Subsurface Exploration Utilizing Large Diameter Borings for the Price Road Drain Tunnel. In Wightman, W.D. & McCarry D.C. (eds). *Proceedings, 1991 Rapid Excavation and Tunneling Conference.* Chapter 1, 3-15. Littleton CO: SME.

Casey, E.F. and Ruggiero, 1981. The Red Hook Intercepting Sewer – A Compressed Air Tunnel Case History. In Wightman, W.D. & McCarry D.C. (eds). *Proceedings 1991 Rapid Excavation and Tunneling Conference*: 179-200. Littleton, Colorado: SME.

Castro, R., Webb, R. & Nonnweiler, J. Tunneling Through Cobbles in Sacramento, California. 2001. In Hansmire, W.H. & Gowring, I.M. (eds). *Proceedings 2001 Rapid Excavation and Tunneling Conference*: 907-918. Littleton, Colorado: SME.

Cording, E.J., Brierley, G.S., Mahar J.W., and Boscardin M.D. 1989. Controlling Ground Movements During Tunneling. In Cording, E.J., Hall W.J., Haltiwanger J.D., Hendron, A.J.

Jr., and Mesri G. (eds.). *The Art and Science of Geotechnical Engineering*: 478-482. New Jersey, Prentice Hall.

Davis R. & Oothoudt T. 1997. The Use of Rotosonic Drilling in Environmental Investigations, *Soil and Groundwater Cleanup*, May 1997, 34-36.

De Pasquale, G. & Pinelli G. 1998. No-Dig Application Planning Using Dedicated Radar Techniques. *No-Dig International*, February 1998, I-12 to I-14. Mining Journal LTD: London.

Dowden, P.B. & Robinson, R.A. Coping with Boulders in Soft Ground Tunneling. 2001. In Hansmire, W.H. & Gowring, I.M. (eds). *Proceedings 2001 Rapid Excavation and Tunneling Conference*: 961-977. Littleton, Colorado: SME.

Essex, R.J. 1993. Subsurface Exploration Considerations for Microtunneling/Pipe Jacking Projects. In *Proceedings of Trenchless Technology: An Advanced Technical Seminar.* 276-287.Trenchless Technology Center, Louisiana Tech University: Ruston, LA.

Essex, R.J. & Klein, S.J. 2000. Recent developments in the use of Geotechnical Baseline Reports. In Ozdemer, L. (ed). *Proceedings Of The North American Tunneling 2000*: 79-84. Rotterdam: Balkema.

Frank, G. & Daniels, J. 2000. The Use of Borehole Ground Penetrating Radar in Determining the Risk Associated With Boulder Occurrence. In Ozdemer, L. (ed). *Proceedings Of The North American Tunneling 2000*: 427-436. Rotterdam: Balkema.

Horn H. M. & Ciancia A.J. 1989. Geotechnical Problems Posed by the Red Hook Tunnel. In Cording, E.J., Hall W.J., Haltiwanger J.D., Hendron, A.J. Jr., & Mesri G. (eds). *The Art and Science of Geotechnical Engineering:* 367-385. New Jersey: Prentice Hall.

Heuer R.E. 1978. Site Characterization for Underground Design and Construction. In *Site Characterization & Exploration.* 39-55. ASCE: New York.

Hindle D.J. 1995. Geotechnical Appraisal. *World Tunneling.* Nov. 1995,371-373. Mining Journal, Ltd: London.

Hunt, S.W. & Angulo. M. 1999. Identifying and Baselining Boulders for Underground Construction. In Fernandez, G. & Bauer (eds). *Geo-Engineering for Underground Facilities:* 25w5-270. Reston, Virginia: ASCE.

Hunt, S.W., Bate, T.R. & Persaud, R.J. 2001. Design Issues For Construction of a Rerouted MIS Through Bouldery, Gasoline Contaminated Ground, In *Proceedings of the 2001 - A Collection Systems Odyssey Conference.* Session 6. Alexandria, VA: Water Environment Federation, Inc.

Mason, D.J. III, Berry, R. S.J., &. Hatem D.J. 1999. Removal of Subsurface Obstructions: An Alternative Contractual Approach. In Fernandez, G. & Bauer (eds). *Geo-Engineering for Underground Facilities:* 1164-1175. Reston, Virginia: ASCE.

Navin, S.J., Kaneshiro, J.Y., Stout, L.J. & Korbin, G.E. 1995. The South Bay Tunnel Outfall Project, San Diego, California. In Williamson, G.E. & Gowring I.M. (eds). *Proceedings 1995 Rapid Excavation and Tunneling Conference*: 629-644. Littleton, Colorado: SME.

Neyer, J.C. 1985. Geotechnical Investigation For Tunnels in Glacial Soils. In Mann, C.D. & Kelly, M.N. (eds). *Proceedings, 1985 Rapid Excavation and Tunneling Conference.* Chapter 1, 3-15. Littleton CO.: SME.

Ozdemir, L. 1995. Comparison of Cutting Efficiencies of Single-Disc, Multi-Disc an Carbide Cutters for Microtunneling Applications. *No-Dig Engineering.* March 1995. 18-23.

Schmidt, B. 1974. Exploration for Soft Ground Tunnels - A New Approach. *Subsurface Exploration for Underground Excavation and Heavy Construction.* 84-96. New York: ASCE.

Smirnoff, T.P. & Lundin T.K. 1985. Design of Initial and Final Support of Pressure Tunnels in the Phoenix "SGC". In Mann, C.D. & Kelly, M.N. (eds). *Proceedings, 1985 Rapid Excavation and Tunneling Conference.* Chapter 26, 428-438. Littleton CO: SME.

Staheli, K., Bennett, D., Maggi, M.A., Watson, M.B. &. Corwin B.J. 1999. Folsom East 2 Construction Proving Project: Field Evaluation of Alternative Methods in Cobbles and Boulders. In Fernandez, G. & Bauer (eds). *Geo-Engineering for Underground Facilities:* 720-730. Reston, Virginia: ASCE.

Technical Committee on Geotechnical Reports of the Underground Technology Research Council. 1997. Essex, R.J. (ed). *Geotechnical Baseline Reports for Underground Construction – Guidelines and Practices.* New York: ASCE. 40p.

Session 1, Track 3

Project / Case histories

North American Tunneling 2002, Ozdemir (ed.)
© 2002 Swets & Zeitlinger, Lisse, ISBN 90 5809 376 X

Predicting boulder cutting in soft ground tunneling

C.M. Goss, PhD
Rocky Mountain Consultants Inc., Longmont, Colorado, USA

ABSTRACT: While much work has been done to predict the presence of boulders, their behavior during excavation remains a mystery. Disc cutters have been effective on boulders on some projects, but on others have plucked out or simply pushed the rocks aside, breaking cutting tools and interfering with the shield steering. In this paper, boulder-ground interaction is modeled for many scenarios using finite elements and the results are compared to 18 case histories. The key result of the research was the importance of boulder to soil shear strength ratio. As the ratio increases, the disc cutter load is transferred from the boulder to the soil. The model showed that when the rock-to-soil shear strength ratio is greater than 600:1, boulders cannot be broken by disc cutters. Of the 18 projects studied, nine validated the model results, seven projects were inconclusive in regards to model correlation and two discounted the model results.

1 INTRODUCTION

In the last few decades, soft ground tunneling has grown in popularity around the world. The growth was spurned by both a greater need and better technology. Successful projects around the world increased the experience and confidence level, leading to more projects.

Soft ground tunneling has enjoyed enormous technical advances during the past 25 years with the dawn of slurry and EPB shields. However, even with all the advances, there are still problems to be overcome in soft ground tunneling. The most significant problem is when boulders (rock blocks too large for the machine to swallow) are mixed in with the soil.

While much work has been done to predict the presence of boulders, their behavior during excavation remains a mystery. Disc cutters have been very effective on some projects, but on others have plucked out or simply pushed the rocks aside, breaking cutting tools and interfering with the shield steering. Boulder removal during construction can be dangerous and costly. If tunnel engineers could predict boulder behavior on their project, they could determine appropriate pre-excavation ground modification.

1.1 *Definition of the Problem*

While a proper site investigation can reveal the likelihood of encountering boulders during a tunnel drive, little is known of boulder behavior. In some projects, tunnel boring machines equipped with disc cutters have broken boulders with ease. In other projects, boulders have simply been pushed to the side, knocking the TBM off the tunnel alignment. In some cases, disc cutters have merely plucked boulders from the face, causing them to roll around, breaking tools and blocking pipes or flood doors.

These problems occur when boulders are too large to be directly ingested by the tunnel boring machine and the ground is too weak to hold boulders in place for cutting. In such cases, it often becomes necessary to enter the face, typically under compressed air, to break or remove the boulders by hand. Such entries are dangerous to both the miners and the surroundings, as the ground can collapse without warning, killing the miners and causing surface settlement and sinkholes. In situations where human entry to the face is not possible, the ground has to be strengthened so that it can hold boulders in place and the disc cutters can do their job. Such a task is not dangerous, but until the equipment is mobilized and has successfully treated the ground, all tunneling stops, causing delays and large cost overruns. A rescue shaft sunk in front of the TBM creates similar delays and costs while disturbing the surface area that the tunnel was supposed to avoid.

Currently, boulder cutting is calculated using simple (Terzaghi) bearing capacity (Navin et al. 1999). The force exerted by a disc cutter is calculated. The force is then applied as a point load to a strip footing (to simulate a boulder). If the resulting pressure is greater than the bearing capacity of the soil, the boulder will move. This calculation is a good

good approach, but it has some shortcomings. Hence, another method of predicting boulder behavior should be considered.

It is known that if the ground is strong, boulders can be broken and if the ground is weak, boulders will move. It is also known that weak boulders will fracture while strong boulders will not. The cut off point for strong soil and boulders or weak soil and boulders is not known. If a relationship between rock and soil, for purposes of breaking boulders, could be quantified, decisions on ground treatment could be made before the tunnel reached the problem area, thereby saving money and increasing site safety. If equipment for ground treatment or human entry to the face is available when the problem occurs, the situation can be quickly remedied, thereby minimizing lost production time.

1.2 *Research Limitations*

It must be considered that the dissertation upon which this paper is based covers original research in an area that has not been the subject of in depth study. Hence, the model developed is a first generation product. It served well to examine the influence of many factors on boulder behavior, but remained rather crude. The results obtained from the model agreed with actual project results in a broad sense, letting general conclusions be drawn. Such conclusions should be seen as trends and guidelines as opposed to hard and fast rules.

To make the study of boulder cutting feasible, assumptions and simplifications had to be made. Most of the simplifications were due to model limitations. Boulders were modeled in two dimensions, thereby assuming plane strain. Rock shapes were kept simple and the mesh relatively unrefined. Only one disc cutter attacking a single boulder was simulated. The model could not apply a pure point load to a boulder and therefore required a correction factor for field correlation. The program also modeled boulder behavior as perfectly elastic-perfectly plastic, which was not realistic since rocks fail in a brittle manner. Finally, the ground was made continuous and isotropic.

Simplifying assumptions also had to be made on site data gathered from the case histories. For some projects, only certain properties and factors were examined. Since data on some projects was limited, parameter values sometimes had to be inferred. In a few case histories, some properties were applied to the entire project if exact cross sections and plans were not available. Since ground parameters were often given as ranges, it was assumed that each value in that range had the same chance of occurring. That is, if rock compressive strength was given as 200MPa to 300MPa, the odds of encountering a 200MPa boulder were identical to encountering a 250MPA or 300MPa boulder.

2 SIGNIFICANCE OF BOULDERS IN TUNNELING

When boulders are expected along a tunnel alignment, a mixed face cutterhead must be used. Such a cutterhead uses picks, scrapers, rippers, and disc cutters to break up boulders too large to fit into the machines muck handling system. Herein lies the problem, while on some projects the disc cutters have been extremely effective, on others they have not been effective at all.

When disc cutters are effective, they roll or glide through the ground until they encounter a hard object (typically a boulder or zone of cemented material). Then, as in hard rock, the disc cutter applies a point load to the boulder, making chips of rock flake off until the boulder fractures. If the pieces are small enough, they are sucked into the machine and excavation continues. If they are still too big, they are further fractured by the disc cutters. Disc cutters can attack and destroy very hard boulders of any shape, as long as the stone is held firmly in place by the imbedding soil matrix (Anheuser 1995).

Sometimes the disc cutters do not work. Instead of fracturing, boulders are pushed aside or plucked out of the face. If the boulders are pushed to the side, they can actually shove the TBM off course. That is, the TBM pushes straight off from the tunnel lining. If there is weak soil to the left and a hard boulder just to the right, the machine will follow the path of least resistance and drift towards the left. If this happens enough, the tunnel alignment will be skewed. If boulders are plucked from the face, they roll around in front of the cutterhead, getting in the way and damaging tools. In such cases, miners or divers must access the face to manually break the boulder apart. Such work is dangerous for the person as well as the project. Face collapse can mean injury or death to the miner and cause excessive settlement on the surface above the tunnel. In situations where face access is not possible, another solution is to increase the shear strength of the surrounding soil. Typically, this is done by pumping in grout from the surface or from the machine through the cutterhead. This procedure has been very effective, but requires the tunnel advance to halt, thus creating very expensive delays and construction claims

In short, boulders often are, quite literally, the obstacles to effective and successful tunneling. If their behavior could be predicted, great benefits could be reaped by the industry and the owners. The prediction of boulder behavior is explored in the next pages.

3 BOULDER LOCATIONS

Determining boulder occurrence has been covered extensively in other literature, thus the following are just a few points to consider. Boulders are commonly found in glacial tills, riverbeds (current and ancient), and at the bottom of slopes. Glaciers may leave boulders randomly scattered or dumped in a pile at a terminal moraine. Rivers, including paleo channels, will tend to leave boulders in distinct beds. Rock slopes, including ancient ones, may leave boulders in piles with some random bits flung further away. Another source of boulders is the bedrock itself. In heavily weathered areas, most of the ground will be weak, but strong blocks of solid rock (core stones) can be expected. This is of particular concern in karstic environments. Finally, there are areas where parts of the soil structure have been cemented or welded together.

It is also important to consider the location of boulders within the face of the TBM. A boulder near the edge of the face has a greater possibility of being plucked by the faster speed of the cutterhead. A boulder near the edge also has a greater chance of simply being pushed aside during the advance. A nest of boulders near the invert can cause the shield to skim over the top while a nest in the crown can cause local loosening and overbreak.

It is imperative that a thorough site investigation be undertaken before tunnel construction. Such a site investigation must include not only local drilling (seismic and conventional), sampling and geophysics, but also the big picture from the general geology of the area. Some work has been done in looking ahead of the face to find boulders and voids. There has been some success in detecting boulders, but the short range means that the information comes after anything can be done about it.

4 CURRENT METHOD FOR BOULDER BEHAVIOR PREDICTION

Currently, boulder cutting is calculated using simple (Terzaghi) bearing capacity (Navin et al. 1999). The force exerted by a disc cutter is calculated. The force is then applied as a point load to a strip footing (to simulate a boulder). If the resulting pressure is greater than the bearing capacity of the soil, the boulder will move. This calculation is a good approach but it has some shortcomings. In the first place, loads on the cutter are difficult to determine. The thrust of the rams is known, but the exact amount lost to skin friction is not known. The amount of force transferred to each cutter is also unknown. In hard rock, only the disc cutters are in touch with the face, but in soft ground tunneling the entire face of the cutterhead is pushing on the ground. Since the disc cutters are mounted ahead of the other tools, it can be assumed that they will apply a larger proportion of thrust to the boulder than their surface area might indicate (i.e. if the disc cutters take up 2% of the cutterhead area, they will apply more than 2% of the total thrust to the boulders).

Plain bearing capacity has some conceptual weaknesses for the boulder situation. Bearing capacity was developed for shallow footings, therefore the effect of confinement on shear strength is not considered. Particularly in deep tunnels, this makes a huge difference. Embedment is accounted for only by surcharge approximations while actual effects are typically much greater. The bearing capacity equations also imply certain characteristics of the foundation and thereby the boulders. That is, the foundations are assumed flat at the base, perfectly rigid, and infinitely strong. The assumption is reasonable for tall concrete walls on soil, but invalid for boulders, since if a boulder were infinitely strong, it would never break. Bearing capacity also does not reveal a fracture mechanism for a boulder. In short, bearing capacity may not always be the best tool for the job. Therefore, other tools are considered in rest of this paper. While they are not perfect, they do address some of the shortcomings of the bearing capacity model.

5 FINITE ELEMENT MODEL

Finite element modeling (FEM) is a numerical method for solving differential equations. The equations are discretised over small elements (lines, areas, or volumes depending on the number of dimensions) that are tied together at nodes. The elements and nodes create a mesh around the area of interest. A set of simultaneous equations ties together the input and output at the nodes. The equations are set up in matrix format (coefficient matrix, variable vector, and solution vector) to allow numerical techniques to be employed in solving them. The matrices from each element are assembled into a global matrix for the entire mesh and then solved. Gauss Points are the locations in each element where integrals are taken and stresses are calculated.

Stress-strain relationships of the elements are described by constitutive laws, (for example, Hooke's Law from elasticity theory). Forces and displacements are applied at the nodes and spread from there to other elements. Changes in one element impact the behavior of neighboring elements as the resulting stresses are spread about. Finite element modeling is an implicit numerical model approach because unknown values are solved for at all points at one time.

Most of this model is based on FORTRAN 90 code developed by Dr. Ian Smith, University of Manchester (UK), and Dr. D. V. Griffiths, Colorado School of Mines (USA). The code can be found in: *Programming the Finite Element Method third edition* by Smith and Griffiths (1998). This code was chosen over commercial products because of its versatility since it was possible to customize the code for the exact modeling needs and output. The code was compiled using the Essential Lahey FORTRAN 90 compiler (ELF90 4.00b).

In finite element terms, the model used in this research features a rectangular mesh of eight node quadrilaterals. Each element has four integration or Gauss points. Element and mesh sizes are user controlled and all calculations are double precision.

The finite element program models a block of material in front of a TBM cutterhead, as shown in figure 1. Disc cutters are simulated by applying a nodal point displacement. A constant displacement was chosen instead of an applied load because unlike hard rock TBMs, soft ground machines are rarely thrust limited. That is, their speed is controlled by rate of muck removal. It can therefore be assumed that the machine will advance continuously until the boulder fractures or is displaced.

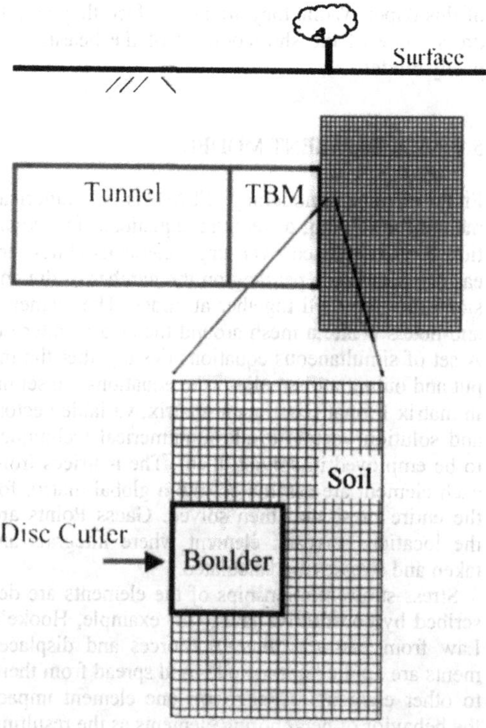

Figure 1. Situation modeled by finite elements.

During a model run, the program first assigns rock or soil properties to each element. It then starts to displace a selected node in a boulder element, simulating the action of a disc cutter. After each displacement, the program runs each Gauss point through the Mohr-Coulomb failure criterion. If its stress level has caused it to fail, the program redistributes stresses to surrounding elements and repeats the failure checking process. When no more elements are at failure and equilibrium has been achieved, the program calculates the reaction forces on the displaced rock node and moves it down another increment. It then returns to checking stresses. The program stops if it cannot redistribute stresses or if the reactions calculated in the previous step are identical to the current reactions (within a specified tolerance). A leveling out of reactions means that the ground has mobilized all of its strength against the displacement. It cannot resist any more and hence, the rock will move under a constant load. Such rock movement indicates that the soil has failed and cannot hold the boulder in place. In the model, all movement results in plastic deformation. The soil deforms plastically under the rock as it moves. The rock deforms plastically in the model as well. Since rock is in reality brittle, significant plastic deformation in the rock is considered a fracture and if a fracture forms in the boulder, the rock is being cut by the disc cutter.

To aid in the boulder behavior analysis, the program creates numerous output files. Figure 2 shows a typical displacement plot in which can be seen the movement of the boulder as well as the deformation of the soil elements around it. Figure 3 shows contours of failed elements. As the boulder is displaced, parts of the boulder and sections of the surrounding soil are forced to yield and fail. These sections appear in white. The dark sections show areas of the soil and boulder that have not yielded enough to fail. The contours in figure 3 show a bearing capacity type failure. Conclusions were drawn from both the numerical and graphical output of the model runs.

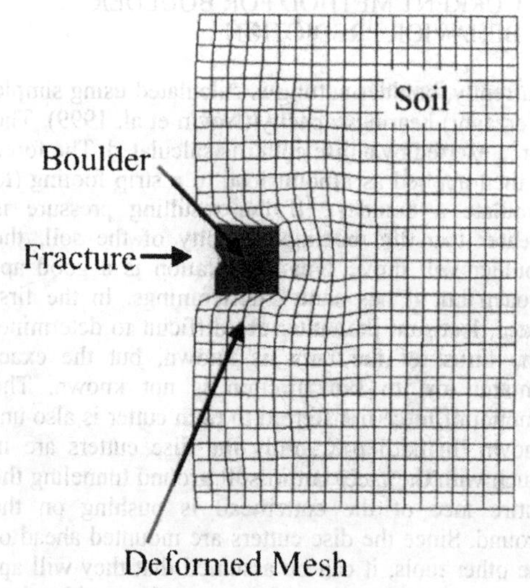

Figure 2. Boulder displacement and mesh deformation.

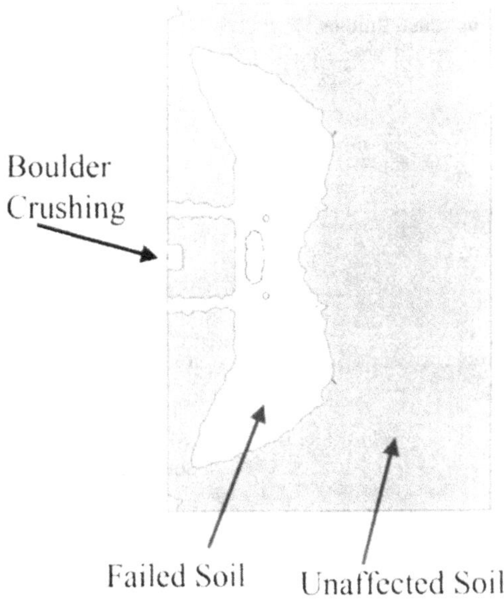

Figure 3. Contour plot of failed elements.

6 CASE STUDIES AND MODEL CORRELATION

The model results cannot be compared directly with field results because of a significant model limitation; point loads could not be modeled. The finite element model distributes loads applied at a single node to the two adjacent nodes. In a very refined mesh, where nodes are very close to each other, the problem is not as severe. However, in the models run for this research, the loaded area changes from a point to an area 0.2 units wide. If one assumes the width of a disc cutter to be around 1cm, the program actually models a disc cutter 20cm wide. Hence, to convert the model results to a real situation, the results have to be multiplied by 20. By modeling, it was found that a boulder 30 times stronger than soil was infinitely strong; it would not break, but move as a solid object. In the field, this would happen when a boulder was 600 times stronger than the soil. While this conversion does not make up for all of the model's shortcomings, it does bring the results closer to reality. For example, a typical soil has a shear strength of around 300kPa. A boulder with a shear strength of 180MPa would therefore be infinitely strong. A shear strength of 180MPa corresponds to a uniaxial compressive strength of 360MPa, a very strong rock.

6.1 *Specific Results*

Figure 4 shows the results of the model to field data comparison. The boulder to soil shear strength ratio runs along the x-axis. The y-axis lists each project by identification number. The horizontal bars show the ranges of possible rock to soil shear strength ratios. Light colored bars indicate that boulders were successfully cut on the project while dark bars mean that boulder cutting was not successful (the ground or machine had to be modified for excavation to continue). The vertical black line indicates the model result for infinitely strong rocks, rock to soil shear strength ratio 600:1. Shear strength ratios to the left of the line indicate that the boulders should be breakable. Shear strength ratios to the right of the line indicate that the boulders are too strong to fracture within the respective soil matrix.

Minimum ratios were calculated by dividing the minimum rock strength by the maximum soil strength. Maximum ratios were calculated by dividing the maximum rock strength by the minimum soil strength. This gave a range of all possible shear strength ratios on the project, from best to worst case.

Project 1, the 4th Tube Elbtunnel in Hamburg successfully fractured boulders. Some boulder movement did occur during two slurry blowouts, but otherwise they were dealt with effectively. The range in figure 4 indicates a somewhat lower chance of breakage, but confirms the field results. It is possible that only weaker boulders were encountered or that the very large TBM was able to ingest boulders instead of relying on disc cutters to break them. The wide range of shear strength ratios is caused by the complex geology where strong soils are found next to pockets of very weak material. Since boulders can be found in any part of a glacial deposit, all combinations of soil strength and boulder strength had to be considered.

Project 2, the Cairo Metro had major boulder problems. Advances were slowed as boulders blocked slurry pipes and rolled around in the face. Since the ground was very weak, instead of breaking boulders, the cutters were heavily damaged. This agrees with figure 4, which shows no chance of boulders breaking. This project also validates the modeled results.

Project 3, the Copenhagen Metro also suffered from boulders. The rocks were damaging cutting tools and knocking the machine off alignment requiring both the ground and the machine to be modified. Figure 4 gives the project an approximately 50-50 chance to cut the boulders. The ground may have been weaker than expected on the drive segment studied. The project is inconclusive in regards to model results correlation.

Project 4, the Kohlbrand-Dueker tunnel in Hamburg had no trouble breaking boulders. The results in figure 4 confirm this. Due to the strong ground, the entire shear strength ratio range is well below the 600:1 line. The case study validates the model.

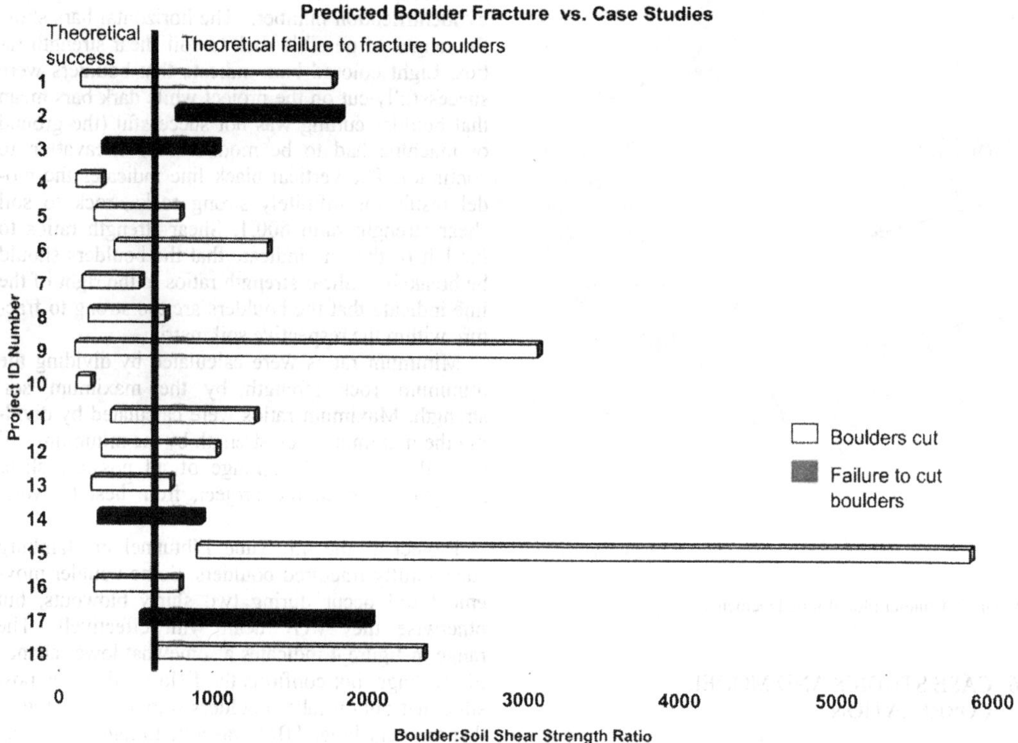

Figure 4. Model and case study correlation.

Project 5, the Fort Lawton tunnel in Seattle had no major boulder cutting problems. Figure 4 indicates that boulders could probably be cut. Only if a stronger boulder was in the weakest material would there be a problem. The project is inconclusive in regards to model results correlation.

Project 6, the Grauholz tunnel in Switzerland ran into several problems, but boulders were not one of them. They were successfully broken by the disc cutters and sent into the slurry pipe via the crusher. Figure 4 shows it somewhat more probable that the boulders would not break. The extremely complex site geology and mixed face conditions made the field results difficult to quantify. The project is therefore inconclusive in regards to model results correlation.

Project 7, the Milwaukee North Shore #9 had splendid success in breaking boulders. This was true for the full size as well as the two pipe jacked units. Figure 4 puts the entire shear strength range to the left of the theoretical line, showing that boulders should be breakable in all cases. The project validates the modeling results more than any other, because the range comes close to the model line, but does not cross it. Projects that have very low shear strength ratios are so far from the line that they do not even challenge the model. The Milwaukee North Shore #9 shows how accurate the model can be.

Project 8, the Sheppard Subway in Toronto had no major problems with boulders during the tunnel drive. While a few made it into the screw conveyer, most were crushed beforehand in the face. Figure 4 shows it most likely that boulders would break. Only in the extreme cases of hard rock and very weak soil, would there be a problem. The project validates the model.

Project 9, the Singapore Lot 706 successfully fractured the boulders in its path. Figure 4 gives it an extremely wide and therefore less useful range. The wide range is due to the geological extremes in the alignment where very weak marine clay covers the top and fills the valleys of very strong alluvial sand. While boulders are much more common in the alluvium, the possibility of encountering boulders in the clay at the bottom of the filled ancient stream valleys must be considered. Hence, the project is inconclusive in regards to model results correlation. If boulders were assumed to be only in the alluvium, the entire range would be well to the left of the line and therefore validate the model.

Project 10, the South Bay Outfall broke boulders with ease. Figure 4 confirms this, placing the entire range far to the left of the infinite strength line. The soil was so strong on this leg of the project that it acted almost like a weak rock. Boulders were easily held in place. The results help to validate the model.

Project 11, the St Clair River railway tunnel between the US and Canada, successfully destroyed boulders. The boulders were weak, but the ground was also weak, leading figure 4 to neither confirm nor deny the results. The range extends to both sides of the model line. The project is inconclusive in regards to model results correlation.

Project 12, the Storebelt tunnel between the Danish isles, was hampered by many difficulties, but not boulder problems. The machines were able to cut the rocks as they encountered them. Figure 4 shows the shear strength ratio range evenly split across the model line. Hence, the project is inconclusive in regards to model results correlation.

Project 13, the U5 subway tunnel in downtown Berlin, was able to break the few boulders it encountered, a fact confirmed by figure 4. The range is almost completely to the left of the model line, indicating that the boulders could be cut. Only if the hardest boulders were encountered in the weaker layers would there be a problem. While it does not make as strong of an argument as other case histories, the project does validate the model results.

Project 14, the VZB Berlin: Fernbahntunnel (long-range railway tunnel), was not able to cut all boulders. Man entry into the face was necessary on four occasions during the eastern tube drive, therefore, the boulder cutting cannot be considered a success. However, many boulders were successfully cut along the remainder of the alignment. Figure 4 agrees with these results as the project bar evenly straddles the model line, leading to the conclusion that some boulders would be breakable, and others not. Hence, the project validates the model results.

Project 15, the Wesertunnel in Dedesdorf, Germany, had no major boulder problems. Documetation described the soil parameters very well, but had very limited data on boulder strength. Given the tunnel's location in the northern German plain, the boulders were assumed to have come from the same glaciers that covered Hamburg. The wide range of boulder strengths thereby assumed gives figure 4 a very wide range for this tunnel. The project discounts model results, due to limited data.

Project 16, the West Seattle Transfer tunnel, was able to cut boulders. Figure 4 agrees mostly with those results. It puts the range mostly to the left of the model line, indicating that boulders could be cut in most cases. Again, detailed boulder data would make the analysis more accurate. The project is inconclusive in regards to model results correlation.

Project 17, the Zimmerberg/Thalwil Pilot Tunnel under Zurich, was unable to cope with the boulders. The entire machine was abandoned and replaced by a hand-mined heading. Figure 4 gives the project a wide range of shear strength ratios, but almost all fall to the right of the model line indicating that the boulders could not be cut. This validates the model results.

Project 18, the John Pier tunnel in England, effectively dealt with boulders, however, figure 4 does not agree. It gives the project virtually no chance of successfully breaking the rocks. The discrepancy is due mostly to limited soil data at the site. The project discounts the modeling results.

Of the 18 projects studied in this dissertation, nine validated the model results, seven were inconclusive in regards to model correlation, and only two discounted the model results. The Singapore project almost validated the results. These numbers are very encouraging given the rather primitive model and limited site data.

One must make a note on the use of figure 4. The range of shear strength ratio assumes that there is the same chance of encountering each ratio. There is no statistical weighting. The figure also looks at the entire project and assumes worst-case scenarios. If one area is weak, the whole project is labeled a failure. Results could be more accurate if each project was divided into sections. Then it could be indicated what areas would have boulder problems. Such action should be taken on individual projects when data is available. However, the results as indicated give a good overview.

While all attempts were made to make the data presented above accurate, in some cases the data simply was not available. Because the author was not on site during the construction of all but one of the tunnels, he had to rely purely on published information. The limited data and associated assumptions led to wide shear strength ratio spreads on some projects. Mostly, the spread was caused by complex geology which had mixtures of very strong and very weak soils filled with boulders of various origins and strengths. Some projects were successful even if the boulder to soil shear strength ratio was very high. While the machines may have been lucky and encountered only weak rocks, it is more likely that the very strong rocks were small enough to be directly ingested by the machine, removing the need to be cut by disc cutters.

6.2 *Sources of Error*

As in most original research, sources of error exist. The shortcomings of the computer model have been discussed in section 1.2. The 600:1 shear strength ratio is an approximation. However, the approximation does not look bad when one examines the range of shear strength values in the field.

The greatest difficulty in making a proper field to model correlation was obtaining exact data. Many projects have dealt with boulders, but most articles and papers about them give only general results which lead to many approximations and assumptions. Geotechnical reports from the site investigation or construction were difficult to obtain and when the reports were available, it was often found that shear strength tests were not carried out, requiring results to be interpreted from other lab tests and ground descriptions.

Other sources of error were also linked to the lack of exact information. Typically, reports and papers would state that boulders of up to a certain strength were to be expected. No mention would be made about the frequency of such boulders or if a high strength was statistically abnormal from the rest of the sample.

When reports included little to no test data on the rocks and soil, numbers had to be surmised from neighboring projects, making the data only useful for general results. For example, if two projects are located in the same geologic region, it could be assumed that boulders came from the same source. That is, if granite boulders were reported on one project and no information was given about the boulders on a nearby project, it could be assumed that the boulders were also granite. However, as shown in the two case studies in Berlin, ground conditions can differ significantly even on adjacent lots.

When no other projects were near, rock properties had to taken from general tables. The main table used was "Rock Substance Properties" on page 867 of Surface Mining 2nd ed. edited by B.A. Kennedy (1990). The table gave many properties for common rocks, including uniaxial compressive strength. The UCS values were given for typical, high strength, and low strength specimens of a given rock type. Such results are good for rough calculations and for checking calculations, but may not apply to specific projects. When rock strengths were taken in this matter, the printed high and typical values became the high and low values in the paper. This was done because boulders tend to be more resistant rock specimens and low strength specimens would have probably eroded. When boulder strength but not type was given, the low value was taken as half of the high.

The documentation also gave only rough estimates for boulder sizes, an unfortunate fact that made correlation more difficult. Figure 4 assumes that the boulders were cut with disc cutters. However, hard boulders small enough to fit into the TBM's muck handling system (screw conveyer or slurry pipes) would not have to be cut, they would be ingested by the machine.

6.3 *Correlation Conclusions*

While correlation between the modeled and field results can be made, the question of usefulness remains. When the boulder to soil shear strength ratio range is small, the results are valid. As shown in figure 4, projects with narrow ranges tended to fall on the correct side of the theoretical success and theoretical failure line. The wide ranges spanned the entire graph and had limited merit.

The key lesson to be learned is that a site investigation must include shear strength properties for both boulders and soil. The strengths for boulders could be obtained by coring and unconfined compressive strength testing (shear strength is half of the UCS) while the soil strengths could be acquired through lab testing, in-situ testing, or through correlation with other index properties such as blow counts. The investigation of tunnel ground conditions must indicate which types of boulder could be expected in each soil type. When just general results are given, the range of boulder to soil shear strength ratios expands and becomes less meaningful. Another factor to be included in the site investigation is the establishment of a narrower range of boulder sizes to determine whether disc cutters or an internal crushing and transportation system would break the rocks.

7 OTHER REASONS DISC CUTTERS FAIL

While the ability to break boulders with disc cutters is mainly a function of the soil behind it, mechanical problems can hamper an otherwise successful boulder break. Even the ideal ground for breaking boulders will stop a TBM if the disc cutters are not working. Abrasive ground can damage discs or bearings so that they no longer roll or cut when boulders are encountered. Bearing seals can fail allowing ground into disc bearings and thereby stopping the discs from rolling. Disc cutter bearings for soft ground must therefore have an overpressure of oil to prevent ground intrusion. The discs also need to move more freely than their hard rock counterparts to keep them from skidding in the soft sediments. Free spinning disc cutters also reduce the chance of plucking boulders from the face. Sometimes not enough discs are mounted or they are spaced improperly. If a TBM can only swallow a 20cm boulder, but the discs are 30cm apart, the cutting cannot be effective. Much thought must go into the placement of tools on the cutterhead. Sometimes boulders are simply too strong to be broken. Disc cutters simply cannot apply enough force to cause them to fracture. It was assumed in this research that TBMs would not be thrust limited. While this is usually true, the possibility of an underpowered machine, particularly in microtunneling must be considered.

8 RESEARCH CONCLUSIONS

The purpose of this research was to determine the behavior of boulders in response to disc cutters in soft ground tunneling. The modeling, while certainly not perfect, yielded many useful results about boulder behavior. The case studies provided a good reference base for future studies and confirmed the model's rock to soil shear strength ratio result. From the modeling and case study correlation, the following conclusions can be drawn. Note that due to space limitations, not all conclusions below could be fully explored in this paper. The reader is asked to reference the dissertation upon which this paper is based for more details.

- During calibration, the program was found to model bearing capacity surface footings in both frictional and non-frictional soils very well. Even embedment runs gave results corresponding with bearing capacity theory.
- Eccentric loads were found to decrease bearing capacity by forcing the boulder to fail out just one side. Therefore boulder cutting can be most effective if the boulder is hit as close to its center as possible.
- Shear strength at the sides of a pushed boulder was found to be fully mobilized. In hand calculations, full theoretical shear resistance can now be applied with confidence.
- Strong boulders increased the bearing capacity of the surrounding soil if they were in the preferred failure path of the pushed boulder. If they were in the shadow of the path or far away, the other boulders had no effect. This was found to be true if boulders were more than 2.5 times the diameter of the pushed boulder away in the direction of tunneling. Boulders were also "far away" if they were more than four pushed boulder diameters away perpendicular to the tunnel alignment.
- Weak boulders were not affected by boulders behind them. Only the boulder underneath the disc cutter was fractured.
- Other boulders, even right behind the affected boulder, were not fractured, only pushed. A concentrated load was found to be the only way to fracture the rocks.
- The ratio of the boulder shear strength to the soil shear strength is the key to determine boulder breakability. As the ratio increases, the crack size in the boulder shrinks and the soil is deformed. The model revealed that if the ratio was greater than 30:1, the boulders would behave like an "infinitely strong" object, moving, but not fracturing. This corresponds to a 600:1 ratio in the field due to the computer modeling a disc cutter as 20 times wider than in reality.
- The actual in situ shear strength is most important to know. The cause of the shear strength (material, water content, overburden) is of secondary importance. The other soil parameters do help in estimating shear strength if direct tests are not done. For example, high overburden will often mean that the soil is stronger since shear strength generally increases with depth.
- Free surface studies showed that the reactions on the disc cutter would be the same as usual (discounting overburden effects) and the shear strength ratio still applied. Test runs were conducted far enough away from the free surface that it had no influence on the bearing capacity. The studies also showed that if the ratio was too high, the boulder would move to the free surface.
- The modeling revealed boulder behavior in frictional soils. Higher friction angles strengthened the soil and weak rock, but had no effect on strong rock. High friction angles were also found to cause numerical instability in the model.
- The lessons learned in this thesis should prove useful in geomechanical modeling in general. In particular, the interaction of soil and concrete in caissons, piles, and deep foundations can use the model as a base.
- The case studies provided a good reference base for future studies and substantiated the model's rock to soil shear strength ratio result (600:1 means boulders will not break).
- The case studies also provided a good qualitative feel for the types of ground conditions that have been overcome using various machines and remediation techniques.
- Of the 18 projects studied in this dissertation, nine validated the model results, seven projects were inconclusive in regards to model correlation, and only two discounted the model results.
- For the results presented in this dissertation to be useful, the tunnel engineer must have knowledge of the ground that will be encountered. A thorough site investigation is necessary to determine both the presence and strength of boulders and what to do if boulders are present.
- Tests must be done to determine the shear strength of the soils along the tunnel alignment. This could be done using traditional lab techniques on bored samples, borehole in situ techniques (probes), vane shear tests, geophysics (shear wave refraction), or correlation with known lab properties such as blow counts.
  The site investigation must also consider what to do if the boulder to soil shear strength ratio is too high (i.e. the rock cannot be cut). The investigators must consider how the ground could be treated to decrease the shear strength ratio and make the boulders breakable. Ground treatment could include different grouting methods from the TBM or the surface. The ground could also be frozen or dewatered. In cases where the shear strength ratio is just slightly too high, allowing the TBM to mine at higher face pressures may be all that is needed.

The knowledge of boulder cutting ability is a key benefit to industry. By determining soil and rock characteristics during the site investigation, potential boulder problem areas can be identified and treated before construction starts. Early treatment is much cheaper because there is no associated down time on the project. If treatment is delayed until problems occur, all tunneling ceases until the ground is strengthened. The early treatment results in lower costs to the contractor and, through reduced claims, to the owner.

9 ACKNOWLEDGEMENTS

I would like to express my sincere appreciation to all of the people, companies, and institutions that have made this work possible. Specifically, I thank Dr. D.V. Griffiths (CSM), Dr. Levent Ozdemir (CSM), Dr. Martin Herrenknecht and Dipl-Ing. Karin Baeppler of Herrenknecht AG, Ray Henn of Haley & Aldrich, and Marco Giorelli of Lovat Inc. For financial support I am most grateful to the CSM Financial Aid Office, Dr. Robert Knecht (CSM), the CSM Coulter Scholarship Committee, AIME for the Henry DeWitt Smith Fellowship, and the CSM Alumni Association. I also thank Golden Software for use of Surfer ® software to help in my analysis. Finally, I would like to express my deepest gratitude to my wife, Tatiana, for her encouragement, support, advice, and understanding during my graduate studies.

REFERENCES

Paper based on:
Goss, Christoph Michael. 2000. *Finite Element Modeling of Boulder Excavation in Soft Ground Tunneling by Earth Pressure Balance and Slurry Machines Using Disc Roller Cutters.* Colorado School of Mines, Golden, CO, USA. Unpublished thesis T-5437.

Anheuser, Dr.-Ing Lothar. Specific Problems Of Very Large Tunnelling Shields. *RETC 1995.* Littleton, CO USA. SME. pp. 467-477.

Burke, Jack. South Bay Outfall. *World Tunnelling* March 1997. London, UK. pp. 47-52.

Call, R. & Savely, J. 1990. Open Pit Rock Mechanics. *Surface Mining 2nd ed.* Kennedy, B.A. (ed). Littleton, CO USA SME. pp. 867.

Craig, R.F. 1995. *Soil Mechanics 5th ed.* London, UK: Chapman & Hall.

Craig, Rodney. Under The Elbe to the Port of Hamburg. *World Tunnelling* April 2000. London, UK. pp. 132-139.

Darling, Peter. Coming Clean Over Storebaelt. *Tunnels & Tunnelling.* October 1993 V25 N10. London, UK.pp. 18-26.

Dowden, P & Robinson, R. Coping With Boulders In Soft Ground Tunneling. *RETC 2001.* Littleton, CO USA. SME. pp. 961-977.

Finch, Alan. St Clair River Tunnel: past & present. *Tunnels & Tunnelling International,* Nov 1995. London, UK.

Griffiths, D.V. 1981. Computation of strain softening behaviour.: Acorn Press, *Proceedings of a Symposium on the Implementation of Computer Procedures and Stress Strain Laws in Geotechnical Engineering.* Desai & Saxena ed. Durham, NC USA. pp. 591-604.

Harrison, Neville. Driving the New St. Clair River Tunnel. *Tunnels & Tunnelling International.* January 1995. London, UK. pp. 17-20.

Heuer, Ronald. 1974. Important Ground Parameters in Soft Ground Tunneling. *Subsurface Exploration For Underground Excavation and Heavy Construction, New England College, Henniker, NH.* New York. ASCE. pp. 41-55.

Heuer, R. & Virgens, D. Anticipated Behavior of Silty Sands in Tunneling. *RETC 1987.* Littleton, CO USA. SME.

Lechman, J. B. 2000. *The Progression of Failure in Earth Slopes by Finite Elements.* Golden, CO USA: Colorado School of Mines Unpublished M.S. Thesis.

Lyon, John. Foams and Polymers for Enhanced Excavation. *World Tunnelling,* March 2000. London, UK. pp. 95-97.

Maidl, Herrenknecht, & Anheuser. 1996. *Mechanical Shield Tunneling.* Berlin, Germany: Ernst & Sohn.

Navin, Kaneshiro, Stout, & Korbin. The South Bay Tunnel Outfall Project: San Diego, California. *RETC 1995.* Littleton, CO USA. SME, pp. 629-644.

Oatman, M. & Lenahan, L. West Seattle Tunnel. *RETC 1997.* Littleton, CO USA:.SME. pp. 749-763.

Ozdemir, L., Miller, R., & Wang, F. 1978. *Mechanical Tunnel Boring Prediction and Machine Design.* NSF APR73-07776-A03. Colorado School of Mines, Golden, CO, USA..

Parker, Harvey. 1996. Geotechnical Investigations. *Tunnel Engineering Handbook 2nd ed.* Bickel, Kuesel, & King (ed). New York, USA. pp. 46-56.

Prandtl, Ludwig. 1921. Ueber die Eindringungsfestigkeit (Haerte) plastischer Baustoffe und die Festigkeit von Schneiden.: *Zeitschrift fuer angewandte Mathematik und Mechanik,* Vol.1, No. 1. pp. 15-20.

Quebaud, Sibai, & Henry. 1998. Use of Chemical Foam for Improvements in Drilling by EPBS in Granular Soils. *Tunnelling and Underground Space Technology,* Vol. 13 No. 2 .Oxford, UK. pp. 173-180,.

Richards, Ramond, & Herrenknecht. Slurry Shield Tunnel on the Cairo Metro. *RETC 1997.* Littleton, CO USA. SME, pp. 709-733.

Ringen, A. & Raines, G. 2000. Ground Improvement For Microtunneling/Pipe Jacking Applications. *Mechanical Tunneling, Raise Boring, and Shaft Drilling Short Course.* Golden, CO USA.

Rostami, Jamal. 1997. *Development of a Force Estimation Model for Rock Fragmentation With Disc Cutters Through Theoretical Modeling and Physical Measurement of Crushed Zone Pressure.* Golden, CO, USA Colorado School of Mines, Unpublished thesis.

Smith, I.M. & Griffiths, D.V. 1998. *Programming The Finite Element Method 3rd ed.* Chichester, UK; John Wiley & Sons.

Wallis, Shani. Elbe Tunnel: Cutting Edge Technology. *Tunnels & Tunnelling International* January 2000 V32 N1. London, UK. pp.24-27.

Williamson, Traylor, & Higuchi. Soil Conditioning For EPB Shield Tunneling On The South Bay Ocean Outfall. *RETC 1999.* Littleton, CO USA. SME.

North American Tunneling 2002, Ozdemir (ed.)
© 2002 Swets & Zeitlinger, Lisse, ISBN 90 5809 376 X

Planning the construction of new particle accelerators

Chris Laughton
Fermi National Accelerator Laboratory, Illinois, USA

ABSTRACT: A new generation of particle accelerator projects is being planned. The accelerator systems will be housed in tunnels up to 230 km in length, and large experimental caverns and access shafts will be placed along the alignment. Early conceptual layouts for these accelerators have incorporated a wide range of tunnel design and construction ideas and addressed some of the more critical end-user requirements. These studies were reviewed at a workshop held at Snowmass, Colorado, in July 2001. This paper summarizes the findings of the workshop and outlines directions for future work.

1 INTRODUCTION

During a three-week meeting of the international particle physics community held at Snowmass, Colorado, a two-day tunnel workshop was convened to review design concepts and layouts for three large accelerator physics projects. The projects reviewed were the Tera Electron-Volt Superconducting Linear Accelerator (TESLA), the Next Linear Collider (NLC), and the Very Large Hadron Collider (VLHC). Various locations in Germany and the United States are being considered for construction of these new accelerators. A site has already been identified for TESLA, adjacent to the Deutsches Elektronen-Synchrotron (DESY) facilities, in the northern suburbs of Hamburg, Germany. US sites are being studied in northern California and northern Illinois; these sites are adjacent to, or within easy reach of the Stanford Linear Accelerator Center, and the Fermi National Accelerator Laboratory (FNAL), respectively. A large amount of underground construction will be required to house these accelerators, and construction costs will account for up to fifty percent of the total budget. Validating key requirements and developing cost-effective designs and layouts have been given a high priority at the early stage of each project. As a start to this effort, tunneling professionals were invited to attend the workshop. The tunnel team consisted of Messrs. Babendererde, Bauer, Frame, Hilton, Lachel, Neil, and Wightman. The team provided feedback on geologic, geotechnical, design and construction aspects of the projects.

2 TECHNICAL REQUIREMENTS

In many respects, the design requirements for accelerator tunnels are not unlike those associated with electrified rail or metro systems, with emphasis placed on satisfying alignment, stability and dryness criteria. However, in the case of accelerators, stability needs may be more stringent; for example, the Next Linear Collider has a requirement that differential floor displacements over a distance of one kilometer be kept below +/- 1.5 mm over a nine-month operating period (Phinney et al., 2001).

In developing the concept designs for these accelerator housings, specific attention is also being paid to the characteristics of ambient ground vibration. Vibrations in key frequency and amplitude ranges can have a deleterious impact on accelerator operation. Meeting this combination of stability and vibration criteria can prove difficult in certain ground units and at certain locations.

3 SUBSURFACE CONDITIONS

At the time of the workshop, none of the projects had performed any site-specific subsurface investigations. An alignment has been identified for the TESLA project in the northern suburbs of Hamburg (Richards, 2001). At this site, confidence in ground conditions is relatively high based on existing regional data sets. This data suggests that site conditions will be similar to those encountered during the construction of the Hadron Electron Ring Accelerator (HERA), an accelerator housing built adjacent to the DESY campus.

The NLC and VLHC projects are studying a variety of sites and layout and design options. However, a modicum of site-specific geotechnical data needs to be gathered before plans and costs for tunnel construction can be developed with confidence. Geotechnical data is particularly important to assist in the characterization of sites where weak expansive shale and sandstone units have been identified.

4 CONSTRUCTION ISSUES

Given the lengths of the accelerator tunnels, Tunnel Boring Machine (TBM) technology has been identified for construction. For the 33km-long, 5.2m diameter TESLA tunnel, slurry TBM technology has been selected as the construction equipment of choice, with single-pass ground support provided by a gasketed segmental lining. This system was successfully employed in the excavation of the HERA facilities through similar saturated soil conditions.

The requirements and designs for the accelerator sites in California and Illinois are still under development, but conceptual planning calls for single bore tunnel lengths up to 230 km for the VLHC and twin-bore tunnel lengths up to 32 km in length for the NLC. Over twenty reference-sites have been studied during the initial conceptual stages of design. A wide variety of layouts for these sites have been drawn-up, including the use of single and twin-tunnel excavations, constructed as cut and cover and mined structures (Atkinson, 2001). In cross-section, twin tunnels have been vertically stacked or placed side-by-side. A more unconventional twin tunnel arrangement has also been conceived, whereby the running tunnels are offset both vertically and horizontally. Ground support concepts for these tunnels range from pattern rock bolting and shotcrete to a segmental lining. Support was selected based on assumed rock mass conditions of the host geologic units. Finished tunnel diameters used in concept studies have ranged from three to six meters. These studies have been used as a basis for developing parametric cost estimates (Lach et al., 1998), C.N.A. Consulting Engineers and Hatch-Mott-MacDonald (2001) and have permitted an initial round of value engineering studies to be undertaken (Laughton, 2000).

5 VALUE ENGINEERING AND OPPORTUNITIES FOR INNOVATION

Although key tunnel design criteria need further definition, it is already clear that the viability of these accelerator projects will be greatly enhanced if the cost of underground construction is reduced.

Early engineering work conducted at Fermilab in conjunction with The Robbins Company, (1999) led to the identification of potential cost-reduction strategies based on the increased use of automation and improvements of TBM sub-systems' performance. Further investigation of opportunities for reducing costs through design, constructability and innovation paths will be explored during the design process as more detailed information on requirements, site-specific characterization and tunneling-related research initiatives becomes available.

6 COMMUNITY ISSUES

In parallel with the tunnel workshop, a workshop was held to discuss community issues associated with the construction and operation of large accelerators. Past experience gained on the construction of accelerator facilities such as the Large Electron Positron, Superconducting Super Collider and most recently the NuMI-MINOS Project has shown the value of public participation in the building of such facilities. Early involvement of the local community is key if the value and impacts of a project are to be properly communicated and fairly evaluated within the community. No tunnel job can guarantee "zero complaints," but if proper explanations are offered at the outset, realistic expectations set during design and shown to be met during construction and operation, complaints can be minimized and good neighbor relations maintained. As a first step in this community dialog FNAL, in conjunction with Northern Illinois University Public Opinion Laboratory, is conducting a community survey (Northern Illinois Public Opinion Laboratory, 2001). Long-range planning strategies will be developed based on the findings of this survey.

7 CONCLUSIONS

The tunnel workshop served to provide a bridge between the physics community and the tunneling industry and offered the projects' engineering staff an early opportunity for an exchange of ideas and evaluation of plans. Feedback from the industry professionals has greatly advanced the maturity of the design concepts and helped provide focus for follow-on work.

As the projects progress, additional input from experienced tunnel designers and builders will be needed to support ground characterization and site selection work, and evaluate the merits of alternate layouts and design solutions. It is hoped that this workshop will be the first of many such interactions between the physics and tunnel communities to ensure that the prospective tunnel owners develop practical and cost-effective solutions to their tunnel engineering challenges.

REFERENCES

Atkinson et al. "Report of the Fermilab Committee for Site Studies". Fermi National Accelerator Laboratory, Illinois, July 2001.

C.N.A, Consulting and Hatch-Mott-MacDonald. "Estimate of Heavy Civil Underground Construction Costs for a Very Large Hadron Collider in Northern Illinois." Fermi National Accelerator Laboratory, September 2001.

Lach J. et al. "Cost Model for a 3 TeV VLHC Booster Tunnel". Fermi national Accelerator Technical Memorandum 2048, April 1998.

Laughton, C. "Value Engineering the Construction of Long Tunnels in the Dolomites of Northern Illinois, United States of America". Milan, June 2001.

Phinney, N. et al. "2001 Report on the Next linear Collider". Stanford Linear Accelerator Center, June 2001.

Public Opinion Laboratory, "Fermi National Accelerator Community Survey". Northern Illinois University, 2001.

Richard F. et al. "The Superconducting Electron-Positron Linear Collider with an Integrated X-Ray Laser Laboratory – Technical Design Report". DESY, March 2001.

Robbins Company. "Tunneling Cost Reduction Study". Fermi National Accelerator Laboratory, July 1999.

North American Tunneling 2002, Ozdemir (ed.)

St. George Tunnel rehabilitation

T.P. Kwiatkowski, P.E.
Jenny Engineering Corporation

D.R. Klug
David R. Klug & Associates, Inc.

D. Raich
Rosewood Contracting Corporation/A.F.C. Enterprises, Inc. (JV)

ABSTRACT: The New York City Transit Authority developed plans to rehabilitate the brick arch St. Georges Tunnel in Staten Island, New York. The rehabilitation would strengthen the existing brick arch with fabricated structural steel members as well as provide weep holes and a piping system in an attempt to eliminate water infiltration into the tunnel.

A Value Engineering proposal was developed and accepted by the Transit Authority, which consisted of a NATM type waterproofing system, (PVC membrane with geotextile fabric), along with shotcrete and lattice girders to provide structural support of the tunnel. The advantages of this system will be discussed along with state of the art survey techniques, which were utilized to optimize each cross section along the tunnel.

1 INTRODUCTION

The St. George Tunnel, constructed in the late 1800's, is located in Staten Island, New York, adjacent to the Manhattan–Staten Island Ferry Terminal. It is owned and operated by the Staten Island Rapid Transit Operating Authority (SIRTOA). The tunnel is a 24-foot wide, twin track, horseshoe. The tunnel arch is constructed of five layers of bricks with approximate thickness of 2'-0". The arch is supported by 14' high stone sidewalls.

The tunnel is located just to the west of the ferry terminal and extends from Station 0+00 at the east portal to Station 5+09 at the west portal. The tunnel between Stations 0+00 and 2+75 is located beneath a U.S. Post Office parking lot and the remainder of the tunnel is below New York City Department of Transportation property. During construction of the U.S. Post Office in 1930, fill placed for the parking lot increased overburden depth from the original 7 feet to approximately 23 feet, which resulted in significant settlement in the tunnel between Stations 0+00 and 2+75. In 1948, a tunnel survey was conducted. The crown was found to have settled by 1'-6" and the springline was found to be displaced outward by up to 5 inches. In the area near Station 0+75, the tunnel crown is essentially flat. In addition, the entire tunnel experienced groundwater infiltration, and during winter, icicles formed creating hazardous conditions for the train operators and passengers.

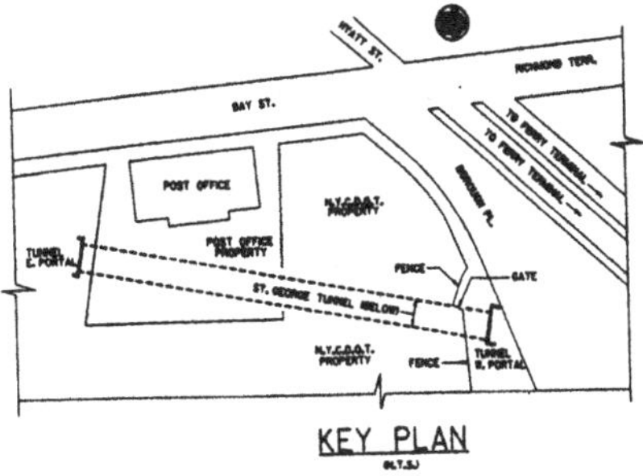

KEY PLAN

2 VALUE ENGINEERING

The rehabilitation of the St. George Tunnel was undertaken to address the issues of tunnel arch deflections and water infiltration, by providing new structural support and a new drainage system. The design of the project was performed by the Metropolitan Transit Authority (MTA) acting on behalf of SIRTOA. The design required installation of steel ribs on 4-foot centers. The tunnel was to be surveyed at each rib location and the steel rib fabricated to the configuration of each arch to provide maximum clearances from the top of rail. The installation of the design also required a drainage system throughout the entire tunnel length, with the installation of three-inch thick shotcrete with welded wire fabric, above the tunnel springline, between the east and west portals.

Rosewood Contracting Corporation/A.F.C. Enterprises, Inc. (JV) of Glendale, New York was the successful bidder for the rehabilitation of the St. George Tunnel. Following award, the Joint Venture contacted David R. Klug, President of David R. Klug & Associates, Inc. to review construction products, methods and techniques, which could be utilized for tunnel rehabilitation. It appeared that a modified NATM support consisting of shotcrete, lattice girders, welded wire fabric, and a proven PVC waterproofing system with a state-of-the-art tunnel survey program would offer a solution to the problems.

The following issues needed to be addressed:

1. Performance of a detailed survey of existing tunnel to establish engineering baseline for a Value Engineering design.
2. Maintenance of at least one live track throughout the entire period of tunnel rehabilitation.
3. Installation of waterproofing and drainage systems, which would completely seal the tunnel arch from springline to springline.
4. Cost savings had to be realized in order to make the Value Engineering alternative attractive to the owner.

Once the advantages of the NATM system became evident, the joint venture retained the services of Jenny Engineering Corporation (JEC), of Springfield, New Jersey to evaluate the feasibility of a Value Engineering proposal for the tunnel zone between Stations 0+00 to 2+75. After a review of the contract documents and evaluation of the existing tunnel conditions, JEC had concluded that utilization of a NATM design approach was not only feasible, but would also result in cost savings. In lieu of the fabricated rib support system with three inches of shotcrete with welded wire fabric, a shotcrete/lattice girder system could be used. The system would provide the flexibility to support varying tunnel cross sections and at the same time provide the structural support necessary of withstanding full overburden loads. The MTA determined that the Value Engineering proposal was technically acceptable in concept, subject to the fulfillment of all conditions stipulated in the specifications.

The specifications required submission of complete design calculations, as well as plans and specifications of the proposed work. These submissions were completed and the MTA accepted the reinforced shotcrete lining as an equivalent support system to that designed.

3 DESIGN

Borings in the vicinity of the tunnel indicated that the overburden consisted mainly of sand and gravel, with traces of silt, with depth reaching 23 feet. The design loading utilized full overburden. The design assumed that the tunnel support system would include lattice girders, shotcrete, and welded wire fabric, and would serve as the final lining of the tunnel. In order to maintain maximum clearances from top of rail, the design thickness of shotcrete was limited to 9 inches, which was equivalent to the depth of steel ribs proposed in the original design. The shotcrete utilized in the design had a compressive strength of 7,000 psi.

A number of limitations had to be considered during the design. They included live rail construction conditions, selection of a lining system that would be structurally equivalent to that originally designed, maintenance of maximum tunnel clearances, and providing a drainage/waterproofing system to prevent water infiltration onto the tracks.

Following a detailed evaluation, a reinforced shotcrete arch with lattice girders, welded wire fabric and waterproofing system was selected for design. The reinforced shotcrete/lattice girder arch was divided into three segments: two leg segments, and one arch section. This permitted construction to be completed while keeping one track in service at all times. The legs were installed in pockets cut into the brick lining on each side of the tunnel. The pockets permitted the reinforced shotcrete lining system to be founded directly on the stone sidewalls of the original tunnel and permitted arch loads to be transmitted directly onto the stone blocks.

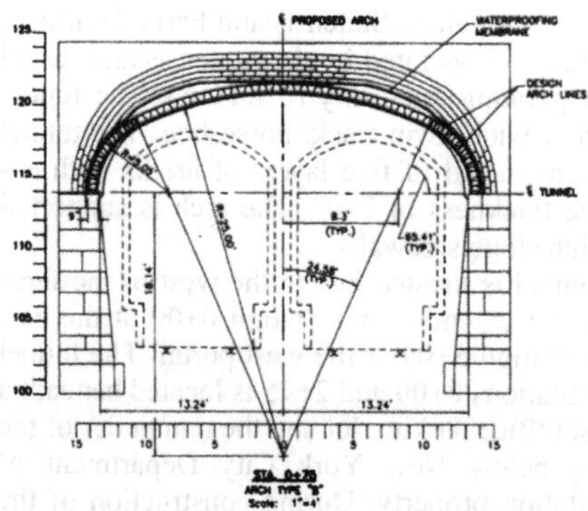

In order to insure the existing tunnel brick arch did not experience any adverse condition from the construction of the pockets, approximately one month prior to the commencement of construction, a test pocket near the east portal was cut. The brick adjacent to the pocket was monitored for movement and after one month of non-movement in the tunnel-lining, the rehabilitation proceeded as planned.

4 CONSTRUCTION

The original tunnel liner was large random size stone to springline with a 5-course brick arch. Most of the rehabilitation work was performed above the springline in the brick arch area. The tunnel provides two-way tracks, eastbound, (EB) and westbound, (WB). Construction began in October 2000 and was completed in November 2001, with one, three-month shutdown due to cold weather and winter conditions. The rehabilitation work was accomplished during non-rush hour periods, allowing one track to stay open and provide uninterrupted rail service. Each working shift allowed six to seven hours track time, and work was generally accomplished during the day, five days per week, with a five-man crew.

SIRTOA provided the Joint Venture with a work train that was comprised of two, 10 foot x 50-foot flatcars, and one diesel engine. To support the work train, a SIRTOA work crew, that consisted of a power man, motorman, break man, and a conductor was provided. The project consisted of four major phases of work: cutting the pockets in the brick, installing the "legs," installing the tunnel waterproofing system, and installing the arch section. Therefore, the flatcars were modified to support each phase and were equipped with sufficient materials to complete the work on each shift. In addition, one of the flatcars was modified to include scaffolding that would allow the work platform on the flatcar to slide against the tunnel wall to ease construction operations. After completion of the work, the platform would slide back past the edge of the flatcar and permit the work train movement.

In addition, to permit the supply of temporary utilities, compressed air, water and shotcrete into the tunnel, two, 6-inch diameter holes were drilled from the surface within the construction yard through the crown of the tunnel. The holes were drilled on the centerline of the tunnel at Station 3+00. One hole was used for compressed air and water, and the other was used for a shotcrete slick line.

5 SURVEY/TUNNEL SCANNING

Prior to the award of this contract, a quarterly tunnel monitoring program had been in effect to detect movement within the tunnel arch. The monitoring program consisted of measurements taken at 7 points transversely across the tunnel arch at 27 stations throughout the tunnel. This monitoring operation was done utilizing conventional survey techniques. Each monitoring session took approximately three days with a three-man survey party.

The contract required that the monitoring program be continued throughout construction. Since conventional survey techniques were relatively time consuming, and only produced information at a specific location or point, the Joint Venture proposed the utilization of a state-of-the-art tunnel survey system called DIBIT to the MTA.

The DIBIT tunnel scanning system was developed and patented in Austria, and is offered in North American by ILF Consultants, Inc., a tunnel engineering firm located in Fairfax, Virginia and Oakland, California. The DIBIT system has been successfully used on major tunnel projects throughout the world. The St. George Tunnel Rehabilitation Project was the first use of the DIBIT system in the United States. DIBIT uses two digital cameras mounted at a fixed distance on a rotatable base producing stereoscopic images of the tunnel surface. Through highly sophisticated software, the images are translated into 3D coordinates with a grid density of 3/8 square inch.

The survey of the entire St. George Tunnel took approximately four hours with two men. The resulting data accuracy is within plus/minus, 1/8 of an inch. The coordinate data can be used to detect movements, deformations, and to calculate very accurate volumes and areas along the tunnel alignment. Plots of the tunnel profile, at essentially any station can be easily obtained. The DIBIT system proved to be a very useful and effective tool for tunnel monitoring, laying out/installing lattice girders, checking the tunnel profile accuracy, and maintaining overall survey control in the tunnel.

In addition, the results from the DIBIT system were utilized to standardize arch geometrics for analysis purposed, and assisted in the installation of the lattice girder legs. After completion of the structural analysis, three arch geometrics were selected along the tunnel. Within each specific geometry defined by stations, lattice girders were interchangeable at any station to insure proper installation of the lattice girder arch section between the two leg sections. Consequently, it was imperative that the legs be installed properly and accurately. The DIBIT provided the accuracy needed for installation.

Therefore, the DIBIT system worked very well, not only for tunnel survey, but also for determination of shotcrete quantities, optimization of tunnel lattice girder fabrication, and accurate installation of the lattice girder legs and arches.

6 TUNNEL REHABILITATION

As discussed previously, limitation and working restrictions within the tunnel not only influenced design details but also dictated the construction sequence. As a result, the optimum construction sequence included the following:

- Cutting the pocket in the brick lining.
- The installation of the lattice girder in the pocket followed by shotcrete.
- Continuation until one side of the tunnel was complete.
- Move to the other side and repeat the process.

- The installation of tunnel waterproofing and drainage system.
- The installation of the arch lattice girders and welded wire fabric.
- Shotcreting one side of the tunnel.
- Shotcreting the other side of the tunnel.

7 LATTICE GIRDER LEGS

The Value Engineering design required 101 lattice girders (CP 95/8/11) to be installed from tunnel Stations 0+00 to 2+75, at a spacing of 2'-6", and 2'-9" center-to-center. The DIBIT survey data was utilized to create a profile at each girder location. From this data, the best-fit girder geometry was superimposed onto the profile and a cut drawing and leg set drawing were produced. The field crew was given these drawings and each cut and leg set was performed accordingly.

The construction process began by installing a flexible cut guide on the brick face, the cut guide provided line and depth control. The vertical cut was made using an ICS hydraulic chain saw. Once the vertical cuts were made, the guide was removed and the brick was spaded out using 15 lb. hand spaders. The lattice girder leg was then bolted to the stone within the pocket utilizing two 7/8-inch pins. Once the leg was secured in the correct position, the leg and pocket were shotcreted using Master Builders, Shotpatch 21 with compressive strength of 10,000 psi. The dry mix process was chosen for this application due to the small quantity of shotcrete required for each shift. Allentown Equipment supplied the admixtures and the mixing and placing equipment used for shotcreting.

The project specifications limited spacing of pocket excavation to minimum 8-foot centers. This required three passes to complete one side of the tunnel. After completion of one side of the tunnel, the procedure was repeated on the other side. However, the second side of the tunnel included an additional step. To insure that the arch section of the lattice girder would fit properly, the arch piece was temporarily set and removed once the leg was secured in the proper position before shotcreting of the leg took place.

8 TUNNEL DRAINAGE SYSTEM

After all lattice girder legs were set and shotcreted, the tunnel drainage system was installed. The tunnel drainage system consisted of waterproofing system (a geotextile fabric (fleece) and PVC waterproofing membrane), collection pipes, drainage gutters, and gutter drains. The system was supplied and installed by WISKO America, Inc.

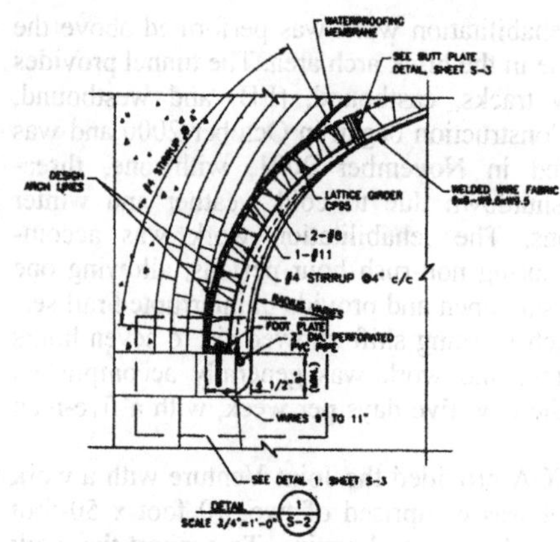

In order to install the drainage system on one side of the tunnel at a time, special scaffolding was erected on the flatcar. The 22 oz. fleece was installed followed by 2.5 mm PVC membrane. At each lattice girder leg location, a window was cut in the waterproofing. This window allowed a structural tie-in to the lattice girder leg for the final shotcrete lining. At the interface between the brick arch and the stone sidewall a horizontal termination strip made out of PVC waterproofing membrane was installed, and a 2-inch perforated PVC pipe was wrapped in the termination strip and welded to the arch portion of the waterproofing membrane. Outlet drains were installed on 50-foot centers.

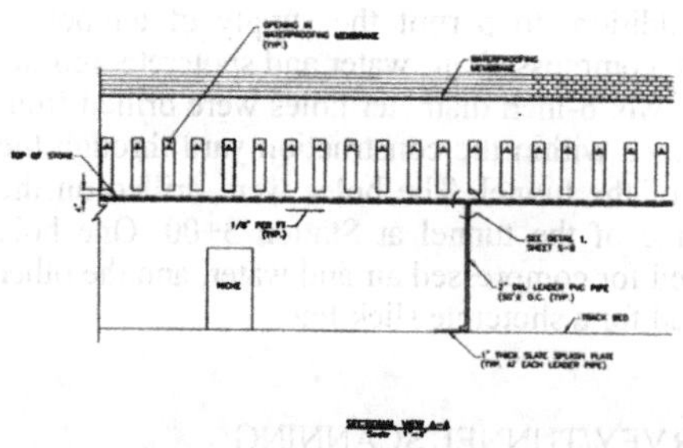

9 LATTICE GIRDER ARCH AND WELDED WIRE FABRIC

Upon completion of the tunnel drainage system, the lattice girder arch piece was then installed. In order to keep trains moving within the tunnel, the arch had to be installed from one side of the tunnel. A hydraulic erection arm was constructed to manipulate the arch piece into place. It was determined during design that the arch section would be installed from the westbound side of the tunnel. This permitted the eastbound leg to be fabricated with a hole in the butt plate for acceptance of a pin welded to the butt plate of the arch piece. The arm griped the bottom bars of the lattice girder and rotated it into position with the leg on the eastbound side of the tunnel. A pin welded to the butt plate of the arch piece was stabbed into a hole on the butt plate of the eastbound leg. The arch was then raised into position and bolted to the westbound leg. Although it was not part of the design requirements, 4 foot x 4 foot welded wire fabric (4x4 – W2.9xW2.9) was installed above the lattice girder arch to help support the shotcrete to be shot against the PVC waterproofing membrane. Once all the arch pieces were in place, the eastbound joint was bolted and the remaining welded wire fabric was installed.

10 FIRST PASS SHOTCRETE

The next phase of construction was first application or pass of shotcrete. The intent of the first pass shotcrete was to cover all of the waterproofing membrane and create a shell that the final fill shotcrete could adhere to. Prior to the shotcrete application, the drainage systems windows were trimmed and a hydrophobic caulk applied around the perimeter of the window. The lower, outside piece of welded wire fabric was also installed at this time. Each sheet of welded wire fabric was approximately 3 foot x 6 foot, and rolled to the radius of the girders. The sheets were tied to the lattice girder legs and pined in the center of the window with ½-inch expansion anchor.

A three-inch slick line was installed from the surface, down the drop hole, and reduced to a 2-inch line, which was run along the tunnel rib. A 7,000 psi wet shotcrete was delivered to the site in transit trucks from a local supplier and pumped from the surface to the nozzle. The maximum pump distance was approximately 400 feet.

At the start of each shift, the flatcars were positioned in the work area and a protective curtain was raised between the rails. Wooden shields were folded out against the tunnel rib for protection of the stone below the arch. Protective tarps were also laid between the rails on the invert to catch any rebound. Shotcreting was performed from a scissors lift located on a gondola flatcar.

All of the admixtures used in the shotcrete mix were provided by Master Builders including the accelerator, Meyco SA160. Rebound and waste was approximately 30% and accelerator dosage ranged from 3% to 6%, and the 7-day breaks of shotcrete samples were all above 7,000 psi. All shotcrete equipment was supplied by Allentown Equipment. Normal operations required 2 - 8 cubic yard trucks per shift, with the majority of shift time spent on setup and cleanup.

11 FINAL PASS SHOTCRETE

Upon completion of the first pass shotcrete, the shotcrete was cleaned off the lattice girders and the upper piece of welded wire fabric (6x6 – W9.5xW9.5) was tied to the girders. Each sheet of welded wire fabric was approximately 3 feet x 10 feet, and rolled to the radius of the girders. The final pass shotcrete was applied in the same manner as the first pass shotcrete. The thickness ranged from 6 to 18 inches. The shotcrete was brought out to the welded wire fabric and trimmed off with shovels to provide a smooth surface for the final 1½-inch layer. To provide a smooth transition from one half of the tunnel to the other, the protective curtain along the centerline of the tunnel was dropped and the joint was shot between trains. The final shotcrete finish was a nozzle finish.

12 SUMMARY

The project was completed under budget and approximately three months ahead of schedule, with no lost time or reportable accidents. The success of this project can be attributed to the acceptance of a Value Engineering proposal, which utilized a shotcrete support system as a final lining system because it was the best suited system for the project.

13 SUPPLIERS

The following companies provided products, equipment, and/or services to make this project a success.

- American Commercial, Inc. – lattice girders
- WISKO America, Inc. – drainage system
- Master Builders, Inc. – shotcrete products
- Allentown Equipment Co. – shotcrete equipment
- KSL Construction Systems, Inc. – tunnel reinforcement
- ILF Construction, Inc./DIBIT Scanning System – tunnel survey/scanning equipment.

North American Tunneling 2002, Ozdemir (ed.)
© 2002 Swets & Zeitlinger, Lisse, ISBN 90 5809 376 X

Monitoring measures for underground rehabilitation of Line 4 Bucharest, Romania

V. Ciugudean-Toma, I. Stefanescu & O. Arghiroiu
METROUL SA Company, Bucharest, Romania

ABSTRACT: Bucharest Metro Line 4 has 4.7 km in length and it is an extension of Line 1, from Nicolae Grigorescu Station to Linia de Centura Station. This line includes two other stations (1 Decembrie 1918 and Policolor). The design for implementing Line 4 started in 1988 that including site investigations, hydrogeological studies and urban constraints (public utilities, buildings) surveys. The construction works began in 1989, as well as the monitoring program, which included surface marks, settlement pins in buildings, tunnel convergence pins and piezometers. The construction works for Line 4 were stopped in 1996, leading to flooding of tunnels and stations. Now, for the completion of works and the rehabilitation of Line 4 structures it is necessary to evaluate the geotechnical and tunnelling conditions, to safeguard the existing infrastructure.

1 INTRODUCTION

The city of Bucharest, capital of Romania, is the largest city, occupying 227 km^2 of urban area and 1571 km^2 including suburban areas, with an average diameter of 21 km. Its population is around 2.3 million inhabitants, which is approximately 10% of the total population of Romania. Considering this, there is a great demand for an effective mass transit system for Bucharest.

Bucharest Metro Company was established in 1975 and began the construction of metro network. Nowadays it encompasses 71.6 km of underground lines, 45 stations and 4 maintenance depots. The Bucharest Metro lines develop basically on the North-South and East-West directions, with some rings in the centre part of the city. This system allows an efficient integration among several neighbourhoods, university and industrial districts of the city and also with other types of transit systems (busses, trams and trains). Considering the line length of all mass transit systems in Bucharest, the length of the metro lines is only 3.7%, but it carries more than 27% of the total mass transit volume.

Presently, the Bucharest Metro network is under expansion. There are 8.4 km and six stations under construction. Line 4 is an extension on Line 1, from Nicolae Grigorescu Station to Linia de Centura Station. It is 4.7 km long and includes two more stations (1 Decembrie 1918 and Policolor). The construction works in Line 4 were stopped, leading to flooding of tunnels and stations. For completion of works and rehabilitation of Line 4 structures, it is necessary to evaluate the geotechnical and tunnelling conditions, to safeguard the existing infrastructure. The map from Figure 1 illustrates the underground lines in Bucharest city, having the approximately area of a circle with a diameter of about 22 km.

2 SPECIFIC GEOLOGICAL AND HYDROGEOLOGICAL CONDITIONS OF BUCHAREST

From the geological point of view, Bucharest is situated between two views in the central part of Moesic Platform, in the so-called Romanian Field, which represents a major depresionary unit that stretches along the Danube River.

The morphology of the area (see Figure 2) on which the town has developed, is a plane with a microrelief resulting from erosion and sedimentary processes, which occurred along two valleys, the valley of Dambovita River in the South and the valley of Colentina River in the North. The two streams are almost parallel with a NW-SE trend, with old large meanders, close to which there are small depressions in, which accumulated the surface waters and the water from the springs situated at the bottom of the slopes. Marshes were formed, that are now hydrotechnically drained. Bucharest has three typical geomorphological zones: 1) the lower meadows on the riversides; 2) the field between the rivers, with altitudes up to 90 m, in the N-W, and up to 75 m above the Black Sea level in S-E; 3) the higher

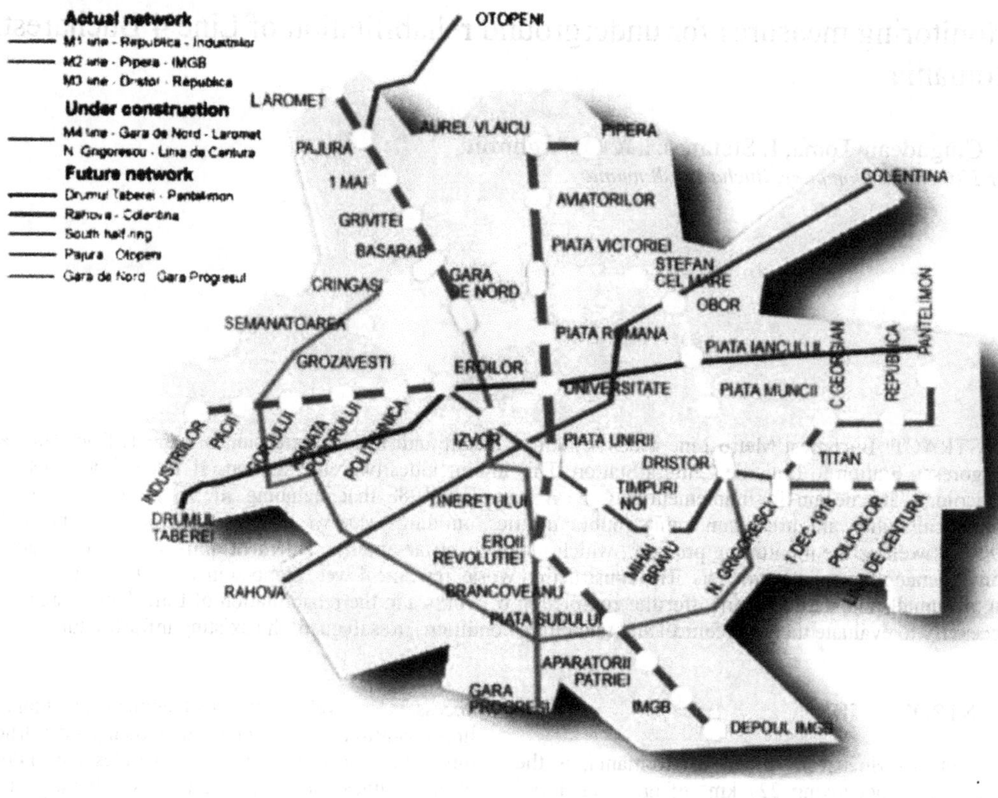

Figure 1. Bucharest Metro Network Map.

plains, extending towards S and N. The city center is developed in the area between the two rivers.

The structural frame specific to the Romanian Plain, due to the neo-tectonical movements, is one of a synclinal with a subsidence character, orientated SW-NE, on which Neogene and Quaternary deposits where accumulated. Bucharest is placed in the axial area of the syncline, where the thickness of the sedimentary deposits is more than 1000 m. This fact explains the seismic character of the region due to the lack of a rigid foundation near the surface, made of hard rocks.

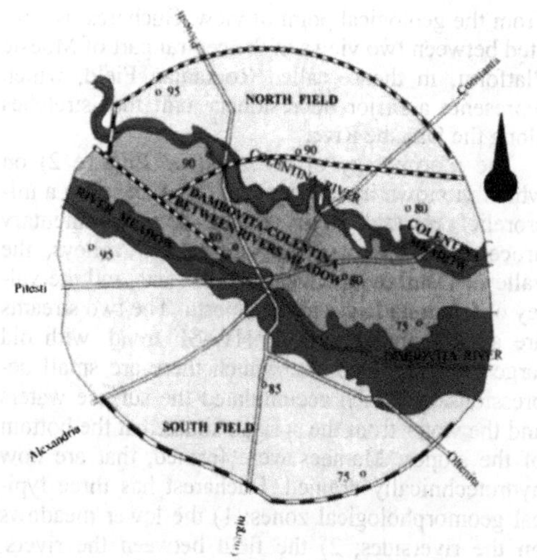

Figure 2. Morphological Map of Bucharest (scheme).

Taking into account the geotechnical information resulted from over 2000 geotechnical borings and using the classification made in our technical literature, the following strata have been identified according to the geological block diagram presented in Figure 3. Type 1 layer is constituted from recent surface sediments, made up of vegetal soil and clay sediments. Type 2 layer named upper clay complex, is constituted of loess formations, often moisture sensitive, with sand layers. Type 3 layer is named Colentina gravel complex, made up gravel and sand (with large variations in grain size) and frequently with water bearing clay layers, with a variable phreatic level. Type 4 stratum is an intermediate clay complex, made up of alternating brown and gray clays, with intercalation of hydrological fine confined sandy layers. Type 5 stratum called Mostistea sand complex, is a confined water-bearing layer made up of fine gray sands with lenticular intercalation of clay.

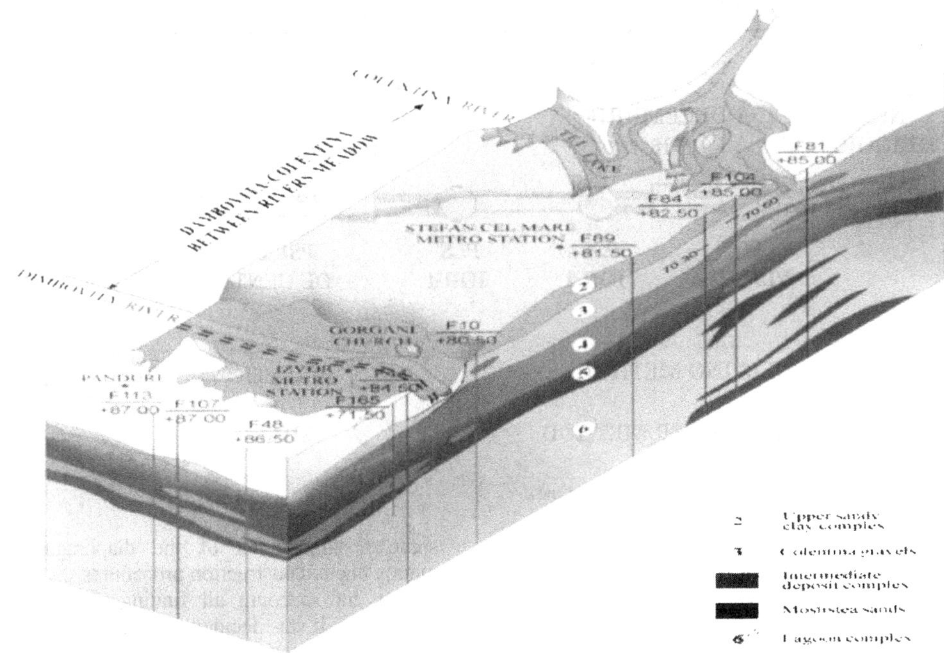

Figure 3. Geological block diagram.

3 CONSTRUCTION PHASES OF LINE 4

Line 4 is located in the South-Eastern region of Bucharest. The line is totally underground, built by cut-and-cover and tunnelling (open-face shield) techniques. Tunnel overburden (distance from ground surface to the tunnel crown) ranges from 5 to 12 m. Design for implementing Line 4 started in 1988, including site investigations, hydrogeological studies and urban constraint (public utilities, building etc.) surveys. These investigations and studies indicated to a water table quite high and geological profile composed by intercalated sand and clay layers.

A simplified view of the geological profile is depicted in Figure 4. The construction works began in 1989, as well as monitoring program, which included surface marks, settlement pins in buildings, tunnel convergence pins, piezometers and water level wells. Considering the total length of 4.7 km, 1.2 km was excavated by cut-and-cover methods and the remaining tunnel by an open-face shield. Construction works continued until 1996, when they

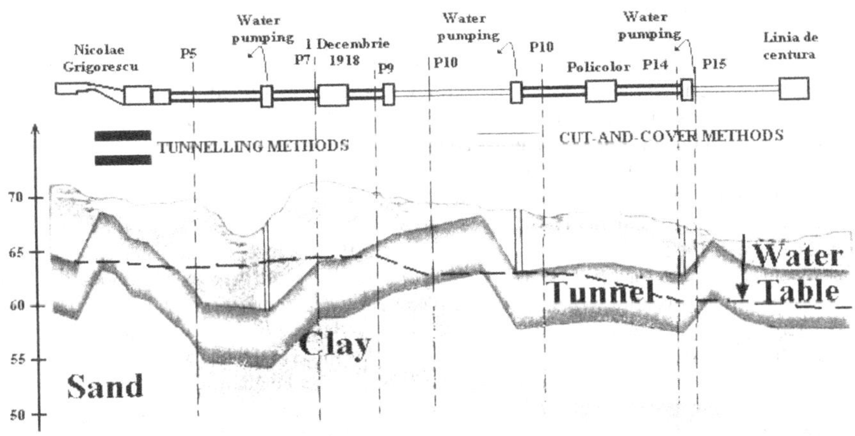

Figure 4. Simplified geological profile of Line 4.

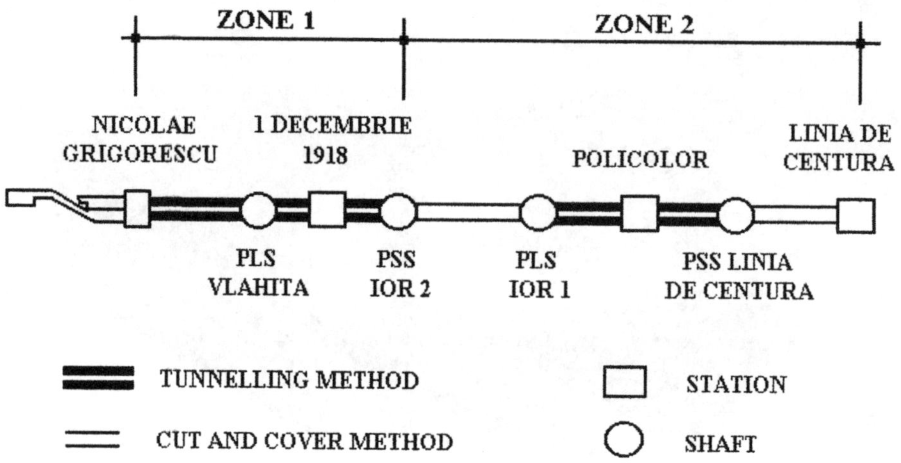

Figure 5. Schematic view and construction methods of Line 4.

were stopped.

The main characteristics of the works when stopped were:

- pre-cast concrete segmental lining of the tunnel already installed;
- filling grouting between ground and tunnel support mostly done;
- water tightness grouting and segment joint stuffing barely done;
- dewatering system turn off and consequent flooding structures.

Since 1996, only a controlled monitoring program continued, including piezometric and water level measures using dewatering wells. The results pointed to water level raising and later stabilization at its equilibrium position (in average 6 to 7 m below ground surface). This means most of the underground structures were flooded, what may cause severe damage to them. Starting middle of 2000, a new geological and geotechnical investigation, based mainly on penetrometer tests, was executed and interpreted for Zone 1 (Figure 5), as well as a more intense monitoring program, in order to evaluate changes in geotechnical characteristics, water levels, ground and building displacements. A similar investigation started at the end of 2000 for Zone 2. The findings of the complementary investigations for Line 4 (both Zones 1 and 2) pointed out to:

- cracks or even crushes of lining segments resulting into additional displacements;
- severe geometry changes from the original circular shape of the support rings, eventually leading to loss of ring stability;
- presence of sand sediments inside the tunnel, indicating leaching through segment joints and consequently ground loosening around tunnels;
- ground settlements measured at the surface;
- variations in the hydrostatical level;
- damages to some station and shaft structures;
- some misalignments of the diaphragm walls mainly due to construction procedures.

Taking into account all findings and probable consequences, it was decided to establish a technical solution for dewatering, in such a way to minimise structural displacements and ensure stability. As the tunnel longitudinal axes present an anticline shape, it would be possible and advisable to work (dewatering) independently in two zones (Figures 4 and 5). Then, for the sake of technical reasons and simplicity during construction works, the total length was divided into two zones:

- Zone 1: from Nicolae Grigorescu Station to PLS IOR2 Shaft (2.1 km in length);
- Zone 2: from PLS IOR2 Shaft to Linia de Centura Station (2.6 km in length).

4 EVALUATION OF THE GEOLOGICAL, GEOTECHNICAL AND HYDROGEOLOGICAL CONDITIONS

Recent site investigations have confirmed the lithologic sequence defined at the beginning of the works in 1989. A simplified geological profile is composed by fill, followed by layers of sand and clay. A typical lithologic sequence is presented below:

- Type 1 layer: non-uniform and heterogeneous fills, in different consolidation stages.
- Type 2 (upper clay) layer: surface clay dust complex, encompassing dusty clays and clay powders. This layer presents low clay activity, confirmed by moderate values of plasticity indexes (IP). It seems to be a normally consolidated clay.
- Type 3 (upper sand) layer: sand and gravel complex (Colentina), composed by large to medium sands and small gravels. Grain size analyses indicated to a high uniformity.

– Type 4 (lower clay) layer: intermediate clay complex, composed by clays with limestone concretions, dusty clays and silty clays. This layer presents a high clay activity, confirmed by high values of plasticity indexes (IP). It also seems to be a normally consolidated clay, only a bit more consistent and stronger than upper clay due to its greater depth.
– Type 5 (lower sand) layer: sand complex (Mostistea), composed by medium to fine sands and dusty sands. This layer also presents high uniformity of grain sizes.

Analysing some borehole logs from penetrometer tests and the longitudinal hydrogeological profile of Line 4 (Figure 4), it can be noticed that clay layers seem to be interrupted in several locations.

This fact justifies the water pressure levels for the two sand formations are approximately the same. In other words, the links between the upper and the lower sand layers establish the water pressure level for these layers.

A geotechnical testing program was carried out at the beginning of the works in order to obtain design parameters. In addition, other reports were analysed, including new laboratory data sheets. A compilation of geotechnical parameters is presented in Table 1.

the ground surrounding tunnels and consequently changes in their geotechnical properties. This suspicion has led to a new site investigation program, based on static and dynamic penetrometer tests (Borros type), and on laboratory tests.

New transversal sections were investigated by penetrometer tests, spaced from 5 to 80 m, depending on present requirements of the underground structures, based on field displacements. Their results were compared to controlled values obtained in areas not affected by underground works (in average 50 to 70 m away from them), which are representative of the local soil conditions. As a general conclusion, there is a decrease in the penetrometer strength in areas where the water table is raised. A broad analysis points out to strength reduction up to 50% in the upper clay and sand layers. The most affected areas also coincide with zones where excessive displacements were measured in tunnels.

Regarding the results for Zone 2 (which were better than those for Zone 1), a decrease between 35% and 50% in the penetrometer strength occurred in some areas of this zone, mostly located in the Colentina sand layer, up to a depth of 5 m and more pronounced in some areas surrounding tunnel portals of shafts and stations. In comparison with data stated in Romanian standards, especially for non-cohesive materials (sand layers), it was observed a tremendous effect of loosening on the main geotechnical parameters (deformability and strength).

Table 1. Range of variations of the geotechnical parameters.

Soil Layer	IP (%)	w (%)	IC	n (%)	ϕ (°) **	c (kPa) **	M_{2-3} (Mpa) ***	ep_2 (cm/m)	k (m/s) *	cu	γ_{sat} (kN/m^3)	E (Mpa)	K_o
Type 2 (upper clay)	14 49	12 26	0.6 1.0	35 45	12 25	30 87	4 15	2.2 6.0	1×10^{-3} 9×10^{-6}		18.4 20.3	10	0.49
Type 3 (sand & gravel)					36	0			1×10^{-3} 7×10^{-4}	1.6 2.6	20	25	0.49
Type 4 (lower clay)	22 63	17 33	0.5 0.9	36 43	14 24	20 87	7 20	1.2 3.4	8×10^{-5} 3×10^{-6}	2.5 3.5	17.7 20.6	15	
Type 5 (sand)					32	0			1×10^{-3}	1.6 8.5	20	25	0.47

(*) the permeability values for strata Type 2 and Type 4 were measured in sand lenses.
(**) other Mohr-Coulomb strength parameters were also reported to stratum Type 2: c = 5 kPa and ϕ = 30°.
(***) M_{2-3} is the oedometric modulus between 200 kPa and 300 kPa.

After stopping construction works and consequent tunnel flooding, some leaching of sand sediments has been observed towards the tunnel, probably caused by high hydraulic gradients in the regions close to segment joints. This phenomenon allied to construction disturbances may cause loosening of

5 STRUCTURAL EVALUATION AND SAFETY ANALYSIS OF THE EXISTING INFRASTRUCTURE

Existing infrastructure includes tunnels (circular shape), galleries (rectangular shape), stations, buildings and public utilities (water supply, sewerage, power, communication cables etc.). For the structural evaluation and safety analysis of these structures, there was performed a complete review of monitoring data and a visual inspection (see Table 2) only on shafts and stations, because of flooding. After dewatering, a detailed inspection will be performed inside the underground structures.

Table 2. Main observations from visual inspections of stations and shafts of Line 4.

Structure	Main observations
PSS Nicolae Grigorescu Shaft	Flooding water contaminated by urban waste. Signs of subsidence zones on two sides, probably caused by shield breaking through.
Station 1 Decembrie 1918	Failure of a crown beam on a side. Pipeline failure (water or sewerage). Clear water on opposite side. Sediment deposits (1 m in thickness) on the tunnel invert.
PLS IOR 2 Shaft	Satisfactory aspects of the structure. Some cracks and exposed reinforcement bars.
Policolor Station	Clear water. Struts to replace temporary tiebacks. Corrosion of exposed reinforcement.
PSS Linia de Centura	Signs of dewatering on walls (1 m), showing brownish and greenish colors. Unidentified bluish color on diaphragm walls.
Other zones	Satisfactory visual aspects.

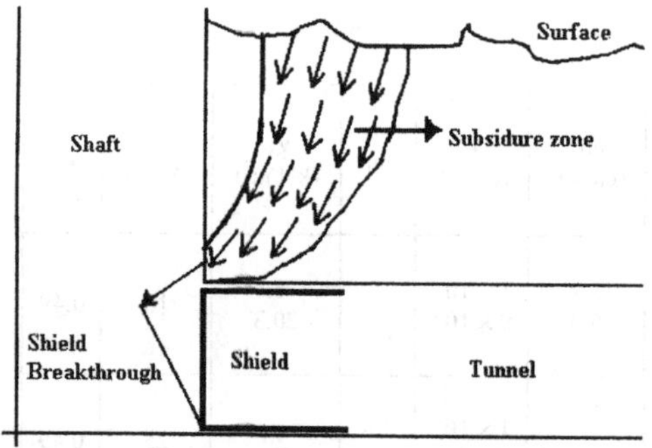

Figure 6. Shield breaking through and subsidence or failure zone.

From Nicolae Grigorescu Station to PSS IOR 2 Shaft and from PLS IOR 1 Shaft to PSS Linia de Centura Shaft (Figure 5), the underground structures include a twin tunnel. Both tunnels were excavated by a semi-mechanized open-face shield, with an external diameter of 6.4 m. The support lining is 0.35 m in thickness, yielding an internal diameter (initial clearance) of 5.7 m. The tunnel depth ranges along Zones 1 and 2, with an overburden varying from 5 to 12 m, crossing all major geological strata. The distance between tunnels also varies from approximately 1.8 to 6.4 m. In general, they are closer in the proximity of stations and shafts and then they separate apart. During tunnelling, the water table was kept lowered (approximately 2 m below the tunnel invert) by dewatering wells, in operation 50 m ahead and 150 m behind the tunnel face. Soil excavation at the shield face was done by shovel when dominated by sands and by pneumatic hammer and shovel when clay appeared in front. When arriving to or departing from stations and shafts, the shield breaking through was facilitated by blasting the concrete walls. This procedure has caused some overbreak, leading to loss of ground and consequently excessive surface settlements or even some failure zones. This mechanism is depicted in Figure 6.

Other important characteristic of the tunnel that also caused some ground displacements is related to shield advances, which were done with the face fully or partially opened. In some occasions, the shield was not immediately jacked towards the tunnel face in order to increase the excavation rate, leaving some span unsupported and consequently leading to loss of ground (ground movements towards the tunnel excavation perimeter). As the layer around the tunnel was composed predominantly by non-cohesive materials (sands and gravels), these ground movements easily propagated to ground surface, leading to settlements and soil mass loosening (Figure 7). This mechanism may propagate throughout the sand layer if this construction procedure remains, leading to disturbances of the whole layer. On the other hand, sandy ground movements tend to close any empty space between the ground excavation perimeter and the lining support, which is very positive for the adequate behaviour of the support system.

Just after the excavation, the shield was moved forwards and the tunnel support installed. Five precast concrete segments and one block compose the support system. They are 0.35 m in thickness and 1.0 wide. The key block was always installed in the tunnel invert.

For waterproofing, segments are transversally and longitudinally surrounded by a rubber ring and gasket (neoprene). The excavation overbreak (difference between ground excavation and external lining diameter) was filled by grouting (cement, sand and bentonite injected at 200 to 300 kPa). This first phase grouting (filling grouting) is very important to distribute loadings to the support, as well as to keep an interaction between the tunnel support and ground. When necessary for improving waterproof-

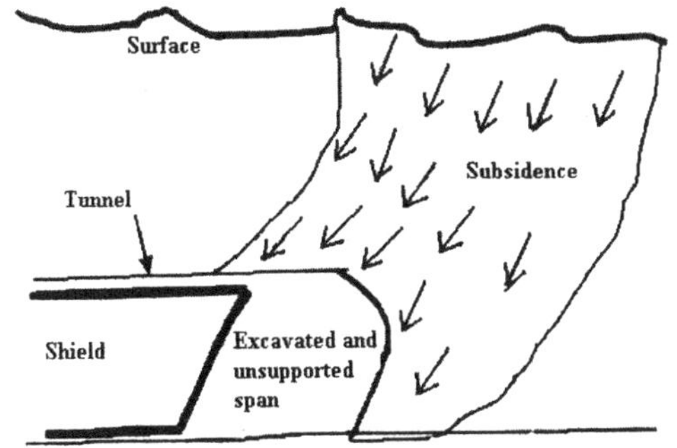

Figure 7. Loss of ground mechanism and consequent disturbance zone due to unsupported span at the tunnel face.

ing, a second phase of grouting was performed, with a mixture of cement and bentonite, injected at 400 to 500 kPa.

Finally, segment joints were stuffed with paste based on rapid and expansive cement. Eventually, after turning off the dewatering system and returning the water level to its initial position, a new grouting phase was executed with resins to improve water tightness.

Moreover, if the segments present some imperfections, such as cracks or deteriorated zones, they would also be treated. Cracks would be filled with resin grouting and deteriorated zones would be firstly chopped out than treated with shotcrete.

The gallery section (rectangular shape), stations and shafts were built by cut-and-cover methods, mainly due to the lack of nearby surface structures and other interferences in that area (green field).

The construction phases included diaphragm walls, tiebacks, bottom slab (invert) and inside structure. For the diaphragm walls, a vertical cut was excavated (for instance, by clamp-shells) and stabilized by bentonite, then placement of the reinforcement mesh and finally concreted (pumped concrete). Cut-and-cover structures also required dewatering during construction.

As the construction works were halted in 1996, some stages of this standard procedure were nor concluded satisfactory. Open space between ground and tunnel support, intense water flow through segment joints, leaching of sand sediments and attack of the structures by polluted water are the main concerns before restarting the works.

In order to establish the best possible approach about safety of the structures, it is important to highlight the main events occurred during construction:

- section 1A (closer to Nicolae Grigorescu Station, 178 m in length): some problems occurred in the dewatering procedure, sediment leaching towards the tunnel happened, leading to loss of ground and causing displacement propagation to the surface;
- section 1B (remain section from Nicolae Grigorescu Station up to PLS Vlahita, 480 m in length): some problems occurred in the dewatering system, long periods of stoppage and maintenance due to the lack of funds, causing great discrepancies in the construction quality; as the standard procedure for waterproofing was not completed, water leakage and sediment leaching occurred. These facts have produced distorted lining geometry, segment breakage and eventually loss of stability. The consequences propagated to the surface, producing subsidence caves.
- section 1C (from PLS Vlahita Shaft to 1 Decembrie 1918 Station, 311 m in length): similar problems occurred, related to dewatering and consequent water leakage and sediment leaching towards the tunnel, but in greater quantities.
- section 1D (from 1 Decembrie 1918 Station to PLS IOR 2 Shaft, 291 m in length): the works in this section occurred satisfactorily in terms of hydrogeology and construction technologies.
- section 2A (from PLS IOR 2 Shaft to Policolor Station; 800 m in length built by cut-and-cover methods up to PLS IOR 1 Shaft and then approximately half of 980 m in length built by tunnelling methods up to Policolor Station): geologically speaking, galleries were excavated either dominated by sand layers or by clay layers. Tunnels were typically based on Type 4 layer (lower clays), but their crowns were always placed on sand layers (Colentina layer). Generally speaking, the works were much less affected by funding shortage, leading to better conditions (hydrogeology and technologies).
- section 2B (from Policolor Station to Linia de Centura Station, approximately half of 980 m in length built by tunnelling methods up to PSS Linia de Centura Shaft and then 810 m in length built up to Linia de Centura Station): tunnels were also based on clay layers and their crowns dominated by sand layers. It seems that the conditions of works were a bit less ideal than in section 2A, leading to more disturbance to the Colentina sand layer.

Table 3 presents a summary of the current status of the structures and their background for Zone 1 and Zone 2 (see Figure 5).

6 EXISTING MONITORING DATA AND FUTURE ACTIONS

Monitoring program started at the beginning of the works. After stopping the works in 1996, tunnel convergence monitoring was halted due to flooding, but buildings have been continuously monitored until now. In 1999, a complementary monitoring program was implemented, focusing on settlement troughs. There were carried out topographic meas-

Table 3. Current status and background of the structures of Line 4 (Zone 1 and Zone 2).

Section	Ground nature	Water table status (hydrostatic level)	Current status of performed grouting	Conclusions from penetrometer tests	Structure deformations
1A	Sedimentary deposits	Level at 7 m below surface (0.5 ÷ 5 m above tunnel crown)	Filling 80% and water tightness 30%.	Parts of the ground with severe damages up to 14 m of depth in approximately 35% of the section length.	Diameter changes up to 60 cm. Foundation of the metro path for tunnel 1.
1B	Sedimentary deposits	Level at 5 ÷ 7 m below surface (4 ÷ 5 m above tunnel crown)	Filling 80% and water tightness 30%.	Current ground status known up to 5 m of depth.	Nothing to be reported.
1C	Sedimentary deposits	Level at 7 ÷ 9 m below surface	Injections for arch outline filling and water tightness in the base: 90% for tunnel 1 and 70% (filling) for tunnel 2.	Current ground status known up to 12 m of depth.	Nothing to be reported.
1D	Sedimentary deposits	Level at 7.5 ÷ 8 m below surface (1 ÷ 4 m above tunnel crown)	Injections for arch outline filling and water tightness in the base: 90% for tunnel 1 and 70% (filling) for tunnel 2.	No ground changes.	Nothing to be reported.
2A	Sedimentary deposits	Level at 7 m below surface (0.5 m above tunnel crown)	Filling 100% and complete grouting on tunnel invert only.	Parts of the ground with severe damages up to 5 m of depth in approximately 20% of the section length.	Minor damages in tunnels close to PLS IOR 1 Shaft.
2B	Sedimentary deposits	Level at 5.5 ÷ 6 m below surface (same level at tunnel crown)	Filling 100% and complete grouting on tunnel invert only.	Loose ground in general up to 3 m to 4 m (locally up to 7.5 m to 9 m)	Tunnel deformations close to diaphragm walls of Policolor Station and PSS Linia de Centura Shaft.

urements for convergence of tunnels, surface settlements by surface marks, and pins on buildings.

A detailed analysis of topographic measurements on convergence of tunnels was done, indicating some lining rings excessive deformation. The current status, specially after flooding, cannot be verified. By this time, only predictions are possible based on numerical simulation. After dewatering, new readings will be taken, besides the in-situ inspection and the guidelines to rehabilitate damage sections, specially those where clearance is not assured.

Monitoring data of building pins have not pointed to any concerns. Most readings float around very few millimetres, what may be caused by temperature changes or accuracy of the reading apparatus. Analysing this kind of data and the surface settlement trough, it was reasonable to conclude that most buildings are quite safe because they are out of the settlement trough width.

Other important fact that reduces these trough width values is the presence of paving, foundations and existing buried structures (water galleries and pipelines etc.). They work as ground reinforcement, diminishing the area affected by the settlement trough induced by tunnelling. Despite this, it is necessary to continue the measurements to identify any failure zone propagation.

Unfortunately, there are no measurements inside the ground (extensometers). These readings would provide a more realistic picture of ground and tunnel movements. In future, it will necessary to install them for next construction phases, which would provide sound bases for decisions taken in relation to the need or not of ground consolidation.

From visual inspections, some actions can be made in future:

- analysis of the water quality, specially in areas clearly identified as contaminated by urban waste and sewerage;
- analysis of the cause of the bluish colour on shaft walls and possible negative reaction of this unknown substance with concrete or reinforcement bars;
- mechanical destructive and electrical non-destructive tests to evaluate the degree of corrosion affecting exposed reinforcement bars.

For flooded underground structures, it will be necessary to elaborate a plan of cleanness and disinfections to be carried out immediately after dewatering, encompassing hot water under pressure, compressed air and high strength brush, in order to remove all fungus, bacteria, calcinations, oxides etc. After dewatering, a new visual inspection will be required to define structural rehabilitation.

7 SOLUTIONS AND TECHNOLOGIES FOR DEWATERING AND STRUCTURE REHABILITATIONS

Some future solutions and technologies for dewatering, ground consolidation, monitoring, structure rehabilitation and structure waterproofing will be carried out, as follows:

- monitoring – for all Zones 1 and 2, including ground settlement, building performance, water level and underground structure geometry;
- dewatering – for section 1A, dewatering the external ground by vertical wells up to the tunnel crown level, and then simultaneously with the inside water level up to the tunnel spring line. For section 1B, water bailing out from inside the tunnel. For section 1C, external ground dewatering by horizontal drillings parallel to tunnels up to 2 m under the tunnel crown level. For section 1D, water pumping out from inside the tunnels. For all Zone 2, dewatering the external ground by vertical wells up to the tunnel crown level and then simultaneously with the inside water level up to the tunnel invert, keeping always the inside level at the same level or higher than external level;
- ground consolidation – for section 1A, grouting from inside the tunnels (depending on penetrometer test results) and from surface. For sections 1B, 1C, 1D and all Zone 2, after dewatering, grouting (clay & cement) from inside the tunnels;
- remote ground consolidation – for section 1A and 1B, from the surface in loose regions indicated by penetrometer tests. For section 1C, grouting from inside the tunnel and from surface in regions indicated by penetrometer tests. For section 1D, not applicable. For all Zone 2, from the surface in loose regions indicated by penetrometer tests, only if necessary as indicated by monitoring results;
- structure waterproofing – for all Zone 1, grouting (clay & cement) behind the arch from inside of the tunnels, with high speed fortify materials from pontoons linked with inside dewatering system. For all Zone 2, grouting (clay & cement) performed from inside tunnels, with quick set (high speed consolidation) materials and water level 3 m above the tunnel invert;
- structure rehabilitation – for all Zone 1 and Zone 2, that will be defined after tunnel inspection.

The problems that can occur may be related to the dewatering process: if this will impose any additional loading or displacements to structures that could be harmful. Also, considering that the ground around tunnels was disturbed during construction and loosened by leaching as demonstrated already by the penetrometer tests to some regions, may be it will be necessary ground consolidation before dewatering.

Trying to answer these questions, it seems reasonable to approach the problem as follows:

- running of a numerical simulation to obtain the general behaviour of the ground and structures, following different paths (sequences) of dewatering, considering different ground conditions;
- establishment of a step-by-step procedure and guidelines for the works, allowing diversions that could not be predicted by the standard behaviour or simulated by numerical analysis. In this case, decisions will be taken heavily base on monitoring (observational method).

Numerical simulation should be done for both typical cross-sections of used technologies on Line 4: cut-and-cover and tunnelling. These numerical simulations should be considered preliminary, only attempting to represent the main construction stages of underground structures.

A preliminary numerical simulation should consider a typical cross-section of diaphragm walls, which are part of the underground structures (galleries) built by cut-and-cover methods (Figure 8).

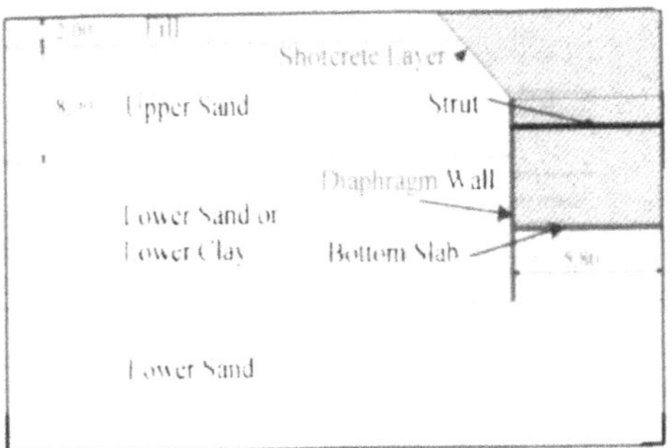

Figure 8. Sketch of the diaphragm wall structure.

Then numerical simulation should consider another cross-section of twin tunnels (Figure 9) built by tunnelling methods (open-face shield).

The analysis should consider the diaphragm walls and tunnels as plane-strain structures (2D analysis), which is a reasonable hypothesis for design, since the longitudinal structure (gallery and tunnel) dimension is much greater than other cross-section dimensions. The soil behaviour should be modelled as a linear-elastic-perfectly media with Mohr-Coulomb failure criterion. This constitutive law requires five parameters: elastic modulus, Poisson ratio, cohesion, friction angle and dilatancy angle. Dilatancy should be taken as null, considering the loose state of sand layers, especially after construction. The diaphragm walls should be considered, firstly, completely surrounded by a sand layer and, secondly, surrounded by a sand layer on the top and by clay layer on the bottom (Figure 8) part of the structure, aiming to evaluate more realistically the clay layer influence upon the behaviour of the dia-

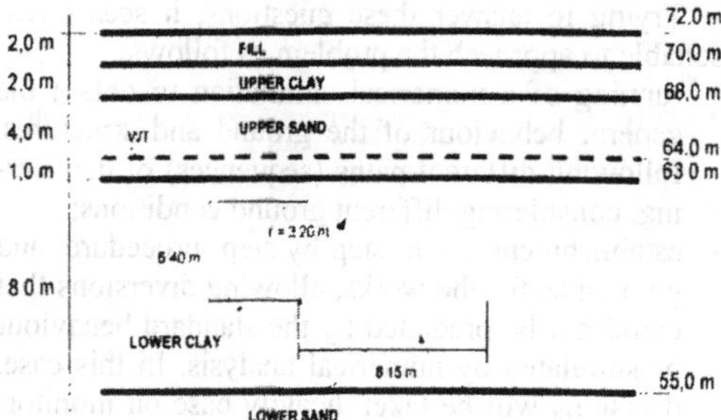

Figure 9. Typical twin tunnel cross-section surrounded by clay layer.

phragm wall and the bottom slab. Then, numerical simulation should be completed with the twin tunnels which will be considered, firstly, surrounded by clay layer (Figure 9) and, secondly, surrounded by clay and sand layers (Figure 10).

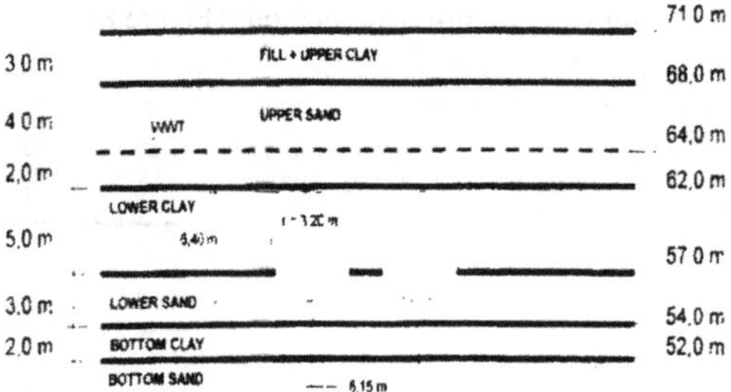

Figure 10. Typical twin tunnel cross-section surrounded by sand layer on the top and clay layer on the bottom part of the structure.

In general, the numerical simulation will encompass the main construction stages since the beginning of gallery and tunnel excavation works to the next imminent stage (dewatering after flooding).

Also, it will be made measurements inside the ground (see Figure 11 for possible scheme of extensometers) during dewatering, which will provide a real view of the ground state and tunnel movements. These will feed with initial data for decisions taken in relation to the need or not of ground consolidation.

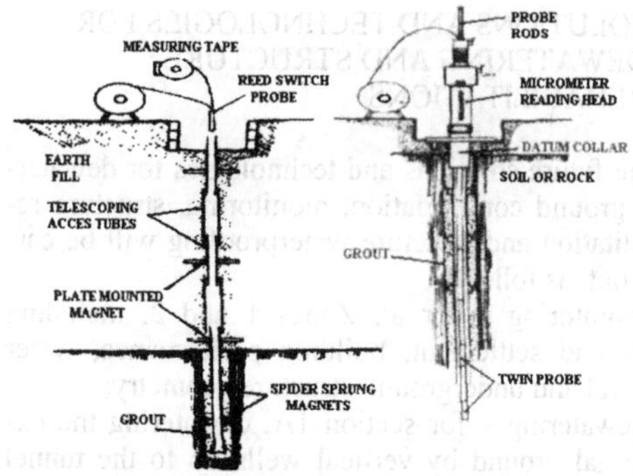

Figure 11. Monitoring scheme for exetensometers.

REFERENCES

Beldean, V., Farmus, R. & Uta, V. 1996. Metro Network in Bucharest – Present and Future. 2nd National Conference for Underground Structure of Romanian Tunnelling Association – Tradition & Development in Underground Construction, ITA-ART. Technical University of Civil Engineering Bucharest, Brasov, Romania.

Assis, A. P. 2000. Evaluation Consultant Report on the Completion of the Construction Works to Safeguard the Existing Infrastructure of Metro Line 4 (from Nicolae Grigorescu Station to Linia de Centura Station). Preliminary Reports no. 1 – Zone 1, Publication G.RE-AA001/00, Department of Civil and Environmental Engineering, University of Brasilia, Brasilia, Brazil.

Assis, A. P. 2001. Evaluation Consultant Report on the Completion of the Construction Works to Safeguard the Existing Infrastructure of Metro Line 4 (from Nicolae Grigorescu Station to Linia de Centura Station). Preliminary Reports no. 2 – Zone 2, Publication G.RE-AA005/01, Department of Civil and Environmental Engineering, University of Brasilia, Brasilia, Brazil.

Calinescu, S. 1996. Computer Programs for Subway Tunnelling Design - Some Aspects of the Romanian Practice. First International Conference, Computer Applications in Transportation Systems, Basel, Switzerland.

Rozorea, G., Fierbinteanu, V., Ciugudean – Toma, V. & Arghiroiu, O. 2000. Underground Work in Bucharest – Geotechnical and Hydrogeological Factors. GeoEng 2000, International Conference on Geotechnical & Geological Engineering, Melbourne, Australia.

Rozorea, G., Assis, A., Calinescu, S. & Arghiroiu, O. 2002. Completion of the Construction Works on Bucharest Metro Line 4 to Safeguard the Existing Infrastructure, World Tunnel Congress, Sydney, Australia.

*** Projects of METROUL SA Company (1975-2001).

North American Tunneling 2002, Ozdemir (ed.)
© 2002 Swets & Zeitlinger, Lisse, ISBN 90 5809 376 X

Matching TBM designs to specific project requirements

E.h. Martin Herrenknecht, Dr.-Ing.
Chairman of the Board of Directors
Herrenknecht AG
Schwanau, Germany

ABSTRACT: Tunnel projects today undertake increasingly difficult geological and site conditions. This paper reviews how TBM designers must adopt and adapt new technologies to meet these ever increasing challenges. Specific projects are highlighted.

1 INTRODUCTION

Picture 1 left

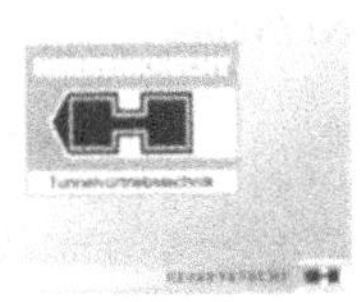

Picture 1 right

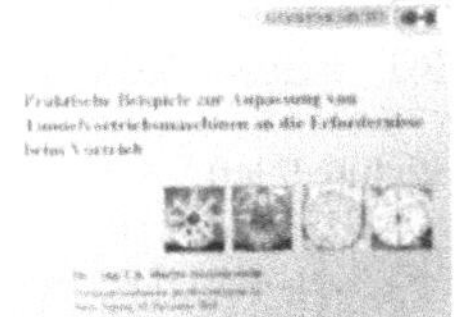

In today's presentation I would like to show the adaptation of TBMs to specific project requirements using a few practical examples.

Since some of the projects have already been discussed in greater detail, I would like to put my focus on the technical area.

The requirements of each project are of great importance for TBMs. Therefore they should be taken into consideration as early as the planning phase.

Picture 2 left

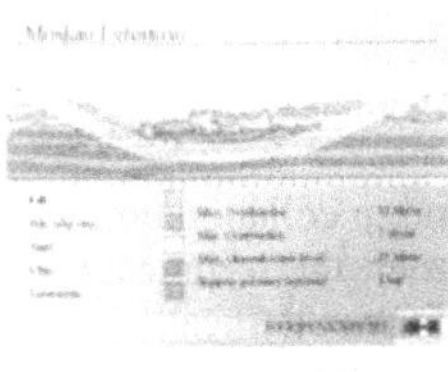

Picture 2 right

- Anticipated geology and hydrology on one hand
- And the designed final lining of the tunnel on the other.

The last aspect is important for the backup design in particular.

Before I discuss the connection based on the following examples, I would quickly like to comment on the specific meaning of geology and hydrology.

2 GEOLOGY AND HYDROLOGY

It is very important to have most precise knowledge of the geology and hydrology in order to design the complete machine system.

In particular for:

Picture 3 left

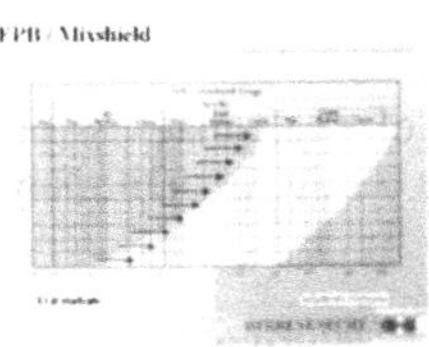

Picture 3 right

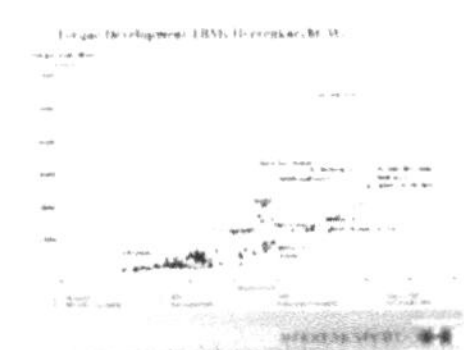

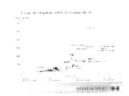

- The choice of operating mode (EPB, Slurry, Hard rock),
- The multifunction of the operating mode and
- The design of the cutterhead, cutting wheel and cutter tools,
- Material transport and backup system design.

Today's trend in soft ground, particularly for machine types with a diameter of up to 10 m, takes the direction towards EPB – Shield with bentonite conditioning. However, starting at the diameter of 12 m, only the slurry shield may be realized based on the high torque.

3 INTERESTING TUNNELING PROJECTS

I would like to use some interesting tunneling projects to present – abbreviated – the adaptation of tunnel boring machines to project requirements.

3.1 *4th Tunnel Elbe River Project (Germany) Lefortovo, Moscow (Russia)*

Picture 4 left

Picture 4 right

After completion of the tunnel in Hamburg, the machine was refurbished at the Herrenknecht AG plant in Schwanau and prepared for its next job in Moscow.

According to the anticipated geology, the cutterhead was re-designed.

Excavation in Moscow started up last week.

Picture 5 left

Picture 5 right

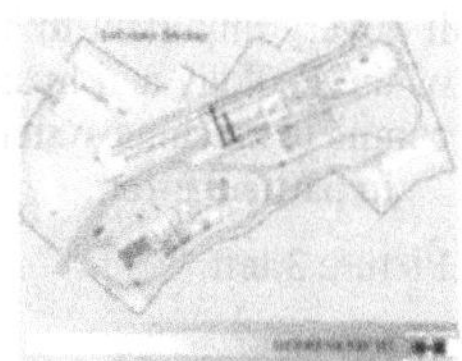

Not only was there a lot of sweat that was produced, but also a lot of vodka consumed in the past 2 years.

3.2 *TBM- Excavation, Zurich-Thalwil (Switzerland)*

Picture 6 left

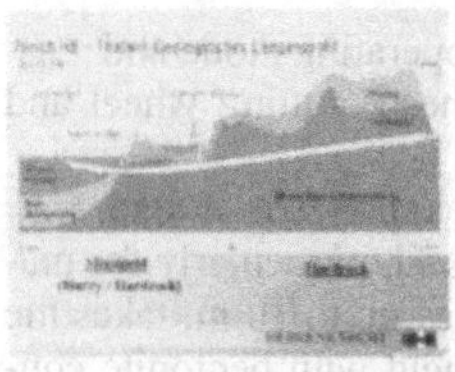

Picture 6 right

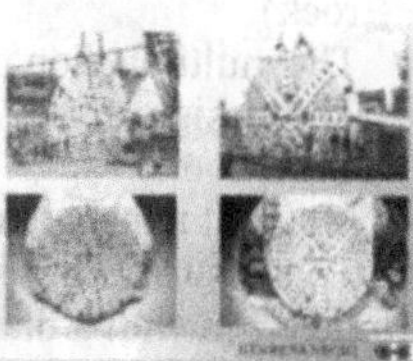

One of the most spectacular inner city railway tunnels was done in Zurich.

The 2.6 km long northern section was excavated with a Mixshield. This shield was converted underground from hard rock to EPB mode and therefore capable of excavating hard rock as well as groundwater containing loose rock.

Picture 7 left

Picture 7 right

The segments, set within mm conditions, should be credited to 2 other factors in addition to the quality of the machine:

- the additional measures of umbrella grouting and a pilot tunnel for grout injections as well as
- The excellent jobsite personnel.

Picture 8 left

Picture 8 right

3.3 *Westerschelde, Terneuzen (Netherlands)*

Picture 9 left

Picture 9 right

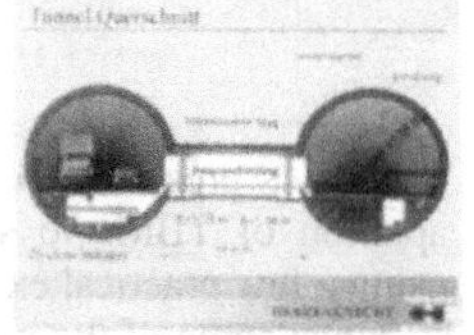

Both tunnels with an outer diameter of 11.33 m and a section length of 6.6 km each present a unique challenge worldwide.

Both Mixshield TBMs drive with a maximum decline of 4.5% under the navigational waterways "Pas van Terneuzen" and "Pas van Everingen", with a maximum support pressure of 6.5 bar.

This impressive world record, however, challenges both, personnel and machine with highest requirements.

During the design phase and in collaboration with the Joint Venture and the owner, the technical team put the focus on the design of the cutterhead in order to avoid blocking due to highly plastischen Boomse Kleis.

Picture 10 left

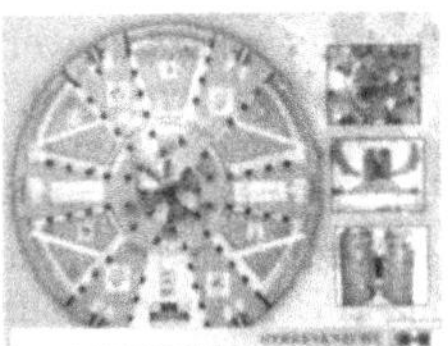

For that reason, the cutterhead was equipped with a center cutter. A vertical crusher was attached in front of the suction stub. In addition, 2 agitators provide the turbulences in the suction pipe.

The highly sophisticated rinse concept, which distributes fresh suspension continuously and evenly along the tunnel face, maintains the density as low as possible and guaratees optimum transport.

This, on the other hand, has a large influence on the economical side of the project.

Picture 10 right

In order to continue the tunnel lining process during excavation of long tunneling projects, the cable channel follows along the backup system, which logistically is another high requirement.

The "Westerschelde" Project is also a good example of how restricted the prediction of geological and hydrological conditions is and how it can lead to new challenges during the tunneling process:

Let us follow the tunneling process:

Picture 11 left (remains)

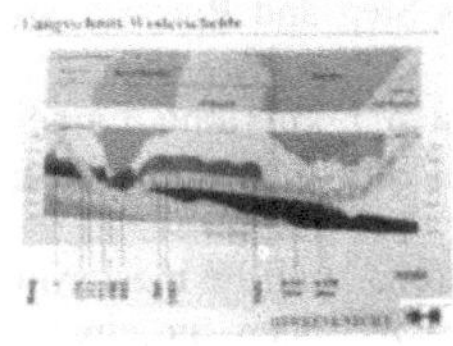

Picture 11 right

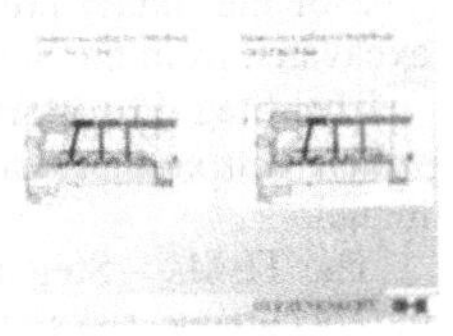

- Damage of the main seal through foreign objects (front seal was exchanged during TM).

Picture 12 right

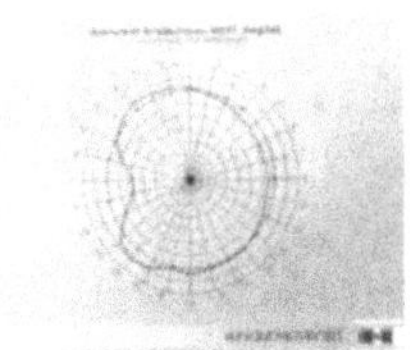

- Deformation of the shield at the lowest point of the tunnel.

Picturc 13 right

- Injection stub was sheared off.

Picture 14 right

- Replacing the seal.

Picture 15 right

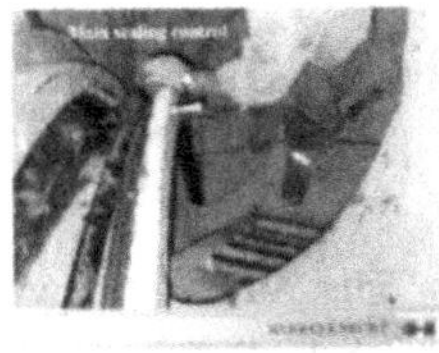

However, these difficulties are the reasons that lead to innovations and new solutions in mechanical excavation and therefore ultimately broaden the applications of this technology successfully.

Picture 12 left

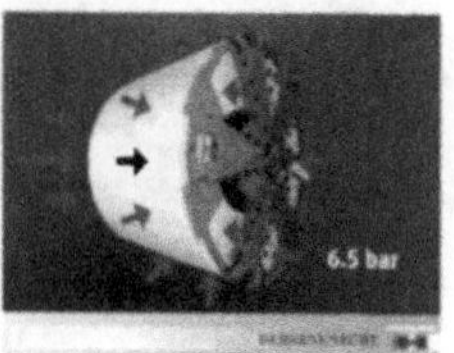

Picture 16 right

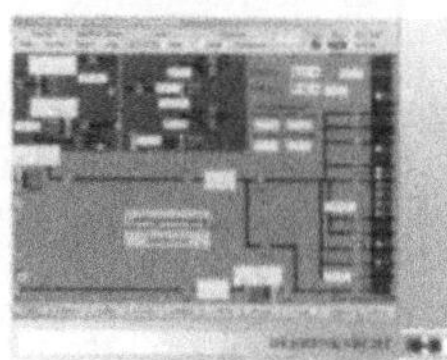

We created the expression:

"Under high pressure, pressure from the font, pressure from above, pressure from below, pressure from all sides; we tunnel builders work successfully."

3.4 *Socatop, Paris (France)*

Picture 13 left

Picture 17 right

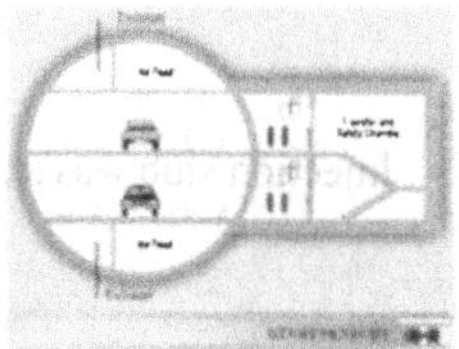

To relieve the traffic load in Paris, a two story freeway tunnel is in the process of being built. 3 lanes on 2 levels each are anticipated to allow fluent traffic.

Picture 14 left

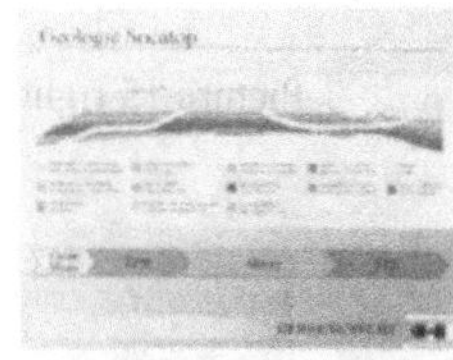

Picture 18 right

The local geology (sand, limestone, sandstone and clay as well as groundwater on three different levels) requires a change of operating modes during excavation. Looking at the economical side of the project, the time required for the conversion of operating modes is also of great importance.

Picture 19 right

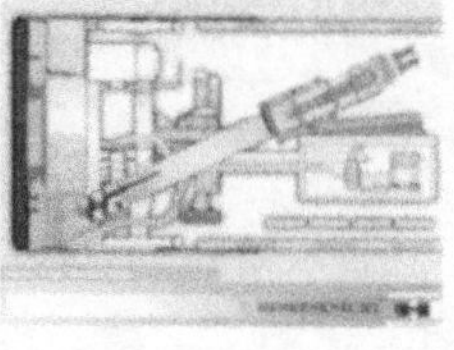

For that reason, a Mixshield / EPB / Slurry / Open mode with a diameter of 11.56 m was utilized. The machine can be operated in slurry, EPB as well as semi-open mode. The conversion between the different modes is possible to be performed within the tunnel. Experiences made in previous projects were also included in the design, whereas for the first time a TBM will be converted from EPB to slurry mode in the tunnel.

3.5 *Lötschberg Basis Tunnel (Switzerland)*

Picture 15 left

Picture 20 right

The planned 34.6 Kilometer long Lötschberg Basis tunnel connects Frutigen located in the Kander valley with Raron and Steg, which are located in the Rhone valley. The complete heterogeneous rock formation of the Alps will be encountered:

- Anticipated from the North portal is Flysch with varying volcanic intermediate layers and schist with limestone faults.
- Following is Quintner limestone of the Doldenhorn cover with highly abrasive granite and grandiorite of the Aar massive, with maximum cover of 2,300 m above.
- This reaches up to the southern area, Steg, consisting of crystal gneiss, schist, phyllite amphibiolite and limestone.

To structurally master the enormous length of the Lotschberg Mountain, three additional tunnels (Mitholz, Ferden, Steg) were built, which results in 11 possible tunnel projects with maximum section lengths of 10.2 km.

2 of the tunnel projects – Steg and Raron – are excavated from the South.

The Steg Lot refers to a 3.3 km long tunnel, which will connect to the 5.2 km line of the basis tunnel.

The TBMs – Steg and Raron – will be mastering the occurring risks, which have been anticipated since the planning period of this centennial project: falling rock, springs, methane, temperature extremes, high abrasiveness – just to mention a few.

Picture 16 left

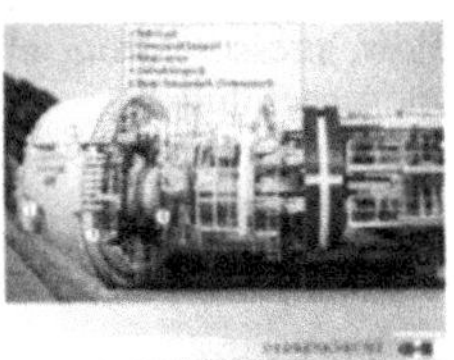

Picture 21 right

At the same time, the TBMs have to provide an optimum work space for the personnel in order to fulfill the conditions. This criterion relies on the so-called L1 area, which is located between the rear of the cutterhead and the first gripper level.

For that reason, the choice fell on a single gripper TBM, which uses the expandable telescopic shield as additional support in the cutterhead area. That helps to maintain the static of the complete TBM system and enables steering during the tunnel drive. Furthermore, the location of the center load of the TBM is elective, which means that the level of the gripper shoe can be as far back behind the cutterhead as possible and provides the space necessary for rock securing measures.

Picture 17 left

Picture 22 right

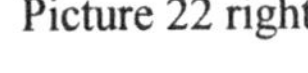

Newly developed for the Steg and Raron machines are the machined rock securing system with wire mesh erector, anchor drills as well as segment erector directly behind the cutterhead in the L1 area.

2 additional drill lances installed on the movable ring erector approx. 38 m behind the tunnel face can perform additional anchor work.

Picture 23 right

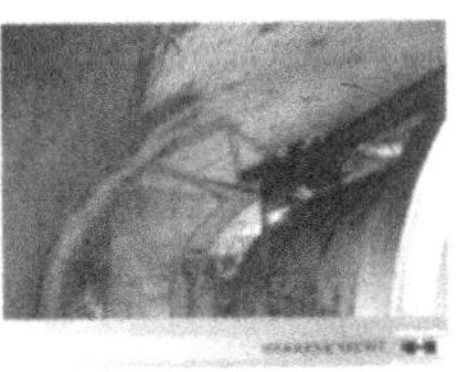

The wet shotcrete system is located behind the drill equipment: 2 spray robots, both capable to spray 3 m horizontally as well as the complete circumference from the sole segment to the roof of the tunnel.

Picture 18 left

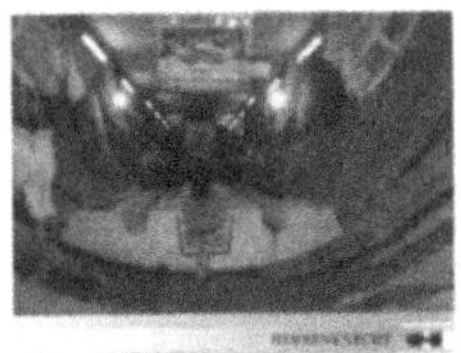

A spaciously arranged sole segment installation site as well as anchor and shotcrete equipment in the L2 area are also important factors for high tunneling performance in rock formations.

Picture 19 left

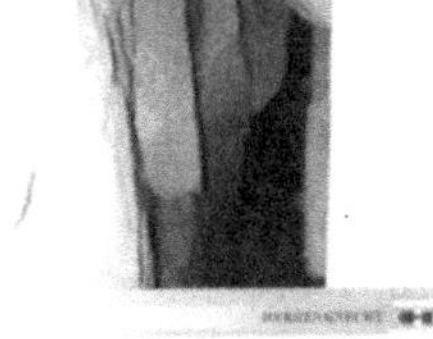

Picture 24 right

Extensive testing at the technical Institute Karlsruhe before manufacturing focused on cutterhead structure, Cutter housing installation as well as the complete structure to optimize these constantly used machine components.

Picture 25 right

The size of the excavated grains shows, that s cutter spacing of 90 mm is an advantage for the recycling process of the excavated material.

Picture 26 right

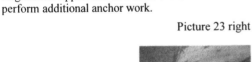

Without doubt, we are grinding our way with difficulties through

- Blocking material in front of the cutterhead
- High temperatures in the cutterhead area – specifically the gauge cutter with temperatures up to 170 °C

– Partially high abrasiveness

Picture 20 left

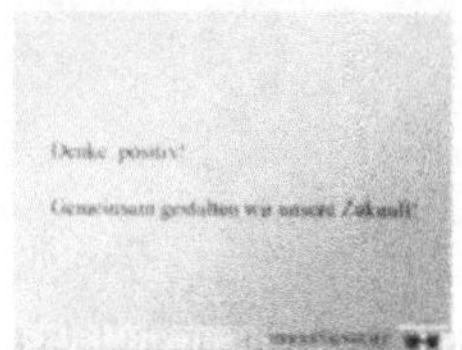

Picture 27 right

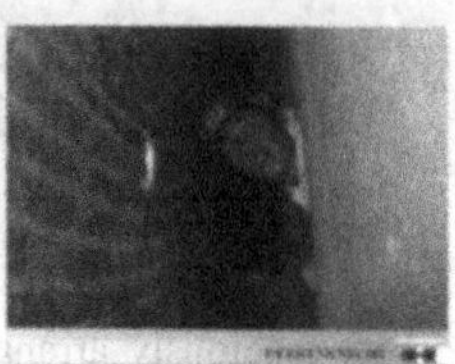

These problems have to be solved according to our motto “Under high pressure; pressure ahead, pressure on top, pressure below, pressure from all sides – we tunnel builders work successfully”.

Let’s take it on! Together we create our future. Think positive!!!

North American Tunneling 2002, Ozdemir (ed.)
© 2002 Swets & Zeitlinger, Lisse, ISBN 90 5809 376 X

Control of contaminated groundwater during tunneling: five case histories

F.J. Klingler & K.M. Swaffar
NTH Consultants, Ltd., Detroit, Michigan, USA

T.S. DiPonio
Jay Dee Contractors, Inc., Livonia, Michigan, USA

ABSTRACT: Tunneling through water bearing soil and rock has always been challenging, requiring specialized methods to control the groundwater and permit safe advancement and adequate support of the tunnel. An added challenge occurs when the groundwater contains contaminants that are potentially harmful to workers or to the public. This paper will provide a discussion of case histories for five projects where tunnels and shafts were driven through contaminated water bearing soil and/or rock. These projects include the Dearborn Retention and Treatment Tunnel, the Downriver Regional Storage and Transport Tunnel, the East Lansing CSO Tunnel, the Weinbach Outfall Tunnel, and the Detroit River Outfall Tunnel. In each case, contamination ranging from gasoline and related hydrocarbons, to naturally occurring dissolved sulfide, presented significant challenges to completion of the project. This paper will discuss the challenges for each project, the solutions employed, and lessons learned.

1 INTRODUCTION

Throughout the history of mining, the presence of groundwater has been a common challenge to the successful advancement of tunnel excavations. These challenges typically revolve around preventing excessive groundwater infiltration, as well as associated loss of granular soils into the tunnel excavation. Numerous methods have developed for controlling groundwater and infiltrating ground, and as we enter the new millennium, almost any groundwater condition can be handled by a variety of modern tunneling equipment and methods.

An added challenge to controlling groundwater involves controlling and treating contamination that is sometimes encountered with it. For the purposes of this discussion, groundwater contamination is defined as the presence of substances in groundwater that could be hazardous to workers and/or that require treatment of the groundwater prior to disposal.

Historically, the most common forms of groundwater contamination have been natural, usually consisting of dissolved methane, sulfide, and/or carbon dioxide. Dissolved methane and carbon dioxide usually come completely out of solution upon exposure to atmospheric pressure. The groundwater is then free of these contaminants, and handling and disposal are not a problem. As a result, however, methane and carbon dioxide gas are formed, which require special air handling methods to prevent explosive and/or asphyxiating conditions from developing. Acceptable air handling methods are well defined in the industry and by the (United States) Occupational Safety and Health Administration, and are not addressed in this paper.

Dissolved sulfide will also come out of solution upon exposure to atmospheric pressure, although not nearly as easily or completely as methane and carbon dioxide. The extent of dissolution of sulfide into hydrogen sulfide gas is dependent on several factors, including the atmospheric pressure, temperature, and turbulence of the groundwater as it enters the tunnel excavation (Doyle, 2001). While the resulting hydrogen sulfide gas can typically be controlled by a well designed tunnel ventilation system, the portion of sulfide that remains dissolved in the groundwater can present difficult challenges during handling and disposal of the groundwater.

While the theoretical saturation concentration of hydrogen sulfide in water is approximately 3900 parts per million (ppm), maximum naturally occurring concentrations in the Midwestern United States are usually in the range of 100 to 200 ppm. Depending on the tunneling methods and dewatering control measures employed, up to 75 percent of this concentration will remain dissolved in the groundwater at the time that it is removed from the tunnel excavation. This groundwater will continue to exsolute hydrogen sulfide gas to the atmosphere, which creates toxic (and noxious) gas, until the dissolved concentration is zero. Disposal of the groundwater into surface waters is not environmentally acceptable, since the sulfide is highly toxic to marine life. Disposal into sewers is not acceptable because of the resulting hydrogen sulfide attack on concrete sewer linings, as well as the potential harm to waste water treatment plant anaerobic digesters. As a result of these conditions, treatment of sulfide bearing groundwater is typically required immediately after its removal from the tunnel excavation, before it is exposed to the open atmosphere.

A separate but increasingly common form of groundwater contamination during tunneling is man-made, resulting from the presence of dissolved and/or free petroleum product within the ground. Such contamination typically results from leaking underground gasoline storage tanks, and can be quite extensive and concentrated. For a moderate to high concentration of the contaminant, a relatively small volume of the liquid portion of gasoline will volatilize to form explosive vapors, which must be properly ventilated to prevent explosive and toxic conditions. This can leave a relatively concentrated mixture of groundwater and gasoline, which must be properly removed from the tunnel and disposed of.

This paper presents discussion of methods used to control, handle, and treat sulfide or gasoline contaminated groundwater for five tunnel projects in the Midwestern Unites States.

2 DEARBORN RETENTION AND TREATMENT TUNNEL

The Dearborn Retention and Treatment Tunnel (DRTT) was designed in accordance with federal requirements to reduce combined sewage overflows (CSOs) to the Rouge River in Dearborn, Michigan. Phase I of this tunnel was designed to be approximately 4,800 m (15,000 ft) long, with diameter of 5.5 m (17 ft). The tunnel system included a 16 m (50 ft) diameter overflow shaft and 11.3 m (35 ft) diameter pumping shaft at the downstream end, and three 3.9 m (12 ft) diameter drop shafts along the alignment.

Ground conditions within the project area generally consist of 26 to 32 m (80 to 100 ft) of variable granular and cohesive overburden, underlain by a relatively thin (less than 3 m thick) layer of highly overconsolidated glacial till (hardpan), further underlain by the Dundee Limestone Formation. The tunnel was designed to be mined through Dundee Limestone, at a depth below ground surface of approximately 52 to 58 m (160 to 180 ft), corresponding to rock cover of 19 to 26 m (60 to 80 ft). Based on artesian water levels in the range of 3 m (10 ft) above ground surface, the groundwater pressure head at the level of the tunnel was in the range of 55 to 61 m (170 to 190 ft).

The Dundee Limestone is a documented producer of hydrogen sulfide gas, and was expected to produce hydrogen sulfide bearing groundwater during the construction. The project design called for controlling inflow of groundwater into the excavations by grouting the hardpan and rock at the shaft locations, and by pumping within the tunnel excavation. The groundwater discharge criteria established by contract required that groundwater be treated to 0 ppm dissolved sulfide (100 percent removal), with the treated water to be disposed into local combined sewers.

Construction of the DRTT commenced in February, 1995, with the first elements of construction being the overflow and pumping shafts at the downstream end of the project alignment. At these locations as well as the other drop shaft locations, the contractor proposed to dewater the hardpan and bedrock, as an alternative to the grouting method of control that was indicated in the contract documents. The engineer accepted this proposal.

The contractor advanced the overflow and pumping shafts through the overburden using the sinking caisson method. This shaft worked well to limit the volume of groundwater seeping into the excavation. During this stage of construction, the groundwater was removed from the excavation from four, 0.3 m (12 in) diameter dewatering wells spaced around the perimeter of each of the shafts, as well as by pumping from within the excavations. Upon landing of the sinking caissons on the hardpan, the shafts were advanced into the limestone by drill and shoot techniques, with the groundwater being controlled through the use of the dewatering wells.

The contractor began advancing the drop shafts between April and August 1995. These shafts were sunk as caissons through the overburden using 3.9 m (12 ft) diameter pre-cast concrete pipe sections. Groundwater at these locations was controlled within the overburden through the use of 0.3 m (12 in) diameter wells spaced around the perimeters of each of the shafts.

In the first three months of dewatering at the overflow shaft and pumping shaft site, the rate of dewatering discharge ranged from 1,100 to 1,900 liters per minute (300 to 500 gallons per minute). Initial dewatering discharge rates at the drop shaft sites ranged up to about 230 lpm (60 gpm). Initial sulfide concentrations within this groundwater were as high as 180 ppm, but stabilized at 80 to 100 ppm once the groundwater levels were lowered somewhat.

As the shaft excavations proceeded into bedrock, the dewatering discharge rated increased dramatically, up to 9,500 lpm (2,500 gpm) at the overflow and pumping shaft site, and up to about 1,900 lpm (500 gpm) at the drop shaft sites. At that time, the project was suspended due to concerns about the increasing rate of groundwater dewatering discharge, and the costs associated with the treatment. At the time that the project was suspended, the overflow and pumping shafts were close to the design depth, although tunnel mining had not started.

Initially, venturi groundwater treatment plants were used to remove the sulfides from the groundwater. In this process, the water is forced under pressure through venturi valves, drawing in air at the same time. The air mixes with the contaminated water, oxygenating the sulfides, and releasing them to the atmosphere. Up to three of these relatively simple plants were used. Each plant consisted of a holding tank, and a centrifugal pumping system that recirculated the water through the venturi. While this system uses electrical energy instead of chemicals to treat the water, it was only successful in removing about 80 percent of the sulfides, which was an insufficient amount according to State and local authorities. Chemical treatment was then used to treat and remove the final 20 percent of dissolved sulfides.

Chemical treatment included the use of a Ferric Chloride-Hydrogen peroxide treatment method. This involved mixing the discharge with ferric chloride, then hydrogen peroxide to oxidize the sulfide. This was successful in reducing the dissolved sulfide concentration to near zero, however the resulting discharge from the treatment system possessed high turbidity, being black in color.

Due to the high volume of dewatering, the method of disposal was changed in October 1995, to involve discharge of the treated groundwater to the Rouge River. This made the highly turbid discharge from the ferric chloride-hydrogen peroxide method unacceptable, and the treatment method was switched to involve mixing with hydrogen peroxide only.

Due to the large volume of water that required treatment during the peak of dewatering, the project sustained costs up to $100,000 per month in chemicals alone.

The greatest challenge in treating the groundwater on this project was probably keeping up the treatment quality with the rapidly increasing dewatering discharge rates. The venturi part of the treatment process was relatively inexpensive compared to the chemical treatment part, but produced a large amount of noxious odors. Most of the project site was well away from residences, however there were some resident complaints.

3 DOWNRIVER REGIONAL STORAGE AND TRANSPORT TUNNEL

The Downriver Regional Storage and Transport System (DRSTS), relieves sewer surcharges and basement flooding for residents of 13 Wayne County, Michigan communities. The DRSTS was designed to capture excess sewage flows and divert these flows into the transport and storage tunnel. Major components of this $120 million system include the 2.2 to 2.4 m (7 to 7.5 ft) diameter, 23 km (13.5 mi) long storage and transport tunnel; the 10.2 km (6 mi) long, 0.97 to 1.45 m (36 to 54 in) diameter Eureka extension; the 16 m (50 ft) diameter, 35 m (110 ft) deep dewatering pump station, and numerous junction and connecting structures.

The DRSTS project was divided into ten construction contracts, with the first contract started in 1996, and all construction completed by 2002. The contract numbers were sequenced from north to south with the depth of the construction elements also increasing accordingly. Contract Nos. 2 through 6 and Contract No. 8 involved tunnel construction, all including at least some dewatering of hydrogen sulfide bearing groundwater.

The project geotechnical investigation revealed that subsurface conditions along the alignment generally consist of 19 to 29 m (60 to 90 ft) of medium to stiff silty clay soils over 1.6 to 3 m (5 to 10 ft) of hardpan, then Detroit River Group dolomite. Both the clay and hardpan deposits contain occasional granular layers. The project design called for the tunnel to be driven almost entirely through the thick clay (and occasional granular deposits), with one relatively small area to be driven through granular hardpan. The geotechnical investigation indicated that artesian groundwater levels in the hardpan and bedrock generally dropped from north to south and from west to east. This corresponded to a pressure head at the tunnel invert ranging from 10 to 31 m (30 to 90 ft). Sulfide levels in the artesian groundwater were measured during the project geotechnical investigation by laboratory analysis and with Drager dip tubes. Dissolved sulfide levels were recorded at concentrations up to 60 ppm.

The contract documents required treatment of any groundwater containing sulfides to less than 1 ppm (dissolved) at the point of discharge, and to less than 5 ppm in the atmosphere at ground level at any point on the project site. The treated groundwater was specified to be discharged in the municipal sanitary sewers along the alignment.

Contract No. 2 required dewatering at one of the shaft locations due to hydrostatic uplift concerns and some leakage into the shaft. The total dewatering quantity was relatively small with rates on the order of 190 lpm (50 gpm) over a 3 week period. To treat this small amount, the contractor elected to use a filter type of treatment system brand named SulfaClean. The system uses a proprietary granular filter containing iron material through which the groundwater flow passes and reacts to form iron pyrite (FeS_2). This cost of this system was about $0.0025 per liter ($0.01 per gallon) to treat, with setup costs of about $10,000.

Contracts 4, 5 and 6 had substantial dewatering efforts with approximately 1,500 lpm (400 gpm) on Contract 4 at a deep sinking caisson shaft and within a hardpan outcropping through which the tunnel passed. Due to long dewatering time period (about 1 year) The Contract 4 Contractor elected to set-up a relatively large hydrogen peroxide treatment plant. The set-up was about $50,000, and treatment costs were about $0.00025 per liter ($0.001 per gallon). The treatment plant consisted of a square metal storage building with internal electronic mixing, sampling, and recording equipment. The injected amount of hydrogen peroxide was automatically adjusted based on groundwater volume and sulfide concentrations. Automatic recordings of all data was easily obtained and submitted to regulatory agencies.

Contracts 5 and 6 also had dewatering rates of 530 to 1,900 lpm (300 to 500 gpm), mainly due to sand layers within the tunnel zone at the southwest corner of the project alignment. An on-site hydrogen peroxide system was also used for these contracts. Minor infiltrations in shafts were treated with burlap bags filled with a chemical similar to the SulfaClean process discussed for Contract 2. These bags were somewhat successful is decreasing the hydrogen sulfide concentrations in the discharged groundwater.

Contract 7 was dewatered by use of sumps within excavations at the base of a large sinking caisson that extended into the bedrock at downstream terminus of the project. The dewatering operation produced groundwater volumes on the order of 680 lpm (180 gpm) for several months. The groundwater was treated with a hydrogen peroxide treatment plant, although much smaller than those used on Contracts 4, 5, and 6.

Overall, the various hydrogen sulfide treatment operations on the project were successful. Problems were mainly limited to public complaints about atmospheric orders. On Contract 2, a dewatering contractor working over the weekend accidentally switched the discharge flow to a nearby surface ditch without going through the treatment system. Residents noticed an odor and contacted the local police who subsequently mobilized the hazardous material team from the State.

Also, occasional complaints of atmospheric odors were received from residents along the other contract alignments. No complaints were received from the municipal treatment plant regarding hydrogen sulfide concentrations in the sanitary flow coming from the project area.

4 EAST LANSING CSO TUNNEL

The East Lansing CSO tunnel was designed as an approximately 2,700 m (8,400 ft) long, 3.1 m (10 ft) diameter concrete lined sewer, with overflow and outfall facilities, intermediate shafts, and appurtenant facilities. The sewer alignment extends along a four to six lane urban boulevard, and in at least four locations, extends adjacent to existing or former gasoline station properties. The tunnel invert depth ranged from about 6.5 to 14.5 m (20 to 45 ft) below grade.

Subsurface conditions along the tunnel alignment are somewhat variable, consisting of interbedded layers of granular and cohesive soils. Soil conditions within the tunnel bore varied from interbedded clay and silt, to an entirely granular profile. Groundwater levels along the alignment varied from 2 to 6.2 m (6 to 20 ft) below grade, which corresponds to levels in the range of 3 to 8 m (10 to 25 ft) above the tunnel invert. During two geotechnical investigations conducted along the alignment prior to the start of construction, groundwater contamination was not encountered, except at the eastern terminus of the project, where low levels of hydrocarbons (0.2 ppm) were detected. In addition, low levels of hydrogen sulfide gas (0.1 ppm) were reported at several locations.

Construction of the tunnel sewer commenced in December, 1994, with mining proceeding from west to east using an open wheel mining machine with adjustable doors at the face. The tunnel excavation was dewatered at selected locations through the use of 0.3 m (12 in) diameter wells, spaced from 16 to 32 m (50 to 100 ft) apart. In other locations, groundwater infiltration was relatively low, and was handled by pumping seeping groundwater from the face of the bore. Initially, the groundwater quality was good (aside from mild turbidity and suspended solids), and dewatering discharge was disposed into the city's combined sewers.

In May 1995, the contractor encountered gasoline seepage into the tunnel at a limited location near the middle of the alignment. Although the condition was a health and safety concern, the volume of seepage into the tunnel was very small, and groundwater treatment was not required at this location.

In November 1995, after the tunnel heading had reached the eastern third of the alignment, gasoline was again encountered seeping into the tunnel at the heading (Doyle, et al. 1997). Following an explosion in which four workers were injured, mining was suspended for approximately six months.

Several subsequent studies, evaluations, and environmental investigations determined that the gasoline contamination was relatively wide spread east of the location of the explosion. As a result, further mining efforts were by hand, with extensive dewatering ahead of the face through 0.1 m (4 in) diameter wells spaced at 6.5 to 10 m (20 to 30 ft). All groundwater was treated using an activated carbon filter treatment plant manufactured by Calgon Corp. This system was self contained within a trailer plant about 8 feet wide by 20 feet long by 10 feet high. The total dewatering discharge rates were typically in the range of 300 to 750 lpm (80 to 200 gpm). Following treatment to remove 100 percent of hydrocarbons from the discharge, the groundwater was disposed of into the city's sewers.

While this system worked relatively well for groundwater rates below 375 lpm (100 gpm), it required frequent maintenance and changing of the carbon filters, which became more of a problem as rates increased beyond 375 lpm (100 gpm). The rental cost of the filter plant was approximately $3,000 per month, not including filters.

5 NORTH WEINBACH OUTFALL TUNNEL

The North Weinbach Outfall Tunnel extends along an urban thoroughfare in Evansville, Indiana. The original project design included an approximately 2,700 m (8,400 ft) long, 2.6 m (8 ft) diameter tunnel, with appurtenant manholes, connecting structures, and junction structures. The planned sewer invert was approximately 8 to 10 m (25 to 30 ft) below ground surface (Henn, et al. 2001).

The regional geologic setting is characterized by a combination of glacial deposits and river valley sediments, generally consisting of silts and clays with some outwash sands. The river-lain sediments are generally composed of sands with some interbedded silt and clay seams. The alignment was divided into three general reaches. The tunnel bore within Reach A was generally through clay; within Reach B was through a mixed face of sand and clay; and within Reach C through sand.

Groundwater occurs within the granular deposits at depths typically in the range of 3 to 8 m (10 to 25 ft) above the design tunnel invert elevation. Hydraulic conductivities of the granular layers were in the range of 10^{-2} to 10^{-3} cm/sec. During the design process, volatile organic compounds (VOCs) were detected, apparently originating from multiple sources along the alignment.

Based on the presence of dissolved VOCs in the groundwater, the City of Evansville initially precluded dewatering as a construction approach, and required tunnel mining using an earth pressure balance tunnel boring machine. However, the contractor proposed an alternative involving dewatering along portions of the alignment through a series of dewatering wells, and mining with a conventional Lovat tunnel boring machine with closable flood doors and muck ring capability. The primary tunnel lining consisted of steel ribs and wood lagging with filter fabric. The proposed secondary lining was precast concrete pipe sections grouted into place.

The contractor's groundwater handling strategy for this project was to design the dewatering program to minimize the spread of contaminant plumes in the groundwater, while achieving adequate drawdown to allow tunneling with conventional mining equipment. The final design involved the use of six 0.3 m (12 in) diameter wells in the southern reach of the alignment. These wells typically discharged at rates of 1,100 to 1,900 lpm (300 to 500 gpm) each, for a total discharge of approximately 9,500 lpm (2,500 gpm).

The wells were installed to draw water below the contaminant plume, and hopefully minimize contamination drawn into the system. A local analytical laboratory was retained to monitor the discharge for volatile organic compounds and other petroleum contaminants, and no positive readings were recorded. Initially, the testing was conducted twice weekly. Since no contaminants were detected in the first part of the project, the frequency was reduced to once per week.

The approach on this project to minimize movement and potential disturbance of the contaminant plume on this project was very successful. As a result, no groundwater treatment was required, and groundwater was disposed into local sewers.

6 DETROIT RIVER OUTFALL TUNNEL

The Detroit River Outfall Tunnel project is currently under construction, and involves mining of an approximately 1,00 m (5,700 ft) long, 6.4 m (20 ft) diameter rock tunnel. The project design includes a 12.9 m (40 ft) diameter drop shaft, a 9.7 m (30 ft) diameter riser shaft, and a river outfall structure.

Subsurface conditions at the site generally consist of 22 to 27 m (70 to 85 ft) of soft clay overburden, underlain by about 2 to 3 m (5 to 10 ft) of hardpan, further underlain by the Anderdon and Lucas bedrock formations. These formations consist of relatively flat-lying dolomitic limestone. RQD values of core samples procured from the bedrock formations ranged from 0 to 100 percent, with an average value of 90 percent. Results of constant head permeability testing during the geotechnical investigation indicated a relatively wide range of hydraulic conductivities, from 10^{-1} to 10^{-5} cm/sec.

The geotechnical investigation revealed the groundwater to contain relatively high concentrations of dissolved sulfide, resulting in hydrogen sulfide gas readings within the borings of up to 710 ppm.

Because the hydraulic conductivity was markedly lower in the rock zone between depths of 64 to 100 m (200 to 310 ft) below ground surface (typically in the range of 10^{-4} to 10^{-5} cm/sec with occasional higher permeability zones), the tunnel was designed with invert at a depth of 290 feet. Due to the extensive presence of high permeability rock above a depth of 64 m (200 feet) within the shafts (as well as occasional high permeability zones below that depth), the project required full penetration rock grouting at the shaft locations. In addition, the contractor was required to advance a pilot hole ahead of the tunnel, grouting as necessary to control groundwater inflow into the tunnel heading. The tunnel lining method was specified as pre-cast gasketed segments, in part due to concern about groundwater leakage through the completed liner during construction.

The contract requirements called for treatment of dewatering discharge to less than 0.1 ppm dissolved sulfide, with maximum atmospheric concentrations at the ground level of 10 ppm. The contract required disposal of the treated dewatering discharge into the local sanitary sewer system.

The construction was begun in March, 1999, starting with installation of the drop and riser shafts. The rock grouting program was very successful in limiting the inflow of groundwater into the shafts, with total volumes of seepage limited to about 100 gpm at each shaft location. Dewatering of the excavation has been limited to pumping from sumps located at the shaft bottoms.

Tunnel mining began in August of 2001, with mining proceeding north from the riser shaft. The initial 300 feet of mining was almost completely dry, apparently due to lateral penetration of grout at the shaft location. At this writing, the mining is proceeding with the use of probe and grout holes at the mining face, mining pushes of about 3 m (10 ft) between grouting intervals. The maximum infiltration at the face has been limited to about 190 to 225 lpm (50 to 60 gpm).

Dewatering discharge has typically contained between 40 and 70 ppm dissolved sulfide (at the location of the treatment plants). Treatment has consisted of turbulance-oxygenation, together with hydrogen peroxide dosing, and has been successful in reducing the dissolved sulfide content to the required level of less than 0.1 ppm. The treatment plants consist of a prefabricated dosing plant (about 3 m square) with adjacent chemical storage.

Following treatment, the dewatering discharge has been disposed into City of Detroit sewers.

7 CONCLUSIONS AND LESSONS LEARNED

Based on the authors' experiences on these projects, several general conclusions are made:

The presence of contamination within groundwater can usually be identified during the subsurface exploration phase of the project. As a minimum, test borings for tunnel projects should include testing of each borehole for the presence of hydrogen sulfide and explosive gas. In addition, in any urban areas, the soil samples obtained should be tested for the presence of hydrocarbons using a photo-ionization meter. Any positive readings from such testing should be followed up with analytical laboratory testing. Draeger tubes may also be used to better measure concentrations of hydrogen sulfide within field samples, which typically oxidize before they can be transported for laboratory testing.

The best system is the most cost effective one based on the amount of water to be treated. Having sufficient data from the getechnical study is key for the contractor to consider the anticipated volume to be treated over the planned construction period and select the most appropriate treatment system.

Contract documents should provide specific treatment and disposal criteria. For sulfide bearing groundwater, this should include levels of dissolved sulfide to treat to, levels of atmospheric hydrogen sulfide allowed, and an approved disposal location (i.e., storm sewer or sanitary sewer). For gasoline contamination, this should include the required discharge/disposal location, together with the maximum contaminant concentrations allowed.

In situations of localized contamination along a tunnel alignment, the dewatering system designer should investigate the potential for limiting the spread and disturbance of contaminant plumes. This may pay great dividends in reduced treatment costs.

Small volumes of contamination may be best handled by relatively inefficient and expensive treatment techniques such as the "burlap bag" method used on the DRSTS project.

Venturi systems may only be effective for removing sulfide down to about 20 ppm, and will result in a great amount of hydrogen sulfide discharge into the atmosphere. However, such systems may be efficient and appropriate in combination with chemical systems.

The public should be educated ahead of time to expect odors during treatment for hydrogen sulfide. All treatment methods produced at least some odor, and all drew at least some public comment.

During construction of any urban tunnel through granular deposits, the possibility of gasoline should be expected and provisions should be include in the contract documents for handling such an occurrence (even if the data does not indicate the presence of such).

REFERENCES

Doyle, B.R. 2001. Hazardous Gasses Underground. New York: Marcel Dekker, Inc.

Doyle, B.R., Hunt, S.W., Kettler, J.A. 1997. Gasoline Explosion in East Lansing Sewer Tunnel. Rapid Excavation and Tunneling Conference Proceedings: p 827-842.

Henn, R.W., Pease, K.A., Rorison, G.J 2001. Weinbach Soft Ground Tunnel: Lessons Learned. Rapid Excavation and Tunneling Conference Proceedings: p 99-111.

North American Tunneling 2002, Ozdemir (ed.)
© 2002 Swets & Zeitlinger, Lisse, ISBN 90 5809 376 X

History and exploration redefine Portland's West Side CSO Tunnel alignment

M.L. Fong, S.L. Bednarz & G.M. Boyce
Parsons Brinckerhoff Quade & Douglas, Inc.

G.L. Irwin
City of Portland Bureau of Environmental Services

ABSTRACT: The West Side Combined Sewer Overflow (CSO) Tunnel, Pump Station & Pipeline Project is currently beginning construction along the downtown waterfront of Portland, Oregon. This 14-foot diameter, 4-mile-long soft ground tunnel is located along the west bank of the Willamette River and crosses the river to Swan Island. A comprehensive historical and geologic assessment along the City's waterfront was undertaken to understand potential obstructions that the tunnel and pipelines mighty encounter. This assessment identified buried streams, lakes, and the ancient Willamette River channel. An extensive geotechnical exploration program was also implemented. Existing and abandoned foundations, the river crossing, and geologic conditions combined to refine both the horizontal and vertical tunnel alignments for the West Side CSO Tunnel. This paper presents a general overview on the information collected and evaluates how useful each exploration method was to tunnel design.

1 INTRODUCTION

The West Side CSO Tunnel and associated shafts are the central element of the Willamette CSO Program. The City of Portland entered into an Amended Stipulation Final Order (ASFO) agreement with the Oregon Department of Environmental Quality to reduce the frequency and volume of combined sewerage overflows from the drainage areas on the west side of the Willamette River. The tunnel will collect and intercept overflows from existing combined sewer and storm outfalls that discharge to the Willamette River from the City of Portland's Central Business District and basins immediately north. The tunnel is approximately 22,000-feet long, extending along the western waterfront adjacent to and beneath Portland's famous Tom McCall Waterfront Park and Port of Portland Terminal 1. The tunnel then crosses beneath the Willamette River to the Confluent Shaft and the Swan Island Pump Station. Figure 1 shows the tunnel and pipeline alignments and the location of the major shafts. Tunnel, shaft, and pipeline construction will be in soft-ground consisting of fill, alluvium (sand, silt, and gravel), and older consolidated gravel deposits. The location of the tunnel passes adjacent to and underneath the Willamette River and therefore groundwater heads in excess of 100 feet will be experienced during construction.

The tunnel design element includes construction of seven shafts along the alignment. The shafts range in depth from 100 to 170 feet, with diameters between 16 to 147 feet. The 220 Million Gallon per Day (MGD), dry-pit submersible Swan Island Pump Station will transfer flows from the tunnel through a new force main system to existing collection system tunnels. The pump station will be designed to accommodate low-flow, dry-weather conditions as well as wet weather flows up to the design capacity.

The third major element of the project is connecting pipelines to the tunnel and force mains from the pump station. The major elements of this work include Segment 3 of the Southwest Parallel Interceptor, Mill Conduit, Jefferson Conduit, Tanner Extension, Balch Conduit, and the Peninsular and Portsmouth Force Mains. Diversion structures and smaller connection pipelines for the abandonment of existing small pump stations are also planned.

2 GEOLOGIC SETTING

The project area is located near the western edge of the Portland Basin in Northwestern Oregon. This northwest-trending structural basin is bordered by the Portland West Hills on the west and the foothills of the Cascades on the east. The Portland Basin is underlain by thick sequences of middle Miocene-age Columbia River Basalt. The upper surface of the basalt was deeply weathered to form a thick layer of decomposed basalt and residual soil.

As the Portland Basin subsided, upper Miocene to Pliocene-age Sandy River Mudstone and the Trout-

Figure 1. West Side CSO Tunnel and pipeline alignments showing major shaft locations.

dale Formation were deposited over the basalt by the ancestral Columbia River. The Sandy River Mudstone consists primarily of overbank deposits of mudstone, siltstone, sandstone, and claystone, with lesser areas of gravel deposits within the ancient channels. The overlying Troutdale Formation contains braided channel deposits of sandy gravel and sand, silt, and clay lenses.

During Pleistocene glaciation, when sea level was ±400 feet lower, Troutdale Formation deposits were weathered and eroded and deep channels were cut by rivers and streams. The bottom of the ancient Willamette River channel was locally 145 feet below current sea level. Catastrophic glacial outburst floods repeatedly scoured and then deposited both coarse and fine-grained material in the Portland Basin during the late Pleistocene. Channels cut in the Troutdale Formation during glacial flooding were subsequently partially infilled with flood deposits and recent alluvium. Significant amounts of interbedded silt and fine sand alluvium were deposited as a result of the post-glaciation rise in sea level.

Artificial fill overlies the alluvium and catastrophic flood deposits along the banks of the Willamette River and within old stream channels. Borehole data and historic maps identified Marquam Gulch, Couch Lake, and Guilds Lake along the west side of the river. Along the eastside of the Willamette River, Swan Island once separated two historic channels of the Willamette River. The eastern, deeper channel was identified as Swan Channel. Between 1920 and 1986, the southern portion of the historic Swan Channel was filled and the remaining channel was renamed Swan Island Basin. The site of the Swan Island Pump Station, once located near the center of the Swan Channel, has been backfilled with up to 71 feet of dredge fill to match the Swan Island ground surface.

During our geotechnical investigation, we identified six material units: Fill, Sand/Silt Alluvium, Gravel Alluvium, Troutdale Formation, Sandy River Mudstone, and Columbia River Basalt. A synopsis of these units is provided below:

2.1 *Fill*

Artificial fill includes a wide variety of materials, but predominately consists of medium stiff to medium dense sand, sandy silt, and silt with organic debris. Fill also contains significant quantities of gravel, brick, riprap, concrete rubble, building debris, buried logs, and wood debris. Piling and portions of structures are often buried in the fill. Localized areas of sawdust and wood chips are present. This material ranges in thickness from 1 to ±71 feet. Fill was placed during the last 100 years along the waterfront, in creek channels, lakes, and on Swan Island. Much of the fill material came from dredging of the Willamette River.

2.2 *Sand/Silt Alluvium*

The recent fine-grained alluvial materials consist predominately of interbedded sandy silt and silty fine sand deposited as recent alluvium and fine grained catastrophic flood deposits. Gravel lenses, wood fragments, woodpiles, and logs are found within this unit. Often the alluvium is stratified with alternating layers of fine sand, sandy silt, silt, and plastic silt. The consistency of the alluvium is most often described as soft to medium stiff or loose to medium dense.

2.3 *Gravel Alluvium*

Gravel Alluvium predominately consists of coarse-grained catastrophic flood deposits of dense to very dense, poorly graded gravel and cobbles in a sand and silt matrix. Scattered boulders (up to 27 inch maximum diameter) and lenses of silt and sand were observed (Fig. 2). The gravel, cobbles, and boulders are predominantly fresh basalt with lesser amounts of andesite and scattered quartzite. Gravel and cobble-sized clasts are generally rounded and are randomly arranged with little reworking. Open zones are common near the upper contact and borehole caving is common in this unit.

Figure 2. Gravel Alluvium Boulders and Cobbles.

2.4 *Troutdale Formation*

The Troutdale Formation generally consists of very dense poorly graded gravel in a matrix ranging from clayey silt to sand. The clayey or sandy silt matrix material sometimes provides very weak cementation within the formation; however, the unit is best described as gravel rather than weak rock. Localized 1 to 2 foot thick weakly cemented gravel lenses were encountered. Sand lenses up to 20 feet thick and lesser silt lenses were randomly encountered within the gravel. Unlike the Gravel Alluvium, the gradual deposition of the Troutdale gravels has resulted in an aligned inclined stacking (imbrication) of the clasts. Generally, the horizontal thickness of the insitu cobble-sized clasts is greater than the vertical thickness. Large cobbles up to 17 inches in maximum diameter were observed. Open graded zones with very high ground water flows are encountered (Fig. 3) and zones where the matrix material is clayey with relatively low permeability are also common. Often the gravel is difficult to drill and sample. The lithology of the clasts is similar to the Gravel Alluvium, but quartzite is more common in the Troutdale Formation. At outcrops along the Sandy River (Fig. 4) and Willamette River (Fig. 5), the Troutdale Formation appears to be coarse gravel and cobble conglomerate with a silty sand matrix. This apparent cementation is most likely a localized effect related to the evaporation of ground water from the surface of the outcrop and the precipitation of the dissolved solids within the matrix material. Significant caving occurred within large diameter boreholes and running sands were encountered.

2.5 *Sandy River Mudstone*

Although not present within the tunnel zone, Sandy River Mudstone was encountered in deep borings conducted on both sides of the Willamette River. Locally, this unit contained gray-green to brown, extremely weak to weak (R0 to R1) siltstone, claystone, and weakly cemented sand. Near the upper contact, the unit resembled a hard clay soil and swelled significantly during sampling. This unit is not anticipated within the tunnel zone, but may be used as a groundwater cut-off during shaft construction.

2.6 *Columbia River Basalt*

Decomposed Columbia River Basalt was encountered below the tunnel zone on the west side of the river crossing. This material generally consisted of orange-brown sandy silt and silty clay that showed relict rock texture (saprolite) at one location and gray, very close jointed, slightly to moderately weathered basalt with fat clay joint infillings below the Nicolai Shaft. Although none were encountered during our investigation, bouldersized fresh or slightly weathered core stones are common in the deeply weathered Columbia River Basalt.

Figure 3. Rotosonic Core Showing Open Zone in the Troutdale Formation.

Figure 4. Troutdale Formation Outcrop, Troutdale, Oregon.

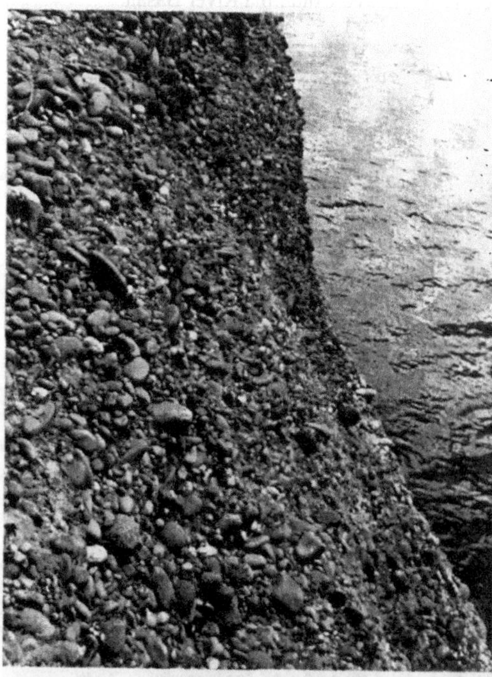

Figure 5. Troutdale Formation Outcrop on East Side of the Willamette River.

3 HISTORICAL DEVELOPMENT

As downtown Portland developed, a wharf was built along the west side of the Willamette River. The wharf consisted of docks, boardwalks, and warehouses supported on closely spaced timber piles. The docks typically extended from near the existing location of Naito Parkway to the current location of the seawall. Pile supported structures were once located along nearly the entire length of the alignment. A seawall was constructed along the riverside of the wharf and the area behind the wall was filled. Much of the fill consisted of sand and silt, but most of the piles and significant quantities of debris remain buried within the fill. Figure 6 is a photograph of the Portland Waterfront during seawall construction.

Further to the north, fill was also added to the western bank of the Willamette River to create current port facilities. Renovation of Port of Portland Terminals began in the 1960s and continued in the 1980's. The existing military port facilities were modified and dredge sand from the Willamette channel was used to backfill old berths and extend the shoreline to the east. Abandoned woodpiles were buried beneath the current facilities.

The Willamette River through downtown Portland is famous for its many historic bridges. The tunnel and pipeline alignments cross under the western abutments or approaches of eight bridges. In addition to existing foundations, numerous abandoned bridge foundations were identified during the historical assessment. The location of both the existing and abandoned foundations influenced selection of the tunnel alignment and profile.

During the urbanization of downtown Portland, the historic western shoreline was filled and moved eastward and fill was placed in low areas, channels, swamps, and lakes to even the topography and increase the usable urban land area. In addition to the

Figure 6. Portland Waterfront During Seawall Construction. (Note closely spaced abandoned timber piles).

construction of additional roads and buildings, early industrialization along both sides of the river saw the construction of a coal gasification plant, numerous warehouses, railroads, and an airport on Swan Island.

4 EXPLORATION PROGRAM

The geotechnical exploration program began by collecting existing subsurface data and then a three-phase investigation was conducted to obtain data necessary for the design of the tunnel, shafts, and pipelines. Initially, an extensive historical assessment along the City's waterfront was undertaken to understand potential obstructions that the tunnel and pipelines might encounter. This information on abandoned buried bridge approaches and wharf pilings, combined with data on existing structures, heavily influenced how the horizontal and vertical alignments of the tunnel were selected. Concurrently, a preliminary geologic assessment was conducted to compile existing subsurface data to produce a preliminary geologic profile of the alignment (Fig. 7A). This assessment identified buried streams, lakes, and the ancient Willamette River channel which had cut down into ancient gravel deposits and were later infilled with fine grained sediments.

4.1 *Phase A Investigation*

The preliminary tunnel alignment was shallow (invert elevation -15 to -35 feet) and was located beneath Tom McCall Waterfront Park. The purpose of the Phase A investigations was to characterize ground conditions, establish the final alignment and profile, and identify geologic contacts between the units described above. During Phase A, mud rotary borings were conducted within city streets to obtain geologic and geotechnical data for the higher profile (Fig. 8). Casing for cross-hole tomography surveys was installed in dedicated borings and converted mud rotary borings. Standpipe piezometers were installed in selected mud rotary borings and groundwater level monitoring was conducted.

Based on borehole data, we were able to define the horizontal reach of the historic lakes along the alignment and characterize the Troutdale Formation as a gravel soil, rather than a conglomerate. Significant concentrations of methane gas (produced during the decay of organic material) were detected within the Sand/Silt Alluvium, Gravel Alluvium, and Troutdale Formation. Methane monitoring of borehole samples continued through all three phases of the investigation.

During the Phase A investigation, we observed that mud rotary drilling was unable to recover adequate samples of the Troutdale Formation gravel deposits. Three-inch ID (Dames and Moore) samplers were seldom driven more than 5 inches during Standard Penetration Testing. Cobbles and larger gravel were not recovered. Larger diameter (PQ-sized) wire line coring was attempted in the Troutdale Formation to improve recovery. However, individual rounded clasts dislodged from the matrix and collided with the core bit, plucking off the diamonds and preventing additional down cutting.

Two cross-hole tomography surveys were conducted which identified the gross outline of abandoned foundations at two of the historic bridges. When the tunnel dropped beneath these structures at the end of Phase A, their impact on construction was negated. Concurrent with the geotechnical investigation, FLAC modeling was conducted at bridge crossings to model how the surrounding soil and bridge foundations would deflect during tunnel excavation.

Based on the result of Phase A investigations, a deeper tunnel with a steeper gradient was selected to avoid existing pile foundations and concrete obstructions. Similarly, the alignment shifted from Waterfront Park into Naito Parkway to avoid the foundation of the demolished historic Morrison Bridge and other abandoned waterfront building foundations. The geologic profile developed at the end of Phase A investigations is shown in Figure 7B.

Shortly after completing the Phase A investigation, the City decided to extend the West Side CSO Tunnel across the Willamette River to Swan Island. The decision was made to build one pump station on Swan Island and combine the West Side and future East Side tunnels at the Confluent Shaft.

4.2 *Phase B Investigation*

The new tunnel profile was deeper to cross under the river. The Phase B investigation required deeper borings to supplement the existing subsurface data. Offshore borings were also required to characterize the new river crossing. Below the Phase A shallow borings, additional contacts between the Sand/Silt Alluvium, Gravel Alluvium, and Troutdale Formation were identified in the deeper Phase B borings.

To provide more accurate data on the grain size, cementation, void space, and weathering of Gravel Alluvium and Troutdale Formation; rotosonic drilling became an important part of the Phase B geotechnical investigation. Representative samples of both fine-grained and coarse-grained deposits obtained during rotosonic drilling (Fig. 9) were supplemented with strength data obtained from mud rotary boreholes. Continuous core recovered during rotosonic drilling improved identification of the geologic contact between the two gravel units. Both rotosonic and mud rotary drilling (Fig. 10) was conducted from barges anchored along the river crossing alignment to defined the gravel contact and the amount of tunnel cover.

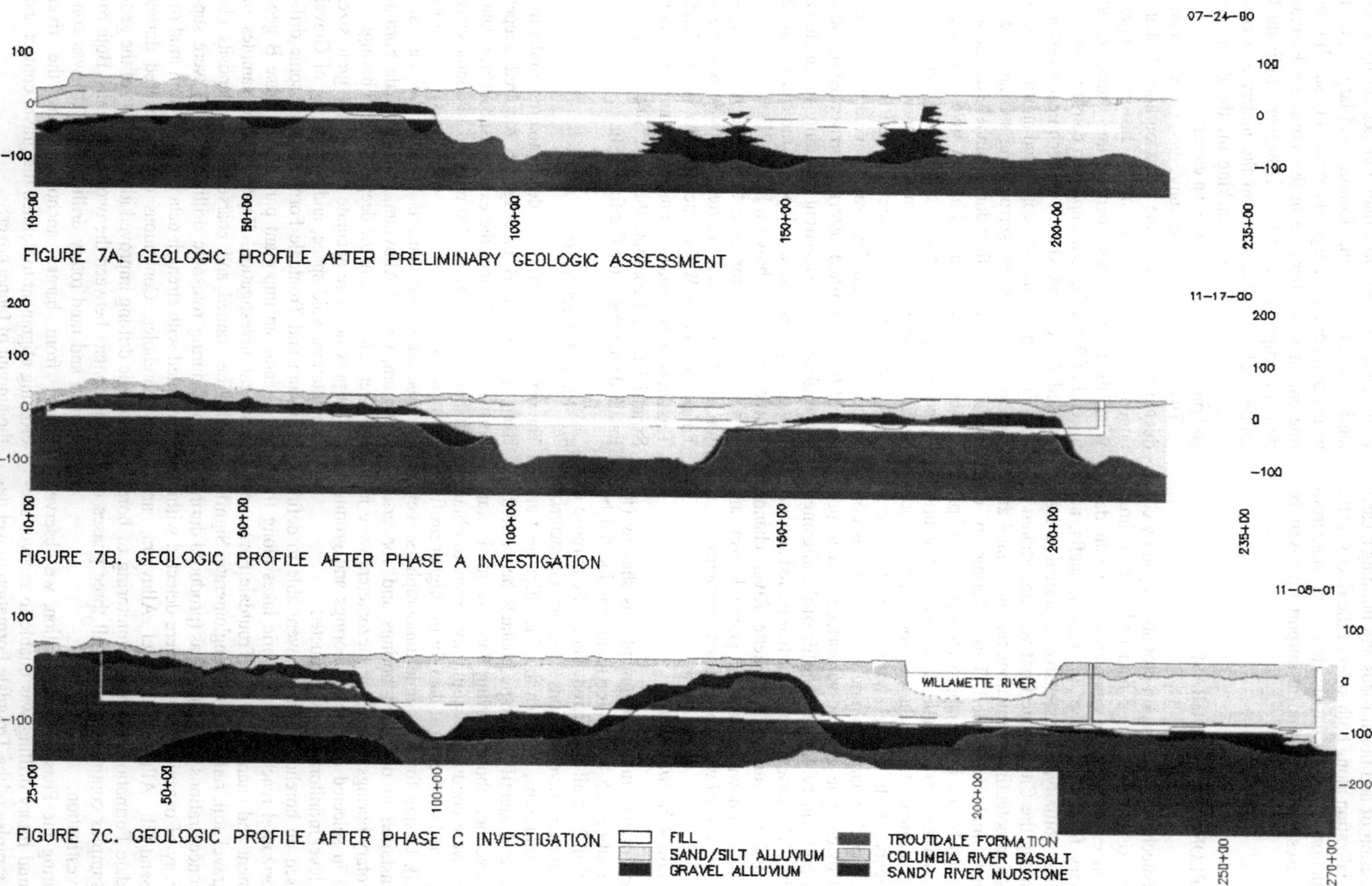

Figure 7. Simplied versions of the three Geologic Profiles.

Figure 8. Mud Rotary Drilling in City Streets.

Figure 9. Rotosonic Drilling in City Streets.

Figure 10. Mud Rotary Drilling in River.

In addition to the rotosonic and mud rotary borings, large (3 foot) diameter borings (Figure 11) were augured in areas of shallow gravel at three locations along the alignment to obtain data to set a baseline for the size and frequency of boulders. Numerous deep cone penetrometer test (CPT) soundings (Fig. 12) were conducted within the Couch Lake area to define the top of the gravel.

Figure 11. Large Diameter Drilling.

Figure 12. CPT Soundings in Deep Fine-Grained Alluvium.

The Phase B investigation was very successful in determining the characteristics of the Gravel Alluvium and Troutdale Formation. Physical samples recovered during this phase were used to develop grain-size distributions of the gravels for selection of the tunnel boring machines.

Geophysical (seismic reflection) and bathymetric studies were conducted along the river crossing to define the depth to gravel and establish bathymetric contours to evaluate tunnel cover within the river channel. Unfortunately, seismic reflectors in organic deposits and gas pockets within the alluvium and channel bottom multipliers masked the gravel contact during the geophysical investigation.

Based on Phase B investigations, the depth of cover within the river crossing became the driving factor in determining the tunnel depth. Thus, the tunnel became deeper, but the slope was not changed. Due to the deeper profile, shafts were either combined or eliminated.

4.3 *Phase C Investigation*

Phase C investigations focused on obtaining data for shaft design and the design of the Swan Island Pump Station. Western Oregon is considered tectonically active and is vulnerable to significant local crustal and deep subduction zone earthquakes. Due to the proximity of two major shafts and the Pump Station to the Willamette River and the Swan Island Basin, the impact of seismic hazards such as liquefaction and lateral spread in Fill and Sand/Silt Alluvium was evaluated at these locations. Additionally, groundwater velocity and gradient data was obtained to ascertain the feasibility of ground freezing to prevent groundwater infiltration during shaft construction. Phase C investigations also identified impermeable layers that could be used for groundwater cut-off in conjunction with ground freezing.

Subsurface exploration methods used during Phase C investigations included rotosonic drilling, mud rotary drilling, and CPT soundings. Rotosonic borings were drilled at shaft locations to better characterize the gravel deposits and identify impermeable materials. Falling head permeability tests were conducted in open gravel zones during rotosonic drilling. Since rotosonic drilling did not provide information on soil consistency and undisturbed samples for strength testing, deep mud rotary borings were conducted at shaft locations underlain by deep Sand/Silt Alluvium deposits. These borings provided SPT N-values to evaluate soil density and obtain samples for triaxial shear testing, unconfined compressive testing, and consolidation testing. Additionally, drilling mud loss during mud rotary drilling allowed us to identify zones of higher permeability.

CPT soundings, including shear wave velocity testing, were conducted at the Pump Station site to obtain data for SHAKE modeling of seismic accelerations. Additionally, FLUSH analysis was performed to model dynamic soil/structure interactions and liquefaction/lateral spread analyses were conducted to estimate anticipated lateral spread during a seismic event.

Permeability, groundwater velocity, and hydraulic gradient data was needed to assess the number and configuration of freeze pipes needed to freeze the ground around the major shaft excavations. To obtain this data, insitu permeability testing was conducted in both Phase B and C to estimate the hydraulic conductivity of the various geologic units. This testing consisted of slug tests in project monitoring wells and falling head permeability tests in open boreholes. Hydraulic conductivity values ranged up to 1.3 x 110 cm/sec, indicating significant permeability in the Gravel Alluvium deposits. Since the Willamette River is under tidal influence along the Portland waterfront, tidal lag time analysis as conducted to determine the delay in tidal fluctuations between water levels in the river and in the monitoring wells installed in borings along the alignment. The magnitude of the delay, combined with hydrogeologic modeling, identified groundwater velocities of 0.3 ft/day. Single well tracer testing, which involved measuring the dilution of a Sodium Bromide dye in a monitoring well over a 24 hour period, provided more accurate velocity data.

Following completion of the Phase C investigation, a Geotechnical Data Report (GDR) and Geotechnical Synopsis Report (GSR) were prepared which summarized both the geotechnical data obtained during all three phases of the investigation and then summarized the anticipated subsurface conditions. The contractor will work with the PB design team to develop a Geotechnical Baseline Report (GBR) that will establish baseline conditions.

No significant changes in the tunnel alignment or depth occurred as a result of Phase C investigations. Based on groundwater data and proposed construction techniques, ground freezing was judged as a viable option for groundwater cut-off at shafts. Based on seismic analysis, ground improvement at two of the major shafts will be required to limit lateral spreading. Figure 7C shows the geologic profile following Phase C investigations.

5 CONCLUSIONS

The phased geotechnical investigation provided an excellent opportunity to tailor an exploration program to best fit the needs of the evolving project. Drilling techniques were selected on a cost versus value basis based on the type of information required for design. Approximately 64 locations were drilled at an overall spacing of 345 ft. Borings include 55 mud rotary holes drilled 10,100 ft, 3 large diameter boreholes drilled 250 ft, and 17 rotosonic borings drilled 4400 ft. Rotosonic drilling was focused on tunnel zones and shaft locations where representative, continuous sampling in gravels was most vital. Mud rotary drilling was about half as expensive than rotosonic and was most useful to obtain strength data, laboratory test samples, and define the top of the gravels. Less expensive CPT soundings were conducted for establishing the depth of the ancient channel in the Couch Lake area and to measure shear wave velocities for seismic design at shaft locations. Thirty-nine CPTs were pushed a total of about 4000 ft.

At each stage of the investigation, geologic profiles were developed that incorporated our understanding of subsurface conditions based on existing borings and those drilled for the project. Simplified versions of three of these profiles are shown in Figure 7. Figure 7A, which shows the geologic profile after the Preliminary Geologic Assessment, is based on interpretation of existing borehole data. Figure 7B, which shows our understanding of subsurface conditions after the Phase A investigation, is based on shallow borehole data. Both profiles identify deep fine-grained alluvium in historic Couch Lake (Sta. 90+00 to 140+00) and shallow gravel on the southern portion of the alignment. Figure 7C, which shows the geologic profile after the Phase C investigation, incorporates the Willamette River crossing into the profile and terminates at the Swan Island Pump Station. The northern portion of the revised alignment follows the ancient river channel and is underlain by thick deposits of fine-grained alluvium.

Session 2, Track 1

Design build case histories

North American Tunneling 2002, Ozdemir (ed.)
© 2002 Swets & Zeitlinger, Lisse, ISBN 90 5809 376 X

Experiences with design construct contract of the "Green Heart" Tunnel, the Netherlands

W. Leendertse
Project Organization High-Speed Line South, Utrecht, the Netherlands

P.S. Jovanovic
Project Organization High-Speed Line South, Project Office North Holland (Holland Railconsult), Utrecht, The Netherlands

ABSTRACT: The Dutch Ministry of Transport and Water Management, Project Organization High-Speed Line South has aimed a clear goal to realize, before 2005, the prestigious project of the world's biggest tunnel under the "Green Heart" in the Netherlands. The 7 km long single tube, 14.8 mdiameter, takes a part of the high speed line between Amsterdam and Paris. Special challenge for the client and contractor was reflected to the unconditional use of the creativity applying innovative solutions available at this moment on the market. This led to the preferable cost reduction, shorter construction schedule and efficient results. The client expectations from the chosen Design-Build contracting form were based on better estimation of the final product considering the possible optimization of the available solutions. To achieve this the client proposed a program of requirements to the contractors pointing the range where the optimum solution could be find. The job has been awarded to only one party, which is now taking care about all phases of the project – from the design to the construction. The contractor has been stimulated to use his full expertise, creativity and innovation to realize the best product getting at the same time a privilege to build the biggest shield driven tunnel in the world for the high speed trains. This paper deals with the client experience concerning contracting issues, advantages and disadvantageous of applied Design Build contract through the risk management control of this unique project.

1 INTRODUCTION

The Amsterdam to Paris will be just a matter of three hours in armchair comfort. Or from central Rotterdam to the centre of London in three hours when the Channel Tunnel Rail Link will be ready. In 2005, when the Netherlands will be connected to the European high-speed rail network, both should be possible. No motorway tailbacks or waiting at the check-in desk at Schiphol. Instead, smooth journeys at up to three hundred kilometres an hour, not only in the European countries but also inside the Netherlands itself. In the construction of the High-Speed Line emphases, wherever possible, will be on the state of the art, technologies, safety, reliability and sustainability with low costs and low risk profile. Much attention will also be paid to architecture of the required infrastructure engineering, including around a hundred and seventy bridges, tunnels and underpasses.

The High-Speed Line is unique project, an interplay of state and private enterprises resulting in a safe and fast transport system that meets both the needs of our times and the needs of tomorrow and beyond. The shield driven tunnel under the "Green-Hart" is going to be constructed as a part of the High-Speed Line South to meet the needs of development and mobility. Since 1991, the high-speed line has gone through all the statutory procedures for of major infrastructure projects, including the procedure for the Kay Planning Decision (KPD) which is embodied in the Infrastructure (planning Procedures) Act. Following the approval of the High-Speed Line KPD by the First Chamber, at the end of April 1997, a start was made on a detailed elaboration of the route. This was done in close consultation with local residents and the relevant local, provincial and national authorities and governmental agencies.

Figure 1. The "Green Heart" Tunnel.

In the meantime the ministers also accepted the Route Decision and at this moment hearing activities are running at the Council of State. It is still possible for local residents, companies, organisations and communities to make objections and to submit them officially. The Council of State will make a decision and will announce in August 1999 when the Route Decision is going to become irrevocable with no more official and formal obstacles to begin with constructing of the tunnel.

2 PROJECT DESCRIPTION

The total length of the tunnel is more than 8 km comprising a 741 m north incline, a 707 m southern incline and 7,16 km of the shield driven tunnel. There will be three maintenance and safety shafts at intervals of about 2,0 km and additional ventilation shafts. Tunnel diameter is 14,9 m and the thickness of the precast concrete segments is 0,6 m. The safety shafts have diameter of 32 m and depth of about 38 m. In the serviceability phase these shafts will be used for emergency purposes and ventilation, and also as a plant building.

Figure 2. TBM of the "Green Heart" Tunnel, (NFM, Lyon – 14,9 m diameter is the largest ever built).

The northern and southern inclines are broken down in three sections: open cutting of 540 m; cut and cover tunnel of 200 m and start respectively exit shafts. The inclines are built inside the containment made of steel sheet piles and diaphragm walls where the underwater concrete has been anchored by driven piles and barrettes.

Soil characteristics of the "Green Hart" location are challenging. There are multiple soil layers consisting of salty and brackish water with water table up to the ground surface. The Holocene Westland geological formation, with 10 m to 15 m thick pit (specific weight of 1100-1300 kg/m^3) and marine clay layer settled on the lose sand, gives a proper picture of difficult soil conditions for any construction works. Under this Westland formation is the Twente and Urk formation 25 m to 30 m from the surface. This formation consists of fine sand layers with density between 10% to 60%. This means a great risk of squeezing and settlements. Furthermore, any construction works that are going to take place under these soil conditions should be carefully considered, taking into account the influence of a soil disturbance on the structures

3 CONTRACTING PROCEDURE

A contract procedure for the shield driven tunnel "Green Heart" started in September 1998 considering an original and unusual way of tendering. There were no traditional advertisements with information, process to record and general specifications but informal, frank and public approach through an organised symposium with domestic and foreign constructors. The main goal of this was to call the market for an intelligent, creative and original solution for the shield driven tunnel project, one of the biggest and expensive parts of the high-speed line rail link. The contract procedure was finished in November 1999 when the "Design-Construct" type of contract has been signed with the awarded constructor – Franco – Dutch consortium consisting of Bouygues Travaux Publics and Koop-Tjoekem.

The contract was awarded on 1st of November 1999 after a second phase of negotiation that began in June 1999 before the bid was submitted. There ware a serious presentations and technical discussions between the Client and contractors and through its complete openness about technical matters, risks, costs and schedule, the Bouygues-Koop consortium succeeded in their approach of "partnering" with the Client. The innovative single tube solution with a central partition wall brought enormous amount of advantages to the Client as well as contractor.

The High-Speed Line Project Organisation is responsible for ensuring a successful construction operation of hundred kilometres of high-speed line. In it, NS-RIB (Dutch Railways Rail Infrastructure Management), Holland Railconsult and DHV (environment and infrastructure consultants) work together under responsibility of the Ministry of Transport, Public Works and Water Management and the Ministry of Housing, Spatial Planning and Environment. In 1998 the structure of the organisation changed from a preparatory phase structure to a structure with emphases on project construction.

The Project Office Shield Driven Tunnel "Green Hart" of the Project Organisation High-Speed Line South has the following tasks:

- To prepare and guide project design and construction, inclusive technical installations

- To establish necessary boundary conditions for realisation (ground purchase, cables and ducts, permissions, etc.)
- Communication activities with local residents and authorities concerning tunnel design and construction
- Preparation of (optional) maintenance plan in a period from five to ten years after completion

By adopting an open attitude and approaching the commercial exploitation of the line, in an innovative way, the Project Organisation stimulated the research and development of the best possible solutions and products that will fit seamlessly into wishes of its users. The contracting process has been realized concerning the following order:

1. Pre-qualification phase
2. Invitation to consultation
3. Consultation phase
4. Invitation for offers
5. Offering phase
6. Offer Submission
7. Evaluation phase
8. Invitation for negotiation
9. Negotiation phase
10. Awarding

The main issue in Design Construct philosophy is the way of risk distribution between the client and contractor where the project management is based on the risk management control (RMC).

4 DESIGN & CONSTRUCT (D&C) BASES

The client's approach to the idea of applying at the first time in the Netherlands the Design Construct contracting form on a mayor tunneling project was based on the statement: Clearness above Reasonableness. This means that the client's responsibilities in Design Construct approach decrease and the contractor's responsibilities increase under the same circumstances concerning the scope of the work with all changes during the project life. On the other hand the risk assessment belongs with its mayor part to the contractor instead to the client. Using the traditional contracting form the client's and contractor's responsibilities are mirrored (see Fig. 3).

Considering the issues of Design Construct assumptions it was necessary for the client to define and chose the most effective and easiest way of the project management control during all project phases from the first idea to maintenance taking care for each phase about risk profile on the one side and the prevision costs, schedule, quality, information and organization issues on the other side.

The client's intention was to keep the risk profile low during the contracting phases establishing the efficient basement of the risk control procedure with the contractor. This basement considers the optimal segmentation of the risks between the client and

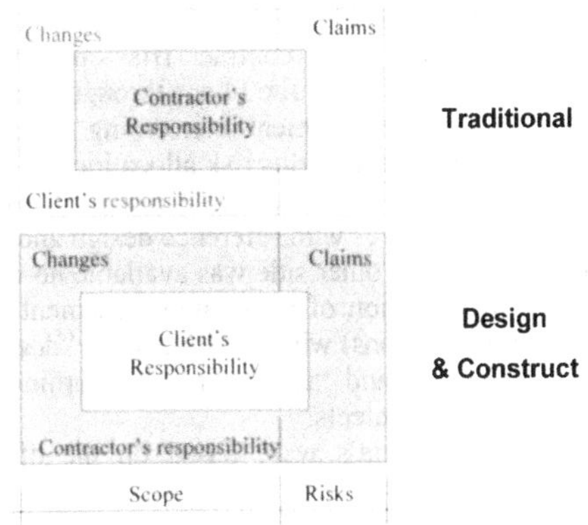

Figure 3. Responsibilities.

contractor avoiding unacceptable risks and an agreement about the risk control with its supervision during the project cycles (see Fig. 4).

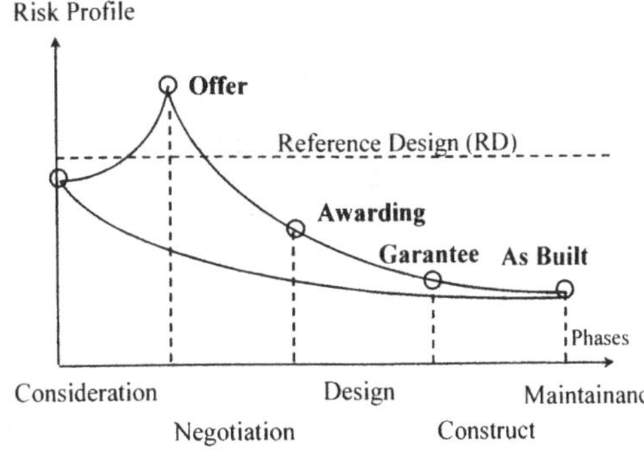

Figure 4. Risk Profile during contracting phase.

Client has chosen for the negotiation procedure based on the Design Construct philosophy intending to establish healthy cooperation between the client and contractor due to their main goal of making the project successful by all means. Contractor on the other hand was surprised with the professionalism of the client granting his full cooperation during the all phases of the project.

5 D&C PHASES

For the consideration phase was important to give a clear picture about the client wishes and procedures. The target was to harmonize the main line with the design (offer). Basic document produce by client ware: the Tender Guidelines with instruction about evaluation; draft contract mentioning the program of requirements and project details; reference design. For the contractors was important that they could submit an offer based on client's reference design (risk profile) and draft contract (risk allocation). This was considering by the client through supervision of design development concerning the offer consistency (risk profile plus risk allocation).

Offers evaluation was based on risk profile definitions: discrepancies with reference design and control measures. The other side was available no-claim bonus with definition of negotiation basement: risk allocation (conditions) with the statements "does not satisfy unless..." and "satisfy if...." regarding the program of requirements.

Awarding criteria's ware based on the offered price realization with offered price for maintenance (15 years / 25 years). Minimum construction nuisance with shorter construction time inclusive recycling possibility of out-coming soil, low risk profile and innovative elements covered with the acceptable quality assessment plan where the additional criteria's consider by the client.

The negotiation phase was important for the client as well as contractor because of the possibility to discuss and solve many issues concerning the control measures "does not satisfy unless..." and "satisfy if...". Furthermore ware the preconditions and the top risks discuss making it clear for contractor what should be considered in their finial offer.

Award went to the contractor who achieved to harmonize the risk profile, risk allocation, control measures agreed in negotiation phase with the design considering the quality assurance in the construction phase. Of course the implementation of the client's external quality assurance has as a main target to control the contractors demonstrability of fulfilling the program requirement. This is based on project risk management where consequences of taking the countermeasures by the contractor should be avoided. This means that "predict and avoid" got the priority before "happened and cured".

6 THE RISK MANAGEMENT PROCESS

The Australian Risk Management Standard [4] is a generally acknowledge standard for risk management. The general method is widely used as a generic risk management process applicable to almost all projects or processes. It consists of 8 phases shown on Fig. 5.

This type of risk management control is applied on the "green Heart" project. The part of risk management involving implementation of procedures and physical changes to eliminate or minimize adverse risks. Risk control includes review of the risk process and review of relevant documents during and after completion of the project.

Risk policy consider the context in which RM is performed; relations/interactions with other management systems and criteria for acceptable risks.

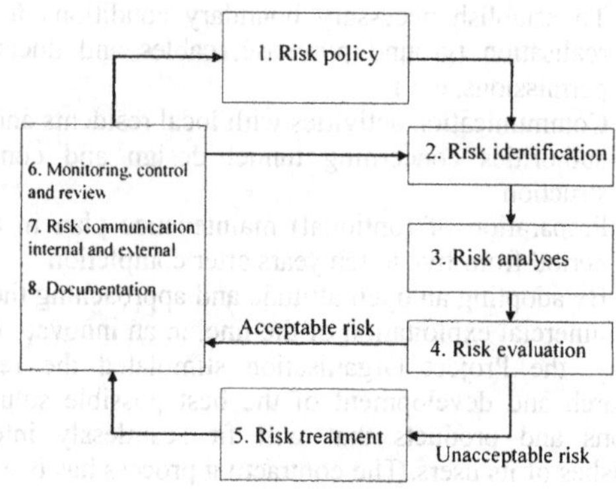

Figure 5. The generic risk management process.

Risk identification consists of what can happen, how and why it can happen. In this stage the client gave to the contractor additional subjects of concern which contractor add to his risk identification file. Risk analyses have been performed from the contractor side for existing risk controls inclusive likelihood and consequences due to the risk level definition. Risk evaluation is very important to establish mayor risk and risk priorities which can be discussed due to measuring risks against the risk policy.

If the risk evaluation shows that the risks are acceptable then the process leads to the monitoring, control and review of the parameters defined in previous steps. If the unacceptable risks occurred then the risk treatment is the next step. This means that the process of identification of possible risk reducing measures shell take place to reduce likelihood or and consequences; risk transfer and the elimination of risks. Risk treatment consider the evaluation of technical economical and administrative measures; evaluation of residual risks against the risk policy; selection of the solutions; implementation of solu-

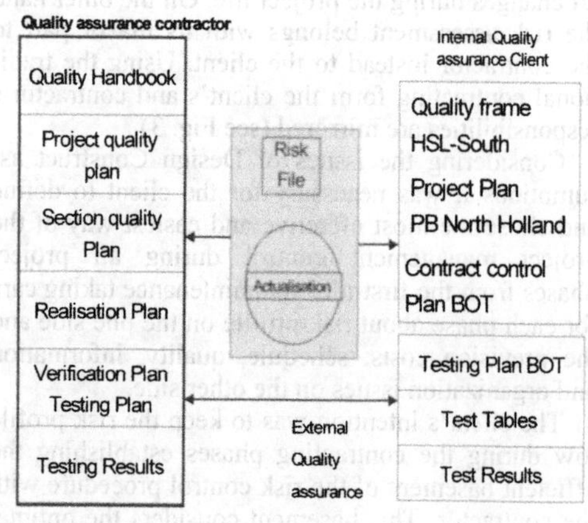

Figure 6. Project control.

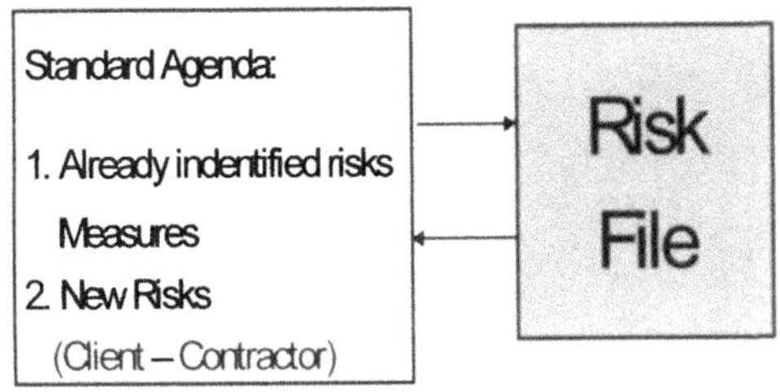

Examples discussions of progress:
- Conditioning
- Healthy & Safety
- Discussion about TBM
- Ramps / strating shaft
- diverse

Figure 7. Discussion of progress.

tions and the transferred risks have to be monitored and reviewed to insure that the conditions are fulfilled. To close the risk loop the last stage in the process is dedicated to the monitoring, control and verifying the risk evaluation and treatment pointing back to the risk identification. Risk communication internal as well as external is recorded to the risk documentation of the contractor always available for the client.

7 CONCLUSIONS

This experience with the risk management control leads to the conclusion that the stimulation of proactive thinking with limitation to the essence under the streamline discussion of progress between the contractor and the client are the key words for the success.

REFERENCES

Leendertse, W.L. (1997), Damages of segments on the west tube The Second Heinenoord tunnel, Bouwdienst Rijkswaterstaat 1933-T-980735, The Netherlands.

Jovanovic, P.S., Veen, C. van der, (1999), Design of lining structure, Conference proceedings, Infratunnel, RAI Amsterdam, The Netherlands.

Blom, C.B.M., Jovanovic, P.S., (1999), Requirements for tunnel design in soft soil, Proceedings Long Road and Rail tunnels, Basel, Switzerland.

Blom, C.B.M., Jovanovic, P.S., Oudejans, W.L., 2000, Recommendations for Tunnel Lining Design in Soft Soil, Conference Proceedings, 16th congress of IABSE, Structural Engineering for Meeting Urban Transportation Challenges, Lucerne, Switzerland.

Bakker, K. J., (2000), Soil Retaining Structures, PhD Thesis, Delft University of Technology, Delft, The Netherlands.

Blom, C.B.M., van der Horst, E.J., Jovanovic, P.S., (1999), Three-dimensional Structural Analyses of the Shield-driven "Green Heart" Tunnel of the High-Speed Line South, Tunnelling and Underground Space Technology, Vol. 14., No. 2, pp. 217-224.

Bakker, K.J., Leendertse, W.L., Jovanovic, P.S., Oosterhout, G.P.C. van, 1999, Stresses and Strains in Segmented Lining, Evaluation of Monitoring and Numerical Analyses, Conference Proceedings, TC 28 Symposium on Geotechnical Aspects of Underground Construction in Soft Soil, Tokyo, Japan.

Blom, C.B.M., 2001, Damage patterns and mechanism of segmented concrete tunnel linings, Delft University of Technology, Delft, The Netherlands.

Bloemhof, K, 2001, Geometrical tunnel model, damage of segmented concrete linings Delft University of Technology, Delft, The Netherlands [in Dutch].

Blom, C.B.M. Oosterhout, , G.P.C. van, 2001, Full Scale laboratory tests on a segmented lining, summary report, Utrecht, Delft, The Netherlands.

Blom, C.B.M., 2001, Lining behaviour -Analytical solutions of coupled segmented rings in soil, Preliminary Thesis "Design Philosophy of Concrete Linings of Shield Driven Tunnels in soft soils", Chapter 1 3, Delft University of Technology 25.5-01-15, Delft, The Netherlands.

North American Tunneling 2002, Ozdemir (ed.)
© 2002 Swets & Zeitlinger, Lisse, ISBN 90 5809 376 X

Design-build of a jet grout access shaft for tunnel rehabilitation

M.W. Oakland, Ph.D., P.E.
CDM Jessberger, A division of Camp Dresser & McKee, Inc., Cambridge, Massachusetts, USA

M.J. Ashe, P.E.
Camp Dresser & McKee, Inc., Cambridge, Massachusetts, USA

J.R. Wheeler, P.E.
Design/Build Geotechnical (Formerly with Hayward Baker, Inc.), Stow, Massachusetts, USA

R.C. Blake, P.E.
CDM Engineers & Constructors, A division of Camp Dresser & McKee, Inc., Madison, Connecticut, USA

ABSTRACT: A design-build approach was used to construct a circular jet grout access shaft as part of the rehabilitation of the 75 year old primary water supply conduit for the City of Providence, Rhode Island. The shaft was needed over the cast-in-place portion to access the entire system for inspection and repairs. The conduit could only be accessed within the footprint of the City's settling basin. Following repairs to the conduit, the shaft would then serve as the excavation for a future valve structure. The circular jet grout shaft was selected for this project due to a number of unique and challenging constraints.

Jet grout was used to form a circular cofferdam access shaft that did not require bracing, reducing the construction time substantially and allowing an unobstructed shaft for access to the tunnel. The jet grout wall formed a stable, erosion resistant surface capable of being disinfected quickly in the event the tunnel unexpectedly had to be put back into service.

This paper discusses the planning, design and construction of the tunnel access shaft. The contracting mechanisms to allow the design-build to be completed are also described. The paper also details the performance of the jet grout system and the contingency measures that were in place to deal with problems. In addition to discussing the engineering challenges of the shaft design and construction, the paper details inspection and rehabilitation of the conduit using polymer modified cementitious mortar.

1 INTRODUCTION

In late 1999, the Providence Water Supply Board (PWSB), the water utility of the City of Providence, RI, became aware of corrosion and deterioration within one of two water supply conduits that are the only connections between the treatment plant and the distribution system. The water system is the sole water supply serving approximately 60 percent of the State of Rhode Island.

The deteriorated water supply conduit was the 2.3 m (90 in) diameter reinforced concrete tunnel extending from the water treatment plant. The supply conduit was constructed in three basic sections, in order of flow from the plant: a 2.3 m (90 inch) diameter cast-in-place reinforced concrete conduit constructed in a cut and cover trench and typically founded on bedrock; a 2.3 m (90 inch) diameter reinforced concrete lined circular rock tunnel and a nominal 2.3 m (90 inch) diameter horseshoe shaped rock tunnel. The rock tunnel sections were constructed by drill and blast techniques. Deterioration of the concrete lining required repairs in both the circular cast in place and tunneled sections of the conduit. A layout of the plant and conduits is shown on Figure 1.

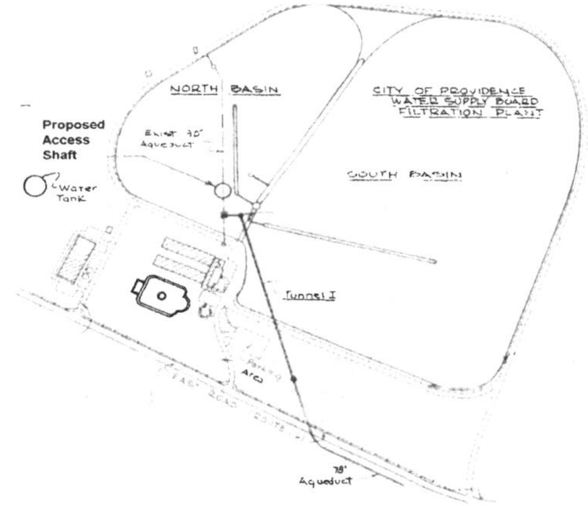

Figure 1. Site Plan.

While the corrosion and deterioration was not excessively severe, it was immediately recognized to be a serious concern because both conduits together are needed to provide an adequate supply of water. Although, either conduit can supply the system under low flow conditions, the conduits were configured such that only a single shutoff valve was provided for both conduits. As a result, the conduits could not be operated individually.

In addition to the critical importance of the water supply conduit to the water system and the State of Rhode Island, concerns were elevated because of the inability operate the conduits individually, loss of one conduit would mean that both would be out of service. Further, there was very limited access into the conduit to conduct inspections and repairs. Only limited inspections and maintenance had been performed on the conduit over the 75 years of its existence. The PWSB had experienced a catastrophic failure of another primary pipeline in their system, a newer prestressed concrete cylinder pipe aqueduct, only a few years earlier. Accordingly, deterioration, as shown in Figure 2, was recognized to be very important and in immediate need of attention and repair.

Figure 2. Typical Conduit Deterioration Exposing Reinforcement.

In response to these concerns, in early 2000, PWSB engaged CDM Engineers & Constructors Inc., the design-build subsidiary of Camp Dresser & McKee (CDM), to perform further inspections, design and repair the water supply conduit. The work was to be performed as a design-build project to take advantage of its early delivery advantages and provide an expedient and cost effective process by which to perform the repairs as the actual conditions were exposed once the conduit was taken out of service.

An immediate task was to determine how best to maintain flows between the treatment plant and the distribution system to allow repairs in the water supply conduit to be performed. By-pass pumping of the treatment plant flows was extensively explored and found to be too risky and costly.

An alternative plan was developed by which the treatment plant would be shutdown and the water supply conduit taken out of service for eight hours every other night during which time the inspections and repairs would be conducted. At all other times during this period, the water supply system would be maintained at maximum levels to service the high demand periods of the day and to replenish the water supply reservoirs in the system. This plan was adequate for the initial inspections, but did not provide sufficient time to conduct repairs.

Another complication, or challenge, of the project, was that the water supply conduit is located beneath the north settling basin at the treatment plant. Access into the water supply conduit was limited to two deep manhole structures located approximately 150 m (500 ft.) apart at either side of the basin. These manhole structures were inadequate to perform the repair with the limitation of the 8-hour shutdowns.

A work plan was developed which included construction of a new shaft at a convenient location, to bulkhead the 2.3 m (90 in) diameter conduit so that the second 2.0 m (78 in) diameter conduit could still continually supply water, maintaining a low flow condition to the system while completing repairs as quickly as possible. However, it was decided that the open access shaft had to be designed such that it could be quickly disinfected and used as part of the conveyance system during a fire emergency or other unanticipated high flow condition. The system had to be capable of returning to service within 2 hours at any point in the construction.

After the conduit repairs were complete and the construction access was no longer needed, the access shaft would house a new valve chamber to allow independent control of the two conduits for better operational control of the water supply system and access into the conduit for routine inspections in the future.

The preferred location of the access shaft was in the cut and cover portion of the conduit where rock excavation could be avoided to reach the top of the conduit. In addition, the access shaft had to be relatively close to the junction of the two conduits for installation of the temporary bulkhead and to be effective as a valve location for future inspections. The ideal location was thus within the north settling basin, very close to the separation berm between the two basins. It was decided to drain the north basin while continuing to operate the south basin at full capacity. Again, making the speed of the construction a primary factor due to the impact on treatment plant operations of having one basin drained.

The access shaft was be about 7.6 m (25 ft.) in depth to reach the invert of the conduit and the valve vault to be constructed within the access shaft was to be about 4.9 m (16 ft.) square. The shaft had to be excavated about 1 m (3 ft.) below the bottom of the conduit, within bedrock, to allow construction of a floor slab for the vault.

2 SITE CONDITIONS AND OPERATIONAL CONSTRAINTS

The site consists of two settling basins contained by earthen berms and separated by a central concrete wall. The location of the proposed tunnel access and valve vault was within the north settling basin. The reservoir is about 6.1 m (20 ft.) in depth, constructed with 3 horizontal to 1 vertical side slopes.

The basin floor and side slopes are generally covered by an unreinforced concrete slab with 0.6 m (2 ft.) wide by 0.6 m (2 ft.) deep drainage trenches in the slab. However, at the northern end of the basin, bedrock was exposed at the surface and was left in place to serve as the basin bottom with the remaining concrete floor formed up to the perimeter of the outcrop.

The proposed valve vault was to be constructed at the southern side of the north basin adjacent to the toe of the separation berm with the adjacent south basin within approximately 1 m (3 ft.) of the drainage trench in the slab.

While it was intended to drain the basin for construction of the new valve vault, several other restrictions created construction challenges for the project. These restrictions included:

- The shaft had to be designed, constructed and excavated within a 3 month period.
- The south basin had to remain full prior to explorations and excavation, creating an apparent artesian pressure below the basin slab.
- The access shaft had to be essentially free of bracing to allow construction of the valve structure.
- The excavation and temporary earth support system had to be capable of serving as part of the pipeline should an emergency require that the tunnel be returned to service. This required that the shaft extend approximately 1.2 m (4 ft.) above the basin slab to provide adequate head. The shaft had to be capable of being cleared of equipment, disinfected and stabilized to allow water flow within a two hour period.
- The excavation system had to be relatively water tight to avoid dewatering in the pervious soils as well as avoid contamination of the water supply if the conduit had to be used in an emergency. The system had to seal the anticipated bedding around and below the existing cast-in-place conduit.

Originally, it was intended that a conventional sheeting system with sump pump dewatering be used as the excavation support system. Similar systems had been used for the original construction of the conduit and subsequent vaults near the location of the access pit. However, in those cases, the basins either were not present or the water in the adjacent basin was likely lowered to facilitate the construction.

3 DESIGN/BUILD ADVANTAGES

PWSB selected a design-build project delivery system to perform the repairs to the water supply conduit because it offered early delivery (schedule) advantages and provided an expedient and cost effective process to perform the repairs as the actual conditions were exposed and design of the repairs were prepared.

Construction had to be completed during winter months when water demands are low. If traditional methods (design-bid-build) were used, there was insufficient time in which to undertake the repairs during the winter months of 1999-2000. Also, a more complete inspection of the disintegration was not possible due to the difficulty in accessing and dewatering the conduit, as well as the amount of debris in the pipe that could not be easily removed through the existing access locations. The unknown conditions complicated the design of the repairs and estimates of material quantities.

Using design-build project delivery, the inspections and design details of the repairs could be prepared as the other construction was mobilizing and getting underway. Previous repairs to conduits for PWSB using design-build methods proved very successful and were believed to be the best way to proceed with the tunnel repairs.

Similarly, due the time constraints of the project and circumstances, design-build of the jet grout access shaft was also to be utilized. Design-build of the access shaft was necessary so the project would not be severely delayed and lose the opportunity to perform the repairs to the water supply conduit during the winter of 2000-2001. In this instance, it was a collaborative effort by CDM and Hayward Baker Inc.

4 SUBSURFACE SOIL AND GROUNDWATER CONDITIONS

Very little was known about the subsurface conditions at the site prior to undertaking the work. A program of test pit explorations was initially attempted to assess the below-grade soil and groundwater conditions. However, boiling sand conditions prevented any test pit from being excavated more than 0.7 m (2 ft.) below the surface. Apparently leakage from the south basin was creating a slight artesian condition. The flow of groundwater into any excavation was substantial.

A test boring was planned as the second phase of explorations. A single, cased test boring was drilled by Geologic Corporation on October 16 and 17, 2000 within the planned footprint of the proposed access shaft. The test boring encountered about 5.2 m (17 ft.) of very dense, gray fine to coarse sand with various amounts of silt and gravel. Granite bed-

Figure 3. Proximity of the South Basin to the Site.

rock was cored to a depth of about 10.1 m (33 ft.) below the bottom of the basin. The bedrock was slightly weathered at the top with a 0.3 m (1 ft.) sand seam approximately 0.3 m (1 ft.) below the top of the rock. Unweathered hard bedrock was encountered in the remainder of the test boring. Rock Quality Designations (RQD) for each of the three core samples was 50, 62 and 92 percent, increasing with depth.

A temporary PVC casing was installed to replace the drill casing and the test boring was later used as a pumping well during construction. Groundwater was recorded at or slightly above the top of the basin slab in the borehole following drilling.

Based on the boring, it appeared that the bottom of the existing conduit was founded about 1 m (3 ft.) within the bedrock at the location of the proposed site of the valve vault. At this location, it was anticipated that up to about 1.5 m (5 ft.) of bedrock would have to be removed around and below the existing pipe to allow construction of the vault that was larger and deeper than the existing conduit. The cofferdam would have to be toed at the top of the bedrock with a relatively water tight seal and allow excavation of the bedrock below the bottom of the cofferdam. In addition, the configuration of the conduit, recessed in a trench in the bedrock, created additional difficulties in forming a seal around the existing conduit.

Conceptual evaluation and preliminary design of the cofferdam proceeded based on the test boring and test pit information. However, to further define the subsurface bedrock conditions at the site for final design and layout, 12 auger probes were drilled by Geologic using solid stem augers. The auger probes were drilled on November 25 and 26, 2000, in part, to identify a potential nearby location that would lessen the rock excavation required.

The probes found the depth of bedrock to increase to the east. In addition, the probes also found a sand seam just below the top of bedrock at several locations. The shaft was moved about 3 m (10 ft.) to the east to reduce the amount of bedrock excavation.

5 DESIGN ALTERNATES

The excavation for a previously installed control structure within the basin near the site of the proposed valve vault had been completed using sheeting and exterior dewatering wells. However, the nature of the soils, amount of dewatering and other construction details were not recorded. In addition, at that time there were no requirements for water containment and the level in the south basin could have been lowered, relieving a major source of the groundwater.

Based on previous experiences, initial concepts were to use a conventional sheet pile or soldier pile and lagging system to support the excavation. A liner or membrane was planned to be used to form the water containment barrier in the event of having to put the conduit back in service during the construction. However, following the initial test pits, it seemed evident that any system selected would likely have to form a relatively effective seepage barrier and be capable of structurally resisting the hydrostatic forces. Alternately, estimates of the dewatering requirements for the project seemed prohibitive to conduct external dewatering.

Further, the work was to be conducted at the downstream toe of what was effectively a dam retaining approximately 6.1 m (20 ft.) of water. Concerns for slope stability of the berm, potential for piping of the foundation soils below the berm and ability to control the artesian condition once a larger opening was made in the slab had to be addressed as part of the design.

The level of the bedrock found in the initial test boring posed several additional challenges to the design of conventional excavation support systems. The first was how to provide an effective groundwater cutoff at the interface with the bedrock. Normally the bedrock surface is not uniform and cobbles and boulders commonly exist at the surface that prevents sheet piles from being driven tight to the surface, leaving a gap that is difficult to seal. This problem was compounded by the sand seam found within the upper portion of the bedrock. The second problem was how to fix the toe of the sheeting at the top of the bedrock. Without being able to drive the toe of the sheeting below the bottom of the excavation, a cantilevered condition exists extending downward. Typically an additional bracing level is required. Often special techniques are required to install this lowest bracing level while still maintaining stability.

An additional consideration related to the wall design was the design requirement to extend the water containment approximately 1.2 m (4 ft.) above grade. This was due to hydraulic considerations should it be necessary to return the line to service while the excavation was open. Ideally, a wall, such as a sheet pile wall, which could remain extended above the ground surface was desirable. However, alternatively, a cofferdam upon which a watertight extension could be added was required.

A second concept was considered for the access shaft using a drilled-in secant pile wall with primary piles reinforced by a soldier beam dowelled into bedrock to provide toe stability. All piles would be drilled through the sand seam that was found within the upper bedrock. The concept for the cofferdam was to be square to match the valve chamber configuration with two interior waler levels. The need for cross bracing was to be determined based on the size of the waler that could be installed without interfering with the wall construction. However, attempts would be made to avoid the cross bracing because the required penetrations in the final chamber walls which would be difficult to seal.

The original design anticipated a set back in the bedrock to maintain the toe of the cofferdam. The setback distance was to be about 2 ft. or about the width of the waler to allow for forming of the valve chamber wall. Splitting would be used to remove the bedrock after the tunnel roof was removed to start repair of the tunnel.

The square secant pile wall was initially selected anticipating that the drilling equipment would be readily available and relatively lightweight for use on the unreinforced basin slab. The square configuration was maintained to limit the amount of wall required to be installed. With this in mind, a site meeting was held with Hayward Baker to discuss the installation in October 2000. At the meeting Hayward Baker suggested considering grouted circular wall. The circular configuration would result in additional wall length, but potentially avoid having to use internal bracing, provided the wall was thick enough to accommodate the compression stress.

6 CONTRACTOR CONSIDERATIONS AND DESIGN/BUILD ENHANCEMENT

Based on input from Hayward Baker, an alternate design was developed using a circular cofferdam. The intent was to develop a configuration that would not require internal bracing. Critical components of a self supported compression ring wall were the compressive force in the jet grout and the compressive loading on the existing 75 year old tunnel in its pre-repaired state.

The specialty geotechnical contractor, Hayward Baker Inc. of Odenton, MD was contracted during the early stages of the design to evaluate potential methods for shaft excavation. Initially, permeation grouting alternatives were considered for these applications as controlled injection of sodium silicate grout, supplemented by polyurethane grouting. However, the significant fines content of the overburden soils, in excess of 15 to 20 percent, and the potential for flowing groundwater to impact the grouting program precluded the use of these alternatives.

The construction of a circular secant pile wall was considered. A secant pile wall offered the benefits of conventional construction techniques and could provide an interior shaft suitable for use in acting as part of the water supply pipeline if the shaft could be constructed to be watertight. However, keying the individual piles into the granite bedrock to provide an adequate groundwater cutoff beneath the irregular bedrock surface would present a time consuming and expensive operation. In addition, a supplemental grouting program would be required to provide a watertight seal between the individual secant piles and the sloping sides of the existing conduit walls and beneath the conduit structure itself.

In subsequent evaluations, the use of jet grouting was considered. In contrast to the secant pile wall, jet grouting is a ground improvement method that employs a high velocity, horizontally directed fluid injection to erode and mix in-situ soils with a cementitious grout to produce a product frequently referred to as soilcrete (Billups, Cavey and Wheeler, 2001). This grout-soil mixture normally achieves an unconfined compressive strength in excess of 0.7 MPa (100 psi) within 24 hours and ultimate strengths typically in the range of 3.3 to 5.3 MPa (500 to 800 psi) in silty sands, sands and gravel. Corresponding soilcrete permeabilities are typically in the range of 1×10^{-5} to 1×10^{-6} cm/sec. Since jet grouting is an erosion based grouting system that is applicable to a wide range of soils, there was little concern for the fines content of the soils to be grouted.

In addition to providing the capability of treating in-situ soils to create a strong and impermeable barrier wall, a series of interconnected jet grout columns could be carried directly to the top of the irregular bedrock surface to create the required groundwater cut-off (Anderson, Chadwick and Wheeler, 2000).

While the jet grout posed several new design challenges, it solved one substantial problem that up to this point had not been addressed. That was how to cut off the groundwater around the existing tunnel within its cut rock trench and any bedding stone. The jet grout could extend around and even under the tunnel forming a relatively positive seal. The conceptual grouting pattern around the culvert is shown in Figure 4.

The jet grout was to be installed through the existing reservoir floor mixing with the existing sand without having to dewater the site. The jet grout was expected to work well to form a seal at the interface with the bedrock as well as around the existing cast-in-place pipe and bedding material. The jet grout wall would form a stable, erosion resistant surface capable of being disinfected quickly in the event the tunnel unexpectedly had to be put back into service.

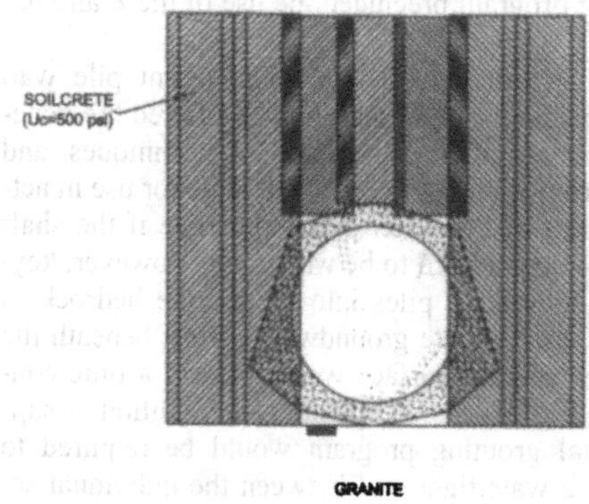

Figure 4. Section Showing Grout Cut-off.

7 FINAL DESIGN

With the advantages offered by jet grouting, the Design-Build team could work closely with the specialty contractor to develop a cost effective and constructible temporary earth support system. Through interactive discussion, a soilcrete strength of 3.3 MPa (500 psi) was selected to make the circular cofferdam wall self-supporting, thereby eliminating the cost and schedule impacts associated with the installation of temporary internal bracing. To construct large diameter columns, Hayward Baker used their proprietary SuperJet grouting technique. As with double fluid jet grouting, the SuperJet system employs a high velocity injection of grout sheathed in a cone of compressed air to erode and mix the soil. However, the tooling for the SuperJet drill monitor utilizes in fluid flow mechanics to create a highly focused jetting stream capable of constructing 3.0 to 4.6 m (10 to 15 ft.) diameter soilcrete columns in lieu of the typical 1.5 to 2.1 m (5 to 7 ft.) diameter columns normally achieved with the conventional double fluid system (Burke and Wheeler, 2000).

The circular shape proposed by the specialty subcontractor for the jet grout shaft provided a structure that resisted external horizontal forces due to soil and groundwater via horizontal compressive forces in the jet grout wall. Conceptually this was a simple and efficient structural system that did not require internal bracing of the excavation or reinforcement of the jet grout. Based on the estimated ultimate compressive strength of the soilcrete of 3.3 MPa (500 psi), the design was based on limiting the maximum compression stress in the jet grout to about 2 MPa (300 psi). The minimum wall thickness was computed based on the compressive hoop stresses due to soil, water and equipment loads plus a 10 percent overstress for out of round conditions.

However, the location, size, and shape of the existing conduit complicated the concept. Where the jet grouted wall crossed the conduit, the compressive forces would be transferred by the existing conduit that occupied a significant section of the grouted cofferdam. Prior inspections of the conduit indicated that as much as 10 cm (4 in.) of the concrete had disintegrated but that the remaining concrete and reinforcing were intact. A structural analysis was conducted to determine if the residual strength of the deteriorated conduit could be relied upon to resist the horizontal compressive forces from the jet grout cofferdam. It was determined that the loads on the conduit, in its current condition had to be limited.

Two options were considered: install horizontal struts across the tunnel to support the compressive loads from the jet grouted wall or widen the wall to distribute the loads over a larger area of the tunnel. The strut option, while less expensive, would have hindered access to the tunnel during the repairs, potentially extending the duration of the repairs. While difficult in the small diameter cofferdam, it was decided to flare the walls and eliminate the need for reinforcement. Given unknowns in the strength and degree of deterioration within the conduit, it was difficult to analyze the reserve strength available. Therefore, the 1 m (3 ft.) jet grout zone was enlarged to 2.4 m (8 ft.) at the points where the jet grout wall intersected the conduit to limit the stresses to only slightly more that the conduit would have experienced under normal conditions with the basin full.

Rather than try to layout individual columns for the contractor, Hayward Baker was provided an outline, shown in Figure 5 of the minimum area to be grouted. He was to select the best grout column diameter and pattern to meet the design.

Hayward Baker elected to use both 1.5 and 3.0 m (5 and 10 ft.) diameter columns to best fit the required area, rather than try to fill in the area using a single size of column. A single row of secant interconnected 1.5 m (5 ft.) diameter columns were used to create the 1.0 m (3 ft.) wide wall, with 3.0 m (10 ft.) diameter columns at the flared sections. All columns were to be constructed with the double fluid jet grouting system, was selected for the design as shown in Figure 6.

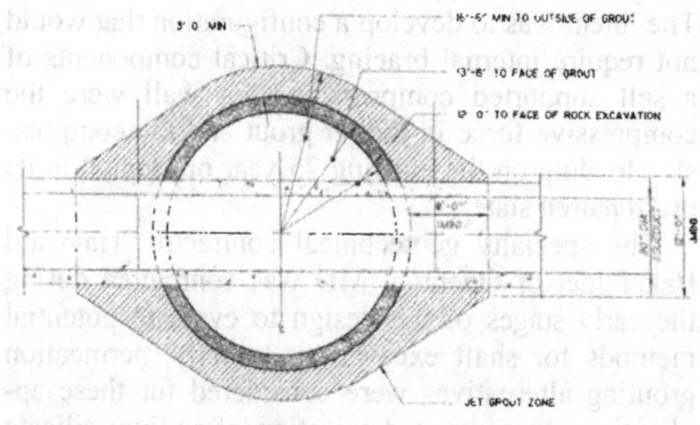

Figure 5. Plan of the Required Jet Grouted Area.

With all of the advantages offered by jet grouting in the difficult subsurface conditions and geometry dictated by the project, there remained concerns about the ability of constructing a watertight barrier wall. Although the specialty contractor could not offer a guarantee as to the volume of water that might flow into the completed excavation the specialty contractor expressed the opinion the groundwater flow into the completed excavation could be held to less then 18 lpm (5 gpm). As a further enhancement to the design, larger 3.0 m (10 ft.) diameter jet grout half columns were added immediately adjacent to both ends of the conduit beyond the minimum outline required for structural support to provide an additional seal against the conduit wall and to extend beneath the wall to stabilize soils beneath the conduit structure.

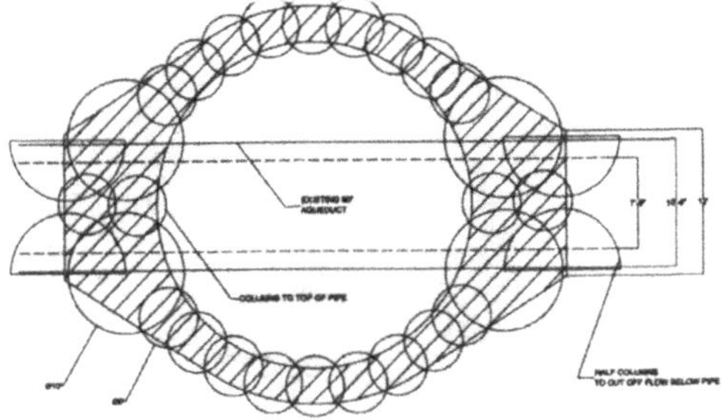

Figure 6. Plan of Jet Grout Column Layout

Consideration was given to doweling each column into the bedrock during grouting to avoid potential for lateral movement due to unbalanced loads from the basin embankment, the toe of which was at about the edge of the cofferdam, and equipment loads. However, it was quickly decided that passive pressure and friction at the base of the wall would be sufficient to resist any lateral movement of the entire structure.

An approximately 1.2 m (4 ft.) high wall was constructed on top of the jet grout ring wall to provided the required head should the conduit have to be returned to service during construction.

8 CONTINGENCY CONSIDERATIONS AND MONITORING INSTRUMENTATION

Two items were considered to be uncertain enough as to warrant a contingency plan. These were the ability of the jet grout to seal around the existing tunnel and potential for the cofferdam to deform the already distressed tunnel. For each of these, a contingency plan and instrumentation programs were to monitor the need for the contingencies were developed.

As discussed above, one of the benefits of the jet grout was that it is capable of sealing around the existing tunnel. However, the effectiveness of the seal is dependent on the as constructed shape of the tunnel, bedding material and backfill used around the tunnel and condition of the tunnel itself. All of these were unknown as well as, to some extent, the performance of the jet grout itself. A program of conventional pressure infiltration grouting was developed to seal leaks that were expected around the tunnel.

Grouting could be undertaken either from the ground surface using vertical and diagonal drill holes or from the subgrade level once excavation began. Grouting from the ground surface would allow the grouting to be conducted with no flow or head on the cofferdam meaning that more viscous grouts could be used that would penetrate farther. While grouting from the subgrade level would allow better identification of the areas of leakage, there could be some difficulty if the flow was too large.

Initially, the use of a pump test and observation wells was considered to assess the water tightness of the jet grouted cofferdam wall. If watertight, the cofferdam could be pumped without recharge. The rate of recharge would be an indication of the amount of leakage in the wall, around the tunnel locations or vertically upward from the bedrock. By using multiple observation well locations, it was hoped that the location of any leakage could be determined. However, given the short allowable construction time and difficulties in angling any grouting from the ground surface below the tunnel, the pump testing was not incorporated into the design. Further, with the widened areas adjacent to the tunnel to help distribute the load, some of the concerns for leakage were mitigated due to more contact surface available to seal the leaks.

After consideration, it was agreed that reasonable water tightness could be expected and that the pump testing would not be needed. Hayward Baker would return to the site during excavation to perform pressure grouting, if excessive leakage was found during excavation.

Concern about the ability of the tunnel to withstand the compressive forces of the cofferdam ring wall also required a contingency plan with monitoring by instrumentation. While the walls were flared where crossing the tunnel to distribute the load on the tunnel, the unknown condition of the tunnel and strength of the concrete forming the tunnel were somewhat unknown. Once again, a contingency plan was developed to relieve the stress on the tunnel should there be indications of unanticipated problems.

The contingency to relieve compressive stress would be to construct a ring beam inside the cofferdam. Approximately 18 inches of space was left around the perimeter of the cofferdam to facilitate construction of the valve chamber. However, if required, that space could also be used for a ring beam around the interior of the cofferdam. The ring beam would take the compressive load that normally would be transmitted thorough the wall and across the tunnel. If needed, one or more concrete ring beams could be readily cast in this area without interfering with the construction of the chamber.

To monitor the stability of the cofferdam wall as well as the tunnel itself, the diameter of each was monitored using an extensometer. Extensometer anchorage points were mounted within the tunnel prior to beginning excavation and on the jet grout wall shortly after the start of excavation and the convergence monitored throughout the construction.

9 COFFERDAM CONSTRUCTION AND PERFORMANCE

Shaft construction work was undertaken during the winter months of December and January. As such, field crews had to deal with freezing conditions that often slow grouting and jet grouting operations as a result of freezing of water within various components of the drilling and grouting equipment. In addition, the contractor's specialty pumping equipment required to complete the double fluid and SuperJet grouting was not available at the start of work. Therefore, a similar pump was rented for use on the project. Unfortunately, this rental equipment was not as reliable as the equipment normally used by the specialty contractor and as such, a full time mechanic was added to the field crew to correct and repair the frequent problems experienced with the jet grout pumping equipment.

Each jet grout hole (Figure 7) was predrilled several feet into the bedrock. In part, this was required since the lowest nozzle is not at the bottom of the stem and to achieve a tight seal at the bedrock, extending the stem into the bedrock so that the nozzle would be flush with the top of the bedrock was necessary. In addition, the predrilled holes were also extended below the sand seam that was observed at most locations around the site within a foot or so of the top of the bedrock to grout any potential seams.

Figure 7. Jet Grout Installation.

Another unanticipated equipment problem became apparent during the use of the SuperJet drill monitor in close proximity to the conduit wall and immediately beneath the conduit. Apparently, the high erosive energy of the jets and the sand/gravel suspended in the soilcrete mixture rebounded off the conduit wall and back into the drill monitor. As a result, this erosive mixture chipped and damaged the eroding jet nozzles that had to be removed and replaced during the jetting operation to preclude permanent damage to the monitor. Following grouting, excavation proceeded as shown on Figure 8.

Figure 8. Excavation within Jet Grouted Cofferdam.

The performance of the jet grouted cofferdam was better than expected. No recordable movement was measured on the extensometers both across the cofferdam and within the tunnel. The cofferdam formed an effective seal against groundwater both along the sides of the excavation and at the bedrock interface allowing excavation without interruption to dewater or install bracing.

The jet grout was very successful in mixing with the soil and bedding materials around and even under the existing odd shaped conduit (as shown in Figure 9) and less than 18 lpm (5 gpm) seeped from any portion of the excavation. The supplemental grouting around the conduit was never needed.

Low flow conditions were maintained throughout the construction and there was no need to use the shaft for water conveyance. However, the mechanisms were in place during all phases of construction to do so if required.

10 TUNNEL ACCESS AND REHABILITATION

The need to repair the conduit and the subsequent construction of the access shaft and valve vault was prompted by a contractor's report of concrete dete-

Figure 9. Grout Seal Around and Below Conduit with Bulkhead seen inside the Conduit.

rioration in the conduit. Divers conducted an inspection and videotaped the condition of the conduit in November 1999. The videotape showed leaking construction joints and disintegration of the invert of the conduit. The disintegration was reported to be as much as 4 inches deep at the invert, gradually decreasing in depth towards the springline and decreasing in severity away from the treatment plant. In December 1999 CDM E&C was contracted to repair the leaking joints and prevent contamination of the treated water by groundwater. While performing this work the disintegration reported by the divers was confirmed and visually assessed by CDM E&C engineers.

The invert of the conduit was deteriorating and only the coarse aggregate from the concrete remained, loosely, along the bottom. The concrete surface above the springline was intact. Observations indicated that the disintegration began just downstream from the point where sodium silicoflouride (flouride) was added to the treated water. There were several slabs, as large as 0.7 m (2 ft.) by 0.7 m (2 ft.) and several inches thick, on the invert and these were identified as flouride.

Flouride is added to the water to prevent tooth decay. When added, the solution breaks down into sodium and silicoflouride ions; the silicoflouride ions react further and produce acidic solutions having a pH of 3-4. Under normal conditions the amount of flouride added is very small and the concentration of the acidic solutions is low; however, it was suspected that there may have been an overdose, allowing the concentration to build up. Concrete in service normally has a pH of about 11, and is susceptible to attack by acids. The acids break down the calcium hydroxide in the hydrated cement paste that binds the aggregates together, but the aggregates are usually more resistant. Conditions in the conduit were consistent with acidic attack of the concrete. Flow conditions in the conduit also supported this hypothesis. The conduit usually flows full, with low water velocities. The flouride solution is more dense than water, which would explain why only the portion below the springlines was attacked, why the lowest portions were the most deteriorated, and why areas further away from the plant were in better condition.

Non-destructive testing (ultrasound and ground penetrating radar) was conducted in June 2000 by NDT Engineering, Inc. The ultrasonic tests showed that the concrete structure was basically intact, with an average compressive strength of 26.7 to 40 MPa (4,000 to 6,000 psi) but that relatively weaker concrete, less than 20 MPa (3,000 psi), existed at or near the surface in areas below the springline. However the results also indicated that the underlying concrete in these areas was stronger. Ground penetrating radar located areas of water infiltration or entrapment, and also indicated the presence of metal waterstops in the vertical construction joints and reinforcing bars.

In July 2000 concrete cores were obtained and a structural inspection was conducted on approximately 300 m (1,000 ft.) of the conduit. In areas where the disintegration of the invert was less extensive and above the springline the pattern of the boards used to form the concrete was evident. This inspection discovered bugholes (voids formed from air trapped against the concrete form) on the concrete surface just below the springline. Below and away from the springline these became larger and deeper and transitioned into the disintegrated area. Wood, ranging in size from splinters to boards, was observed on concrete surfaces and embedded in the concrete underlying the disintegration. The locations and extent of cracks, efflorescence, leakage, exposed reinforcing and exposed waterstops were noted. Four concrete cores, 10 cm (4 in) in diameter and about 20 cm (8 in) long, were taken for visual examination, strength testing, and petrographic examination.

Strength tests showed the compression strength of the concrete was about 36.7 MPa (5,500 psi), confirming the range suggested by the ultrasonic testing. Visual examination of the core surfaces forming the inner surface of the conduit showed exposed coarse aggregate, consistent with acid attack. Visual examination along the length of the cores showed some segregation of coarse aggregate, i.e., settling of the coarse aggregate down and away from the invert of the conduit. Petrographic examination revealed leaching of cement hydrates and alteration and destruction of the cement paste as deep as 2.5 cm (1 in.) from the inner surface of the conduit. The interpreted water to cement ratio ranged from moderately low to moderately high. However, X ray diffraction and chemical testing did not clearly identify the cause of the disintegration. Strength tests and petrography were conducted by Construction Technology Laboratories, Inc.

The conduit was built around 1924-1925. The record drawings show vertical construction joints but do not show any horizontal construction joints and none were observed during the inspections. Photographs of similar structures built then suggest that an inner cylindrical form may have been used, requiring that the concrete be placed from above and moved laterally under the form to create the invert. Given the technology available then this may have been accomplished by using a "wet" mix – a concrete mix with a high water to cement ratio. Observations of the form pattern, bugholes, segregation of aggregate, and weak concrete near the surface would be consistent. Concrete with a high water to cement ratio is more porous and more permeable to water and aggressive chemicals than concrete with a low water to cement ratio, and is also more susceptible to attack by aggressive chemicals. Embedded wood would provide additional paths for water and chemicals to enter the concrete. It was concluded that the most probable cause of the disintegration was poor quality concrete that was attacked by the acidic solution, and that only the concrete near the inner surface had been affected.

Approximately 300 m (1,000 ft.) of the invert was restored with a polymer modified cementitious fast setting structural repair mortar that was safe for use in potable water. Surface preparation consisted of removing the remains of the disintegrated concrete (aggregate), followed by high pressure water blast to remove weakened concrete and wood. After removal of debris and water the mortar was troweled onto the invert. In areas where the repair was greater than 2.5 cm (1 in) in thickness the specifications allowed the addition of 0.9 cm (3/8 in) stone to the mortar (Figure 10).

The drawings and specifications contained provisions to repair severely corroded reinforcing.

A portion of the conduit within the new valve chamber was removed by sawcutting to allow insertion of the valve in the conduit. The sawcut pieces were inspected to verify our observations and conclusions. The valve structure was used as an entry port for the repair of the conduit and also as an entry port to inspect approximately 3.5 miles of the concrete lined horseshoe shaped rock tunnel downstream of the repaired section. Current plans are to improve access to this portion of the tunnel and make repairs in 2002.

Figure 10. Conduit Repairs.

11 SUMMARY OF SCHEDULE AND COST

The site conditions that required the jet grout access shaft were exposed in November 2000, shortly after construction operations were mobilized on site and immediately after the reservoir was drained and exploratory excavations were made to expose the water supply conduit. It was at this time that the saturated flowing sands were found and an alternative approach to the excavation and access shaft was found to be necessary. Over the next 3 weeks, the jet grout construction plan for the access shaft was developed and the engagement of Hayward Baker by CDM was finalized. As the specialized equipment and workers were mobilized to the site, the details of the design were prepared. Construction of the jet grout access shaft got underway in December and was completed in 3 weeks, early January 2001. Had the project not been plagued by very cold winter conditions, it may have been done earlier.

Overall, the jet grout access shaft took approximately 2 months, from conception through testing of the jet grout columns and start of excavation. Following the excavation within the shaft to bedrock, including full exposure of the exterior of the water supply conduit, which took approximately 2 weeks, the interior of the jet grout access shaft was coated with shotcrete to seal minor leaks and provide a clean containment vessel in the event of an emergency in which the shaft would need to be flooded and made part of the water supply conduit during the project. In any case, the jet grout access shaft was completed in sufficient time for the interior of the water supply conduit to be accessed and the repairs to be completed prior to the end of April 2001, when the water supply conduit had to be put back in full-time operation (See Figure 11).

The cost of the jet grout access shaft was $260,000. The cost of the excavation within the shaft and the clean-up and disposal of the grout waste was in addition to this cost.

12 CONCLUSIONS

Design-build should be viewed as a process of optimization the combined cost of design and construction, not necessarily just lowering the designers cost. The design-build approach is popularly misconceived that because fewer drawings are required or

Figure 11. Completed Conduit Repairs.

that the drawing are less complete, the costs of the design will be less and that that will be the primary savings. Conversely, in this particular case, the design costs were probably higher then would have been otherwise needed to bid a conventional design that put the responsibility for cofferdam design on the contractor. However, in being able to tailor the design to the specific methods available to a particular contractor, the combined costs could be optimized while still meeting the aggressive schedule to meet the overall goals of the project.

Traditional project delivery systems often cannot reflect the most efficient means and methods of construction. Particularly in the public sector, designs put out to bid must be able to be constructed by a wide range of contractors, each having their own individual approaches to a particular problem, their own experience, equipment, processes, and staff. Designs are prepared in advance of contractor selection and inherently cannot reflect the selected contractor's capabilities, thoughts and preferences. The design-build system allows interaction between the designer and builder to produce a design that best meets the owner's needs – one that meets design requirements for function, strength, and durability and that also meets cost and schedule limitations.

The creative design experience and the flexibility offered by the jet grouting system provided the design team with the ability to create a watertight shaft in difficult subsurface conditions. Initially, the specialty contractor had to demonstrate his applicable technical experience to the design team and the variety of construction alternatives that he could provide to solve the numerous specific subsurface problems anticipated on the project. Once the design team was comfortable in working with the specialty contractor, it was agreed that any specific design alternatives developed with the specialty contractor would not be shared with others. With the assurance that the specialty contractor's creative and innovative ideas would not be shared with other contractor's and subsequently competitively bid, the contractor was free to work closely and with the design team to explore all possible concerns and to work interactively with the team to develop a responsive and cost effective solution.

From the start, this was clearly a project that would have been difficult to execute in any manner other than design-build. The unique constraints and difficult soil conditions required close coordination between the designers and contractor, especially the specialty jet grout subcontractor. While it is likely that the circular jet grouted cofferdam concept could have been developed as part of the conventional design, bid, build approach, other aspects such as the use of semi-circular seepage cutoff columns, multiple sized columns and penetrations into the bedrock would have been difficult to envision by the designers who are not extensively aware of the unique capabilities of the contractors specialized equipment.

In addition, to have more open access to the contractor's anticipated soilcrete strength estimations, production rates and other factors critical to the design, the design could be tailored to meet the condition without the usual time consuming back and forth of shop drawings that takes place in the conventional construction sequence. While it is likely that other specialty contractors could have approached this project in a similar or possibly alternate manner with equal success, the key was to work out the details particular to that contractor's equipment, techniques and experience which may be unique to his means and methods so that they can be incorporated into the design.

This project was clearly successful, largely due to the cooperation and interaction of the designers, contractor and owner.

13 ACKNOWLEDGEMENTS

The authors want to recognize and express their appreciation to Jason Herrick, PWSB's Project Manager, and Paul Gadoury, PWSB's Director of Engineering, for the opportunity afforded CDM and Hayward Baker to undertake this challenging project. Their foresight in selecting the design-build method of project delivery is also to be recognized as it proved to be successful and the most cost effective and expedient means by which to execute the project. They were key players in the decision to use a jet grout access shaft for the project and are credited with the primary leadership and overall success of the project.

The authors would also like to acknowledge the assistance and work of Mr. Joseph K. Cavey, Hayward Baker Inc. Project Manager, for his interactive work in developing the final jet grout column design and layout in addition to his consistent attention to details and trouble shooting efforts throughout construction.

REFERENCES

Anderson, A., Chadwick, K.R., and Wheeler, J.R., "Reconstruction of the Whitehall Street Subway Station", Proceedings of the 8th Annual Great Lakes Geotechnical and Geoenvironmental Conference, Detroit, MI, May 2000.

Billups T., Cavey, J.K. and Wheeler, J.R., "Grouting Program Bridges New Building Loads Over Unused Railroad Tunnel", Proceedings of 2001: A Geo-Odyssey, sponsored by the Geo-Institute of the American Society of Civil Engineers, Blacksburg, VA, May 2001,

Burke, G.K. and Wheeler, J.R., "SuperJet Grouting for Tunneling Applications", Proceedings of the North American Tunneling (NAT 2000) Conference, Boston, MA, June 2000.

North American Tunneling 2002, Ozdemir (ed.)
© 2002 Swets & Zeitlinger, Lisse, ISBN 90 5809 376 X

Application of design-build contracts to tunnel construction

Robert A. Robinson
Vice President, Shannon & Wilson, Inc.

Michael S. Kucker
Senior Associate, Shannon & Wilson, Inc.

Joseph P. Gildner
Construction Manager, Sound Transit Light Rail

ABSTRACT: Several owners/agencies, unhappy with cost overruns and schedule shifts associated with standard design-bid-build contracts, have proceeded with design-build contracting in the hopes that there will be less risk of claims and legal entanglements with the single source responsibility that is part of the design-build process. Other owners are assessing design-build as a means for accelerating the tunnel project delivery process while potentially reducing owner risks and ultimate project costs. Recently, the Federal Transportation Administration (Tunnels & Tunneling 1999) has elected to implement design-build contracting approaches on five major demonstration projects. Several other European owners have experimented with design-build formats on major tunnel projects. This paper will present case history data for several design-build tunnel projects, noting changes in cost and schedule, during the course of the project. This paper will also discuss likely pros and cons for the owner, contractor and designer in a design-build contracting approach for the construction of tunnels.

1 WHAT IS DESIGN-BUILD

The Engineer's Joint Contract Documents Committee (EJCDC 1995) provides these definitions of the design-build method of contracting:

- A single entity capable of providing both the design professionals and construction services with its own forces.
- A joint venture of an engineering firm(s) and a contractor(s).
- An engineer or a contractor, providing either design professional or construction services respectively, itself, and providing the other through an appropriate subagreement.

The civil industry has utilized design-build contracting to successfully construct a wide variety of buildings, plants, and other facilities. Design-build contracting accounted for about 15% or an estimated $70 billion of the U.S. construction industry in 1999 and 2000, this is an increase of $16 billion from the $54 billion reported in 1990 in Engineering News Record. Design-build contracting procedures are best suited to design and construction projects that are repetitive in nature and where the opportunity for unforeseen conditions (geotechnical or subsurface utilities) or impacts on adjacent property owners (third party issues) are minimal or nonexistent. At the other end of the construction spectrum, design-build has been used successfully for a variety of complex or high-tech projects requiring highly specialized construction techniques and specialty design and construction firms. These include "clean rooms," petrochemical facilities, hydroelectric plants, and steel mills (Loulakis 1992), all with well-defined final performance requirements and conditions.

Typically, the design-build process is thought to provide these benefits:

- *Owner Developed Preliminary Design* - Ideally, the owner's engineer will develop a prelimiary design that incorporates the essential project requirements, owner's preferences for design details and configurations, and provides sufficient assessment of ground conditions and third-party impactsA single entity capable of providing both the design professionals and construction services with its own forces.
- *Unified Design and Construction Team* - The teaming of the final designer and contractor provides a single point of contact and responsibility to the owner, and permits the short-cutting of the procurement phase for the final design team, as would be required for traditional design-bid-build, thus shortening the overall schedule by several months. It may also reduce the time and effort required to administer the work by reducing the number of contracts.
- *Accelerated Schedule* - Reduces the length of the overall design and construction process by combining the competition and selection phases for

the final designer and the contractor into one unified selection process for the design-build team. Overlapping design and construction efforts allow simultaneous design and construction, thus reducing overall schedule. Procurement of long-lead items such as tunnel boring machines can proceed in parallel with the final design.

- *Award Using Best Value Procurement Process* - The owner may chose to award the contract to the most qualified team based on objective selection criteria focusing on the talent and experience of key team members, work programs the team will develop and implement, and the cost of their work.

Design-build gives the contractor more control over the work and provides increased opportunities for the contractor to be innovative with respect to strategies for both reducing cost and accelerating schedule (Brierly et al., 1998). It also allows the designer and contractor to interact closely during construction and react quickly to the excavated and exposed ground conditions or perceived cost and schedule savings with alternate construction means and methods.

However, the design-build submittal and selection process is considerably more costly for the proposing teams than traditional design-bid-build, due in part to the large amount of information and the level of design detail generally required from the project owners to assess the various design-build teams. To encourage a larger number of proposing teams, some owners have elected to include a stipend or honorarium payment for the losing bidders in a design-build contest. This stipend is intended to partially reimburse the losing bidders for the large amount of effort that is necessary for the bidding teams to develop the extensive submittal documents that are required for most design-build procurements. Stipends have generally ranged from 0.05 to 0.3% of the estimated total design and construction cost of the project. Most design-build teams would argue that these stipends, although helpful, do not come close to covering all of their costs (let alone providing a profit) for developing their proposed design and construction approaches, as required for most design-build proposals.

2 TRADITIONAL DESIGN-BID-BUILD CONTRACTING

Most major public sector tunnel projects have been contracted using traditional design-bid-build contracting. In this approach, the owner procures the designers to develop feasibility studies, preliminary design, and final design. A comprehensive site investigation is developed to assess ground conditions, followed by the development of a design and set of plans and specifications. In the process of developing the design drawings, most of the third-party issues, such as permit requirements, construction impact agreements, and public meetings and interaction, are revealed, developed and refined.

Following this iterative design process, the owner then competitively bids the fully explored and designed project. The contractor generally sublets portions of the designed project to a variety of specialized subcontractors. The subcontractors can review the bid documents to fully ascertain the nature and extent of their involvement, as well as conflicts, schedule impediments, and other issues that might affect their costs.

Under the traditional design-bid-build approach, the contractor's primary goal is to economically implement the plans and specifications and complete the project on schedule and under budget in order to receive a reasonable profit for his team's efforts. The contractor is not obligated to interpret or second-guess the intent of the designer, nor the condition of the ground. Ambiguities or errors in design or interpretation of the explorations will likely result in additional compensation for additional work and "changed conditions".

The design engineer's goal is to produce a functional and satisfactorily completed project within the owner's budget, with minimal changes or claims, while meeting the intent and goals of the design. This may at times be at apparent odds with the contractor's goal of successfully completing a project while earning a reasonable profit. The contractor's meeting his goals should in fact be a significant concern of the designer and owner, since only contractors who can successfully and satisfactorily complete a project while earning a reasonable profit can afford to go after further tunnel work.

There are numerous advantages to the traditional design-bid-build contracting approach, which include: 1) strong owner control of all aspects of the project though construction; 2) clearly divides and defines the roles of the designer versus the contractor; 3) resolves most of the third-party (utilities, buildings, highways) issues prior to construction; 4) attempts to define the geology and environmental issues and impacts prior to final design; 5) provides abundant opportunities for public involvement; 6) promotes a higher quality of design and construction for better long-term performance of a completed project; 7) resolves many of the possible design questions and owner design concerns; and 8) is familiar territory for the owner, designer, and contractor. Therefore, an owner must weigh and compare the risks of working with a relatively new and untried contractual approach versus refining his use of the traditional design-bid-build approach. The owner's level of risk in design-bid-build contracts can be managed and reduced by employing risk sharing strategies developed over the last 30 years to suit the tunneling industry (USNCTT 1974) and that include dispute review boards, escrowed bid documents, geotechnical baseline reports, as well as thorough alignment investigations (USNCTT 1984).

The traditional design-bid-build contracting method has been repeatedly promoted by the American Consulting Engineers Council as a preferable contracting approach to design-build, as follows: "The American Consulting Engineers Council strongly believes that the use of the traditional design and separate bid/built project delivery system is in the best interest of the owner as well as protecting the health, safety and welfare of the general public. This traditional system provides the checks and balances necessary to give the owner and the public the greatest degree of assurance that the project is the most appropriate solution for the circumstances through the use of a qualifications-based selection process and direct owner representation by the design professional" (ACEC 1992).

3 DESIGN-BUILD – PROS & CONS

Due to a general lack of industry-wide experience with design-build contracts, there are few owners or contractors familiar with the pros and cons of this contracting method. A few of the more salient issues, concerns and perceptions are discussed in the following paragraphs.

Owners often resist the loss of control inherent in the concept of design-build and react throughout the project as if it is standard design-bid-build. However, the owner's ability for continuous review and full control over all of the nuances of the final design in the design-build process is greatly reduced. One of the most significant drawbacks to the owner is that in joining the designer and contractor together, many of the checks and balances present in the traditional contracting method are eliminated. These checks and balances help to protect the owner from an inferior project.

The level of design appropriate to design-build is uncertain. Some design-build RFPs are based on a conceptual design level of maybe 15%, whereas other RFPs have been carried to a 60% design level, with accompanying plans and specifications. Likewise, the level of geotechnical investigations and third-party property and utility assessments varies considerably from next to nothing, to an 80 to 90% evaluation. There are pros and cons to both of these approaches. Certainly, the minimalist approach provides the greatest potential for innovation by the contractor, although it also provides the greatest risk for changed conditions, and the attendant cost and schedule impacts. Whereas, the maximalist design approach significantly reduces the potential for changed conditions, but reduces the opportunity for innovation on the part of the contractor, while maximizing the owner's control over the nature of the end product.

Many of the third-party and geotechnical issues that impact the outcome of an urban tunnel project are resolved during the iterative phases of intermediate and final design. However, under the design-build format, these issues may not even be recognized, let alone resolved during the preliminary design phase. If they are discovered and resolved during the design phase of a design-build contract, then they may involve some additional costs for changed conditions. However, if significant differences in geotechnical or third-party conditions are not revealed until the construction phase, then major project cost and construction schedule issues may occur.

3.1 *Owner's Perspective*

A perception common to many owners seems to be that the design-build process will shift much of the cost and schedule risk to the contractor, and significantly reduce project schedule and possibly the overall project cost. Some owners have come to associate traditional design-bid-build with extended project design and construction times, disputes over changed conditions, and prolonged and costly litigation. In traditional design-bid-build, the owner has been held responsible for unanticipated changes in ground conditions, unknown third-party utilities, and often for construction delays due to third-party permitting requirements. For design-build, the owners appear to retain much of this same risk for third-party issues and changes in geotechnical conditions.

Abundant case histories from design-build applications to above ground facilities indicate that the owner also does not realize any significant cost savings, but some possible savings in schedule, provided that the owner can restrain their desire to actively control the project. Design-build was never intended for nor has it proven capable of routinely reducing construction costs. In fact, as shown in later sections, the preponderance of the evidence from tunnel projects to date is that design-build generally results in increased costs and schedule time beyond that contracted by the owner.

Most design-build, at least as it applies to tunnel work, appears to involve a negotiation of the final cost of the project after the final design-build team has been selected. At this point, there is very little incentive for the design-build team to minimize their prices, since they are the selected team. However, there may be a tendency to increase prices as the owner adjusts the design during the negotiation period, based on ideas provided by the various competing teams.

Owners and their design engineers are used to frequent and detailed reviews of several iterations of the design and construction documents. It is through this process that many of the problems and issues of a project are revealed and resolved during the various stages of design. However, unless the design-build documents detail out the levels, frequency, and general scope of submittals and reviews, it will be difficult to enforce such frequent submittals and reviews. Furthermore, if the reviews result in substantial changes to the design-build team's approach, then the owner will likely be expected to pay additional costs. This could easily lead to significant disagreements and conflicts between the owner and contractor team.

In design-build, the owner now finds himself on the opposite side of the table from the final designer, who now works for the contractor. This often makes it extremely difficult for the owner to find out exactly what the final design will look like, except in a piecemeal fashion as various portions of the design are being developed to support the construction schedule of the contractor. The owner may have very little opportunity to directly interact with the designer without the intervention, and control of the contractor. Naturally, the contractor is interested in reducing, as much as possible, the complexity and cost of a project, whereas the owner may have design goals, possibly not adequately detailed and explained in the contract documents, that he wishes to impart to the designer as the project progresses (as would be the case in normal design-bid-build).

The Owner may believe his site inspection staffing needs can be significantly reduced, since the design-build team now has a greater responsibility for quality assurance (QA) and quality control (QC). However, yielding this activity to the design-build team further removes the owner from the project. In tunneling, the best opportunities for observation and recording of ground conditions and quality of the work are afforded during the various phases of construction, rather than any observations of the completed tunnel.

3.2 *Contractor's Perspective*

Many contractors perceive that design-build offers an opportunity to maximize profits while retaining the current balance of risk with the owner. Many contractors also appear to relish the idea of having significantly more control over the progress, details, construction means and methods, and character of the final product. There may also be a greater tendency for contractors to try new and innovative construction approaches that could save money and time, and for a more productive partnership between the contractor and designer to develop these innovative construction approaches.

Design-build teams may attempt to shed much of the risk for unanticipated or changed ground and third-party conditions back to the owner by including, as part of their proposal, a long list of assumptions, guidelines, issues, concerns, exclusions and understandings. This list may serve as "baselines" for establishing changed condition claims or for limiting the contractor's liability exposure. Several pages of baseline assumptions from each competing team make comparison of the various proposals and costs extremely difficult, and have resulted in several design-build projects never going to contract.

Contractors that are setting up and leading a design-build team must interact directly with the designer. Due to the presence of numerous uncertainties, and the wide range of possible design and construction solutions, the contractor must often learn to cope with frequent design uncertainty and revisions. This is particularly true if there is considerable latitude in the design requirements and/or a scarcity of information on geotechnical conditions and adjacent third-party utilities and structures. Contractors are often not aware of the amount of lead time, labor hours, and labor costs required to complete a final design and prepare the specifications and plans.

The contractor must also provide a more extensive proposal submittal than is normally required for traditional contracting. A good deal of this proposal information must be provided by the designer, who may have been provided with very limited geotechnical and third-party information. If a stipend is provided, then the contractor may elect to pay the designer's base salaries with a partial multiplier.

The proposal submittals often contain unique construction means, methods, and materials. Consequently, the design-build bidders have often expressed concern over the possible release of their proprietary approaches to possible competitors. Public owners and design-build proposers may find it difficult to prevent release of proprietary proposal information through various forms of the Freedom of Information Act.

For some design-build proposals, utilities and overlying or adjacent property owners have had little or no contact with the project owner, so initial contacts with third-parties made by the design-build team to test out alternative construction options has resulted in considerable consternation for all parties. Unfortunately, where a design is only carried to a 15 to 30% level by the owner, third-party issues are usually not fully addressed by the owner at bid time.

Design-build often shifts a greater proportion of the quality control and quality assurance issues to the design-build team. However, the contractors are often ill-prepared and ill-disposed toward the extensive inspection, testing and reporting necessary for a thorough QA/QC program. A design-build team's goals of constructing a project quickly, efficiently, and cheaply do not generally mesh well with the normal goals of a comprehensive QA/QC program, or the normal agency requirements for thorough QA/QC documentation.

3.3 *Designer's Perspective*

The designer has traditionally worked for and developed the design through its various phases to support the requirements of the owner. This has traditionally been a collaborative and iterative approach, with the design developing and maturing as additional data is obtained and the goals and needs of the project are refined, often under public scrutiny.

Under the design-build approach, the designer, geotechnical engineer, and other design related subcontractors now work for the contractor. This change in allegiance requires that the designer and his subcontractors now promote and protect the interests of the design-build team. These interests include: 1) the development of the cheapest construction approach (not necessarily the most cost-effective from a long-term operation and maintenance approach); 2) minimization of the exploration and design phases of the work to accommodate rapid turn around of construction documents; 3) reduction of the amount of construction monitoring and review; and 4) minimization of the amount of design and construction documentation. Furthermore, the owner has now lost direct access to the final designer, and may have very little control, except through the preliminary specifications, on the final product.

Design engineers and contractors need to develop formulas for equitable sharing of the risk from a problem project and profits from a successful project. Often there is little or no owner payment for design work and, consequently, the team has little cash to reimburse the engineers for their design efforts. There is also a strong push to segment the design so that portions of the project can get under construction to enhance the teams cash flow. Contractors may require that the designer reduce their billing rates in the early phases of the project until cash flow improves by meeting construction milestones.

The designers and contractors in the design-build team recognize that they may face significantly increased exposure to liabilities that were not present in more traditional design and construction methods. Designers may be required by the owner or the contractor to guarantee that their design will meet defined performance criteria, resulting in increased liability potential over and above normal negligence definitions (Loulakis 1992). Some design-build projects have even been expanded to include a period of maintenance and operation. The I-15 project in Salt Lake City, Utah, included an extended maintenance requirement. The Tolt River water treatment project near Seattle, Washington, requires the team to maintain and operate the plant for several years before turning it over to Seattle Public Utilities.

4 DESIGN-BUILD CASE HISTORIES

Very few design-build tunnel project case histories have been published in the literature. However, the few relevant published histories of tunnel projects do shed light on the successes and failures, pros and cons of the design-build contracting approach. From these projects we can glean several lessons to incorporate into future design-build contracts. Recent tunnel projects that have been contracted utilizing design-build methods include:

4.1 *Channel Tunnel, England to France (Lemley 1991)*

The Channel Tunnel project was conceived over a century ago, but it took nearly a full century to realize that dream as tunneling techniques caught up with the needs and requirements of the project. The project was implemented with the award of a concession to design, build, equip, and commission the cross Channel project in 1986. The tunnel was completed in 1993, with full operation commencing in the summer of 1995. The design-build team, Transmanche Link JV, consisted of a consortium of five British and five French firms to construct a total of 57 km, portal to portal, of 4.8 m diameter service tunnel and 7.7 m diameter twin railway tunnels in a scheduled 33 months. Over the course of the design-build construction project, the estimated price of tunnel construction climbed from $3.8 billion in 1981, to $8 billion in 1988. This upward creep was related to inflation, improved geotechnical data, and changes in owner requirements. All parties recognized that there were major risks, not the least of which were geotechnical. Consequently, the risk was shared between the owner and contractor using a target cost reimbursable contract for the tunnel portions of the project.

An extensive geotechnical program (Varley et al., 1996) was undertaken in three phases prior to letting of the contract and in two phases during construction. Prior to the contract, there were 97 over-water and 34 land borings. During construction, an additional 19 over-water borings and 34 land borings were drilled. Over 4,000 km of geophysical surveys were performed on land and sea.

Despite the large number of very costly over-water borings and extensive geophysical surveys, there were geologic surprises that resulted in delays, schedule adjustments, claims, and additional compensation. In the design-build contractor's proposal, a comprehensive discussion of anticipated ground conditions and behavior was presented. This listing basically served as a baseline for comparison with actual ground conditions and ground behavior for the resolution of claims for differing ground conditions as allowed under the contract. Claims revolved around: significantly larger amounts of rock overbreak which contributed to major difficulties in erecting the segmental liner system, local areas of high groundwater inflow, and unidentified zones of folding and fracturing that adversely impacted tunneling (Mansfield 1996).

4.2 *Whittier Tunnel Rehabilitation, Alaska (Moses et al., 2000)*

The 4.2-km-long Anton Anderson Tunnel was originally constructed during World War II to transport goods and people to a military base established at Whittier, Alaska. However, it has been a bottleneck for car traffic in and out of Whittier, since cars could only enter or leave the town via flatbed railcars through the tunnel. Consequently, HDR Inc. developed a preliminary design for Alaska Department of Transportation for conversion to the first U.S. tunnel to provide dual highway and railroad use, making it the longest rail and highway tunnel in North America, and the longest highway tunnel in the U.S. Four design-build teams proposed, but two dropped out prior to the final submittal and selection phase. Although the owner's estimate was $35 million, after negotiations the project was awarded on the basis of "best value" for $57.3 million to a design-build-operate contract team consisting of Kiewit Construction Co., with design provided by Hatch Mott McDonald and five local Anchorage firms. Final cost of the project is estimated at about $61 million, a little more than a 5% increase over the negotiated price, and the contract was completed on schedule despite weather delays and tunnel access scheduling complications with the operating Alaska Railroad.

No geotechnical explorations were necessary since the existing tunnel was more than 99% unlined, leaving little opportunity for a claim for changed geologic conditions.

Discussions with various team members suggest that although this was a successful project overall, there were some problems with implementation of the design-build contract approach. The very linear nature of the project made staging and organization a critical issue for the success of the job. Difficulties in coordinating with the Alaska Railroad's freight and passenger service schedule resulted in irregular short and long work windows. Quality control was often a point of contention between owner and contractor. Several iterations of various design components were required and the owner, by contract, had a very long review period (up to 3 weeks) which fortunately was rarely used completely. Environmental permitting with the U.S. Forest Service caused some delays.

4.3 *Copenhagen Subway System, Denmark (Reina 2000)*

The current project, begun in 1996 and scheduled for completion in 2000, consists of 14 km of rail line bid at $350 million. However, the project has $175 million in claims and will be two years over schedule. The project includes 8.3 km of twin 4.9 m tunnels, 4.2 km of track elevated on embankments and structures, 0.9 km of at-grade track, and numerous stations including six stations reaching nearly 20 m deep. The winning team, CONMET is an international consortium including: Carilion plc (U.K.), SAE International (France), Astaldi SpA (Italy), Bachy Soletanch Ltd. (England), Ilbau GmbH (Austria), NCC Denmark A/S, and ILF Consulting Engineers (Austria).

Reportedly, the owner was reluctant to relinquish control, adding to construction delays and costs, lengthy submittal reviews, and an estimated 300% increase in drawings over the contractor's original estimate. Stringent zero settlement criteria; excessively abrasive silts, sands and gravels; and stringent controls on locomotive exhaust have contributed to increased costs and schedule delays. Also, the international flavor of the team created difficulties due to cultural differences in their approach to the work and the distance between the various team members.

4.4 *Tren Urbano Transit System, Puerto Rico (Gay et al., 1999)*

The new system consists of 17.2 km of track alignment, 16 stations, and 1.5 km of underground alignment. The Rio Piedras Section 7 includes open cut, cut-and-cover, earth pressure balance machine (EPBM), NATM, and stacked drift excavation methods. The design-build team selected for the underground section consists of Kiewit Construction Co., Kenny Construction Co., H.B. Zachary Co., CMA Architects and Engineers, Sverdrup, Jacobs Associates, and Woodward Clyde. The KKZ/CMA Team won the project for $225,600,000, which, although not the lowest price, was determined to be the best value of the three bidders.

A geotechnical program consisting of 22 borings and two pumping tests were undertaken by the owner prior to award of the contract. The selected design-build team accomplished 15 more borings to better define the nature of the alluvial clays, sands, and silts, and the groundwater conditions on the project.

Many lessons were learned on the Rio Piedras design-build project by the various participants. These included: 1) improved awareness and documentation is needed of each member's role and responsibilities including quality control and team management, 2) a shift is needed in the designer's approach to coordinate and segment design with construction for earlier construction progress payments, 3) improved awareness of the owner and designer that design is on the critical path and that quick efficient document control is essential to maintain schedule, 4) a recognition by the owner and design-build team that design costs for design-build are very comparable to traditional design-bid-build, 5) the designer must adjust his specification writing and quality control (QC) requirements to expedite construction, and 6) the normal payment on the basis of constructed elements of the project can lead to serious cash flow problems for the team during the early phases of the project.

Since this project is not yet complete, the actual cost of the project and its success in meeting the schedule and in providing the owner with the quality of work that was desired won't be known for a year or more. However, discussions with various project personnel have indicated that there have been problems with localized excessive and damaging settlements, old undocumented utilities, greater levels of design effort to meet the owner's needs, and a much greater level of QC labor than was anticipated. All of these issues will likely lead to increased cost. As of September 2001, contract costs for the entire 17.2 km alignment had increased by nearly 70% to $1.9 billion and the schedule has slipped by 2 years, due to increases in scope, higher than estimated construction costs, and differing site condition claims. A representative of the U.S. Department of Transportation has suggested that "A traditional design-build contract would not have had so many claims" (ENR 2001).

4.5 *High Speed Rail Link Germany*

A 177 km long high speed rail link is now under construction from Nurnberg to Munich in Germany. Cost for the entire line is estimated at $2.04 billion (DM 4,320 million) and completion is scheduled for late 2000 or early 2001. This alignment includes nine tunnels, totaling about 25.6 km and ranging from 650 m to 7,700 m long. NATM construction methods will be used to excavate the 12.9 m I.D. double-track tunnels. The owner, Deutsche Bahn AG, is experimenting with the use of design-build contracts for this project and divided the alignment into several design-build sections. Negotiations with the winner of each section took up to 6 months, due to price differences, design approach differences, and baseline issues presented in the proposals.

In Germany, the owner is generally acknowledged to retain ownership of the ground. All geotechnical explorations were undertaken by the owner and the data was collected along the alignment and presented to the various bidders in data reports. Borings were typically spaced about 300 m apart. In describing ground conditions, care was taken to avoid discussing construction means and methods in relation to the ground, since the construction approach was to be determined by the design-build team.

The German owners are experimenting with and experiencing many of the same difficulties with design-build that have been experienced in the U.S. Difficult ground conditions have been claimed to be partially unanticipated and have contributed to claims; however, delays in gaining and granting permits have caused the greatest delays to the project. The owner's engineers voiced concern over the quality of the design submittals, quality of the construction, inadequate time for thorough submittal reviews, lack of owner access to and interaction with the designer, and higher numbers of claims. The contractors indicated concern over excessive time for review by the owners, excessive involvement in the project details by the owner, and insufficient preliminary design details. For the traditional design-bid-build, claims were generally within 3% of the bid price, but for design-build the claims appear to be about 25% of the negotiated bid price. Some of the owner's representatives suggested that this should be the last time that design-build be used for tunnel projects in Germany.

4.6 *"Link" Light Rail, Seattle, Washington (Gildner, 1999)*

The entire $3.6 billion project consists of 40 km of light rail alignment including 10 km of twin-bore tunnel. The north 8.4 km of the alignment will consist of up to 6 m diameter tunnels, up to 70 m deep along with three deep (up to 65 m) mined stations, one 40 m deep cut and cover station, a shallow crossing beneath Interstate 5, a deep crossing beneath Portage Bay, and passage beneath numerous streets, utilities, and buildings up to 30 stories high. The north corridor tunnel portion of the project was designated to be design-build. Three teams qualified, but one withdrew during the formal proposal and cost estimating phase. The Modern Transit Constructors (MTC) team was selected as providing the "best value" (a combination of technical qualifications and cost). The MTC team consisted of Modern Continental Construction, S.A. Healy Construction/Impregilo of Italy, Dumez of France, Parsons Transportation, D2 Consult, and Goldberg Zoino Associates.

When Sound Transit selected the design-build contract approach they also established a philosophy for risk allocation based on the fundamentals dealing with "ownership of the ground" as distinguished from "ownership of the means and methods". This approach is reflected in a substantial geotechnical exploration program. A phased exploration program, consisting of 137 borings, spaced an average of 100 m apart along the tunnel and with 6 to 9 borings at each station location was undertaken during the feasibility and preliminary design phase of the project. An additional four borings were drilled at the suggestion of the two design-build competitors. After selection of MTC, an additional 15 borings were initiated based on discussions with the design-build team and minor revisions to the alignment. All exploration data was presented in a Geotechnical Data

Report (GDR) and interpreted in a Geotechnical Characterization Report (GCR). A Tender Geotechnical Baseline Report (TGBR) was prepared, which included definitions of baseline conditions such as boulder quantities, nature of soil unit conditions, but not ground behavior, since this will largely be determined by the contractor's selection of means and methods for tunnel and station excavation and support.

Unfortunately, the selected design-build team could not develop a cost estimate and construction approach that met the goals of Sound Transit. The tentative agreement between the selected team and Sound Transit was terminated.

4.7 *Case History Conclusions*

Of these six projects, the Whittier Tunnel Rehabilitation project appears to have been the most successful from a cost and schedule perspective. This success likely relates to the roughly 99% exposed geology in the mostly unlined tunnel, and relatively few third-party impacts. The most critical third-party issue was working around the scheduled freight/auto carrier train service during the enlargement of the tunnel, and installation of a roadway surface and safety and traffic control facilities to permit both rail and automobile traffic in the tunnel.

5 WHERE'S THE RISK?

Over the last 20 years the tunneling industry (owners, contractors and designers) have evolved a unique risk sharing program to help to reduce the number and severity of claims and disputes (USNCTT 1974). This risk sharing approach included the development of: 1) a geotechnical baseline report (GBR) to establish a level "playing field" for bidders and for reference during construction; 2) escrowed bid documents to provide information on the contractor's assumptions for unit prices, quantities, and means and methods of construction; and 3) a 3-person Disputes Review Board (DRB) consisting of reputable peers from the construction industry. This risk sharing approach has proven to be successful, when correctly applied by experienced and knowledgeable owners, designers, geotechnical engineers, and contractors for traditional design-bid-build tunnel construction. However, no testing and calibration of this system of risk sharing has been accomplished for design-build contracting. Consequently, it is questionable as to whether and how best to use these established risk sharing principles.

A critical and central issue for any tunnel contract is the question of who "owns" the site conditions. It is of critical importance that ownership of the risks and benefits of a particular site be clearly distributed between the owner and design-builder in the early stages of the project (Jaffe and Goode 1992). Some owners have attempted to shift all responsibility and ownership of the project environment to the contractor. This harkens back to the pre-risk sharing days of the heavy construction industry when owner's attempted to abrogate any responsibility for ground conditions, or changes therefrom, that might occur in the course of a project. This was often accomplished by including disclaimers on all exploration boring logs. For these projects, minimal interpretation was provided and consequently no geologic profiles, sections or other interpretations were commonly provided.

Loulakis (1992) notes that a design-build contract should include a differing site condition (DSC) clause that provides for the owner retaining the financial risk for unanticipated site conditions. The incorporation of a DSC clause allows the bidders to submit a minimal bid that does not incorporate contingencies for unexpected conditions. In implementing a DSC clause, it behooves the owner to assess the site conditions to a level necessary to reduce the risk for encountering unanticipated conditions during the design or construction phases of the work. One of the highest areas of potential risk is associated with variations in the ground conditions. It has become a basic tenant of tunnel construction contracting that the project owner retains overall ownership of the site conditions, including geologic conditions and variability, soil and groundwater contamination, and imprecisely located or describe third-party utilities.

A modified GBR was used on the Sound Transit North Corridor project, that established baseline geotechnical conditions utilizing data from borings spaced about 100 m apart. Baselines were established for the geologic and groundwater conditions, soil properties, and boulder and fracture frequency. The selected contractor was required to amend this Tender GBR with the agreement of the owner.

A Disputes Review Board (DRB) is most effective when a well defined set of contract documents, and GBR is in place. Without these two prerequisites, it will be difficult for the DRB to establish the bidding parameters that provide a baseline from which to determine if a changed condition has occurred.

The bid documents of the selected design-build team may mimic the role of escrow bid documents, provided that enough detail is presented in the documents. The owner and contractor will need to thoroughly assess and review these documents, keeping in mind that they may be used by a DRB to assess the validity of a claim. Contractors have expressed a concern over the possible accessibility of escrow documents to competing teams. After review by the owner, these documents should be left in the possession of a third-party, such as a bank, and the contractor should retain ownership of the documents. The documents of losing bidders should remain their property and be returned to them after the selection and negotiation with the winning design-build team. However, when a stipend is paid to the loosing bidders, then some owners may maintain that they now "own" the losing bidders documents, even though the owner has not paid full labor rates for their preparation.

6 THIRD-PARTY AND GEOTECHNICAL INFORMATION FOR DESIGN-BUILD

Very little information has been published on the expected or anticipated levels of site evaluation and exploration that an owner should provide to prospective bidders for a design-build project. For traditional design-bid-build contracts, experience has shown that for both the owner and the contractor, the risk is significantly reduced by presenting the bidders with all of the geotechnical information collected (USNCTT 1984). However, very little has been published regarding the amount of geotechnical, utility, foundation, and historical construction information that should be provided to design-build bidders. A wide range of levels of geotechnical and site utility and foundation investigations have been utilized for design-build projects, ranging from little or no investigation for projects such as the Whittier Tunnel Modification, to a thorough, nearly 100% complete design exploration program for the Channel Tunnel, Tren Urbano, and Seattle Light Rail projects.

6.1 *Failure Without Exploration Data*

Some owners have attempted to let design-build contracts with little or no information on existing utilities, foundations, permits, and/or subsurface investigations, arguing that the design-builder assumes responsibility for all conditions relative to the project site. A recent design-build procurement for the rehabilitation of an approximately 100 year old tunnel presented very little data on subsurface utilities and relied entirely on subsurface data from adjacent properties collected more than 30 years ago. Since the owner's engineer prepared less than a 10% level of design, none of the third-party issues, such as utility realignments and impacts on an adjacent freeway, had been resolved prior to the bidding phase. The lack of utility data prompted several of the bidders to contact local utility owners, who were as surprised as the bidders that numerous relocations were required and that utility companies had not already been contacted by the project owner. Due to the very preliminary nature of the owner supplied design, a lack of third-party agreements, and lack of current geotechnical data, the various proposed approaches and bids were very diverse and included extensive lists of bidder's concerns and baseline assumptions. The very diverse bids and approaches made it difficult to compare the various proposals. The owner elected to cancel this project, pending more design work and agreements with third-party stakeholders.

6.2 *Suggested Level of Explorations*

For design-build, the level of exploration should be similar to the traditional design-bid-build except that the program would be phased to permit input and suggestions for additional explorations from the design-build teams. Experience has shown that any information not disclosed to the bidders will eventually be revealed if any form of claims litigation is pursued. Therefore, as with traditional design-bid-build, the exploration data should be presented in a GDR and summarized and interpreted in a GBR. A Geotechnical Interpretive Report may be prepared for the owner's designer to provide them with the necessary geotechnical interpretations to develop their preliminary design, engineer's estimate, and preliminary plans and specification guidelines. The content of these various documents have been well described and documented by a number of authors (USNCTT 1984 and UTRC 1997). However, the use of these formats for design-build contracts has not generally been addressed in the literature.

There are some significant differences and considerations in presenting geotechnical exploration data for design-build versus design-bid-build. Since a design-build team is fully responsible for final design and for selecting their construction means and methods, it is not appropriate nor in the owner's best interest to present interpretations of construction behavior that might define or limit construction means and methods. Ultimately, the design-build teams should be asked to provide their own interpretations of the exploration and laboratory data, since the contractor has a much greater responsibility for selection of the means and methods of construction. The owner should have an opportunity to comment on and amend the contractor's baseline interpretations.

Also during the proposal phase, the various bidders should be asked for their review comments and input on the content and extent of the explorations. Additional explorations may be suggested by the bidders, dependent on their design and construction concerns. Furthermore, the selected design-build team may, through their further assessment of the project conditions, propose additional borings and tests. When possible, and depending upon schedule, it is desirable to perform these additional explorations prior to the design-build teams final design scope and construction bid. This will help to reduce claims for changed conditions later in the project.

7 CONCLUSIONS

Design-build has only been used for the construction of a few large tunnel projects in the United States and Europe in the last 10 to 15 years. Consequently, this contracting technique is still being refined and tested in the underground environment. The primary advantage of design-build is a potential savings in schedule, which the owner pays for with reduced control over the final product and potentially higher negotiated final bid prices. Thus far, many of the major design-build tunnel projects have been significantly over schedule and have experienced cost increases of 25 to 100%.

7.1 *Murphy's Laws Live Underground*

Tunnel construction frequently reveals the unexpected and few, if any, tunnels have been constructed without some surprises. Many of the same risks exist for both the design-bid-build or design-build approaches. These risks include: 1) the potential for changed geologic and groundwater conditions, 2) potential occurrences of contaminated groundwater or soil along a long linear alignment, and 3) third-party impacts such as unknown utilities, settlement induced damage to structures, and delays for property procurement and permit acquisition. Since it is a general tenet of U.S. construction practice that the project owner retains ownership of the ground (including third-party issues), then these risks are not easily transferred or eliminated by the design-build process. A fourth potential risk in design-build projects is the risk that the owner's 10 to 30% design may contain significant design flaws that will be exposed by the more thorough design process of the design-build team and result in significant additional costs and schedule delays.

7.2 *Risk Reduction Contracting*

To reduce these risks, whether a tunnel project is constructed using design-build or the more standard design-bid-build approach, the owner is encouraged to undertake a thorough geotechnical investigation, combined with a complete assessment of utilities and structures along the alignments and a determination of the permit requirements that will impact design and construction approaches, schedule, and cost. The geotechnical data should be presented in a factual GDR. The interpreted conditions upon which the bidders are to rely along with other geotechnical baseline issues should be presented in a GBR, which should avoid discussion of means and methods for design and construction. The contract documents should contain a DSC clause in recognition of the possibility of unrecognized ground conditions that may warrant a change in construction approach and the associated project cost and schedule. Ideally, the bid documents of the successful bidder should be retained and stored with a third-party, such as a bank, and referred to when resolving disputes. Disputes should be presented to a DRB established jointly by the Owner and Design-Build Team. Utilizing these up-front precautions, should help to appreciably lower the risk in using either contracting approach.

REFERENCES

American Consulting Engineers Council (ACEC) 1992. Board of Directors. October.

Brierley, G. and Smith, G. 1998. Going Under? How about Urban Design/Build! *World Tunneling*. November. p. 42-46.

Engineer's Joint Contract Documents Committee (EJCDC). 1995. *Guide to Use of EJCDC Design/Build Documents*. Issued by American Consulting Engineers Council, National Society of Professional Engineers, and American Society of Civil Engineers.

Engineering News Record (ENR) 2001. Rising Costs, Slipping Schedule Have Tren Urbano in Hot Water. September 17. p. 15.

Gay, M, G. Rippentrop, W.H. Hansmire, and V.S. Romero 1999. Tunneling on the Tren Urbano Project, San Juan, Puerto Rico. *1999 Rapid Excavation and Tunneling Proceedings*.

Gildner, J.P and Borst, A.J. 1999. Sound Move-Seattle's Light Rail Continues Underground. *1999 Rapid Excavation and Tunneling Proceedings*.

Jaffe, M.E. and C. Goode 1992. Chapter 9 – Allocation of Risks Between Designer and Builder. *Design-Build Contracting Handbook*. Edited by R. F. Cushman and K.S. Taub. Wiley Law Publications.

Lemley, J.K. 1991. Channel Tunnel – Overview and Contractural Arrangement. *1991 Rapid Excavation and Tunneling Conference*. p. 739-749.

Loulakis, M.C. 1992. Chapter 1 - Single Point Responsibility in Design Build Contracting. *Design-Build Contracting Handbook*. Edited by R.F. Cushman and K.S. Taub. Wiley Law Publications.

Mansfield, A.J. 1996. Chapter 31- Disputes, Arbitration and the Geotechnical Engineer. *Engineering Geology of the Channel Tunnel*. C.S. Harris, M.B. Hart, P.M. Varley and C.D. Warren, Editors, Thomas Telford House. p. 467-471.

Moses, T., P. Witt, and F. Frandina 2000. Two-In-One Tunnel. Civil Engineering. April. p. 48-53.

PSMJ Resources, Inc., 1995 PSMJ's Book of Design Build Contracts.

Reina, P. 2000. Tunnel Vision. *Design-Build*. McGraw Hill Construction Information Group, December.

Tunnels & Tunneling, North America 1999. Design-Build to Control Costs?. Vol. 2. November. p. 5.

Underground Technology Research Council (UTRC), Technical Committee for Geotechnical Reports 1997. *Geotechnical Baseline Reports for Underground Construction*, ASCE.

U.S. National Committee on Tunneling Technology (USNCTT) 1974. *Better Contracting Practices for Underground Projects*. National Academy of Science.

U.S. National Committee on Tunneling Technology (USNCTT) 1984. *Geotechnical Site Investigations for Underground Projects*. National Academy of Science.

Varley, P.M., C.D. Warren, W.J. Rankin, and C.S. Harris 1996. Chapter 8 – Site Investigations. *Engineering Geology of the Channel Tunnel*. C.S. Harris, M.B. Hart, P.M. Varley and C.D. Warren, Editors, Thomas Telford House. p. 88-117.

Session 2, Track 2

Underground spaces and policies for infrastructure network

North American Tunneling 2002, Ozdemir (ed.)
© 2002 Swets & Zeitlinger, Lisse, ISBN 90 5809 376 X

Road, rail and subway tunnels by the Reichstag, Berlin, Germany

Ole Peter Jensen
Chief Project Manager, COWI A/S, Denmark

ABSTRACT: After reunion of West and East Germany, there has been a great effort to re-connect and upgrade the infrastructure between the two former German States. The project consists of three tunnels, constructed as cut & cover tunnels. The eastern tunnel is a two tracks U-Bahn (Metro) ending in an underground station close to the new parliament for Germany (Reichstag). The central tunnel contains tracks for the ICE-railway, starting with eight tracks at the north end and narrowing down to four tracks at the south end of the project. The western tunnel is a four-lane highway tunnel. All tunnels were constructed in one single construction pit.

1 GENERAL DESCRIPTON

In 1995 the contractor Spie Batignolles GmbH together with COWI as consultant won the project "Projektlos 2 - Spreebogen" for a contract sum of 0.2 billion US$.

COWI was responsible for the detailed design and planning of the construction pit and the tunnel structures. Furthermore, for the planning of the diversion and the following re-establishing of the river Spree. Other temporary works were included in the design.

The Projektlos 2 is a part of a concept to strengthen the infrastructure of the "Central Area" in Berlin with the Lehrte Bahnhof (Projektlos -1) at the north end of the development zone continued in cut & cover tunnels under River Spree to the front of the Reichstag (Projektlos - 2). The tunnels then continues in bored tunnels (Projektlos - 3) under the Tiergarten to the new station at Potsdamer Platz.

2 PROJEKTLOS 2 - SPREEBOGEN

The project consists of three tunnels, constructed as cut and cover tunnels. The eastern tunnel is two tracks U-Bahn (Metro) ending in an underground station close to the new German parliament ("Reichstag").

The central tunnel contains tracks for the ICE-railway, starting with the possibility for eight tracks at the north end and narrowing down to four tracks at the south end of the project.

The western tunnel is a highway tunnel with two bores, each with two traffic lanes.

Special construction methods were implemented for construction of the tunnels, as groundwater lowering was not permitted.

The tunnels were therefore constructed in a special construction pit made watertight in order to minimise the water volumes to be pumped.

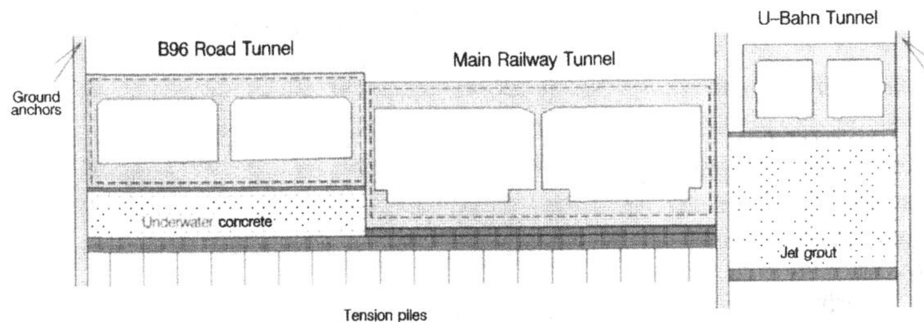

Figure 1. Typical cross sections of the three cut and cover tunnels.

3 CONSTRUCTION PIT

3.1 *General layout*

The underground in Berlin consists mainly of sand. In order to avoid lowering of the water table under the entire city, an agency was established to co-ordinate and balance the water volumes that were pumped out with the water volumes re-injected into the ground.

The construction pit was divided into four sections surrounded by 1.20 meters thick diaphragm walls. The diaphragm walls were designed with one layer of ground anchors at the top of the wall and a free standing height of up to 20 metres. After construction of the diaphragm walls and installation of ground anchors the pit was excavated wet.

On completing the excavation, vertical tension piles in the form of H-piles were installed, underwater concrete was placed and the pit was finally dewatered.

The construction pit was 500 meters long and varied in width from 120 meters to 60 meters. The pit was divided into four parts, where part E (95 x 120 m) was placed in the river, hereafter followed in land part H1 (112 x 90 m), H2 (253 x 80 m) and last part I (40 x 60 m).

Figure 2. Ariel photo over Projektlos 2 – Spreebogen.

3.2 *Groundwater requirements and excavation of construction pit*

The underground in Berlin consists mainly of highly permeable sand layers. Lowering ground water in order to construct the cut and cover tunnels would have impact on the water levels over a wide area and therefore major ground water lowering schemes were not permitted.

Due to the great number of construction sites working at the same time in Berlin, the city had established a water management company which took care of the water pumped up and the re-injection of water to keep a balance of the volumes and levels in the city.

Excavation of the different parts of the construction pit was done wet excavation. The ground water level was close to the surface and by excavating a hole close to the diaphragm wall a basin was created. A small cut and suction dredger was lifted into the excavated basin from where the dredger starts to dredge the closed part of the construction pit by pumping the materials to barges in the river.

At all time during the dredging work, the water level inside the pit was kept higher than the ground water level outside the pit.

The maximum allowable volume of water intruding the construction pit, when it was finalized and empty for water, was 1.5 l/sec per 1,000 m².

3.3 *Construction of diaphragm walls*

The diaphragm walls were constructed in a traditional way with guide beams and bentonite slurry to keep the overpressure.

Figure 3. Excavation of diaphragm panel.

The diaphragm walls have a thickness of 1.20 meters and a length varying from 25 to 30 meters. The walls were divided into a general panel width of 7 meters.

The diaphragm walls are temporary walls that do not have any structural function after the constructions of the tunnels are finalized.

The walls are designed for several load cases depending on different construction phases. When the construction sequence for the different parts of the construction pit were agreed, the amount of reinforcement in each wall panel were optimized.

Along the inner side of the diaphragm wall was excavated a trench for allowing installation of ground anchors in the top of the wall. The holes for the ground anchors are drilled through a pre-installed pipe in the diaphragm wall, fixed to the reinforcement cage. There were placed two anchors in each reinforcement cage and with two cages in each diaphragm panel give four anchors per wall panel.

Figure 4. Installation of ground anchors.

The ground anchors are installed inclined and staggered with an angle of 20 deg and 35 deg from horizontal. The anchors consist of 8 lines diameter 0.6 inch and with an average length of 20 to 30 meters and are stressed to a maximum of 1,200 kN per anchor. The anchors are in general stressed twice depending on the construction sequence of the pit, excavation, empty pit for water etc.

3.4 *Diversion of river Spree*

For construction of the three tunnels under the river Spree it was necessary temporally to divert the river, as it can be seen on the Ariel photo, Figure 2. The river diversion made it possible to construct a construction pit for the part of the three tunnels, which finally will be placed below the river.

A new channel was excavated and bonds were constructed out in the old channel to allow for construction of diaphragm walls forming the construction pit.

Measures were taken to protect the construction pit for impact from the river barges.

Figure 5. Fender structure along the river diversion.

The fender structure consists of steel H-profiles both as vertical piles and as horizontal beams between the piles. The horizontal beams were also prestressed by cables that would absorb the energy from a barge impact during large deflection of the fender structure. The fender system was designed for a barge impact force of 4 MN over a width of 2 meters.

The construction pit in the river was demolished after the tunnels were constructed and back filled with sand before the river was re-established.

3.5 *The Swiss embassy*

The construction pit passes close to the corner of the Swiss embassy. The Swiss embassy is an old, heavy, three story building that is protected by a preservation order. Settlements of the building, which could cause cracking or damage, were therefore of great importance.

At this particular location the standard plane diaphragm panel was replaced with T-shape panels to increase the stiffness. The tops of the T-shape panels were connected to a 3 x 3 meter horizontal prestressed concrete beam. At the ends of the beam were placed 18 numbers of ground anchors with a length of 50 to 70 meters and were reaching to the opposite site of the building.

By using a stiff wall and by placing the grouted part of the ground anchors opposite the building, the soil volume under the building was pre-stressed, thereby reducing the settlements.

There was a program monitoring the behavior of the building and there were no critical settlements at any time during the construction period.

Figure 6. Installation of T-shape diaphragm panel at Swiss embassy.

Figure 7. Ground anchors position in pre-stressed beam.

3.6 *Pit for U-Bahn at station*

The U-bahn vertical alignment was rising toward the station placed in pit "H2". The part of the pit, which contains the U-bahn, was separated with a longitudinal diaphragm wall. This was doing to avoid excessive excavation and re-filling before construction of the U-bahn tunnel could commence.

To seal off from intruding ground water, a jet-grout layer was installed between the diaphragm walls and deep under the U-bahn tunnel. The other section of "H2" was constructed in the usually way, with wet excavation and underwater concrete with tension piles. This arrangement can be seen on Figure 1.

Figure 8. Construction pit "H2" with embraced pit for U-bahn tunnels. Reichtag can be seen in the background.

It was necessary to use temporary steel bracing between the diaphragm walls in the pit for the U-bahn tunnel due to the water pressure from the other part of "H2 ".

3.7 *Underwater concrete and tension piles*

The bottom of the construction pit was sealed off with a concrete slab cast under water, anchored with tension piles.

When the dredging in an enclosed part of the construction pit was finalized the bottom was cleaned and prepared for installation of tension piles.

The tension piles consist of H-profiles, HEB 220, with shear brackets welded on the toe of the piles and a steel plate welded on the top of the piles forming the head. At the inside corners in the H-profile were installed two injection pipes running down to the pile toe. The piles had a length of 19.8 meters.

The piles were vibrated down into the bottom of the pit from a floating barge. The piles were placed in a pattern of 3 x 3 meters. After the piles were installed the toe of the piles were grouted by injection of mortar through the pre-installed injection pipes.

A 1.5 meters thick concrete slab was casted under water and the head of the piles were embedded in the concrete. The water in the pit was pumped out after the concrete had reached its strength and the pressure below the slab were then taken over as tension in the piles. The concrete slab was not reinforced the forces were carried by a compression arch between the pile heads.

3.8 *Test of tension piles*

An extensive full-scale field test of the piles was carried out before any of the piles were installed. A test area was excavated down to a level just above ground water level, so the test piles would be placed in water. An area of 20 x 54 meters was prepared for testing of five single piles with different lengths and a group test of five piles. Two geotechnical boring were carried out in the area together with SPT tests, all to give a good knowledge of the ground conditions.

Figure 9. Plan over test area

The necessary anchor lengths of the piles were calculated to 18.20 meters and the necessary working load was calculated to be 1,000 kN. The single pile test were carried out with pile lengths varying from 13.76 to 17.30 meters and the test load of 700 kN, 1,200 kN and 2,000 kN were applied to the piles in three steps with full load release between the steps. Some of the single piles were even tested up to 2,900 kN. The results from the tests were plotted for each pile in two diagrams load versus time the load was applied and deformation of pile head versus load. The distribution of the axial force, based on strain, was measured by strain gauges placed along the pile.

A group of five piles were likewise tested and for any group effect. Besides monitoring each pile the soil in between the piles were monitored in three level for vertical and horizontal stresses and with rod extensiometers the vertical movement of the soils.

Based on all the results from the pile tests the allowable shear stress to be used in the calculation of the pile bearing capacity was fixed to τ = 150 kN/m^2.

4 TUNNEL CONSTRUCTION

4.1 *General*

The U-Bahn tunnel is constructed as one cross section with two tubes and a constant overall width of 15 meters. The cross section separates into two independent tunnels close to the underground station to make room for a central platform in the station area.

The ICE-railway tunnel varies in width from 69 meters at the north end under river Spree to 24 meters at the south end. Under river Spree the tunnel is tied to the vertical tension piles used for the underwater concrete in order to get enough safety against uplift; those piles will work as permanent tension anchors for the tunnel.

The highway tunnel (road B96) has a constant width of 24 meters and the cross section is divided into two tubes with two lanes in each.

4.2 *Construction*

The tunnel construction was traditional cut and cover tunnel construction, with the tunnels divided into 10 to 20 meters casting segments. The concrete used was dense B-35 concrete also classified as watertight concrete. By casting in segments of 10 to 20 meters it was possible to control the crack development from the concrete temperature and keep the crack width below the required 0.15 mm.

The construction sequence was simply to construct bottom slab, walls and last the roof slab. For tunnels along the pit walls, highway and metro, the pit walls were utilized as outer formwork. In this case the pit walls were cleaned and straighten out and a thin drain layer was placed before the tunnel wall was casted.

Figure 10. Construction of tunnel bottom for the main railway.

In the final condition, when ground water rises the drainage layer ensures that the tunnel wall is subject to the full water pressure. Special measures were taken to ensure that the tunnel bottom and roof slabs were supporting the pit wall when the ground anchors were released. The effects of this arrangement is that in final condition the tunnel wall carries the water pressure and the pit wall carries the soil pressure and transfers the load to the tunnel bottom and roof slab.

The joint between the tunnel segments were equipped with waterstops and shear keys which allows the segments to contract and expand, and that accommodate small rotations.

4.3 *Special detail*

All of the tunnels had sufficient safety against uplift by their weight, except for the part of the main railway tunnel under the river where the available space was limited.

The tension piles below the railway tunnel were extended and were cast into the bottom slab of the tunnel.

In the final condition when the river is back in position the uplift from that part of the railway tunnel will be taken as permanent tension in the piles. All other tension piles will not have any effect in the final condition.

5 REMARKS

The project was very complex with special construction pit and three tunnels side by side.

Especially the construction of the pit had several obstructions. Bunkers from world war two, unexploded ammunition, cellars, pieces of tunnels etc. which could not be located due to registrations and drawings have been lost or burned during the war.

But, also the tunnel construction calls for an experience contractor to handle huge quantities of concrete, reinforcement and steel.

The contractor has in professional way and with high standard finalized this complex project in time.

Figure 11. Extended tension piles.

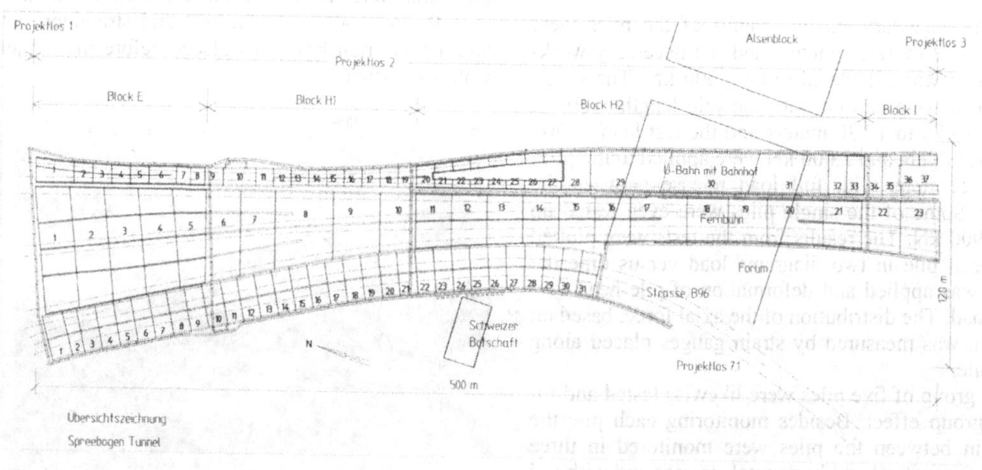

Figure 12. Plan over tunnel construction segments.

North American Tunneling 2002, Ozdemir (ed.)

Trends in tunnel contracting and execution risks, a light moving toward you

H.G. Dorbin
The Nielsen-Wurster Group, Inc., Houston, Texas, USA

J. Dignum
The Nielsen-Wurster Group, Inc., Seattle, Washington, USA

ABSTRACT: Risk Management has become an essential tool for every Tunneling Contractor as Contractors take on more financial and performance risk than before. This paper addresses the Project Risk perspective of the Owner, the shared objectives of the Owner and the Contractor, and the detailed steps the Contractor must take to accept both Owner perspective and Tunneling Industry realities in order to manage Risk successfully on Tunneling Projects. For the Owner, the Tunneling Project is an investment. For both, cost and schedule estimates must be accurate; the plan must be appropriate for the Project; Contract Risks must be identified; Action Plans and a Risk Management Performance System must be utilized. The Contractor's Risk Management Campaign must define its Project Success, acknowledge its realities, understand the contract relationships and scope of work, define the project deliverables, identify and evaluate the risks, and more to prevent large disputes and successfully manage Project Risk.

1 INTRODUCTION

The Tunneling Industry has existed for hundreds of years. As with many mature industries, it has evolved through innovation and advances in technology and efficiencies. These advancements and innovations have led to the successful achievement of Tunneling Projects of proportions unimaginable a generation ago. (Examples: chunnel etc.)

Technological advancements and the mature nature of this industry also have imposed certain commercial realities on Tunneling Contractors. These include increased international competition, higher Project Owner/Developer expectations and shifting of government contracting methods toward allocating Risk to the Contractor.

Because of these new realities, Risk Management has become an essential tool for every Tunneling Contractor's tool belt. Until recently risk was handled simply by a Contractor adding a contingency factor for the "unknown" to a bid submitted for a specific scope of work. Contingency was intended to cover those physical obstacles which were impossible to identify, or those items of work which may have been inadvertently left out of the pricing of the project scope of work. Contingency was never intended to cover costs arising out of changes in contractual allocation of risk between Owners and Contractors. The new risks are often those which Contractors do not identify, or even know exist, until the economic impact of the situation hits.

Today, Contractors who fail to identify and address contractual Risk beyond that traditionally covered by contingency almost certainly will suffer a project financial collapse at some point.

This paper addresses the Project Risk perspective of the Owner, the commercial and transactional realities of the Contractor, and the detailed steps the Contractor must take to accept both Owner perspective and Tunneling Industry realities in order to manage Risk successfully on Tunneling Projects.

2 THE OWNER'S PERSPECTIVE

To the Owner, the Tunneling Project is an investment. As with every investment, the Owner evaluates three things to weigh its merit:
- Capital Investment (Cost)
- Return on Investment (Private Sector – Future Profits/Public Sector – Social Benefit)
- Risk to Success (Project Risk)

From this simple model, the Owner will frame its Project Goals of budget, cost, schedule and quality. Achieving these Project Goals defines Project Success for the Owner.

The Owner has the ability to control Cost and Project Risk. The Return on Investment will rely on events and forces outside of the Project and outside of the Owner's full control.

A number of market forces have greatly enhanced the Owner's ability to control Project Cost. Bidding for large Tunneling Projects is highly competitive between large international firms. As the scale of Tunneling Projects has increased in the past few decades, and as the desire of Owners – public and private – to have single source responsibility for the Projects has grown, large international tunneling firms have evolved which now compete globally for Projects. Advancements in telecommunications, electronic/computerized Project tools and other Project Management methods have allowed these mega-firms to use company resources and infrastructure – wherever those resources are located – to compete for and execute work anywhere in the world. This global competition between big firms, often similar in capabilities, has made cost a key distinguishing factor between competing contractors, thus increasing price competition and lowering Contractor margins.

With price competition high, Owners have increased their focus on addressing Risks to Project Success. In evaluating these Success Risks, the Owner must make a decision for each Risk:

- Keep Control and Responsibility for the Risk: Maintain control of Risks which are too critical to Project Goals to insure or transfer.
- Insure Against the Risk: Pay to transfer the Risk to a third party insurer.
- Transfer the Risk to the Contractor: Develop the Project delivery method and Contract form to hold the Contractor accountable, and compensate the Contractor for accepting this Risk.

While Owners keep control over many critical Risks and procure insurance to address Risks when that is the best economic solution, there is a growing tendency to transfer more Project Risk to the Contractor. From the Owner's Project Investment perspective, this transfer of Project Success Risk to the Contractor is the *Essence of the Deal* between the Contractor and the Owner. State, Federal and foreign governments have adjusted their rules and/or their interpretations of existing rules to add greater flexibility to the Project delivery formats and Contract forms used, including use of Lump Sum Turnkey and Guaranteed Price Contracts in situations where those were not allowed or even thought of a decade ago. While in the past, where Contractors were compensated accordingly, heightened Owner expectations and high international competition in the Tunneling Industry have shifted this Risk/Reward relationship so that in today's marketplace Contractors are accepting more Project Risk for less Reward than ever before.

3 FIVE GOALS FOR RISK MANAGEMENT

The principal objectives of Risk Management for both Owner and Contractor can be stated in five goals:

1. Cost and schedule estimates are correct to an acceptable accuracy.
2. Execution Risks are identified and evaluated, beginning with the question: Is the plan appropriate for the specific Project?
3. Contract Risks are identified and evaluated, including the analysis of:
   - Compliance with corporate legal requirements;
   - Contract structure, Project delivery method an Contract organization Risks; and
   - Scope of Work, performance standards and deliverables.
4. Action Plans are developed to avoid or mitigate impacts from the Risks that were identified.
5. A reliable Risk Management Performance System is in place to monitor and report on Action Plans during each stage of the process: Project planning, bid tender, Contract award and execution of the Project.

4 THE CONTRACTOR'S RISK MANAGEMENT CAMPAIGN

The Contractor has many of the same goals as the Owner in managing Risks. However, because of the Contractor's role in the Project and later point of entry into the Project cycle, the Contractor's Risk Management steps are somewhat different and are performed in a different order from those of the Owner.

4.1 *Step 1. Define Contractor Project Success*

The benchmark used by most Contractors for Project Success is achieving the "as-sold margin." This benchmark accepts the notion that different profit margins are acceptable in different industries serviced by a Contractor and even on different projects within the same industry. A variety of business reasons may also cause a Contractor to intentionally reduce its margins to enhance its chances of being awarded a particular Project. Even the nature of the undertaking will influence the margin expected, since Contractors expect to be paid to accept Risk and, from a Contractor's perspective, the more Risk there is to an undertaking the more reward it should receive. Thus a Project that generated large profits, but only a fraction of the as-sold margin, may not have managed Risk as well as a Project generating its full as-sold margin even though its profit contribution was smaller.

4.2 *Step 2. Acknowledge Contractor Realities*

The Contractor typically enters the Project after the Owner has already defined the Project delivery method, the Contract form and other contracting variables. Bids are often estimated by the Owner prior to sending out bid packages. Depending upon the Project, Owners have a great deal of success forecasting the bid amounts either by performing the estimate themselves or by hiring a contractor to perform the estimate. The methods used to estimate cost and schedules for Tunneling Projects are well established and widely understood. This estimating practice also facilitates a higher degree of competition between Contractors and a knowledgeable Owner, which can negotiate price and schedule based upon its own analysis.

Contractor Risk Management is also affected by the fact the Contractor almost always has considerably less time to evaluate a Project and prepare a plan, schedule and cost estimate than the Owner had. The Contract can be thousands of pages long when the referenced standards and specifications are included. These Contractor Realities, as shown in Table 1 below, must be acknowledged in the process of identifying and managing Project Risks.

Table 1. Contractor realities.

Contract variable	Owner	Contractor
Scope definition	Decided	Accept/negotiate clarifications
Cost	Estimated	Offers in competition
Schedule	Estimated or fixed	Offers in competition
Contract form	Decided/may accept alternative offers	Accept/propose alternative
Terms and conditions	Draft contract included in bid package	Accept/negotiate changes
Planning period	2–10 years	60–180 days

4.3 *Step 3. Understand the Contract Relationships*

In order to identify, evaluate and manage Risks in development of a competitive but achievable bid and in execution of a successful Project, the Contractor must understand the relationships between all parties executing work in connection with the Project.

The Owner will have developed the Project delivery method and will have parceled the tasks required to design and construct the tunnel infrastructure to its optimum financial benefit. Contract liability requires having a Contract with the party alleging breach (contract privity). Thus understanding the Project contracting structure completely (who has a Contract with whom) is critical for Contractors to evaluate where claims or execution difficulties might arise. A Tunneling Contractor sometimes relies heavily upon information and Project deliverables from a third party with whom it has no Contract remedy because it has no Contract.

4.4 *Step 4. Understand the Scope – the Entire Scope of Work*

The Scope of Work is literally thousands of pages in some complex Projects. The first time the Scope of Work is carefully examined by either the Owner or the Contractor can be when there is a dispute. Since Scopes are most often written by Owners, not understanding the entire Scope is a Risk that a Contractor cannot afford to accept.

The mantra promoted by the management of many contractors, "*Read the full Contract*," is based upon the useful concept of full understanding but, in reality, often cannot be achieved by any one person due to the volume of the documents. While some parts of the Contract should be read by everyone, most of the Contract needs to be dismantled into smaller manageable elements for the Contractor team to read and understand. This is done by:

- Identifying the parts of the entire Contract, including referenced standards and procedures;
- Collecting copies of all referenced documents, applicable codes, etc.;
- Using a matrix to organize the Scope elements, identifying who needs which parts; and
- Distribution to designated parties for careful reading and understanding.

In addition to furthering the understanding of the Project Scope and required deliverables, reading and understanding the entire Contract allows the Project costs associated with any Owner-driven (or location-drive) codes, specifications or performance requirements to be identified and included in the bid.

4.5 *Step 5. Define the Project Deliverables*

Contractors are very Project oriented and typically think in terms of deliverables. All of the Execution Deliverables and Contract Deliverables need to be identified by the Contractor and carefully detailed, showing from whom the deliverables are due and to whom they are due.

This exercise shows the Contractor clearly:

- What the Contractor needs in order to perform the Project, which may or may not be stated in the Contract as Owner-provided items, including approvals and reviews (i.e., Contract Deliverables).
- What Deliverables the Contractor has to rely upon third parties to provide, meaning that there may be no entitlement to impacts if they are not delivered.
- What the Contractor is to provide to third parties in order for those third parties to perform their Work Scope, meaning that a claim may originate through the Owner for failure to coordinate with third parties.

4.6 *Step 6. Communicate the Contract Form*

The technical staff of the Contractor sometimes justify not reading the Contract because it is full of legal terms and is unclear to them. Forcing them to read it will only mean they have read something they do not understand. Some Contractors provide a plain language dissertation of the Contract to the Project

Team. Unfortunately, this is often produced by an overburdened legal staff after the Project is awarded, which is likely to be too late from a Risk Management standpoint. As a consequence of not reading the Contract, the Contractor Project Team can be surprised to discover what it must perform within the time and budget it agreed to.

A review of key entitlements and liabilities in simple English is important to the staff executing the Project and administering the Contract. Without understanding how the Owner can access financial guarantees or withhold payment, the technical staff manning the Project cannot protect against such Risks.

4.7 *Step 7. Identify "Soft" Scope*

Many claims result from 1) one party not understanding that it is to perform a task, or 2) one party not understanding the standard that applies to its performance of a particular task. Words such as "verify," "coordinate," "collaborate," "confirm" or "quality," for instance, create an immeasurable standard for performance that does not clearly state the level of responsibility for the task. These "Soft Scope" issues are often included in Owner Contract Forms and Scopes of Work. However, when there is a dispute or a question of entitlement, these "soft" phrases are looked upon to validate or invalidate a party's position.

The Contractor's job is to identify Soft Scope issues and seek or provide clarification as to their meaning *before* the Contract is executed. Otherwise, the estimate for cost and schedule cannot be relied upon.

4.8 *Step 8. Identify the Risks*

The previous steps have helped identify the Execution and Contract Deliverables of and to the Contractor, the liabilities connected with the Contract and how they are triggered when Deliverables are not provided as required, and the Soft Scope. Using that combined set of Project-specific information, the Contractor can identify a particular Project's Risks.

Risk identification seems to be the most difficult step for Contractors. Like the Owner, the Tunneling Contractor strives to avoid or mitigate impacts to Project Success and/or its as-sold margin. However, several factors make Risk identification more difficult for Contractors than for Owners. These include:

- Contract Form and terms already prepared and are voluminous
- High Turnover in staff
- Project Teams typically not the same personnel each time
- Limited time frame to review, as bid preparation period is prescribed by Owner
- Lack of familiarity with Project site
- Stiff competition, and the knowledge that frequent clarification requests to Owner may be viewed as irritating

With these realities facing Contractors, the Risk identification system must be simple and efficient. At the same time, because of the competitive nature of the bidding marketplace and the transparency of the bidding process, key Risks cannot be overlooked, so the system needs to be thorough and accurate.

Typically Contractors use one of three methods or a combination of those methods to identify Risk. Each method has its own pros and cons, as shown in Table 2 below:

Table 2. Risk identification methods pros and cons.

Identification method	Pro	Con
Checklist	Thorough; review consistent issues	Being thorough usually takes a long time; cumbersome to use alone during execution
Brainstorming	Fast, dynamic; can change which items are being reviewed	Relies on knowledge of participants to raise risk concerns
Computer models	Provide detailed quantitative data	The long time required to input data may affect application; projects are specific collections of risk, and models rely upon other projects to predict outcomes of specific risks

The method employed by Nielsen-Wurster is a mixture of techniques called Structured Brainstorming, which uses checklists to orient and structure the thinking of the Brainstorming team. While computer models offer interesting prospects, the variety of specific Risks a Contractor faces on a Project and the timetable the Contractor is typically provided in which to tender a bid generally limit the computer models' usefulness to Contractors.

4.9 *Step 9. Evaluate the Risks*

After the Risks have been identified, the potential impact of each Risk needs to be evaluated so the most attention and resources can be directed to the Risks with the most potential impact, and so the cost and schedule estimates will reflect those extra measures.

Various techniques are used by Contractors to evaluate Risk, all leading to the same two parameters – likelihood of occurrence and severity of impact upon the Project.

A critical mistake often made by Contractors is struggling to develop quantitative potential impact numbers. The time spent developing hypothetical quantitative exposure numbers for identified Risks at the bid tender stage often means sacrificing the breadth of the Risk identification and evaluation of other Risks. It is seldom possible to precisely determine the potential financial impact of the majority of Risks on the Project at the bid stage. A qualitative method recognizes the time limitations and other objectives that the Contractor covers in its Risk Management Process.

4.10 *Step 10. Develop Action Plans and Continue Risk Management*

At this stage the Contractor has performed the same Project Risk Management steps as the Owner, although in a different order and from a different perspective. The Contractor has prioritized its Risks based upon potential impact to the specific Project. Action Plans are developed to best address the nature of each Risk identified.

However, the Contractor also is faced with changing Risks and new Risks during the Project's execution. The Contractor must review its profile of Risks periodically to validate and revise current Risk priorities, groups of identified Risks and Action Plans.

Contractors on large Tunneling Projects perform the same Risk identification, evaluation and management steps as the Owners do when developing their subcontracting plan. The Risks identified from the Risk Profile developed for the Prime Contract relationship serve as inputs to define the appropriate Project delivery method, Contract forms and terms for the subcontracts.

5 CONCLUSION

Contractor Risk Management techniques, even on the most complex Tunneling Projects, are straightforward and direct, but demand discipline and a realistic acknowledgement of the realities of the contracting process. Dedication to a Risk Management system with objective criteria is vital to make Project Risk Management results predictable and consistent, and to allow Contractor management to rely upon them. These systems are most effective when adapted to acknowledge the commercial and transactional realities of both Owner and Contractor in the cycle of developing, bidding and executing engineering and construction projects. Effective Risk Management can prevent development of dispute situations and increase the opportunity for achieving Project Success on Tunneling projects of all scales. Remember: the light at the other end is moving toward you.

North American Tunneling 2002, Ozdemir (ed.)
© 2002 Swets & Zeitlinger, Lisse, ISBN 90 5809 376 X

Environmental impact assessment of underground multi-functional structures

N. Bobylev & M. Fedorov
St. Petersburg State Technical University, St. Petersburg, Russia

ABSTRACT: Nowadays underground multi-functional structures development is a popular solution for improving urban environment. A method of environmental impact assessment of those structures is discussed in the present paper. The method is based on multi-criteria optimization of a number of particular environmental parameters, that comprehensively characterize underground multi-functional structure environment. Results of calculations using the method are represented by figures, which allow to compare environmental quality before, during, and after underground structure construction.

1 INTRODUCTION

According to UN secretary-general Kofi Annan in 2025, two-thirds of the world's population will live in cities. As urbanisation advances, underground facilities becoming an integral part of a modern, sustainable city. Underground multi-functional structures are developments that fit central part of a city within lack of space and various environmental problems in the most proper way. Of cause, underground space development has its environmental benefits and drawbacks, proposed method of environmental impact assessment has the aim to compare this benefits and drawbacks for archiving the best solutions for underground multi-functional structures construction.

2 ENVIRONMENTAL IMPACT ASSESSMENT

Environmental impact assessment (EIA) is a sizeable part of a feasibility study, preliminary and define design of underground facilities. Impact assessment – simply stated, is the identification of future consequences of a current or proposed action. (International Association for Impact Assessment). More detailed definition of EIA was given by the German Association for the Assessment of Environmental Impacts: "To assess environmental impacts means to early, systematically, transparently, and reasonably investigate, evaluate, and consider the relevant future consequences of a current or proposed action".

3 CONSIDERED OBJECT AND THE ENVIRONMENT

3.1 *Considered object*

Underground multi-functional structure (UMFS) is defined in the present paper as an underground development within a high-density built urban territory. Those developments can encompass traffic and car parks, motor ways, public utilities, emergency systems, waste disposal systems. UMFS are usually quite sizeable, deep, and widespread objects. UMFS supposed to be highly implicated into existing city above- and underground infrastructure and eventually be a source of environmental quality improvement.

A lot of research on the development of UMFS has been done within such organisations as ITA and ISSMGE.

3.2 *The environment*

As it was stated above, high-density built urban territory is considered as an environment of an underground structure. From an EIA study standpoint, the most complicated example of underground structure environment looks as follows:

- Central, historical part of a big city;
- High-density built territory that contains valuable historic buildings;
- Considered territory faces heavy traffic loads;
- Considered territory faces various ecological problems (e.g. air and soil contamination);
- Ground conditions are unfavorable.

One of the main difficulty engineers are facing while conducting EIA of underground structures are uncertain ground conditions. Urban territory, which has been used for many centuries, most likely contains abandoned and even unknown artificial underground structures. For example in the city of St.Petersburg among such structures were found storage, shafts, conduits. It is well-known, that it is impossible to restore an underground opening to its original condition. An underground cavern will continuously affect vast under- and above ground surroundings. Therefore overburden caverns and channels, organic layers appear to be a certain difficulty in ground conditions estimation.

The other difficulty for conducting EIA is time variation of environmental characteristics. Traffic loads and transport streams distribution can be a good example. Transport streams distribution modelling is a sizeable problem for environmental assessment of UMFS. For predicting traffic loads at a period of 100 years it is necessary to posses data even on a socio-economic development of a region.

4 THE METHOD

4.1 *General description*

Typical UMFS is a subject for detail and all-embracing considerations form a social, economical, political, and environmental points of view. Considered exploitation period of an UMFS is usually near to 300 years, thus it is especially important to estimate long-range effects of interaction between underground structure and the environment.

It is proposed to proceed EIA of UMFS in four steps:

1. defining characteristics of a UMFS and the environment;
2. defining possible environmental effects of UMFS - environment interaction;
3. assessing variation of the environmental parameters;
4. integral assessment of envirnmental parameters variation.

Initial data is collected on the first stage of the method. This data is analyzed on the second stage during consideration of environmental effects. On the third stage environmental conditions are characterized using a number of environmental parameters. Several mathematical methods are proposed on the fourth stage for integral assessment of environmental parameters variation (see Figure 1).

4.2 *Initial data*

Initial data for EIA is presented by characteristics of the object - UMFS, and description of the environment.

Characteristics of UMFS are functional, architectural, lay-out, and technological solutions. As a rule, this data is an essence of an underground development project, and is quite clear. But, it is useful to mention, that it is important to consider supplemental functions of UMFS and predict possible alterations or changes of the object functions (e.g. new infrastructure instalment).

Main characteristics of UMFS are:

- function,
- dimensions,
- technology of a cavern creation,
- reliability,
- output of pollutants (including heat).

Initial characteristics of the affected biosphere area are rather difficult to be defined. Here are some of characteristics of the environment:

- ground,
- hydrological,
- landscape (including existing buildings),
- transport,
- atmospheric,
- animals and plants,
- social.

UMFS and its environment also have a joint characteristic – ground stress-strain conditions.

4.3 *Types of interaction – environmental effects*

It is appropriate to divide environmental effects of creating an underground development into two groups: the first group – environmental impact of underground structure during a construction period, the second – environmental impact of underground structure during an operation period.

Among possible effects during construction period could be found:

- landscape deterioration due to construction site creation,
- transport of goods to a construction site,
- transport availability changing (most likely negative),
- water ingress to the cavern,
- barrage effect,
- surface settlements,
- waste from construction site transporting,
- atmospheric pollution,
- heat (frozen) impact (most likely negative),
- sound and vibration impact (most likely negative),
- increasing risk of accidents.

Among possible effects during construction period could be found:

- water in/out put,
- surface flow changing,
- barrage effect,
- atmospheric pollution (positive/negative),
- heat impact (most likely negative),
- sound and vibration impact (most likely positive),
- landscape changing (most likely positive),
- transport availability changing (most likely positive),

- reduction risk of accidents.

It appears that in case of an advanced (from environmental standpoint) UMFS development one could found out unbeneficial environmental effects during a construction period and salutary environmental effects during an operation period.

4.4 *Assesment of environmental parameters variation*

For conducting EIA it is proposed to characterize environment of UMFS by a number of environmental parameters. Those parameters should have a multi-level structure, which is presented bellow:

1. Earth related parameters:
- geological structure;
- groundwater level;
- groundwater chemical composition;
- groundwater hydrodynamic characteristics (speed and direction of filtration);
- groundwater conditions (rate during seasons changing);
- hydrology characteristics (composition, dynamics, and conditions of surface waters);

2. Human related parameters:
- landscape;
- transport availability of a city facilities;
- recreational and esthetical value of the territory;
- noise level;
- reliability and safety;

3. Ecosystem related parameters:
- chemical composition of atmospheric air;
- moisture of atmospheric air;
- animals and plants;
- soil composition;
- temperature.

These environmental parameters, their number and structure is the result of possible environmental effects analysis (see Section 4.3). The environmental parameters consideration is necessary and sufficient for conducting EIA of typical UMFS.

Defining variation of each parameter is an independent issue and should be done by an expert (experts) in a relevant area of knowledge.

4.5 *Integral assessment of environmental parameters variation*

Integral assessment of envirnmental parameters variation is the most difficult and discussion problem. In the present work it is proposed to estimate integral environmental parameters variation using methods of system analysis, and in particular multicriteria optimization methods.

The following methods could be used: main criteria, linear accumulation, selected maxims accumulation, Pareto. Multi-criteria function can be presented as following:

$$\sum_{j=1}^{n} \varepsilon_i J_i(x) \xrightarrow[x \in X]{} \max \qquad (1)$$

where $J_i(x)$ = function reflecting environmental parameter variation; e_i = ponderability coefficient; and i = environmental parameter number.

Maximum of the multi-criteria function in Equation 1 corresponds to the best quality of the environment. Function $J_i(x)$ and ponderability coefficient e_i are both being assessed by experts.

It is proposed to assess variation of environmental parameters using a system of marks. Scale of marks is identified from 1 to 5. Mark 5 corresponds to the best value of environmental parameter, mark 1 – the worst. As it was stated above, assessment of environmental parameter variation is proceeded by an experts in a relevant area of knowledge.

Ponderability coefficient of environmental parameter reflects importance of this parameter for EIA of a particular project. While setting up a ponderability coefficient it is useful to take in the account the following aspects:

- Goals of the project;
- Environmental parameter dynamic;
- Space variation of an environmental parameter;
- Social, cultural issuers, and public opinion.

Ponderability coefficients are stetted up by experts, experienced in managing underground projects and EIA. Deep scientific knowledge of each environmental parameter is not required for this stage of expertise.

Ponderability coefficient should be a positive number. Usually value of a coefficient is near to 1. It is useful to select one special environmental parameter first. Ponderability coefficient of this paramiter could be set as 1, thus ponderability coefficients of the other parameters should be compared to the parameter selected.

4.6 *Calculation procedure*

Equation 1 could be presented and calculated as a following table (see Table 1).

Column 4 of the Table 1 should be developed according to the stages of EIA. It is appropriate to carry out EIA of UMFS at the following stages of a project:

1) "zero version" – without any structure;
2) construction period;
3) operation period;
4) liquidation period (mainly for industry-included UMFS).

After Table 1 is filled in it is possible to proceed calculation of integral assessment of environmental parameters variation at the above mentioned stages of a project. For better environmental quality achieving Function 1 should be maximised, therefore inference about environmental benefits of a UMFS

Table 1. Calculation for integral assessment of environmental parameters variation.

1	2	3	4
#	*Parameter*	Ponderability coefficient	Mark reflecting the value of a parameter at a certain period of time
i	J_i	e_i	$J_i(x_i)$
1	Geological structure		
2	Groundwater level		
3	Groundwater hydrodynamic characteristics		
4	Groundwater chemical composition		
5	Groundwater season conditions		
6	Hydrology characteristics		
7	Soil composition		
8	Air chemical composition		
9	Air moisture		
10	Noise level		
11	Temperature		
12	Animals and plants		
13	Transport availability		
14	Landscape		
15	Territory recreational value		
16	Territory reliability and safety		

project implementation is made by comparison of figures corresponding to different EIA stages.

Evidently, worse environmental situation we have within a territory, better results are shown by different stages of a project.

5 THE METHOD IMPLEMENTATION

5.1 *Aggregation of the environmental parameters*

As it was described in Section 4.4 there are 16 basic environmental parameters in the present method. In most cases it is useful to aggregate those parameters. The main reason of aggregation is to allow expert consideration of interaction between environmental parameters of similar nature outside a frame of the method. The method, undoubtedly, is simplified by this issue. Aggregation also allows comparison between upper-level environmental parameters. For example, it is possible to compare Earth-related parameters and Ecosystem-related parameters, but comparison between geological structure and chemical composition of atmospheric air looks unacceptable.

Method of aggregation of environmental parameters depends upon initial data for a project, project value, number and specialisation of experts involved.

5.2 *Detailed elaboration of the Table of integral assessment of environmental parameters variation*

Detailed elaboration of the Table of integral assessment of environmental parameters variation (see Table 1) depends upon requirements for EIA of a particular UMFS project. Typical detailed elaboration is carrying out within Column 4 of Table 1. It is possible to develop stages of EIA using time factor. For example, different stages of UMFS construction could be assessed (bored piles installment, cavern creation, motor-transport ways relocation etc.).

Detailed elaboration also involves assessment of different project versions of UMFS. Various construction and transport logistics methods are surely subject for EIA.

5.3 *Comparison of the EIA method results*

Results of the EIA method are presented by figures. Therefore it is easy to transmit figure value to percents, assume environmental quality before project started as 1 (see brief example in Table 2).

Environmental decision making on UMFS project implementation should be based on comparison between the figures of assessment and the time duration of assessed stage.

Table 2. Calculation for integral assessment of environmental parameters variation. Brief example.

1	2	3	4		
#	*Parameter*	Ponderability coefficient	Mark reflecting the value of a parameter at a certain period of time		
i	J_i	E_i	$J_i(x_i)$		
			"zero version"	Construction period	Operation period
1	Earth related parameters	1	5	3	4
2	Human related parameters	1.2	2	1	5
3	Ecosystem related parameters	0.9	3	2	3
	Sum $J_i(x_i)$		10.1	5.8	12.7
	%		100	57	126

6 CONCLUSION

EIA of UMFS proved to be complicated and discussion problem. The proposed method has its aim to proceed integral assessment, involving as much as possible inter- and multi-disciplinary studies. Method was used for assessing UMFS projects in St.Petersburg, Moscow, and Hamburg. Accuracy of the method is comparable to other ones, (verbal,

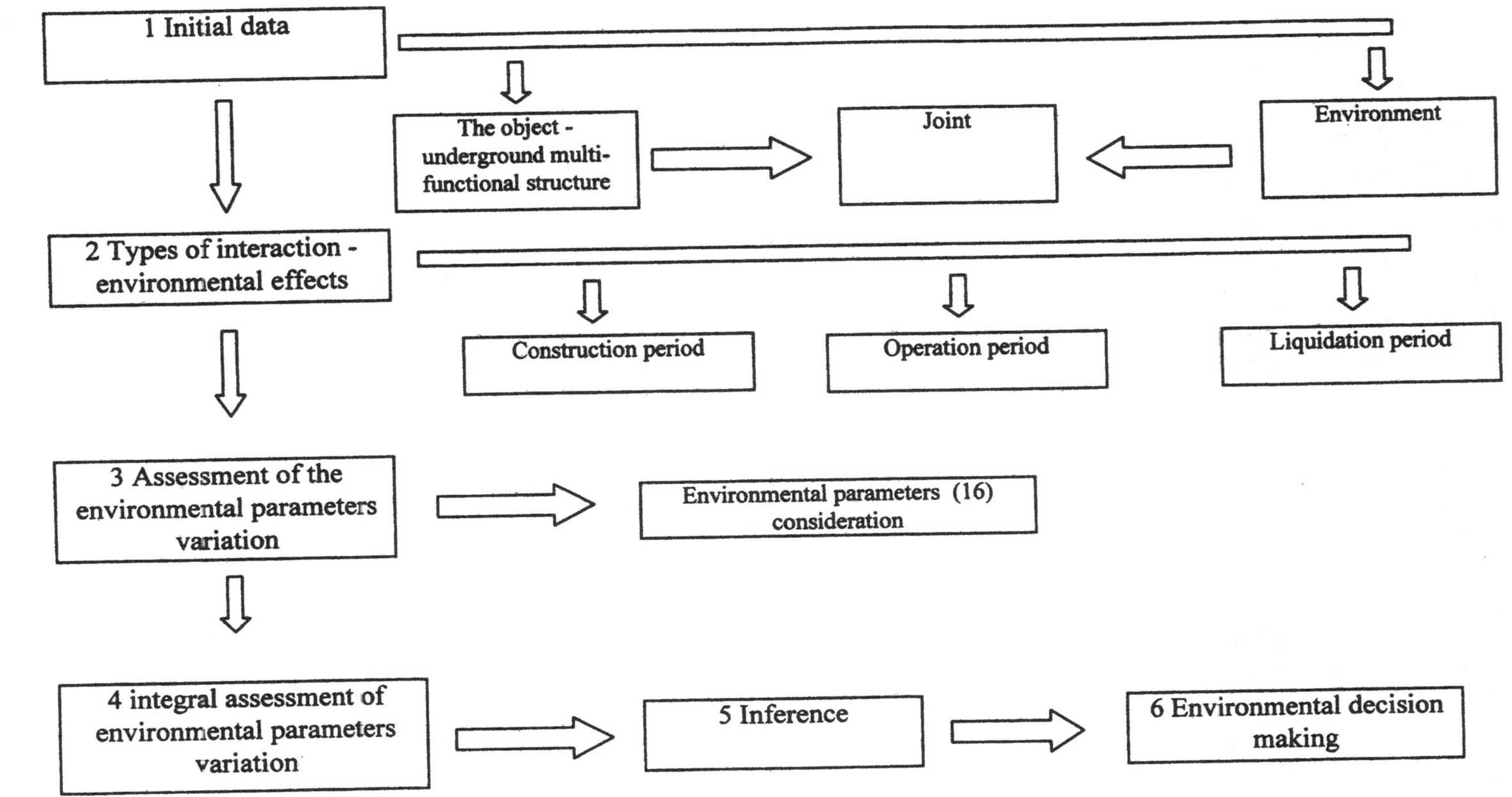

Figure 1. Environmental impact assessment of underground multi-finctional structure. The method scheme.

Leopold matrix), in certain cases the method proved more clear and precise results.

Further development of the method is based on multi-criteria analysis implementation improvement. In future the method could be used for EIA of other types of underground structures.

7 ACKNOWLEDGEMENTS

The authors would like to thank colleagues from St. Petersburg State Technical University (St. Petersburg, Russia) and Hamburg-Harburg Technical University (Hamburg, Germany) for help and inspiration.

REFERENCES

Bobylev N. 2001. EIA of underground motor-transport structures within historic city: St.Petersburg experience. *Proceedings, Fourth Symposium on Straight Crossings, Bergen, Norway, 2 – 5 September, 2001.*

Carmody, J. & Sterling, R. 1993. *Underground space design: a guide to subsurface utilisation and design for people in underground spaces.* New York: Van Nostrand Reinhold.

Iliichev V., Uchov S., Zarezkiji J. 1997. Construction experience of underground structure at Maneznaja square. *Gornij Vestnik #4, 1997.*

Yufin S. (ed.) 2000. *Geoecology and computers – Proceedings of the third international conference, Moscow, Russia,* 1-4 *February 2000.* Moscow.

Session 2, Track 3

Tunnel lining techniques

North American Tunneling 2002, Ozdemir (ed.)
© 2002 Swets & Zeitlinger, Lisse, ISBN 90 5809 376 X

Behavior of a single gasket system for a precast concrete segmental liner subject to high external hydrostatic pressure

Howard Lum, David Crouthamel & David Hopkins
Jacobs Associates, San Francisco, California

Ed Cording
University of Illinois, Urbana-Champaign, Illinois

John Shamma
Metropolitan Water District of Southern California, California

ABSTRACT: High external hydrostatic pressure is anticipated during the construction of the Inland Feeder Arrowhead Tunnels in Southern California. A bolted and gasketed precast concrete segmental liner will be used as the primary support for the tunnels. The performance of the gasket is critical for the ability of the segmental liner to meet stringent water inflow criteria under the maximum design groundwater head of 275 meters during construction. Since limited data is available for the performance of gaskets at pressures anticipated for this project, testing at the University of Illinois at Urbana-Champaign and Metropolitan Water District of Southern California (MWD) was conducted to study the performance and compatibility of various gaskets designs. The testing focuses upon the following issues: ability of proposed gaskets to seal effectively under the maximum hydrostatic conditions; failure mode of the gasket; correlation of contact pressure to sealing pressure and joint gap; compatibility of the gasket line loads with the segment joint design; and stress relaxation of the gasket after initial compression. Design recommendations and testing procedures specific for gaskets subject to high hydrostatic head will be presented, as well as how these findings were implemented for the Inland Feeder project.

1 INTRODUCTION

1.1 *Project description and purpose*

The Inland Feeder Project is part of MWD Capital Improvement Program to increase its water delivery capability to Southern California. The Inland Feeder spans approximately 71 km from the San Bernardino mountains above the city of San Bernardino to the San Jacinto Valley in Riverside County, consisting of 31 km of 3.7-m diameter tunnel and 40 km of 3.7-m diameter pipeline. The Arrowhead West Tunnel is the first leg of the project, consisting of a 6.4 km long tunnel that travels in a southeasterly direction through the San Bernardino Mountains beginning at the Devil Canyon Power Plant above California State University, San Bernardino, and ending at Waterman Canyon near Highway 18. After a 1.22 km pipeline segment under Waterman Canyon, the Arrowhead East Tunnel continues southeasterly for about 9.7 km reaching depths below the ground surface of more than 610 meters, exiting at City Creek Canyon. The finished project will connect the California and Colorado River Aqueducts in Southern California. A general project map is shown on Figure 1.

Construction began on the Arrowhead East Tunnel of the Inland Feeder in early 1997. After mining about 2.44 km of this tunnel from the City Creek Portal site, construction was halted due to elevated groundwater inflows. Both the Arrowhead West and East Tunnels have now been redesigned to incorporate a primary liner system that substantially reduces inflows. Construction is expected to resume in mid-2002.

Figure 1. Inland Feeder Project Location.

Groundwater along the tunnel alignments is located within the bedrock and flows primarily through the rock mass discontinuities (i.e., joints, faults, and shears). External groundwater heads can reach significant levels. Some faults along the project also act as groundwater barriers with up to 60 meters of differential groundwater head across the faults.

1.2 *General requirements*

The Arrowhead Tunnels of the Inland Feeder project will be subject to high external hydrostatic pressure during tunnel excavation. A bolted and gasketed precast concrete segmental lining system is specified as the primary support for the tunnels. In addition to the high water pressure, stringent external water inflow criteria during construction impose additional demand on the gasket system. After the primary support is in place, a permanent final lining will then be installed for water conveyance.

Based on the available exploration data, the maximum hydrostatic head during construction for the Arrowhead East Tunnel is 275 meter (2.70 MPa) and a maximum construction head for the Arrowhead West Tunnel of 168 meter (1.64 MPa). A single gasket is specified for the 330 mm (13-inch) thick concrete segment placed both at the circumferential and longitudinal joints. Because the gaskets are inherently stiff to resist the anticipated hydrostatic pressures, adding redundancy of the gaskets (e.g. double gaskets) would complicate the feasibility of adequately compressing the gaskets during erection.

The performance of the gasket in the concrete liner is one of the critical elements that determine whether the primary lining system will function satisfactorily. Limited research data are available on gasket performance under the anticipated hydrostatic pressures for the Arrowhead Tunnels. Specific testing was necessary to assure that a gasket could be developed to meet project requirements. Three gasket manufacturers volunteered to participate in the tests, namely Construction Polymers/Vertex Rubber (CP) of United States, VIP-Heinke of Great Britain, and Phoenix of Germany. Tests on furnished samples of their gaskets were conducted between December 2000 and March 2001. The gasket profile dimensions from the furnished samples are described in Table 1.

The following gasket characteristics were studied through testing to ensure that a gasket can be developed for the Inland Feeder project:

1 Load deformation characteristics of the gasket
2 Correlation of gasket contact pressure to sealing pressure and joint gap
3 Ability of the gasket to seal effectively under the maximum hydrostatic pressure

Table 1. Gasket Profiles.

Gasket Manufacturer	Groove Depth (mm)	Bottom Groove Width (mm)	Top Groove Width (mm)
Construction Polymers (CP)	12	35	40
Heinke (CS002B)	14	33	40
Phoenix	8.5	36	40.6

4 Development of the gasket line load on the concrete segment joint
5 Failure mechanism at the concrete joint due to high gasket line load

Tests were performed both at the University of Illinois at Urbana-Champaign and MWD La Verne Laboratory. Tests conducted at the University of Illinois will be summarized without detailed data analysis. Tests conducted at MWD La Verne Laboratory will be addressed in details in this paper.

The general hydrostatic design criteria of the gasket are to reliably withstand the anticipated hydrostatic heads over the period during tunnel excavation. After tunnel excavation is completed, an impervious steel or reinforced concrete cylinder pipe will be installed as the final lining. The gaskets are therefore required to retain a residual contact pressure capable of resisting the maximum hydrostatic heads discussed above during construction. In addition to withstanding the hydrostatic heads, the gaskets must exhibit compression loading characteristics that are compatible with the structural design of the segments. The gaskets must be fully compressed without extrusion from the segment groove and without overloading the groove itself, causing localized failure in the segment.

1.3 *Tests conducted at University of Illinois*

The University of Illinois performed load deformation, relaxation and water pressure tests on the CP gaskets. The purpose of these tests is to gain a better understanding of gasket performance under high water pressure. For the load deformation tests, steel rails were used to load a pair of gaskets at three different rates of loading. For the relaxation test, a constant gap at room temperature was maintained, and the applied load was measured over time. For the water pressure tests, the gaskets were tested in a steel "picture frame" device at a range of joint gaps that are anticipated during construction. Test results from the load deformation tests and water pressure sealing tests will be discussed in this paper.

1.4 *Tests conducted at MWD La Verne Laboratory*

Two different types of test were conducted at MWD La Verne Laboratory, namely the gasket sealing tests in a concrete T-section and the flaking analysis tests at the concrete joint.

From past experience, it has been shown that leakage through a gasketed joint can initiate at the intersection of perpendicular gaskets forming the T-joint. The concrete T-tests are intended to determine the feasibility of a gasketed lining system and to gather information on gasket behavior in concrete under high hydrostatic loading. The geometry of the tests was selected to reflect the dimensions proposed for the Arrowhead Tunnels, including the concrete edge distance and the T-intersection at the circumferential and longitudinal joints. The tests are intended to evaluate the general characteristics of gasket and not intended to provide a comprehensive research and analysis of gasket-concrete joint interaction

Because the initial gasket contact pressure is critical and can be high in order to seal at high pressures, full scale joint testing was developed to verify that the concrete section could withstand the gasket line loads produced by compressing two gaskets until the mating segment faces are in contact. Excessive line loads could potentially cause a shearing failure at the corner of the segment. Flaking tests were developed to determine the concrete edge strength against shear failure as the compressive line load increases as the gap between the concrete segments closes. Tests were performed relative to the maximum probable line load from the compression of two gaskets with no segment joint offset and with an offset of 13 mm to complete gap closure

1.5 *Purpose of the test*

The first step to evaluate the gasket performance is to determine its load-deformation curve. Factors such as gasket material stiffness, gasket/groove volume, and gasket/groove shape determine the performance of a gasket. Laboratory testing can be performed to obtain the load-deformation curve at various gap widths. To prevent leakage, the initial contact pressure of the gaskets between segments must be high enough so that the contact pressure exceeds the applied water pressure. A gasket confined in a gasket groove, when it is approaching full compression (very small joint gap), is operating on the steep portion of the load-deformation curve. As the gap is opened by a small amount, a rapid decrease in contact pressure will occur. This implies a limited range of sealing capacity when the gasket is operating in this steep portion of the curve. Ideally, the gasket should operate along the flat portion of the load-deformation curve to maintain sealing capacity at a wide variation of joint gap widths. In practice, the gasket material behavior and high water pressure requirement may not allow its performance be limited to the flat portion of the curve.

1.6 *Test setup and procedures*

Gasket pairs were installed in top and bottom steel plates with properly machined grooves. The ends of the grooves in the 457 mm (18-inch) long test fixtures were confined longitudinally with steel end plates. Tests were conducted on both polyisoprene (EPDM) gaskets and neoprene gasket with and without top plate offset. Plots of the applied load verses vertical displacement (joint gap) were generated.

1.7 *Results and conclusions*

For CP-EPDM (polyisoprene) gasket without offset, the load deformation curve changed from a flat slope to a steep slope at a gap width of 3.2 mm. The final gap was 1.6 mm with a maximum line load of 350 N/mm. The load-deformation curve becomes very steep as the holes in the gasket are closed and the total volume of the gasket approaches that of the material and the volume of the groove.

For CP-EPDM gasket with offset, the load-deformation curve is similar up to 6.9 mm. The maximum load at 1.6 mm gap was 232 N/mm, which is less than that of the case without offset. The difference is due to the total volume of the gap and groove is greater when the gasket is offset. Thus, for a given gap, the contact loads are lower for the offset gasket.

The CP neoprene gasket has the same geometry as the CP-EPDM gasket with similar load-deformation curve. The neoprene gasket has a higher line load than the EPDM gasket due to its increased stiffness. For comparison, at a gap width of 6.4 mm, the contact force for the neoprene gasket was 35 N/mm while the contact force for the EPDM was 27 N/mm.

A typical load-deformation curve is included in Figure 2. From the load-deformation curves, it is evident that the high water pressure requires the gasket to operate at the steep portion of the load-deformation curve. As a result, the gasket will only be efficient to seal against leakage for a limited range of joint gaps.

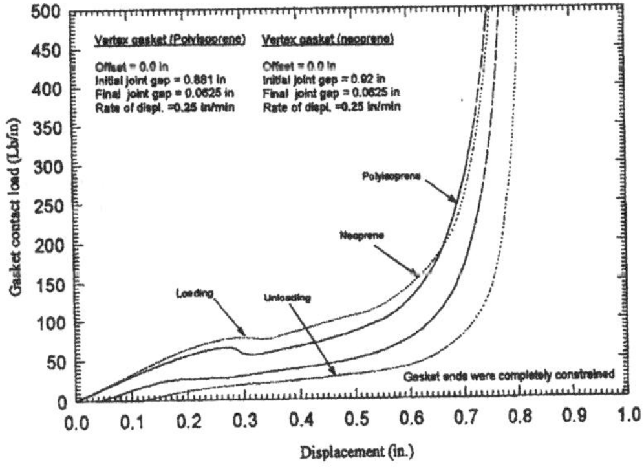

Figure 2. Typical Load vs. Deformation Curve.

2 SEALING TESTS IN STEEL FRAME

2.1 *Purpose*

Due to the difficulty to test the sealing pressure with concrete sections, a simplified approach using steel "picture frame" was performed at the University of Illinois. The gaskets were tested under the anticipated hydrostatic pressures with 13 mm offset and with no offset. The tests studied the behavior of gaskets compressed to an initial gap and contact pressure, the subsequent changes in gasket contact pressure as water pressure is increased, and the final gasket contact pressure and water pressure at leakage.

2.2 *Test setup and procedures*

The water pressure tests were performed by applying water pressure to the interior volume of two gaskets in a picture frame configuration placed in gasket grooves in steel plates. The picture frame dimensions of the gasket grooves were 102 mm by 457 mm (4-inch by 18-inch) measured inside to inside. The test setup is shown in Fig. 3

The total load on the gaskets was measured with the two load cells placed between the top steel plate with the gasket groove and the top plate with the threaded rods. Extrusion of the gasket into the gap was measured with dial gage with the plunger inserted horizontally into the gap. Change in the gap height was measured with two dial gages.

2.3 *Results and conclusions*

The CP-EPDM and CP Neoprene gaskets were capable to withstand water pressure up to 4.14 MPa (600 psi) without leakage. The sealing pressures did not change whether the gaskets were offset or not.

The gap width of 6.4 mm (0.25-inch) was the allowable limit for minimizing extrusion and leakage of the CP-EPDM gasket at high water pressure. It is preferred to have a smaller initial gap of 4.8 mm (3/16-inch) to ensure that the gaskets will perform satisfactorily. The extrusion of the neoprene gasket was smaller when compared to the EPDM gasket under the same loading.

Leakage develops between the gaskets when the water pressure approaches the gasket contact pressure. Measurement of contact loads during testing permits the determination of the relation among the initial gasket pressure, initial gap, and final water pressure at leakage and final gasket contact pressure. From the results of the pressure test and the gasket load-deformation curve, it is possible to obtain the sensitivity of the drop in water sealing pressure as a result of gap increase.

3 SEALING T-TESTS IN CONCRETE

3.1 *Purpose*

The following factors related to gasket performance are focused in the concrete T-tests:

1 Feasibility of manufacturing a gasket that can seal at high pressure based on the maximum specified gasket groove dimensions in the contract documents. Given a maximum pressure of 2.70 MPa, it was difficult to justify the sealing capacity of a gasket without testing.
2 The potential leakage pattern through the T-intersection of the gaskets (where the circumferential joint meets the longitudinal joint).
3 Installation procedures of the gasket in the gasket groove. It is important to understand the manufacturer's required procedures for installlation.

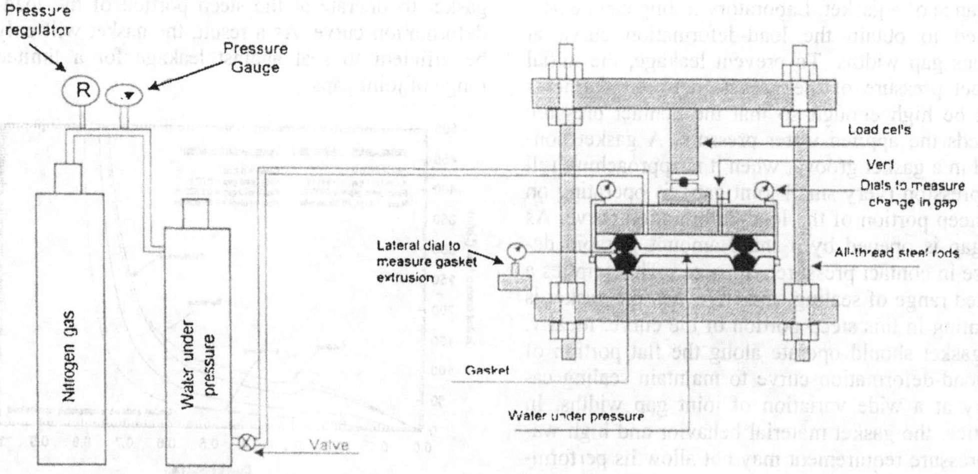

Figure 3. Setup for steel picture frame sealing test.

Surface preparation and curing procedures as required by the gasket manuacturer were followed in the testing laboratory.

4 The deformation characteristics of the various gaskets during compression at various pressures. Gap widths at the joint were measured and recorded at each stage to monitor the changes.
5 Leakage paths that could develop during pressurization. Potential seepage paths include leakage through the gaskets interface, leakage between the gasket and the concrete groove and leakage through the concrete section were studied.

3.2 *Test procedures*

The T-test configuration (Fig. 4 and Fig. 5) consists of one top concrete section and two bottom split-sections. The three-block design simulates a T-joint similar to the segment joints at the circumferential and longitudinal interfaces. Each block was cast with the exact groove dimensions for the respective gasket design. In addition, the design dimension (330 mm segment) from the gasket groove to the extrados of the segments was used in each test block. This configuration ensures that the edge condition between where the hydrostatic pressure is applied and the gasket was duplicated.

The entire concrete blocks assembly measured 508 mm by 508 mm (1'-8" by 1'-8") in plan dimensions with an overall height about 508 mm (1'-8"). A pressure chamber measured 203 mm by 203 mm by 102 mm (8" by 8" by 4" tall) was cast at the upper block. Two steel pipes of diameter 9.5 mm (3/8") each were embedded in the upper block for water inflow and outflow control. The concrete blocks were formed using steel formwork. The concrete strength was specified as 55 MPa (8,000 psi) per ASTM C39. The concrete mix design was identical to the one required in the Specifications for the segments. The constituents of the concrete mix included locally available aggregates that met design mix criteria specified for the actual concrete segments. The actual compressive strengths of the test blocks during testing were in the range of 59 MPa to 62 MPa (8,500 to 9,000 psi).

Reinforcement was not used in the concrete blocks based on the following reasons:

1 The dimensions of the blocks are small in comparison to the development length of reinforcement. This means that the strength of the reinforcement cannot be developed to be effective to resist any tensile stress. In addition, the limited space would make it difficult to achieve adequate consolidation of concrete if reinforcement is used.
2 Providing reinforcement will increase the ultimate strength of the section, but the limit condition for the concrete blocks to perform satisfactorily is the appearance of the first crack. Once the first crack (minor hairline fracture) is formed, a water seepage path will develop and testing cannot proceed. Reinforcement was judged to be of limited value to solve this problem.

Prior to the gasket shipment, a plaster mould of the concrete gasket groove was sent to each gasket manufacturer to alleviate any gasket fit up problems. The gasket manufacturers also provided MWD with

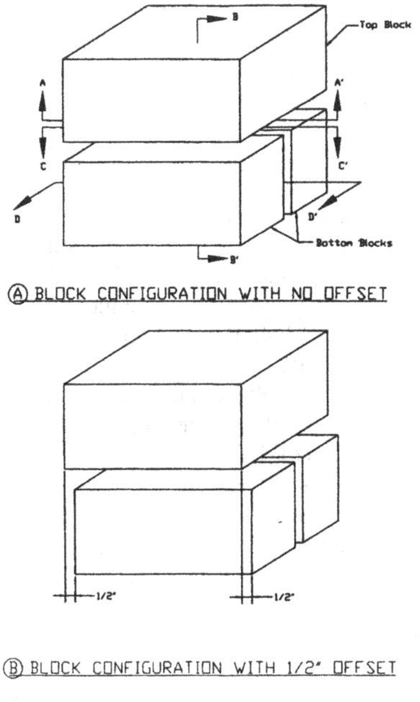

Ⓐ BLOCK CONFIGURATION WITH NO OFFSET

Ⓑ BLOCK CONFIGURATION WITH 1/2" OFFSET

Figure 4. Schematic drawing for layout.

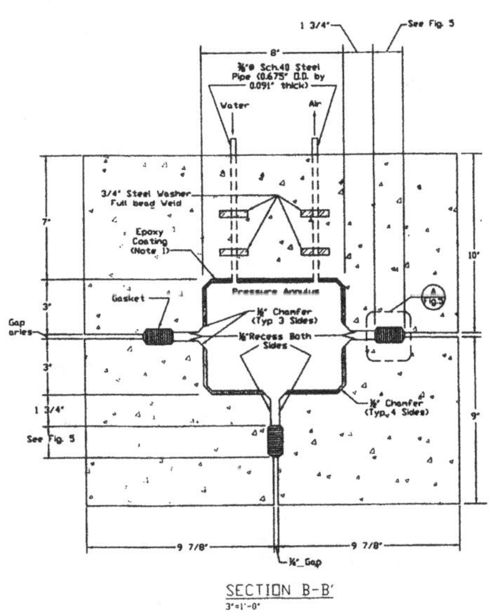

Figure 5. Typical cross section of concrete blocks.

their recommendations for the type of bonding agent between the gasket and concrete, and the adhesive application procedures.

To provide confinement of the blocks during testing and uniform compression of the gaskets, a confinement frame shown in Figure 6 consists of steel plates and bolts was constructed. The bolts and structural steel plates were used to compress the gasket and confine the concrete blocks in the direction of gasket compression. The purpose of the steel plate and the bolts are to limit deflection of the blocks, to uniformly compress the gaskets during closure of the gaps, and to maintain confinement of the concrete blocks during hydrostatic pressurization. Bolts were manually tightened, in sequence, using wrenches to compress the gasket to the specified gap width.

Attentions were given to assure uniform concrete bearing and torquing, but some of these challenges could not be overcome due to limitations of the confinement system using only bolts and plates with manual torquing sequence. The limitations of the test frame and the method of load application constituted to some of the premature concrete failure observed.

It is crucial to torque the bolts in a manner to result in uniformly compression of the gasket. Uniform compression of the gaskets should result in a more uniform contact pressures along the gasket grooves and in turn limit unbalanced loads in the test blocks. The procedures to torque the blocks were revised and improved during the tests but it was evident that nonuniformity in the compression and torquing remained. In addition, because the geometry of the test gaskets contains several right angles (horizontal and vertical planes), this results in a highly variable gasket line load. As a result, high stress concentration was developed at the corners for the stiff gasket. The inability to provide a uniform compressive force for the gasket accounts for most

Figure 6. Assembly of the concrete blocks.

of the premature failures occurred in the concrete blocks.

The assembly was set up with an initial gap at the gasket joint between concrete blocks. Bolts were tightened to achieve the prescribed gap width. The gap widths at different locations around the blocks were then monitored during pressurization. Leakage was checked by visual inspection and through pressure gauge readings at incremental pressure values. After some initial trial runs, an initial gap width of 3/16" was set for the sealing tests.

Once the blocks are torqued to the required gap, the concrete blocks would be pressurized through the internal chamber using a controlled pumping device. The water pressure was then gradually increased incrementally and pressures were held at specified intervals for gap widths to be measured. The gaps were also inspected for any potential leakage. Any signs of leakage through the gaskets were continuously monitored and recorded. Pressure was held constant for a minimum of 10 minutes before the next incremental pressure was applied. The goal was to determine maximum water pressure allowed before leakage occurred.

3.3 *Results and conclusions*

The Construction Polymers (CP) gasket was significantly stiffer than other gaskets tested, making it more difficult to compress the gasket uniformly using a manual torquing method. Because of the uneven distribution of the high line loads due to the various transitions (corners) of the gasket geometry applied in the test, high torsional moments were transmitted into the blocks causing concentrated stresses to exceed the concrete strength.

The pressure test on the CP gasket was limited by localized micro-cracking of the concrete blocks from the stresses mentioned above, but this gasket exhibited no leakage up to 3.0 MPa (430 psi). The gasket should withstand higher pressures, and other tests conducted at University of Illinois had verified their sealing capacity at higher head. In testing of the less stiff gaskets, the concrete block assembly remained intact and the gaskets from Heinke and Phoenix developed a leakage path along the gasket to gasket interface at many locations for pressures between 1.5 MPa and 2.0 MPa (222 psi and 290 psi) respectively. In all cases, the leakage path did not develop at the T-intersection of the gasket joint. Table 2 summarizes the results of the tests.

The observed difficulties of the test assembly included the followings:

1 Limited capability in compressing the gasket uniformly using only bolts and nuts. Torquing sequence became very crucial to determine success.
2 The line load (restoring force) of the gasket can be very high for gasket with high sealing capacity (e.g. CP gasket). It becomes a practical limit to

Table 2. Results for Sealing Tests.

Gasket Manufacturer	Offset	Maximum Pressure Tested		Remarks
Construction Polymers	0 mm	3.0 MPa	430 psi	Top concrete block failed.
Construction Polymers	0 mm	2.1 MPa	300 psi	Initial bolts torquing caused hairline crack in bottom concrete block
Phoenix	0 mm	1.5 MPa	222 psi	Leakage through gaskets
Heinke (CS002B)	0 mm	2.0 MPa	290 psi	Leakage through gaskets

compress the gasket uniformly in a laboratory environment using only manual torquing tools.

3 The size of the test assembly was limited so several 90-degree corners were introduced to the test blocks. The 90-degree corners exhibited extra stiffness and caused unbalanced forces for the concrete block sections.

4 Full contact bearing between the concrete blocks and the load frame was difficult to maintain. Stiff steel sections were used for confinement, but it approached the practical limit for such a test frame given the high water pressure and gasket line loads experienced.

Even the tests were not entirely successful to determine the maximum pressure that the CP gasket can withstand; several important conclusions and recommendations can be drawn:

1 Gasket groove dimensions were shown on the contract drawings. Based on the tests performed, it is concluded that providing a gasket based on the specified gasket groove dimensions can seal against the high water pressure.

2 The CP gasket will most likely withstand the given high pressure. Prior to the concrete failure at 3.0 MPa pressure, no signs of leakage were observed. Other tests conducted at the University of Illinois verified that CP gasket would seal at higher pressure using steel test frames.

3 Given the maximum design pressure of 2.70 MPa required for the Arrowhead tunnels, two out of three gasket manufacturers did not provide a gasket that could withstand this pressure. These two gaskets were off-the-shelf items and not originally designed for the high head applications required at the Arrowhead tunnels. For the actual tunnel construction, special design considerations for high pressure must be considered in gasket selection and design.

4 For the two gaskets (Heinke and Phoenix) where the gasket failed due to a continuous leakage path, the leakage did not occur at the T-intersection of the gasket joint. Due to the stiffness of the gaskets at the T-joint, high contact pressure was achieved and it was unlikely that initial leakage will occur at a properly installed T-joint.

5 The gasket that is most likely to seal against high pressure will be composed of stiff material with a high profile (CP gasket), since the other two gaskets (Heinke and Phoenix) with lower capacities were made of softer (less stiff) material with a lower profile.

6 The stiff CP gasket used in the tests demonstrated that precautions should be used to ensure that the gasket could be properly installed and compressed. The difficulty in applying manual torquing methods in the laboratory proves that the contractor should apply special measures to properly compress the stiff gasket with mechanical means during erection.

7 It was difficult to perform T-intersection testing using concrete sections, as demonstrated by the premature concrete failure of the test blocks. This failure mechanism, however, is unique to the test setup and will not apply to the actual tunnel construction. In the actual tunnel segment construction, the joints will be reinforced and plywood packing will be used for load distribution. Furthermore, the design of the segment ring is such that load sharing and re-distribution will occur in the actual tunnel construction. In comparison, the concrete test was very vulnerable to minor cracking because of the use of a non-redundant load frame in a laboratory situation.

8 Due to the difficulty to perform a concrete T-test and the variable results obtained, such a test in concrete should not be required as an acceptance criterion of performance. Other tests such as T-tests in steel sections and flaking tests should be required to verify the sealing capacity and line load of the selected gaskets.

4 FLAKING TESTS

4.1 *Purpose*

Due to requirement to center the packing between the concrete segments, the gasket cannot be centered at the joint. Flaking tests on the concrete section were performed to investigate the concrete edge shear capacity relative to the maximum probable line load from the compression of two gaskets in both the no offset and offset positions. It has been shown that the initial gasket contact pressure is critical and can be very high in order to seal at high hydrostatic pressures. Therefore, full scale joint testing was developed to verify that the concrete section could withstand the gasket line loads produced. Excessive gasket line loads could potentially cause a premature shearing failure at the corner of the segment face, rupturing the gasket groove. Since the failure mechanism is localized at the corner, the reinforcing steel in the segment body is not effective to resist this type of failure. An unreinforced concrete block was used for testing to verify the concrete shear strength against the maximum line load developed.

4.2 *Test procedures*

Two materials were used to manufacture the gaskets: Polyisoprene/EPDM and Neoprene. Three manufacturers submitted the following four gaskets for testing: Phoenix (EPDM), VIP Heinke (EPDM), Vertex-Construction Polymers (Polyisoprene) and Vertex-Construction Polymers (Neoprene).

Each gasket manufacturer supplied MWD with their respective gasket specimens in rolls. The gaskets were cut into 12-inch straight lengths to fit the test blocks constructed at MWD La Verne Laboratory.

The 55 MPa (8,000 psi) concrete mix for the test blocks was identical to the design mix required by the Contract Specifications. The concrete test blocks were steam cured for a minimum of seven days in a moist environment. A minimum of three compressive tests and three tensile tests were performed at 28 days after casting the test blocks and cylinders. The testing requirements stated that if both test results exceeded the required design strengths, the concrete blocks could be used for testing at that time. If any of the cylinders failed in either tension or compression below the required strengths, additional cylinders were to be tested later until the required strengths were achieved. For the tests performed at the La Verne Laboratory, the cylinder compressive strengths ranged between 55 MPa to 62 MPa at 28 days.

Both 305 mm (12-inch) and 330 mm (13-inch) segment joint configurations were tested. The 330 mm segment adds 13 mm (½-inch) to the distance from the edge of the gasket groove to the outside of the block. The additional distance increases the localized shear capacity for the segments. Figure 7 shows a sketch of the configuration of the blocks without offset.

4.2.1 *Offset of top block from the bottom block*

During construction of the segmental liner, it is possible that the two gaskets do not line up at a joint. The Contract specified a maximum offset of 13 mm (1/2-inch) during segment installation, so the effects of this offset would be required. Gaskets were compressed using zero offset and an offset of 13 mm normal to the running length of the gasket. Figure 7 shows a sketch of the configuration of the two blocks in the zero offset position. To obtain the offset, the top block was shifted by 13 mm relative to the bottom block in one direction.

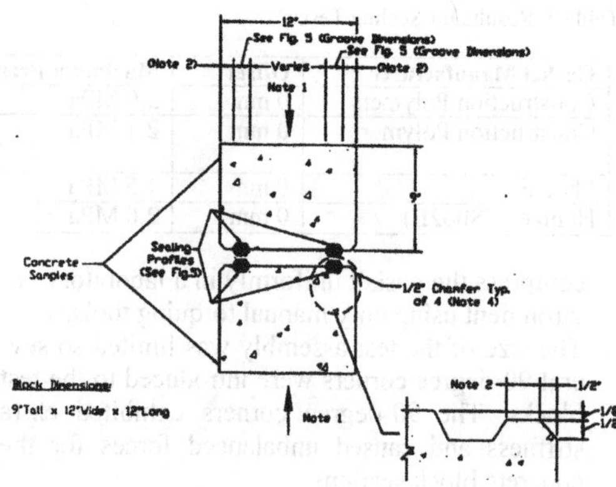

Figure 7. Flaking Test Setup.

4.2.2 *Lateral restraint*

Since plywood packing will be used at the joint along one side of the gasket, the effects of the packing material was studied. A steel shim plate (1.6 mm thick) was placed at the intrados edge of the gasket groove to simulate the plywood packing restraining the gasket from extruding during compression. This test was performed using both 305 mm and 330 mm blocks with the gasket mating surfaces offset by 13 mm.

4.2.3 *Longitudinal ends confinement*

Load-deflection tests conducted at the University of Illinois resulted in stiffness values on the order of two times the values obtained from the tests performed at La Verne using the stiffest gasket. The increase in stiffness was attributed to the fact that the University of Illinois test fixture included steel plates at the ends of the gasket grooves to confine the gaskets longitudinally, similar to situations in the field. Steel plates were added to the ends of the concrete blocks used in previous tests as shown in Figure 8. Using this modified test fixture with the CP Neoprene gasket, tests were run to investigate the effects of longitudinal confinement and whether the segments could handle the increased line load.

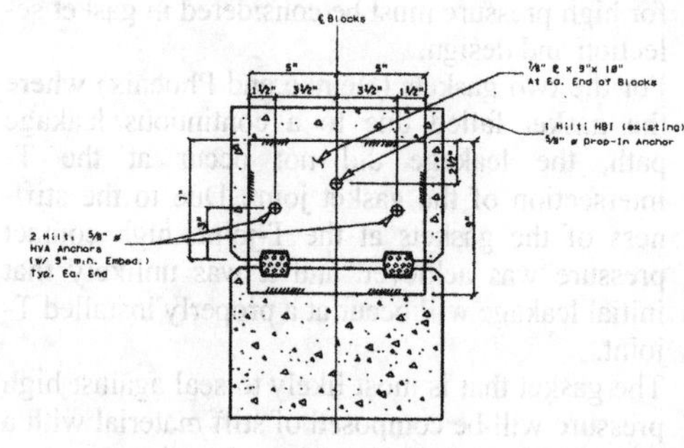

Figure 8. Confinement End Plate Detail.

Figure 9. Steel rails test for gaskets.

4.2.4 *Steel rail test*

As a final test on each gasket, a 330 mm concrete block was arranged with the gasket faces aligned on a steel rail and compressed until failure. The load was applied at a constant rate through a steel bar towards the gasket until concrete failure occurred. The load at failure was recorded, and the orientation and dimensions of the concrete failure plane were measured and recorded. Figure 9 shows a concrete block placed on a set of steel rails for testing.

4.2.5 *Summary of tests*

For each set of gaskets, a total of seven tests were performed using the setup shown in Table 3.

Table 3. Types of test for flaking analysis.

Segment Thickness (mm)	Offset (mm)	Remarks
305	0	
305	13	
305	13	Lateral Restraint
330	0	
330	13	
330	13	Lateral Restraint
330	0	Longitudinal End Confinement
330	0	Steel Rail Test

4.3 *Results and conclusions*

Table 4 summarizes the results of these tests by showing the average line load per unit gasket length at a joint gap of 3.2 mm (1/8-inch). The stiffness of Phoenix and Heinke gaskets is about 20% of the stiffness of Construction Polymer gaskets. Neither the Heinke nor the Phoenix gasket line loads resulted in any distress or shear failure of the concrete blocks. However, the relatively low stiffness exhibited by these gaskets did indicate that the sealing capacity would be less than that of the Construction Polymer gaskets. Confining the ends of the gaskets increased the inherent stiffness of the gasket material. Since the gaskets will be continuous in the tunnel, it is prudent to consider the stiffness exhibited by the Construction Polymer gaskets with longitudinal confinement as the most critical load case to develop the ultimate strength of the concrete flaking capacity.

Table 4. Summary of Load-Deflection Results.

	Average Line Load at 3.2 mm gap (N/mm)			
	Phoenix	Heinke	CP-EPDM	CP-Neop
No Offset	29.4	20.1	94.2	118.7
13 mm Offset	24.2	17.5	74.4	89.3
13 mm Offset with Lateral Restraint	21.9	16.6	73.5	84.9
No Offset with Longitudinal Confinement	N/A	N/A	N/A	214.4

CP-EPDM: Construction Polymer Polyisoprene /EPDM
CP-Neop: Construction Polymer Neoprene

Figure 10 shows a typical shear failure at the corner of the concrete segment due to gasket line load. Failures occurred for the CP-EPDM gasket at the 305 mm concrete block with no longitudinal confinement and for the CP Neoprene gasket at the 330 mm block with longitudinal confinement.

For the 305 mm thick concrete segment, both the CP EPDM gasket and the CP Neoprene gasket caused a shear failure at a line load of 165 N/mm. At failure, the gap width was 1.8 mm for the CP-EPDM gasket and 2.5 mm for the CP Neoprene. These results confirmed that the CP Neoprene gasket was stiffer than the EPDM gasket. The Neoprene gasket was designed to be stiffer in order to improve its sealing capacity.

For tests conducted on the 330 mm blocks without end confinement plates, none of the gaskets

Figure 10. Flaking (shearing) failure of concrete.

caused a shear failure. The increase in edge distance (distance between gasket groove and concrete edge) provides additional shear strength against flaking. At a maximum line load of 179 N/mm for CP Neoprene gasket with full closure, no signs of cracking or failure were observed.

For tests performed on CP Neoprene gasket with end confinement plates and 330 mm blocks, shear failure occurred at a gap width of 2.5 mm. The results of applied loads and gap widths for the two gaskets are summarized in Table 5.

Table 5. Line Loads with Longitudinal End Plates Confinement.

Gap (mm)	Line Load Test 1 (N/mm)	Line Load Test 2 (N/mm)
6.4	39	42
4.8	81	78
3.2	218	211
2.5	251*	250*

* Concrete failure occurred at these loads

5 CONCLUSIONS

This study has led to the following conclusions:

1 The gasket must have an adequate initial stiffness and confinement to withstand the applied hydrostatic pressures. This can be developed by proper selection of the gasket material stiffness, height, geometry and groove depth. If properly confined and with adequate initial stiffness, the gasket contact pressure is a function of both the initial compression and the water pressure applied laterally to the gasket.

2 The line load exerted by the gasket on the concrete segment must not cause a localized structural failure. A shearing failure mode was verified through testing, and suitable laboratory testing procedures were established as gasket selection criteria. This failure can be prevented by proper selection of the gasket material properties and geometry to prevent the gasket from transitioning to a very stiff mass and limit the maximum contact pressure to an acceptable level as the gap approaches full closure.

3 The steel picture frame provides a positive method to measure contact pressure, water pressure and gap width. The mode of failure was leakage through the interface between the gaskets. The concrete T-tests confirmed that seepage through the concrete-gasket interface should not govern the sealing capacity, and leakage through the T-intersection should not control the design because of the inherent gasket stiffness at these transitions.

4 T-tests performed in steel for subsequent tests should yield the control exhibited by the steel picture frame tests and confirm the sealing capacity at the T-intersection.

REFERENCES

ACI 318 and 318R, Building Code Requirements for Structural Concrete and Commentary. 1995. American Concrete Institute (ACI).

Cording, E., Paul, S., & Shalabi, F. 2001. Performance of Concrete Segment Gaskets for the Inland Feeder Tunnels. *Jacobs Associates Report for Inland Feeder. 2001.*

Hansmire, W. 1984. Example Analysis for Circular Tunnel Lining. *Tunneling in Soil and Rock, ASCE.*

Crouthamel, C., Swartz, S., Curtis, J., Warren, S., & Shamma, J. 2001. Evaluation, Monitoring and Design Concepts for Pre-Cast Concrete Segmental Primary Support Under High Hydrostatic Pressures. *American Rock Mechanics Association Conference, 2001.*

Swartz, S., Lum, H., McRae, M., Curtis, J., & Shamma, J. 2002. Structural Design and Testing of a Bolted and Gasketed Pre-cast Concrete Segmental Lining for High External Hydrostatic Pressure. *North American Tunneling Conference, 2002.*

North American Tunneling 2002, Ozdemir (ed.)
© 2002 Swets & Zeitlinger, Lisse, ISBN 90 5809 376 X

Structural design and testing of a bolted and gasketed pre-cast concrete segmental lining for high external hydrostatic pressure

S. Swartz, H. Lum & M. McRae
Jacobs Associates, San Francisco, California

D.J. Curtis
Mott-MacDonald, London, England

J. Shamma
Metropolitan Water District, Los Angeles, California

ABSTRACT: Restrictions on water inflows into the tunnel excavations for the Inland Feeder Arrowhead Tunnels require the utilization of a bolted and gasketed pre-cast concrete segmental primary lining system. The bolted gasketed segments have been designed for a maximum expected groundwater head of 275 meters, in addition to other applicable loads. Each of the segment's critical components, including the joints, segment body, joint edges at the gasket groove, and load-transferring packer, are designed to assure segment performance under the design loading conditions. This paper discusses the load conditions anticipated for the tunnels, provides an overview of similar projects, discusses the design procedure for the segments, summarizes the design verification testing program, and discusses other important issues that impact performance.

1 INTRODUCTION

1.1 *Project Description, Purpose, and Geology*

The Inland Feeder Project spans approximately 71 km from the San Bernardino mountains above the city of San Bernardino to the San Jacinto Valley in Riverside County, consisting of 31 km of 3.7-m diameter tunnel and 40 km of 3.7-m diameter pipeline. The **Arrowhead West Tunnel** is the first leg of the project, consisting of a 6.4 km long tunnel that travels in a southeasterly direction through the San Bernardino Mountains beginning at the Devil Canyon Power Plant above California State University, San Bernardino, and ending at Waterman Canyon near Highway 18. After a 1220 m pipeline segment under Waterman Canyon, the **Arrowhead East Tunnel** continues southeasterly for about 9.7 km reaching depths below the ground surface of more than 610 m, exiting at City Creek Canyon. The finished project will connect the California and Colorado River Aqueducts in Southern California. A general project map is shown on Figure 1.

Construction began on the Arrowhead East Tunnel of the Inland Feeder in early 1997. After mining about 2440 m of this tunnel from the City Creek Portal site, construction was halted due to elevated groundwater inflows. Both the Arrowhead West and East Tunnels have now been redesigned to incorporate a primary liner system that substantially reduces inflows. Construction is expected to resume in mid-2002.

The Arrowhead Tunnels are located in the southern foothills of the east-west trending San Bernardino Mountains, uplifted in Quaternary time along and north of the San Andreas fault. The southern portion of the range rises abruptly from the San Bernardino valley floor and is characterized by steep rugged slopes, sharp ridges and deeply incised canyons. The mountain range is underlain by crystalline igneous and metamorphic rocks. Metamorphic rocks are primarily gneiss, and also include calc-silicate gneiss and calcareous and dolomitic marble in thin to thick layers. The granitic rocks consist primarily of quartz monzonite, with lesser amounts of quartz diorite and granodiorite. Both the igneous and metamorphic rocks are generally relatively fresh to slightly altered at tunnel depth except for portal areas, around fault zones, and within about 100 meters of the ground surface.

The largest and most significant fault in the project area is the active trace of the northwest trending San Andreas fault, which is very linear across the mountain front and defines the southern boundary of the San Bernardino Mountains. Other notable faults crossing the project area trend east-west, and appear to merge with the San Andreas fault system at their western terminations.

Groundwater along the tunnel alignments is located within the bedrock and flows primarily through the rock mass discontinuities (i.e., joints, faults, and shears). External groundwater heads can reach significant levels, as discussed in more detail below. Some faults along the project also act as groundwater barriers with up to 60 meters of differential groundwater head across the faults.

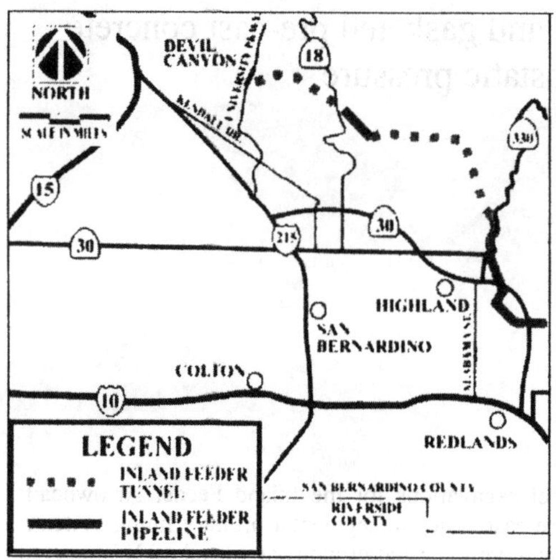

Figure 1. Project Location.

1.2 *General Requirements*

The final lining of the Inland Feeder will have an inside diameter of 3.7-m and will consist of steel pipe or reinforced concrete cylinder pipe. The anticipated inside diameter of the concrete segmental primary lining is 4.9-m to provide the necessary clearance for the installation of the final lining.

Given the required water inflow restrictions of the segmental primary lining, high hydrostatic pressures will develop on the primary lining after it is installed and sealed. The maximum hydrostatic heads are 275-m for the Arrowhead East Tunnel, and 175-m for the Arrowhead West Tunnel. In addition to the hydrostatic loads, it is anticipated the segmental lining will be subject to ground loads and grout pressures during the initial stages of construction.

1.3 *Historical Perspective of Design Approach*

Three methods adopted in the past to control water inflows into deep tunnels where precast concrete segmental linings have been the primary lining of choice are:

1 The use of bolted and gasketed segments. One example of this method is the French designed and built part of the Channel Tunnel.
2 The use of hexagonal segments with very detailed and onerous grouting requirements. An example of this method is the Evinos-Mornos Tunnel in Greece.
3 The use of grouting to limit inflows to the extent possible, and with ungasketed segments. One example of this method is the first 2440 m of the Arrowhead East Tunnel.

The Evinos-Mornos Tunnel, completed in 1994, had a lining which was designed to accommodate a maximum hydrostatic pressure of 20 bars, this being the highest pressure for which grouting and sensible lining design were presumably considered practicable. For this method to work in conditions where the potential hydrostatic pressure is greater than 20 bars, as at Evinos-Mornos, the ground surrounding the tunnel must be made effectively impermeable by grouting to a distance large enough to ensure that hydrostatic pressure upon the lining cannot exceed the design value. The Evinos-Mornos Tunnel was the first, or one of the first, tunnels constructed by this method. This tunnel was built with up to 800 meters of ground cover and at least 70 meters of perched water head. Many difficulties were encountered during construction, but water inflows were not reported ever to have been a difficulty.

The initial portion of the Arrowhead East Tunnel, built between 1997 and 1999, utilized a segmental ring without gaskets, and a combination of pre-excavation and post-excavation grouting to reduce the permeability of the surrounding rock. However, it was found that the nature of the rock in certain areas made grouting to the extent necessary to limit inflows below allowable limits very difficult.

Based on past difficulties with grouting in general, it was felt that implementing the second or third methods listed above would prove to be unreliable in controlling groundwater inflows.

Table 1. Projects Utilizing Gasketed Segmental Linings.

Project	Year	Hydrostatic Head m	Internal Diameter m	Lining Thickness m
South Bay Ocean Outfall, San Diego	1996-1999	55	3.30	0.23
Storebælt Eastern Tunnel, Denmark	1988-1994	80	7.71	0.40
Channel Tunnel, (French Side)	1987-1993	110	7.62 (x2) 4.79	0.40 0.32
St. Clair River Tunnel, USA/Canada	Through 1995	35	8.41	0.40
El Salam Siphon, Egypt	1993-1996	45	5.73	0.30
Lesotho Delivery Tunnel North	1992-1995	100	4.57	0.25
Hastings Tunnel, UK	1997-2000	65	6.49	0.42

The French part of the Channel Tunnel, completed in 1993, employed technologies dating back several decades. The tunnel is 120 m below sea level and 70 m below seabed level. This project comes closest to the Arrowhead tunnels in terms of loading conditions, and the design and testing of radial joints to carry the very large circumferential (hoop) stresses. The segmental concrete lining was designed to carry the full hydrostatic load plus a substantial proportion of the rock load. Table 1 summarizes a number of projects completed within the last decade that utilized gasketed segmental linings.

2 DESIGN REQUIREMENTS

2.1 *Design Load Cases*

Two primary load cases are considered for the structural design of the primary lining. The first is the initial loading conditions on the lining, soon after installation, before the segments are fully sealed. These loads include ground loads, equivalent to a *non-buoyant* rock height of 14-m, and backfill grout loads behind the segments. These combined loads produce a low ring thrust-moderate moment condition.

The second load case is for conditions after the segments are fully sealed. These loads include ground loads, equivalent to a *buoyant* rock height of 14-m, and a maximum hydrostatic head of 275-m. Ground loads are considered buoyant due to the magnitude of the expected hydrostatic pressure, and grout loads are neglected, as it is assumed that the grout would have set, and subsequent slight movements of the lining would relieve the initial stresses caused by the grout loads. The combination of hydrostatic and ground loading produce a high ring thrust-low moment condition in the segment. This load case also produces localized high line loads along the gasket groove associated with the stiff gasket necessary to seal the segments.

2.2 *Components of Structural Design*

The structural design of the segments involved the assessment of three modes of failure. The first mode considered the design capacity of the longitudinal joints between segments. The longitudinal joints must be designed to withstand the high ring thrusts associated with the second load case discussed above. These joints have often been found to be the critical component for design of segments under similar load conditions. The primary failure mechanism is a tensile splitting failure, often described as a bursting failure, slightly below the bearing surface of the joint. In addition, the performance and behavior of the packer installed between the segments is critical to transfer and distribute the load across the bearing surface of the longitudinal joint.

The second mode considers the behavior of the segment body. Both load cases discussed above are accounted for in designing the body of the segment. The expected failure mechanisms are similar to wall panels exposed to combined bending and axial loading, either a yielding of the panel in bending (for high-moment, low axial thrust loads), or a compressive failure (for high axial thrust loads).

The third mode considers the capacity of the joint at the longitudinal gasket groove to withstand the high line loads generated at the gasket, necessary to seal the segments against the external hydrostatic pressures. The primary expected failure mode is shearing of the corner block of the segment, propagating from the gasket groove to the outside surface (extrados) of the segment. The shear plane is likely to occur just outside of any effective reinforcement.

2.3 *Load Fact ors and Factors of Safety*

A load factor of 1.7 is used throughout the design for ground loads and grout loads. This load factor accounts for some uncertainty in the magnitude of the loads, and follows the recommendations of ACI 318-95 for live loads. For hydrostatic pressures on the lining, substantial work was done to conservatively determine the maximum anticipated hydrostatic heads for the tunnels. Therefore, a load factor of 1.0 for hydrostatic pressure is considered appropriate to eliminate design redundancy.

For factors of safety on materials, or strength reduction factors per Load Factor and Reduction Design (LFRD) criteria, the structural component under consideration dictates the values used. For the longitudinal joints, it is possible to assign two separate strength reduction factors for the two materials contributing to the strength of the joint. A reduction factor of 0.7 was used on the concrete strength and a reduction factor of 0.95 was used on the strength of the steel. The tensile splitting, expected at the joints, is more reliably resisted by steel than concrete, which has well-documented behavior in tension. These factors have been used on previous projects with success, including the Channel Tunnel, the Storebælt Railway Tunnels, and the St. Clair River Tunnel. Note however that use of these lower strength reduction factors requires that the segments be factory produced high quality, well-cured segments that meet tight tolerances.

For the second mode of failure considering the body of the segment, a ϕ factor (strength reduction factor applied for the composite strength of the reinforced concrete) of 0.7 was used, as recommended in ACI 318-95 for permanent concrete design.

For the third component, factors of safety were based on full scale testing using different gasket materials to ensure compatibility between the gasket and concrete. Depending on the stiffness of the gasket material and the gap width (level of compression), factors of safety varied from 1.2 to 2.1.

2.4 *Other Design Criteria Issues*

The Contractor shall be required to submit proposed methodology for handling, transporting, and storing of the segments, as well as loads associated with the installation of the segments and the loads imposed by the TBM. The design of the segments will then be augmented as necessary to assure the segments can resist the expected loads during construction based on the criteria discussed in Section 2.2.

The segmental lining was not designed specifically for seismic loading conditions because the primary lining is essentially a temporary structure. In addition, it has been found that similar linings in rock and soil tunnels have typically performed well in earthquake events, with only minimal cracking and other minor damage.

Concrete cover is designed per ACI 318-95 code for permanent structures to assure adequate durability.

3 DESIGN METHODOLOGY

3.1 *Longitudinal Joints*

The design of the longitudinal joints is based primarily on an empirical procedure, in conjunction with theoretical and physical testing validation. This methodology has been used successfully on a number of projects including the Channel Tunnel. Comparable procedures include those for the design of prestressed concrete end blocks (British Standard (BS) 8110, for example). The primary difference is that the procedure used herein allows for the benefit of the tensile strength of the concrete. The design is based on a contact width of approximately 0.45 times the thickness of the segment. The basis of this procedure is to calculate the ultimate capacity of the joint against two types of failure mechanisms, tensile splitting (ultimate capacity designated N_{uc}) and low-angle shearing (ultimate capacity designated N_{us}). Several factors dictate the ultimate capacity of the segmental lining, including the thickness of the segment (designated h), the tensile strength of the concrete (designated f_t), the width of the segment (designated b), and the steel capacity that is capable of resisting the expected modes of failure (designated either F_{st} for tensile splitting, or F_{sts} for low-angle shearing, and is equal to the amount of applicable steel to resist failure multiplied by the allowable stress in the steel). The ultimate capacities calculated can then be reduced by the partial strength reduction factors discussed above to arrive at acceptable design capacities.

The equations used for this design procedure are:

$$N_{uc} = 4.45 \cdot f_t \cdot h \cdot b + 4.0 \cdot F_{st} \quad (1)$$

$$N_{us} = 6.75 \cdot f_t \cdot h \cdot b + 2.0 \cdot F_{sts} \quad (2)$$

Equation 1 is illustrated graphically on Figure 2, a plot of the ultimate splitting capacity normalized by the quantity (A_c*f_t) on the y-axis, and the steel reinforcement capacity normalized by the same quantity, called the reinforcement ratio, on the x-axis. The resultant plot of the target load is a straight line with a y-intercept at 4.45 and a slope of 4.0, as shown on Figure 2. Based on past large-scale tests, Equation 1 has been found to be an a lower-bound to ultimate capacities for segments within the reinforcement ratio range of 0.375 to 1.2. Below 0.375, behavior becomes more erratic, as the joint behavior can be more susceptible to load eccentricities. Above 1.2, the increase in strength of the segments does not necessarily increase with additional reinforcement. Also, the testing used to develop this procedure typically maintained reinforcement ratios within this range. Therefore, the design should use a reinforcement ratio between 0.375 to 1.2.

For tensile splitting, effective reinforcement has to be through-thickness ties that are adequately anchored at each side of the joint, and are located within a distance of about 85% of the thickness of the segment from the face of the joint. Only reinforcement on one side of the joint is used in Equation 1, as tensile-splitting can occur at one face only.

For low-angle shearing, effective reinforcement has to comprise similar ties as for tensile splitting reinforcement, but the ties must also be anchored on each side of the expected failure plane in order to be effective. For Equation 2, the total amount of steel intersecting the expected failure plane is counted, and thus steel on both sides of the joint can resist low-angle shearing. The expected low-angle shear failure plane has been typically observed at about 26 to 27°. Dependent on the amount of eccentricity in the thrust across the joint, the capacity of the reinforcement against low-angle shearing (F_{sts}) can be up to about twice the value of the capacity of the re-

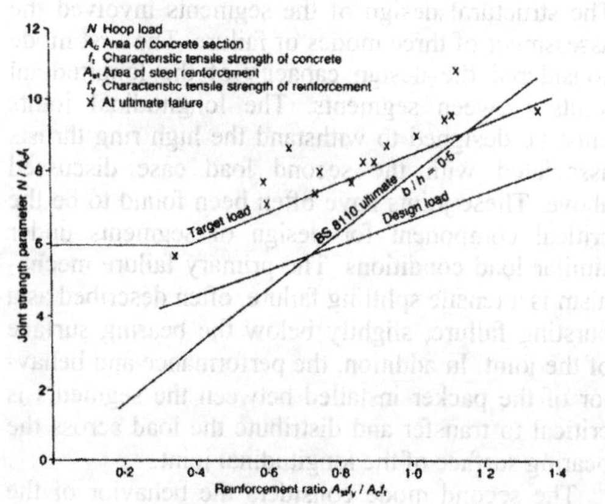

Figure 2. Normalized Splitting Capacity versus Reinforcement Ratio (Based on Equation 1).

inforcement against tensile splitting (F_{st}). Therefore, based on Equations 1 and 2, all else being equal, tensile splitting is typically the more critical mode of failure for the joints. This conclusion has been well supported by past large-scale testing, and is also supported by the tests run for this project.

3.2 *Segment Body*

The design of the body of the segments is based on wall-panel design for combined bending moments and axial loads. The procedure for design is outlined in the ACI-318-95 code, and is graphically represented by a moment-thrust interaction diagram (Figure 3). The capacity of the segment is a function of the thickness of the segment, compressive concrete strength, amount of both tension and compression reinforcement, and location of reinforcement within the segment body. The combinations of moments and thrusts that fall within the envelope are below the capacity of the segment whereas those combinations that fall outside the envelope exceed the capacity of the segment.

3.3 *Edge of Joint at Gasket*

In order to seal against external hydrostatic pressure, the gasket develops significant closure forces which could cause the edge of the segment to shear off. These forces are dependent on the stiffness of the gasket material, the geometry of the gasket groove (width and depth), the amount of confinement (both longitudinal and radial), and the amount of closure of the gap between the joint faces, limited by the stiffness of the packer. Since the location of the gasket groove is near the outside edge of the joint, the loads are applied at a point along the joint where reinforcement will be inefficient to resist the load. The design of the edge of the joints at the gasket groove are therefore reliant on the geometry of the gasket groove, the distance from the gasket to the edge of the segment, and the shear strength of the concrete.

Design of the joint edge was based on full scale joint testing conducted at the Metropolitan Water District (MWD) La Verne Laboratory. This testing was used to verify that the concrete section could withstand the gasket line loads anticipated during construction. Further testing was also performed to evaluate the ultimate shearing capacity at the corner of the segment face by rupturing the gasket groove.

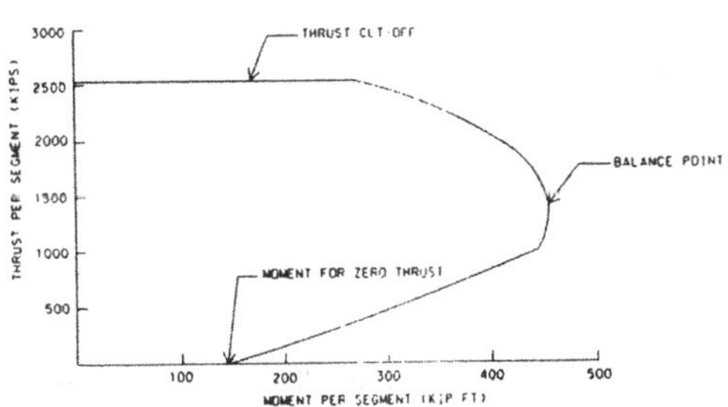

Figure 3. Moment-Thrust Interaction Diagram.

3.4 *Behavior of Packer*

A packer is used to transfer the ring compression across the longitudinal joint. The behavior of this packer is critical in assuring that loads are transferred properly, without creating large load eccentricities or stress concentrations. In order to adequately transfer the loads, the packer must deform enough to allow early eccentricities to diminish to create a fairly uniform pressure across the bearing surface, yet retain enough stiffness to prevent concrete edges from coming into contact under the high compression stress acting on the packer.

Based on past large-scale testing programs, in order to properly transfer compression loads across the joint, the packer needs to provide bearing over approximately 45 to 50% of the joint surface. If the packer is widened, compressive stresses will be transferred to the edges of the segment, where it is difficult to place reinforcement to resist these loads. For narrower packers, tensile stress concentrations can become very high, resulting in a greater likelihood of bursting failure.

Pressure distributions in the packer often result in eccentric loading across the packer. To avoid the development of additional eccentricities, the packer should be aligned on the centerline of the segment, to the extent possible. Offsetting the packer away from the centerline of the segment will not significantly impact the splitting capacity of the joint, provided sufficient tensile reinforcement is provided at the joint. Tensile stresses produced in the segment remain about the same for offset loads. However, shear planes through the joint will be offset from the centerline due to eccentric loading, thus potentially causing joint reinforcement to not effectively resist shear failures. In addition, the offset loading will cause a moment in the segments equal to the thrust multiplied by the eccentricity between the centerline of the packer and the centerline of the segment.

3.5 *Other Design Issues*

Although durability of the segments is not a critical factor in the design as the lining is essentially a temporary structure, it is important to limit cracking in order to assure the segmental ring remains watertight. Significant cracking is not expected, as long as handling and installation of the segments is performed carefully. However, crack control can be improved by addition of fiber reinforcement to the concrete mix. Although difficult to ascertain, the addition of fiber reinforcement may also increase the tensile strength of the concrete (f_t), thus increasing the capacity of the joints against splitting and shearing. Based on the results of the large scale testing discussed below, it was decided that adequate capacity of the segments could be developed without the addition of fibers to the mix.

4 VERIFICATION TESTING

4.1 *Large-Scale Joint Testing*

4.1.1 *Purpose and Scope*

The purpose of the large-scale testing on the joints is to validate the design procedure for the longitudinal joints. Although the procedure has been successfully used for several projects in the past, the magnitude of the expected loads warranted additional testing. A series of tests on large scale specimens was conducted to assure that design procedures adopted were applicable for the higher load conditions.

The tests were conducted at the University of Illinois, Champaign, under the direction of Professor William Gamble. These tests were conducted on approximately 70% scale specimens sized to fit the available testing equipment. A total of fifteen tests were run, with nine of the fifteen tests assumed to be applicable to the design used for the segments. The other six tests were run on segments with either different joint configurations or for segments with different strength requirements than the present design. These six tests were therefore disregarded for purposes of verifying the design procedure.

4.1.2 *Test Procedure*

Each test consisted of two segment blocks loaded across a common joint, with packer material of the same type and thickness as proposed for the final design. Reinforcement was schematically the same to the expected design reinforcement, taking into account the smaller scale of the test segments. The concrete mix used for the segments was the same as anticipated for the final design, using materials shipped from the project site that are anticipated to be used in the final design mix. Strain gages to monitor performance were installed at multiple locations in the test segments, including two each on two of the layers of the joint reinforcement to measure tensile strains, on two of the vertical rebars to measure compression strains, and on two through-thickness gages installed within the concrete in close proximity to the joint surface to monitor tensile strains in the concrete.

Of the nine test results deemed applicable to the current design, three were performed on standard blocks, and served as a control group for the other tests. Three were performed on the same blocks, but with one block offset by 12-mm with respect to the other one. The third set of three tests was performed on similar blocks as the first two sets, but with steel fiber reinforcement added to the concrete mix. It was felt that the fibers would prevent early cracking, but would not significantly improve the ultimate capacity of the segments.

4.1.3 *Test Results*

Figure 4 shows the results of the nine applicable tests on a graph of normalized tensile-splitting capacity versus reinforcement ratio, a similar graph as Figure 2 above. Figure 4 shows that all but one of the nine tests falls very close to the Ultimate Capacity line, which is based on Equation 1. The one test that plotted below the line was observed to fail in the body of the test segment in a combination shear-ing/spalling failure, not observed in the other eight tests. Load eccentricities from apparent tilting of the segments may have contributed to a premature failure. In addition, strain measurements at the joint indicated much lower tensile strains at failure than those for the other eight tests, indicating that the failure mode was not tensile splitting. Therefore, it seems reasonable to disregard the one low test as a product of the testing procedure, and not representative of tensile splitting. The other eight tests failed in tensile splitting, as evidenced by the large tensile strains observed in the joint reinforcement, and characteristic crack patterns observed proceeding the ultimate failure of the segments. Figures 5 to 7 show the test segments in various stages of load development and subsequent failure.

The results of this testing procedure confirm the design equation for splitting capacity. Although some early cracking of the segments was observed, this cracking did not adversely affect the segment capacity. Furthermore, the cracks appeared to lack continuity from the intrados side of the segments to the extrados side, which would suggest that the cracks would not provide a conduit for large quantities of water to flow through the segmental lining. It was found that addition of steel fiber reinforcement reduced the occurrence of pre-failure cracking and post-failure spalling, but did not appear to greatly increase the overall capacity of the segments.

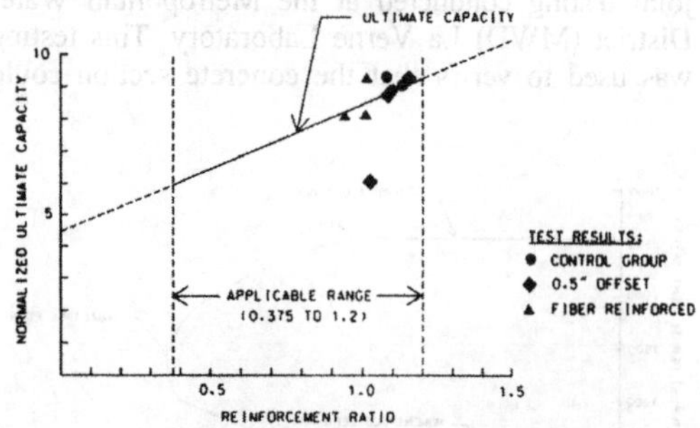

Figure 4. Applicable Test Results Plotted as Normalized Ultimate Capacity versus Reinforcement Ratio, Compared to the Ultimate Capacity Predicted by Equation 1.

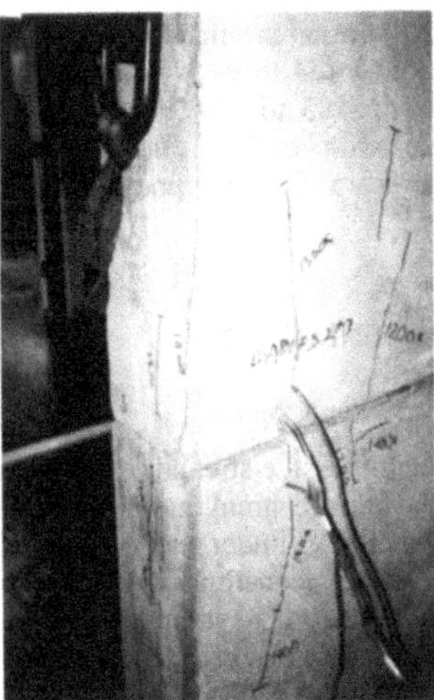

Figure 5. Development of Tensile Splitting Cracks Before Failure (Control Group, Load of 6200 kN (1400 + kips)).

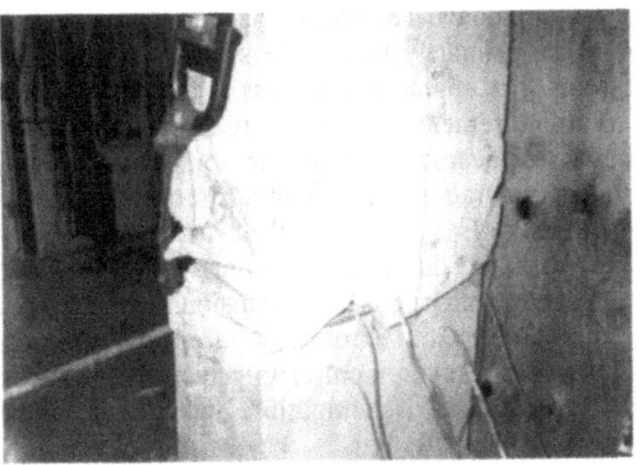

Figure 6. Damage to Segments at Ultimate Failure (Control Group, Load of 6700 kN (1500 kips at Failure)).

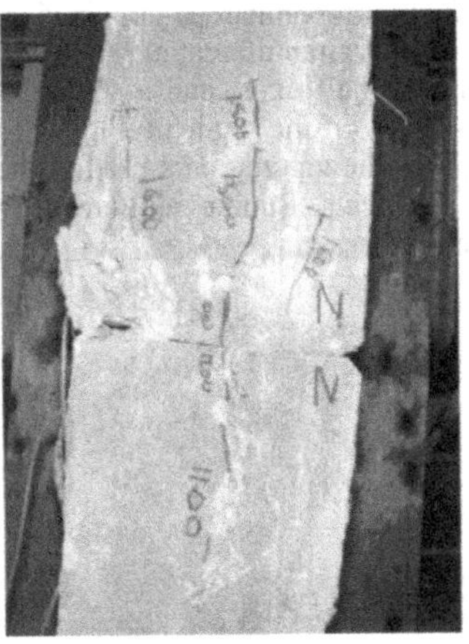

Figure 7. Crack Development and Damage to Segments at Ultimate Failure (Fiber-Reinforced Segments, Load of 7600 kN (1700 kips)).

4.2 *Gasket Groove Testing*

Three manufacturers submitted the following four gaskets for testing: Phoenix (EPDM), VIP Heinke (EPDM), Vertex-Construction Polymers (Polyisoprene) and Vertex-Construction Polymers (Neoprene). Each gasket manufacturer supplied MWD with their respective gasket specimens in rolls. The gaskets were cut into 305 mm (12-inch) straight lengths to fit the unreinforced test blocks built at the MWD La Verne Laboratory.

The 55 MPa (8,000 psi) concrete mix for the test blocks was identical to the design mix required by the Contract Specifications. Both 305 mm (12-inch) and 330 mm (13-inch) segment joint configurations were tested. The 330 mm segment adds 13 mm (½-inch) to the distance from the edge of the gasket groove to the outside of the block. The additional distance increases the localized shear capacity for the segments. Figure 8 shows the configuration of the blocks without offset.

Several load cases were tested at the laboratory for all the submitted gaskets to account for installation deviation during construction. The load cases and test results are discussed below.

4.2.1 *Offset of top block from the bottom block*

During construction of the segmental liner, it is possible that the two gaskets do not line up at a joint. The Contract specified a maximum offset of 13 mm (1/2-inch) during segment installation, so one of the primary goals of the gasket testing was to evaluate the effects of this offset. Gaskets were tested using zero offset and an offset of 13 mm normal to the running length of the gasket.

4.2.2 *Lateral Restraint*

Since plywood packing will be used at the joint along one side of the gasket, the effects of the packing material was studied. A steel shim plate (1.6 mm thick) was placed at the intrados edge of the gasket groove to simulate the plywood packing restraining

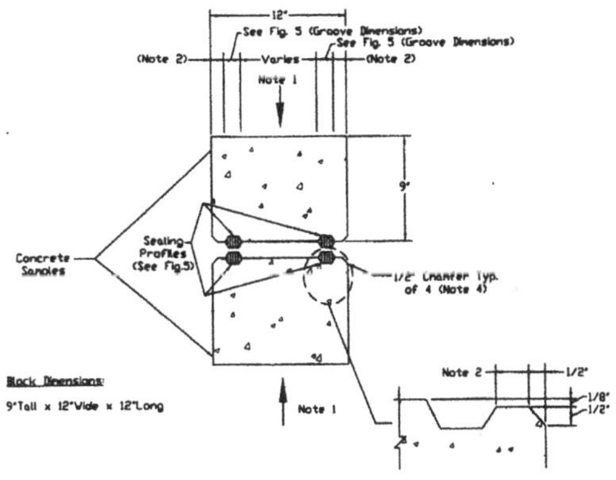

Figure 8. Flaking Test Setup.

the gasket from extruding in the radial direction during compression.

4.2.3 *Longitudinal Ends Confinement*
The increase in gasket stiffness from end confinement was simulated with the ends of the gasket confined longitudinally by steel plates. Tests confirmed that the gasket line load can increase by up to 100% when the ends are fully confined. The confined length in the laboratory (305 mm) is significantly smaller than the anticipated width of the segments (1.2 m). Therefore, increased line loads for the laboratory tests are considered conservative for field conditions.

4.2.4 *Steel Rail Test*
As a final test on each gasket, a 330 mm concrete block was arranged with the gasket faces aligned on a steel rail and compressed until failure. The load was applied at a constant rate through a steel bar towards the gasket until concrete failure occurred. The load at failure was recorded, and the orientation and dimensions of the concrete failure plane were measured and recorded.

4.2.5 *Results*
For the 305 mm concrete segment, shear failure occurred only when the stiff gaskets from Construction Polymer were tested, as shown on Figure 9. For the 330 mm concrete segment, shear failure did not occur for all the gaskets tested without longitudinal confinement. When steel confinement plates were used for the CP-neoprene gasket (stiffest gasket in the group), shear failure occurred at a line load of 250 N/mm for a 2.5 mm gap width.

For a normal gap width of 3.2 mm that is likely to be used during construction, none of the gaskets caused a concrete flaking failure. For the unconfined load case, the line load for the Phoenix gasket varied from 22 to 29 N/mm; the line load for Heinke gasket varied from 17 to 20 N/mm; the line load for CP-EPDM gasket varied from 74 to 94 N/mm; and the CP-neoprene varied from 85 to 119 N/mm. Under the worst load case with end confinement, the maximum line load from CP-neoprene was 214 N/mm without failure.

Figure 9. Flaking (Shearing) Failure of Concrete.

5 DESIGN RECOMMENDATIONS

5.1 *Concrete Strength Requirements*

A minimum 28-day compressive strength of 55 MPa and a minimum 28-day tensile strength of 4 MPa are required. Based on concrete cylinder tests performed on concrete mixes using construction materials available at the project site, these strengths can be consistently met using a number of alternative mix designs.

5.2 *Segment and Joint Requirements*

A 330-mm thick segment is required to meet the design requirements. The required segment thickness is controlled primarily by tensile splitting capacity of the joint. The segment thickness is also affected by space requirements to accommodate the gasket groove and necessary clearances on each side (towards the extrados, and towards the packer) to accommodate offsets and provide adequate load capacity to withstand the gasket compression forces. The joint layout with required dimensions is shown on Figure 10. The Contractor is required to coordinate with the gasket manufacturer to select the gasket so that the line load is compatible with the concrete segment.

5.3 *Reinforcement Requirements*

Reinforcement in each segment is shown schematically in Figure 11. Reinforcement is provided in both the longitudinal and circumferential (hoop) directions, at the longitudinal joints, and as through-thickness ties in the body of the segment. Figure 12 shows a photograph of the zig-zag bars used in the large scale testing segments. Similar reinforcement will be used for the final design.

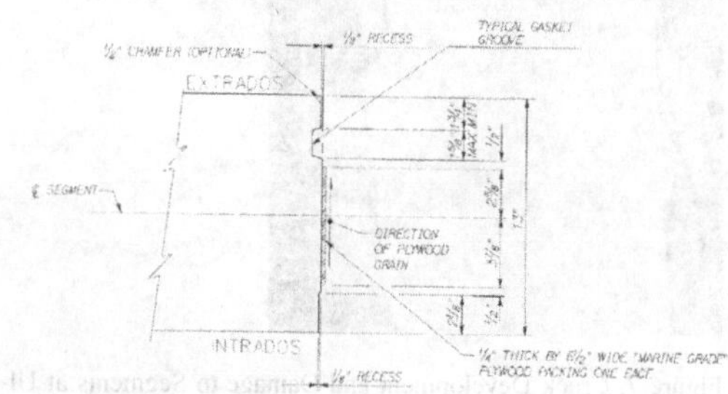

Figure 10. Joint Configuration.

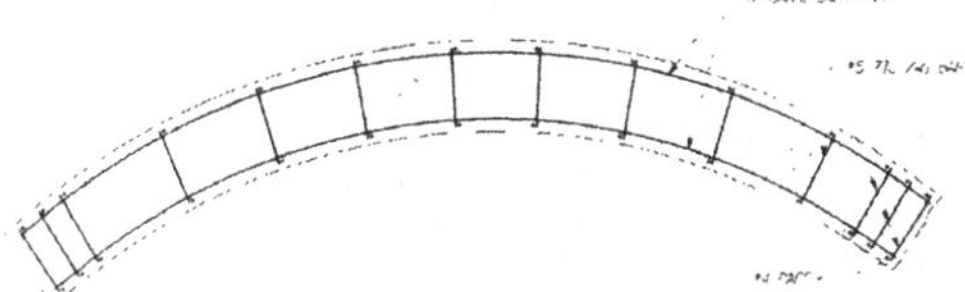

Figure 11. Layout of Reinforcement in Segments.

Figure 12. Zig-zag Bars Used in Test Segments.

The following general guidelines are important for optimizing the performance of the reinforcement:

1 Hoop reinforcement should be placed within the bends in the zig-zag bars, to the extent possible.
2 The ends of the zig-zag bars should be bent 90 degrees back into the segments.
3 For the three layers of zig-zag bars at the joint, the direction of the bends should be alternated.
4 Longitudinal rebar should be either tied or tack-welded to the zig-zag bars.

5.4 *Packer Requirements*

The packer is required to be 165-mm wide and 6-mm thick. Based on compressional/deformation testing on various materials, and trial tests carried out by others, a Marine Grade AB plywood packer is recommended to achieve the necessary behavior for the expected range of stresses. In addition, the lateral strain under compression in one direction for a plywood packer is minimal, thus limiting the development of shear stresses along the joint face in this direction. The packer should be oriented such that strains are limited toward the intrados and extrados of the joint surface. Based upon the results of the compression testing used to validate the joint design, the performance of the plywood packing used for these tests was as predicted and met the criteria of distributing the load across the joint surface without causing undue shear stress along the joint surface.

To provide adequate space and clearances for the gasket groove, the packer needs to be offset 16-mm from the centerline of the segment.

5.5 *Additional Design Requirements*

To assure initial gasket compression, bolts will be needed across the joints. The design of these bolts and the number used will depend on the Contractor's choice of gasket, segment geometry, and other issues. As an alternative to bolts, longitudinal dowels or bars could also be used to maintain compression across the circumferential joints.

Thrust capacity based on TBM requirements, loads associated with handling, and other construction induced loading will be based on the Contractor's means and methods. The design will then need to assure that these loads will not exceed the capacity of the segments.

In order to assure performance of the segments, as discussed above, the segments need to be factory produced high quality products. It is imperative to maintain tight manufacturing tolerances, segment casting tolerances, and requirements for curing to prevent cracking.

6 CONCLUSIONS

The design of the bolted, gasketed segmental lining for the Arrowhead Tunnels used a combination of analytical and empirical methods that were validated by large scale laboratory testing. These procedures verify that the 4.9-m I.D. segmental lining can carry the maximum external head of 275-m. The design requires the combination of a 330-mm thick lining, a compressive strength of 55 MPa, a tensile strength of 4 MPa, and extensive reinforcement, especially in the portion of the segments adjacent to the longitudinal joints. The design also requires the use of specific gasket materials and gasket groove geometry to prevent localized shear failures at the longitudinal joints.

The design procedures documented in this paper provide insight for future designs under similar conditions. Potential restrictions on inflows for environmental reasons during all phases of construction could result in the development of significant hydrostatic pressures, especially for deep tunnels under bodies of water, or under mountainous regions. For these situations, the design of the joints become critical, and additional design procedures, as implemented in the design presented in this paper, need to be applied.

Further study is required in the following areas to refine the design procedures for segmental linings subject to high external loads:

1 Interdependency of the gasket design, the strength of the concrete, and the spatial requirements for accommodating the gasket groove, with the purpose of improving capacity of the gasket groove edge.

2 Use of steel fibers to improve the overall performance of the lining system.
3 Required build tolerances and impacts upon constructability in the field.

REFERENCES

ACI 318 and 318R, Building Code Requirements for Structural Concrete and Commentary. 1995. American Concrete Institute (ACI).

Biggart, A. R.. 1993. Storebælt Railway Tunnel - Construction. Proc. Int. Symposium: Technology of Bored Tunnels under Deep Waterways. Copenhagen: International Tunneling Association and the Danish Tunneling Society.

Craig, R. & Mazen, A. 1993. El Salam Syphon under the Suez Canal. Proc. Int. Symposium: Technology of Bored Tunnels under Deep Waterways. Copenhagen: International Tunneling Association and the Danish Tunneling Society.

Crouthamel, C., Swartz, S., Curtis, D. J., Warren, S., & Shamma, J. 2001. Evaluation, Monitoring and Design Concepts for Pre-Cast Concrete Segmental Primary Support Under High Hydrostatic Pressures. American Rock Mechanics Association Conference, 2001.

Curtis, D. J. et al. 1991. The Channel Tunnel: Design, Fabrication, and Erection of Precast Concrete Linings. Tunneling '91. London: Institution of Mining and Metallurgy.

Curtis, R. 1993. St. Clair River Tunnel. Proc. Int. Symposium: Technology of Bored Tunnels under Deep Waterways. Copenhagen: International Tunneling Association and the Danish Tunneling Society.

Finch, A. P. 1996. The New St. Clair River Tunnel between Canada and the USA. Proc. Instn. Civ Engr: Civil Engineering. London.

Gotfredson, H. H. 1993. Storebælt Project. Proc. Int. Symposium: Technology of Bored Tunnels under Deep Waterways. Copenhagen: International Tunneling Association and the Danish Tunneling Society.

Grandori, R. et al 1995. Evinos-Mornos Tunnel, Greece. Proccedings of RETC, 1995. San Francisco, California.

Lemley, J. K. et al. 1993. The Channel Tunnel – Cooperation in a Transfrontier Environment. Proc. Int. Symposium: Technology of Bored Tunnels under Deep Waterways. Copenhagen: International Tunneling Association and the Danish Tunneling Society.

Lum, H., Crouthamel, C., Hopkins, D., Cording, E., & Shamma, J. 2002. Behavior of a Single Gasket System for a Precast Concrete Segmental Liner Subject to High External Hydrostatic Pressure. North American Tunneling Conference, 2002.

Norie, E. H. & Curtis, D. J. 1990. The Channel Tunnel: Design for UK Tunnels and Related Underground Structures. Strait Crossings. Rotterdam: Balkema.

Odgård, A. et al. 1993. Storebælt Railway Tunnel - Design. Proc. Int. Symposium: Technology of Bored Tunnels under Deep Waterways. Copenhagen: International Tunneling Association and the Danish Tunneling Society.

Pearse, G. 1999. Coastal Cleaning at Hastings. World Tunneling. Vol 12, No 6.

Robinson, B. & Jatczak, M. 1999. Construction of the South Bay Ocean Outfall. Proc. of RETC. Orlando, Florida.

Sharp, J. C. et al. 1993. Lesotho Highlands Water Project. In Burger, H. (ed.), Options for Tunneling, 1993. Elsevier.

The Channel Tunnel, Parts 1, 2 and 3. 1992-3. Proc. of the Inst. of Civil Eng. London.

Vigl, L. & Purer, E. 1997. Mono-shell Segmental Lining for Pressure Tunnels. In Golser, Hinket and Schubert (eds.), Tunnels for People. Rotterdam: Balkema.

North American Tunneling 2002, Ozdemir (ed.)
 ISBN 90 5809 376 X

Upper Narrows Tunnel – Innovative support and lining system

Kent Pease
Haley & Aldrich, Inc., Denver, CO, USA

Michael McKenna
Jacobs Associates, Los Angeles, CA, USA

ABSTRACT: The Upper Narrows Tunnel presents the most recent innovations in the evolution of highway tunneling technology in North America. Innovations include the use of shotcrete by itself as the tunnel final lining, the use of a spray-on membrane for water control, and escape adits for emergency egress. The spray-on membrane and shotcrete work well in combination for an economical and effective lining system. To further economize the lining, the contractor is allowed to use fiber reinforced concrete as an alternative to welded wire fabric. With this system, voids behind the lining are eliminated, negating the need for contact grouting. Although components of the final lining system have been used on other highway tunnels in North America, this is believed to be the first use of shotcrete by itself for a final lining, and the first use of a spray-on membrane as the primary means of water control in a highway tunnel.

1 INTRODUCTION

1.1 *Project Overview*

The Upper Narrows Tunnel is part of a larger project with the goal of improving the safety of highway users. The US 160/Wolf Creek Pass East Roadway Reconstruction Project consists of approximately eight miles of existing US 160 southwest of the town of South Fork as shown in Figure 1.

The Upper Narrows Tunnel is within a 1-mile long segment at the western end of the overall project. It is approximately seven miles below Wolf Creek Pass and six miles below the Wolf Creek Pass Ski Area.

The project owner is the Colorado Department of Transportation (CDOT), Region V, with partial funding from the Federal Highways Administration (FHWA). Project design was led by Carter Burgess with Haley & Aldrich responsible for geotechnical and tunnel design.

Design and construction of the entire project is planned to be in several phases over the next few years. As a result of available funding and contracting requirements, tunnel construction is divided into two phases with two independent contracts bid at different times. The first phase is for the exterior rock cuts, and excavation and initial support of the tunnel.

The second is for final lining, mechanical and electrical, and roadway improvements. This contracting situation presents certain requirements on the transition between contracts and contractors, and also has some influence on project design and economics.

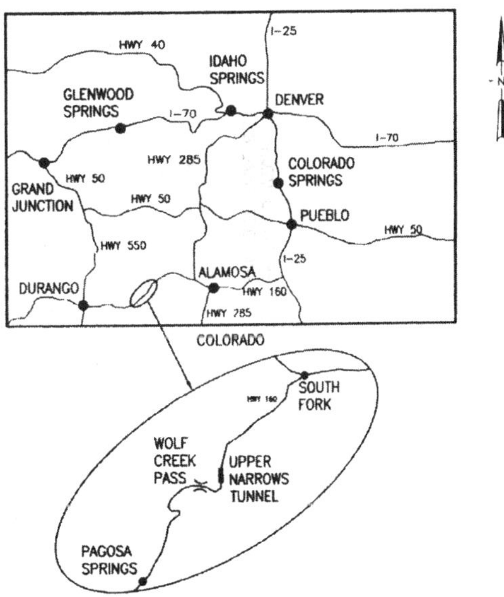

Figure 1. Project Location.

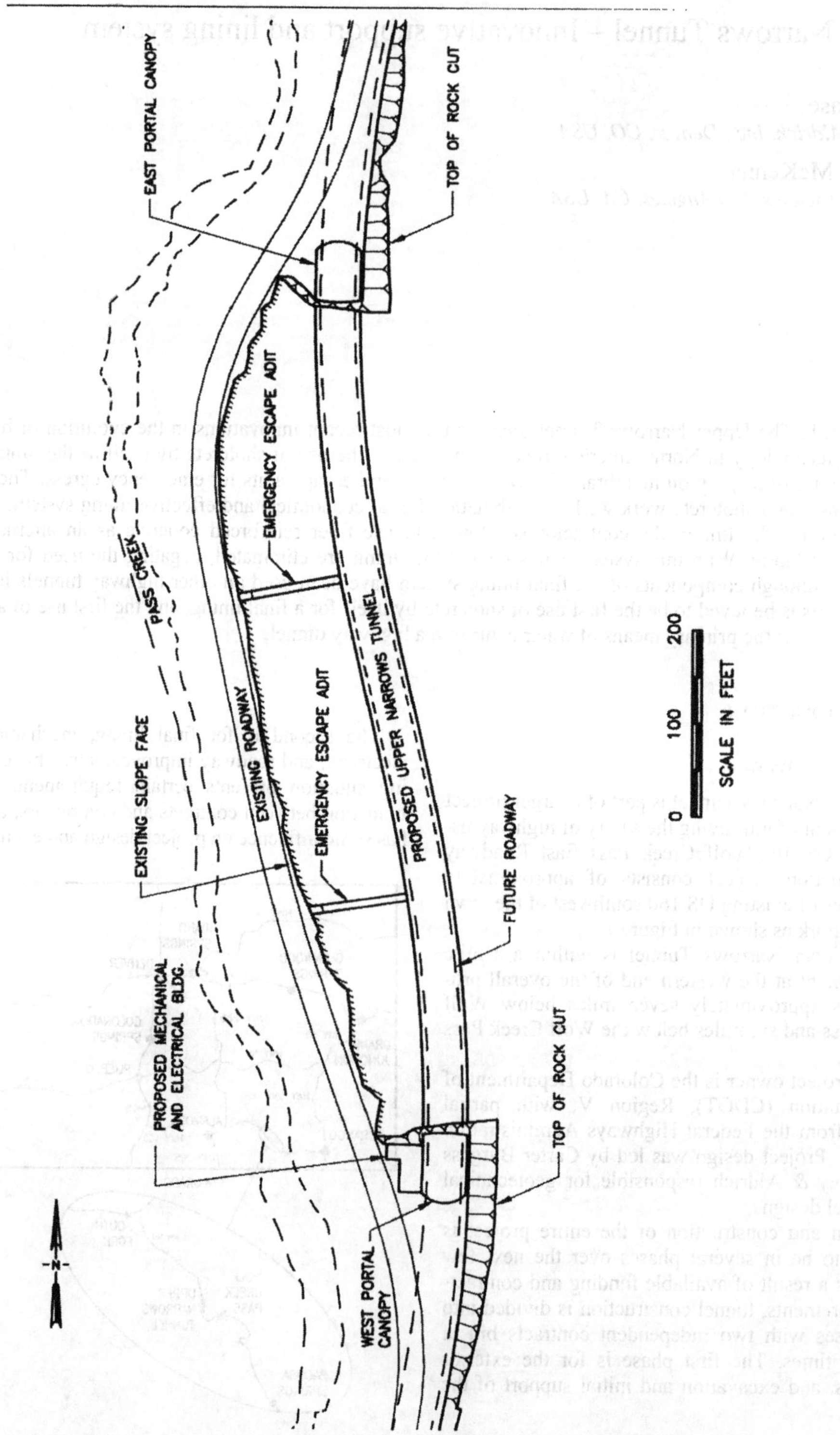

Figure 2. Plan view of tunnel alignment.

1.2 *Geometry and Layout*

Within the Upper Narrows project area, US 160 is within the canyon for Pass Creek, which is a tributary to the South Fork of the Rio Grande. The tunnel will be will be 916 ft long and will penetrate a lobe of rock cliffs to avoid tight curves. It will parallel the existing road with a gentle S-shape in plan as shown in Figure 2. The existing alignment goes outside the lobe and will be abandoned after the tunnel is opened, with all traffic in both directions flowing through the tunnel. The tunnel slopes downward to the east at a grade of approximately 5.6 percent. For the purpose of this paper, compass directions are consistent with the east/west orientation of the highway, although the true geographic orientation of the tunnel is north/south.

Finished internal dimensions will be 44 ft wide by 25.7 ft high in a modified horse shoe shape, with excavated tunnel dimensions two to three ft larger in both directions as shown in Figure 3. The finished tunnel dimensions are slightly wider than for most two-lane highway tunnels, and were dictated by the need to maintain two-directional traffic even in the event of a stalled vehicle, plus sidewalks for safety and emergency egress.

The project is perched 150 ft above Pass Creek with steep talus slopes and cliffs below the road down to the Creek, and cliffs and steep rock outside and above the tunnel as shown in the cross section in Figure 4. The tunnel is approximately at the same level as the existing road. Horizontal cover between the cliff face and the tunnel ranges between 40 and 120 ft. Vertical cover of the tunnel ranges from approximately 50 ft at the portals to more than 200 ft., The minimum cover occurs diagonally from the tunnel and is approximately 25 ft.

Two emergency escape adits are included in the design to provide egress of people from the tunnel in potentially dangerous or life threatening conditions, such as may result from an accident or fire. The adits are at the tunnel third-points, and will exit onto the shoulder of the existing road. The adits are horseshoe shaped with finished internal dimensions of 10-ft high by 8-ft wide.

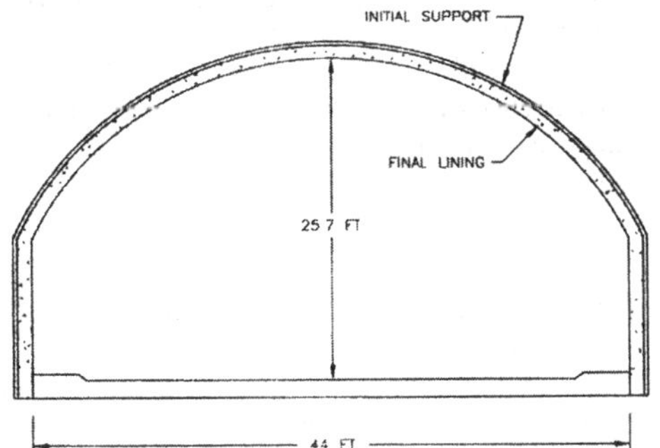

Figure 3. Tunnel cross section.

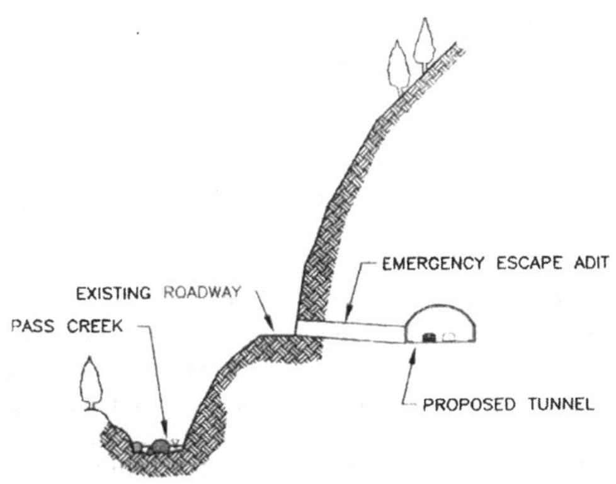

Figure 4. Overall project cross section.

1.3 *Project Goals and Site Challenges*

The overall project goal is to improve highway safety without significant environmental or visual impacts, and to minimize disruptions and delays to traffic during construction. The project is situated at an elevation of 9,000 ft in scenic and harsh alpine environment, in a steep canyon, and adjacent to a protected stream. These conditions place special demands on project design and construction. A tunnel was chosen over an exterior alignment primarily to facilitate construction, and to reduce environmental and visual impacts. The exterior alternative, which would have been superimposed on the existing highway, would have required more extensive rock cuts, a bridge over Pass Creek, and a longer construction period impacting the active highway.

Some of the construction challenges are the remote location with inconvenient access, limited construction season, harsh winter conditions, and tight work and staging areas. Additionally, the existing highway is heavily used and must remain open with strict adherence to temporary closure schedules. With these challenges, construction methods that are rapid, easily mobilized, and flexible are beneficial. Therefore, tunnel support and lining favors shotcrete, simple rock bolts, and a spray-on membrane, rather than other materials and methods that are more difficult to mobilize or slower to construct.

The reality of project funding requires strict consideration of costs and adherence to project budgets. This is not a project where new technology is purchased, but rather where new technologies are used to reduce project costs. It is in this environment that technology innovations were developed and implemented.

2 EVOLUTION OF LININGS FOR HIGHWAY TUNNELS IN ROCK

Ground support and lining systems for tunnels has evolved over time. To follow is a brief review of technologies that have been used for highway tunnels in rock situated in North America.

In the early to mid 1900's support, when used, was primarily timber or steel ribs and lagging. In good rock there was often no final lining. Where a final lining was necessary, it was usually reinforced, cast in place (CIP) concrete. In most situations no water control membrane was used. In the late 1900's rock bolts and shotcrete were used for initial support, especially in moderate quality rock. Final linings were almost exclusively CIP concrete, often with the use of a PVC membrane to control water infiltration. The next state in the evolution was the incorporation of rock bolts as part of the final lining. The Reverse Curve and Hanging Lake tunnels (Glenwood Canyon in Colorado) constructed in 1988 and 1991, respectively, were reported to be the first use of rock bolts as part of the final lining for a highway tunnel in the United States. However, a CIP concrete lining was still installed for structural reasons and to support the PVC membrane.

The Upper Narrows Tunnel near Wolf Creek Pass in Colorado is the next step in the evolution of linings for highway tunnels. In this tunnel, the final lining is designed as reinforced shotcrete combined with a spray-on membrane for water control. Rock bolts and a thin shotcrete layer will be utilized for initial support, but assumed to be sacrificial relative to the final lining. Although the rock bolts could have been utilized as part of the final lining, this would have required more expensive bolts and more difficult installation to protect against corrosion, and it was more economical to assume that they were sacrificial. The lining system designed for the Upper Narrows Tunnel is practical because of recent advances in products, a history of successful shotcrete use, and because of the high quality rock mass and low level of seepage expected in the tunnel lining. It is anticipated that this system, or components of this system, will be used as the lining for future tunnels.

3 SAFETY AND EMERGENCY PROVISIONS

Design for the safety of people in the tunnel in potentially dangerous or life threatening emergency conditions, such as may result from an accident or fire in the tunnel, was a prime consideration in tunnel design. Most highway tunnels recently designed provide for the emergency safety with a robust ventilation system to remove smoke and heat from the tunnel, and provide breathable air. Some tunnels also provide safe houses, which are side rooms insulted from the main tunnel with supplied or filtered air. For this tunnel, the relatively thin wedge of rock outside the tunnel allowed for the use of escape adits exiting to the pre-existing highway outside the tunnel as the primary emergency safety provision. These adits provide a higher level of safety than could have been possible with a ventilation system, and were used instead of such a system. Although an emergency ventilation system is not part of the design, a minor ventilation system of four jet fans is included to improve air quality in normal (non-emergency) operating conditions. For details reference Gonzalez & Pease (2001).

Two emergency escape adits are used and are located at the tunnel third-points as shown in Figure 2. The adits, will exit onto the shoulder of the existing road as shown in Figure 4. The adits will have a horseshoe cross-section with a finished inside dimensions of 10 ft high by eight feet wide.

4 GROUND CONDITIONS

4.1 *Geologic Conditions*

The Upper Narrows Tunnel is at an elevation of 9000 in Colorado's San Juan mountains. The canyon is steep and over 1500 ft deep at the project site. Rock outcrops and cliffs are common in the canyon, forming stair-stepped palisades on the canyon sides.

The tunnel is completely within bedrock of the Fish Canyon Tuff formation consisting of welded volcanic ash flows occurring in near-horizontal beds typically 20 to 100 ft thick. In the vicinity of the project the formation is over 1000 ft thick. The rock mass is generally competent, fresh to slightly weathered, and massive, with moderately- to widely spaced joints. The majority of the rock mass is considered to be of fair to excellent quality, having RQD values greater than or equal to 75 percent. However, less competent rock exists, both in terms of weathering and rock jointing. Colluvium overlies the bedrock in small discontinuous deposits, primarily in shallow talus wedges between the cliffs. The only occurrence of colluvium relative to the tunnel is thin deposits in the upper areas of the cuts at the two portals.

4.2 *Intact Rock Properties*

The rock is grey, welded, dacitic tuff that is slightly vesicular and has lithic fragments from ¼- to 2-in. It is hard, strong, and abrasive. A summary of material properties based on laboratory testing of intact specimens of the rock from core holes is presented in Table 1.

Table 1. Properties of Intact Rock.

Property	Range	Average
Unconfined Compressive Strength	2,000 to 16,000 psi	10,000 psi
Elastic Modulus	$2x10^6$ to $3x10^6$ psi	
Brazilian Tensile Strength	300 to 900 psi	700 psi
Cerchar Abrasivity	2.7 to 3.4	3.2 (highly abrasive)
Specific Gravity		2.4

4.3 *Rock Mass Jointing*

Three main joint sets have been identified in the project vicinity; one primary, two secondary, and one random. The terms "primary" and "secondary" does not necessarily correlate with age, prevalence, or frequency. Rock outcrops in the canyon generally occur in benches and cliffs that are structurally controlled by the primary and secondary jointing, respectively.

The primary joint set is defined as parallel or subparallel to the bedding surfaces between different ash flows and deposits. On a wide scale the primary jointing is near horizontal to subhorizontal often with a shallow dip to the west. On a smaller scale the joints have an undulating character with a wavelength on the order of several ft to tens of ft, and localized dips up to 30 degrees. These joints are relatively continuous, and are traceable for tens of ft to hundreds of ft. Spacing ranges from 10 to 50 ft, with more frequently spaced joints at the bedding contacts. Most primary joint surfaces are tight, and unweathered or lightly weathered with a moderately rough texture. However, some primary joints exhibit a thin zone of tight fracturing with weathering and some clay infilling.

Two secondary sets were identified. These are near vertical, and orthogonal to each other and to the primary joint set. Dips of 80 to 90 degrees are common. Spacings typically ranged from 3 to 15 ft, with some jointing occurring in narrow swarms as tight as several inches. On a project-wide basis the joints have a consistent orientation; however, the orientation varies greatly at each location. At the west portal the secondary sets strike nearly parallel and perpendicular to the tunnel, and at the east portal they strike at angles between 30 and 45 degrees to the tunnel. Secondary joint continuity varies with some joints traceable for the full height of exposed cliffs; other joints extend for only ten feet. Although individual joints may or may not be continuous, the trends of the major joint sets are relatively consistent. These joints are typically tight to open less than ¼-in. with no in filling, fresh to slightly weathered, and with a moderately rough texture. However, some joints have a clay coating, or are open by several inches and infilled with soil, especially joints near the ground surface.

Occasionally a third set of randomly oriented joints is also present. These joints are infrequent and can have any orientation. The character of these joints is similar to the secondary joints.

4.4 *Rock Mass Characterization and Behavior for Tunneling*

Descriptions of the rock mass as used herein are based on a combination of the parent rock, jointing, weathering, and ground water conditions relating to tunneling. To describe these conditions and present them in contract documents, the rock mass is described as either Common or Poor and further subdivided into the three categories of A, B, and C for contracting purposes. Descriptions of each based on common tunneling classification systems are summarized in Tables 2a and 2b.

Table 2a. Rock Mass Descriptions.

Ground Description & Category	Terzaghi[1]	ROD (%)
Common Rock, A and B	*Massive to moderately jointed*	75 to 96
Poor Rock, C	*Moderately blocky and seamy to stratified vertical*	24 to 75

Table 2b. Rock Mass Characterizations.

Ground Description & Category	Geomechanics System[2] RMR[3] Range	Geomechanics System[2] RMR[3] Description	The Q-System[4] Q[5] Range	The Q-System[4] Q[5] Description
Common Rock, A & B	52 to 59	Fair	15.75 to 23.75	Good
Poor Rock, C	23 to 52	Fair to Poor	0.5 to 15.75	Good to Very Poor

Notes: 1. Terzaghi (1946), Huer (1974)
2. Bieniawski (1973)
3. RMR ranges from 0 to 100 in a linear scale
4. Barton, Lien, and Lunde (1974)
5. Q ranges from 0.001 to 1,000 in a quasi-log scale

It was estimated that 75 percent of the tunnel will be Category A rock, 20 percent in Category B rock, and 5 percent in Category C rock.

4.5 *Groundwater Conditions*

The ground water table is below the tunnel invert, and is assumed to mimic topography in a muted fashion with flows toward Pass Creek. The project groundwater system relevant to the tunnel consists of an unconfined aquifer within joints in the bedrock fed by rain and snowmelt from the ground surface.

Observations of cliffs and rock cuts along the existing highway outside the proposed tunnel revealed some seepage from joints. The magnitude of seepage changes seasonally, with the maximum seepage during spring runoff and summer thunderstorm seasons. Some of the joints are damp seeps but some discharge enough water to visibly drip or rarely, to flow. Flowing conditions were observed only from open joints near the ground surface that were visibly traced to the ground surface and a water source. Ground water conditions in the tunnel are expected to be primarily dry or moist joints and occasional seeps from joints, with the possibility of flowing water at a few locations.

5 DESIGN

5.1 *Lining Overview*

Tunnel stability and lining will be achieved by using two systems consisting of temporary initial support and the final lining. The function of the initial support is to stabilize the opening for one to two years prior to installation of the final lining. For overall economics and efficient excavation, it was decided that the initial support system would be installed quickly and assumed to be sacrificial. A sacrificial initial support system is less expensive than an initial support system that has long-term durability and is incorporated into the final lining. The decision to use a sacrificial initial support system was also based, in part, on the desire to complete tunnel excavation as quickly as possible, and to be consistent with available funding for the first phase of the work.

Although it is possible to design and construct an initial support as part of the final lining system, this approach would require different bolting systems, as well as substantial quality assurance and quality control during installation. With a sacrificial initial support system, the bolts could be installed quickly, without long-term corrosion protection, and a relatively low level of QA/QC could be employed during construction. This system was determined to be less expensive than a more robust initial support system incorporated into the final lining. Additionally, if it was assumed that the initial support contributed to the final lining, the design of the final lining would not have been substantially different. Therefore, there would have been an overall higher cost for the combined initial support and final lining system. Based on these factors, independent initial support and final lining systems were used.

The initial support is designed to be primarily rock bolts and shotcrete (Category A and B rock) with steel sets used only in poor rock conditions (Category C rock). The final lining is designed to be primarily 9 inches of reinforced shotcrete, with a few areas of CIP concrete at the portals and in poor quality rock. Integral with the two lining systems is the water management system. Although discussed separately, the three systems are interdependent and have to work well together.

A key consideration in design of the lining systems was to maintain good contact between the linings and the ground. This is to maintain the integrity of the rock mass and prevent the rock mass from relaxing, thereby using the rock rather than the lining to support ground loads. In consideration of this phenomenon, a sheet type water proofing membrane in combination with shotcrete for a final lining was considered, but not used because of potential voids generated behind the lining. Although it is possible to grout these voids, this extra step is not desirable and the system was estimated to be more expensive than the spray-on membrane system.

Because the use of these systems is somewhat novel and contractors may not have a high degree of experience or comfort with their use, the use of a sheet membrane and CIP concrete will be allowed as an alternative. Also, by allowing the traditional alternatives, the project was assured of the lowest price, and avoiding paying a premium for new technology if the existing technology is less expensive.

5.2 *Excavation*

Tunnel excavation will be with drill-and-blast methods using multiple drifts. There will be three drifts in the top heading and two or three in the bottom bench (or heading), with initial support installed in each drift as excavation proceeds. The central drift of the upper heading will be excavated first and will in effect be a pilot tunnel. Multiple drifts are beneficial; 1) to control the excavation and support, and 2) to provide better relief for side slashes resulting in a neater cut line and less damage outside the excavation limits. Additionally, precision blasting methods are required at the final cut lines to result in a low level of damage to the remaining rock.

5.3 *Initial Support Description*

The function of the initial support is to provide temporary stability and safety for the tunnel opening prior to installation of the final lining. This is expected to be one to two years. As presented above the initial support was assumed to be sacrificial and was not designed for long-term performance. Three levels of initial support will be used consistent with the three categories of ground, A, B, and C. Levels A and B consist of rock bolts with shotcrete, and level C consists of steel ribs with shotcrete as presented below in Table 3. Shotcrete will be reinforced with either welded wire fabric or steel fibers at the contractor's option. Determination of design details

Table 3. Initial Support.

Support Level	Bolts	Shotcrete	Other
A	10 ft #8 @ 6 ft both ways	3 inches	NA
B	12 ft #8 @ 4 ft both ways	4 inches	NA
C	Spot as necessary	4 inches	W10x65 Steel Ribs @ 4-ft center to center

of each level of support was based on evaluations and calculations as presented later in this paper.

5.4 *Final Lining Description*

The function of the final lining is for long term stability of the opening and to support the water management system. As presented above, the final lining system was assumed to act independently of the initial support. Two different types of final lining systems are used: 1) shotcrete for most of the tunnel, and 2) Cast-in-Place (CIP) concrete in the portal zones and in areas of poor ground.

The use of shotcrete by itself for a final lining has not previously been used for a major highway tunnel in North America. A key factor in the use of shotcrete in this case was the good quality of the rock mass, which is basically self-supporting, provided it remains intact and does not loosen over time. The shotcrete provides this long-term protection and restraint. Final lining shotcrete will be a total of nine inches thick consisting of a 1-in. flash coat, at least 6-in. of steel fiber-reinforced shotcrete, and a coverage coat of 2-in. of shotcrete without the fiber reinforcement to create a smooth finished surface. Alternative reinforcement consisting of welded wire fabric or reinforcing steel can be used at the contractor's option.

The ability to apply shotcrete against the spray-on membrane was a key consideration in the choice of shotcrete as a final lining. There was a concern that the shotcrete may not stick to the membrane, especially overhead at the tunnel crown. However, a test program and similar project applications have demonstrated that it is feasible with careful control of shotcrete properties and application. After the shotcrete has taken an initial set it will be self-supporting, and eventually the membrane to shotcrete interface will go into compression as the rock mass relaxes inward. Therefore, there is no need for a long-term bond between the membrane and the shotcrete.

CIP concrete will be used as the final lining at the portals and in areas of poor rock. At the portals, CIP concrete is used to provide a higher level of support due to the lack of confinement, the likely more open character of the joints, and to provide a transition from the portal canopies. In areas of poor rock it is necessary to support the expected higher ground loads. The concrete will be 12-in. thick, with one layer of #6 bars at 12-in. spacing both ways positioned 2-in. from the inside surface. This reinforcement position will give the best resistance to bending caused by inward movement of the rock.

5.5 *Groundwater Control*

Although the flow rates into the tunnel are expected to be quite small, any water penetrating the tunnel lining could cause damage to the lining, and could result in ice on the inside of the lining or on the road surface. Therefore, the water control system is designed to control seepage and to be compatible with the lining systems chosen. A key consideration in designing the water management system is that the tunnel is above the ground water table and that seepage is expected only from isolated joints. A second key consideration is that the water management system should be integrated with the chosen initial support and final lining. The water control system design is based on products and features used in other tunnels and for other applications that are combined in a new and innovative way to work with the unique ground water conditions and lining systems for this tunnel.

A three-part water control system has been designed for this project consisting of: 1) vertical drainage wicks, 2) a water barrier membrane, and 3) formation drains to collect and discharge the water. This system is illustrated in Figure 5. Strip drains will be used to intercept seepage prior to reaching the membrane and to reduce the buildup of water pressure outside the membrane, thereby reducing the burden on the membrane. Strip drains are specified to be a commercially available composite drainage material consisting of a dimpled sheet with a non-woven geotextile on one side. They will be placed in two layers; against the rock, and between the initial support shotcrete and the final lining, both exiting into drainage pipes cast into the invert below the sidewalk. The strips will be will be primarily vertical and one foot wide at approximate six-foot intervals around the tunnel perimeter from invert to invert. Supplemental strip drains will be placed directly over seeps and observed wet joints. The discontinuous pattern allows for good contact and adhesion of the shotcrete with the rock (exterior layer of drains), or membrane with previously placed shotcrete (interior layer of drains). A formation drain at the bottom corners of the tunnel will intercept the strip drains and transmit seepage out of the tunnel. This drain will be a perforated pipe embedded in porous concrete below the outside edge of the sidewalks.

The membrane will be a spray-on product sandwiched between the initial support and the final lining. In this position it will be easily applied to a relatively uniform surface and will serve as the final line

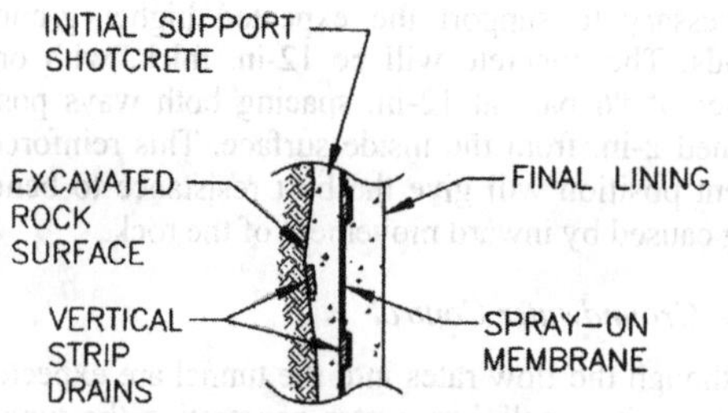

Figure 5. Cross-Section of Water Control System.

of defense inside the strip drains. Also, at this location it will be protected from physical damage and weathering by the final lining, and will be supported by the final lining.

This will be the first use of a spray-on membrane for this application in North America. A spray-on membrane is economical and provides an appropriate level of protection for this application, which is discontinuous seepage above the water table. Advantages of the spray-on membrane are that it is easy to install, works well in combination with shotcrete, does not have seams, and can easily cover protrusions into the lining such as rock bolts. In comparison, a sheet membrane is more labor intensive to install, seams are required, and the application of shotcrete inside the membrane requires installation of reinforcing to support the wet shotcrete. Additionally the use of shotcrete with a sheet membrane results in a void behind the membrane; negating good ground/lining contact and interaction, or requiring a contact grouting program. With a spray-on membrane, there is good contact and a grouting program is not necessary.

5.6 *Tunnel Support and Lining Evaluations*

Tunnel initial support and final linings were designed based on several different types of evaluations including:1) empirical correlations, 2) kinematic analyses, 3) ground arching, 4) two-dimensional continuum finite element analyses, and 5) beam-spring finite element frame analysis. Not all analyses were used for each support or lining type; however, more than one type of analysis was used for each. The intent of this section of the paper is to provide a brief overview of design methods for the shotcrete final lining, and to note methods used for other tunnel support and lining systems for future reference. To follow is a brief description of the analysis and design methods used:

* Empirical Correlations – These are based on rating systems of rock mass quality and associated correlations with previously used support types. Both the Q and RMR rock mass classification systems were used. Calculated rock mass quality ratings are presented earlier in this paper. Correlations were used to determine likely initial support and final lining based on what has been used on other projects in similar ground conditions. Specifically these analyses were used to obtain a preliminary assessment of initial support types, to estimate the sizing and spacing of bolts, and determine initial support and final lining shotcrete thicknesses. Results indicate that pattern bolts and shotcrete are appropriate for initial support in most areas of good rock, and that steel ribs and lagging are necessary in poor rock.

* Kinematic Analysis - Kinematic analyses address stability of openings and individual blocks based on the spacing and orientation of joints relative to the opening. Evaluations used typical joint patterns and evaluated stability of the entire opening, and stability of individual blocks, wedges and slabs that may be freed by joints and combinations of joints. Results were used to set rock bolt patterns of initial support, and for punching shear loading of the shotcrete final lining.

With the primary and secondary joints approximately perpendicular and square, rectangular blocks are expected to predominate. The primary mode of movement is from horizontal joints near the crown that will tend to form unstable slabs, especially in combination with secondary joint sets. Additionally, wedges and triangular blocks are expected to result from variations in joint orientation and from random joints.

* Ground Arching Analysis of Steel Ribs – Ground arching was used to design initial support steel rib size and spacing. Analyses were based on the methods presented in Proctor & White (1946) in which the ribs are loaded with a given rock volume that is transferred into thrust and moment in the rib. Thus the ribs are loaded by a combination of active and passive forces at evenly spaced blocking points. Based on rock mass conditions, the rock load was chosen to be one half-tunnel diameter (23 ft of rock at an average unit weight of 150 pounds per cubic foot). Based on these results the rib size and spacing was chosen.

* Continuum Finite Element Analyses – The behavior of the ground and interaction of the ground and the lining were analyzed with a continuum two-dimensional finite element model incorporating beam elements, continuum elements, and interface elements. The model was run with the finite element program PLAXIS. This model was used to evaluate stresses and deformation in the ground in different excavation and support phases, especially in uniform loading. Model output provided estimates of deformations and ground stresses, that were compared with allowable values. These analyses indicate that the stresses in the rock were well below allowable material strengths.

* Beam-Spring Frame Analysis – The CIP concrete final lining and interaction of the ground and the lining was analyzed with a beam-spring model using a

frame analysis. Beam elements were used to model the lining, and radial and tangential springs used to model the ground. An external load, based on the geologic conditions around the tunnel, was then applied to the lining. The frame analysis was run by the computer program STAAD III. This analysis allowed a detailed evaluation of the lining with different loading conditions with ground/lining interaction. A uniform, vertical rock load of 12 ft of rock was applied based on the same methodology described in the previous section on ground arching analysis for tightly spaced steel ribs. Nonuniform loads were based on blocks and loose areas determined from the kinematic analyses. Results of thrust and moment were used to design lining thickness and reinforcing for the CIP concrete final lining using the ACI working stress design method.

The shotcrete final lining was evaluated and designed for both uniform and nonuniform ground and water loading using a combination of four of the above methods of analysis. First, empirical correlations were used to determine and approximate design, and to compare the design with other tunnels in similar ground conditions. Second, the two-dimensional continuum finite element model was used to determine deformations and the distribution of stresses in the ground and the lining, primarily under uniform loading conditions. Third, kinematic analyses were used to determine the size and configurations of blocks and loose areas likely to place nonuniform loads on the lining. Fourth, the beam-spring frame finite element model was used to determine the deformations and distribution of stresses in the lining, in both uniform and nonuniform loading. Results of the kinematic analyses were used as an input for nonuniform loading on the lining using the beam-spring frame model. One of the nonuniform load conditions was for two-dimensional punching shear of the lining caused by loose blocks directly loading a limited area of the lining. Output from the two computer models included the thrusts and moments in the lining around the tunnel perimeter in different loading conditions were then used to design lining thickness and reinforcing.

6 SUMMARY

The Upper Narrows Tunnel presents the latest in tunnel support and lining technology for Highway Tunnels in North America. These innovations include the use of a spray-on membrane and strip drains for water control, and the use of shotcrete for the tunnel final lining. The innovations were made possible by advances in membrane technology combined with successful use of shotcrete and fiber reinforced shotcrete in tunnel lining systems throughout the world. Additionally, the project owner, the Colorado Department of Transportation, and the Federal Highway Administration were instrumental in the design with their support of new technologies. Finally, lining design and water control for the Upper Narrows Tunnel are tailored specifically for the ground conditions present and suitable for the level of service required of the tunnel. It is the combination of all these factors that allowed for the innovative lining and support systems used for the Upper Narrows Tunnel.

7 ACKNOWLEDGEMENTS

The authors appreciate the project support and willingness to use innovative technologies of CDOT and FHWA. Individuals who were instrumental in project design include Steve Long/Project Manager, Tom Livingston/Carter Burgess, John Schneider/CDOT, and Anthony Caserta/FHWA.

REFERENCES

Barton, N., R. Lien, and J. Lunde. 1974. Engineering Classification of Rock Masses for the Design of Tunnel Support, *Rock Mechanics* 6, pps. 183:236.

Bieniawski, Z.T. 1973. Engineering Classification of Jointed Rock Masses, *Trans.* Southern Africa Institute of Civil Engineering 15 pps. 335:344.

Bieniawski, Z.T. 1989. *Engineering Rock Mass Classifications*, published by John Wiley & Sons, Inc. Wiley, New York, 251 PP.

Gonzalez, Joseph and Pease, Kent. 2001. Escape Adits Provide for Tunnel Safety. *Proceedings of Transportation Research Board, 80th Annual Meeting, Washington, DC*, 12 pp., January 7-11, 2001. Also to be published in TRB Record, 2001.

Hoek, E. and Brown, E.T. 1980. *Underground Excavations in Rock*, Institute of Mining and Metallurgy, London.

Heuer, R.E. 1974. Important Ground Parameters in Soft Ground Tunneling, *Subsurface Exploration for Underground Excavation and Heavy Construction*, New England College, Henniker, New Hampshire, American Society of Civil Engineers, New York, pps. 41-55.

Proctor & White, *Rock Tunneling with Steel Supports*, Commercial Shearing, Inc., 1946 (revised 1968, reprinted 1988).

Terzaghi, K. 1946. Rock Defects and Loads on Tunnel Supports, in *Rock Tunneling with Steel Supports*, by Proctor and White, pps. 47-85, Commercial Shearing.

North American Tunneling 2002, Ozdemir (ed.)
© 2002 Swets & Zeitlinger, Lisse, ISBN 90 5809 376 X

Selection and design of corrosion protection liner for precast concrete segment tunnel liner

D.R. Chapman
LACHEL & Associates, Inc., Morristown, NJ, USA

ABSTRACT: Microbially induced concrete corrosion requires major rehabilitation efforts in many municipal sanitary sewer systems, leading to more frequent selection of corrosion protection liner (CPL) systems for new projects. CPL system design for precast concrete segment tunnel linings presents different challenges than for concrete pipe linings because of the large amount of additional joints. Detailed screening of several systems for a new project in Columbus, Ohio left two for detailed consideration. These were 1) Ribbed or studded PVC or HDPE liner cast onto segments with welded joints, and 2) Linabond Co-Lining System, a proprietary chemically attached liner system. This paper describes the selection process, technical and cost comparisons, and describes the characteristics of the final candidate CPL system. This paper should be of interest to utility owners, consulting engineers, and contractors concerned with design and construction of tunnels for sanitary sewers in geologic environments requiring pressurized-face machines and precast concrete segments.

1 INTRODUCTION

Major rehabilitation costs are being incurred by wastewater utilities in dealing with corrosion-damaged concrete pipe and structures. Protection of new concrete sanitary sewer facilities from such corrosion will prolong infrastructure life and reduce future maintenance costs. Selection and design of a corrosion protection liner (CPL) system has strongly influenced the design concepts for a new sanitary sewer tunnel in Columbus, Ohio. This paper describes the CPL selection process and results and presents liner test results for a promising liner system.

2 BASIS FOR PROVIDING CPL SYSTEMS FOR SANITARY SEWER TUNNELS

Concrete has long been used for sanitary sewer system infrastructure, in the form of concrete pipe for conveyances and cast-in-place structures for junction chambers, manholes, sumps, treatment basins, etc. Corrosion of concrete in this aggressive environment has been known for some time, particularly for manholes and structures at treatment plants. Much effort and expense has been applied to protect against and remediate corrosion damage.

This problem has traditionally been associated with warmer climates and higher ambient temperatures thought to better support the microbial growth that gave rise to the particular mechanism of concrete attack. It may indeed be more severe in such climates. Houston, Texas is known to have great problems with concrete deterioration in sewers, and the City of Houston has studied many alternatives to provide protection against such corrosion. The University of Houston has a significant research program under the Civil Infrastructure System Initiative sponsored by the National Science Foundation. University of Houston, (1997). Large testing chambers allow performance of research on corrosion mechanisms and means of preventing or mitigating concrete attack. However, problems have increasingly been observed and required attention in more northern locations with cooler climates as well, including in Canada, Joyce, (2001).

Mechanisms for concrete corrosion have been thoroughly documented, and a detailed treatment of this subject is beyond the scope of this paper. Simply stated, sulfides can be generated in sewers under certain conditions, namely: low dissolved oxygen content, high strength wastewater, long detention times, extensive pumping, and high wastewater temperatures. Part of the equilibrium that results in the sewer environment involves the generation of hydrogen sulfide gas. Bacteria that colonize surfaces above the water line in sanitary sewer pipes and structures can consume hydrogen sulfide and oxidize it to sulfuric acid. This process can result in the cultivation of colonies of bacteria that can live in progressively lower pH environments. Surfaces of sanitary sewer pipes have been found to exhibit conditions with pH values as low as 0.5.

The sulfuric acid attacks the cement matrix of the concrete and this attack is not prevented by the use of sulfate-resisting cement of Type II or Type V. Sewer flows can then remove softened cement, exposing aggregate particles to removal by sewer flows. This successively exposes new surfaces to attack, and eventual progression to the steel reinforcement that is also attacked by the acidic environment.

Certain factors may exacerbate the corrosion mechanism in today's sanitary sewer conveyances. First, regulations require pretreatment of sewage for metals removal. Metals are toxic to bacteria and it is thought that their removal has allowed bacteria that were previously kept in check to flourish. Second, regionalization of sewage treatment through closure of older treatment plants that cannot meet current effluent standards and conveyance of sewage to larger modern treatment plants significantly increases residence time, contributing to greater sulfide generation, Joyce (1995).

3 BWARI PROJECT BACKGROUND

The BWARI sewer project will provide over 37,000 feet of 4.27 m (168") and 3.66 m (144") finished diameter sewer on the south side of the City of Columbus, Ohio to augment and existing 2.74 m (108") interceptor, to provide 227,125 cu m (60 million gallons) of in-line wet weather storage to minimize bypassing, to eliminate existing pump stations and open new growth areas, and to provide future service capacity throughout the tributary area. Uhren (2001). All but about 1219 m (4,000') of the BWARI sewer that will be open-cut will be tunneled through layered glacial soils consisting of dense silty and clayey glacial tills and of stratified sand and gravel from glacial outwash, with boulders. These soils display high permeability in some zones and measured groundwater levels indicate the need to assign design groundwater pressures up to 210 to 240 kPa (30 to 35 psi). Geologic conditions and their influence on project development have been discussed in previous papers. Frank (2000); Frank (2001); and Uhren, (2001).

The decision to provide a CPL was made early in the project, based on the considerations expressed above. At that time, it was planned to tunnel using a two-pass system with initial support of steel ribs and timber lagging followed by placement of concrete pipe lined with Ameron's PVC (poly-vinyl chloride) T-Lock lining material as a CPL. This PVC sheet liner has T-shaped ribs on the back, which are cast into the inner wall of the concrete pipe in the casting yard. After placement in the tunnel, the joints are covered with capping strips and heat welded on both sides of the joints, using additional weld strips.

Through the geotechnical investigation, it became apparent that the presence of impermeable glacial till soils at and above the tunnel invert would make effective and complete dewatering very difficult. Additional concerns existed regarding dewatering because of potential impacts on domestic wells in the area and on the nearby City well field. Accordingly, it was decided to alter the design to use a onepass method with bolted and gasketed precast concrete segments serving as initial and final support. This meant that the CPL would need to be applied directly to the inside face of the concrete segments. The selection process, the two leading CPL system candidates, and testing on one system are the primary subjects of this paper.

4 BWARI PROJECT DESIGN CONDITIONS

As discussed above, the BWARI sewer will serve multiple purposes within the City of Columbus' long term sewer infrastructure planning. The following considerations dictate the project design conditions:

- The wastewater storage function results in longer residence times than for a standard flow conveyance.
- Existing hydrogen sulfide levels in the connecting sewer and at the Southerly Wastewater Treatment Plant indicate strong potential for microbially induced concrete corrosion.
- Recent inspection of the existing 2.74 m (108") diameter outfall sewer indicates significant corrosion damage with exposed aggregate and reinforcing steel and very low pH values on concrete surfaces. These findings validate the envisioned need to provide a CPL.
- With the precast concrete segment tunnel liner, each 1.22 m (4') advance of the tunnel requires a ring constructed from five larger segments and a smaller trapezoidal key. This results in six longitudinal (radial) joints as well as the circumferential joint, a total of 20.7 m (68') of joint for each 1.22 m (4') advance of the tunnel. Bolt pockets and grout holes are additional sites that will require CPL protection.

5 REVIEW OF PREVIOUS CPL EXPERIENCE AND CURRENT CPL DESIGN PRACTICE

The focus of this paper is not to exhaustively review the history of CPL liner provision for sanitary sewer conveyances. However, experience with various CPL systems that have been applied in the past is an important component of the selection process for new projects. Therefore, selected information from performance data that has been gathered is presented here.

CPL systems for new construction involving concrete pipe have most often utilized Ameron's PVC T-lock lining system, applied at the casting yard, with joints welded after pipe placement. As the corrosion process does not really take place below the water flow line, practice has frequently involved placing the liner material over the entire pipe circumference, but allowing a section at the bottom to be cut out, removing the slippery pipe invert which exists with this liner in place. The coverage is frequently specified to be 270°, although it is important to assess the level to which typical low flow conditions will fill the pipe and design to ensure against exposure of unlined concrete above the water flow line. Besides Ameron's T-Lock material, several manufacturers of high-density polyethylene (HDPE) offer lining sheet goods with integral anchors of various configurations that perform essentially the same function as T-Lock. These materials are somewhat stiffer, and have more traditionally been used to line various basins and structures. Peggs (1999). They are clearly applicable to lining concrete pipe as well.

Other CPL systems have been developed primarily for remedial construction, and often involve HDPE or PVC liners with integral anchors placed over an annular form and grouted in place. Another system involves a lining system that is spirally wound into the tunnel or pipe, with a snap-together joint detail, also involving annular grouting.

A mechanically-anchored system utilizing HDPE sheet lining material fastened in place with stainless steel battens and concrete anchors has also been employed in several applications, with mixed results as indicated below.

5.1 *Houston, Texas*

The Northside Sewer Relief Tunnel was the largest of Houston's deep sewer tunnels, and was constructed beginning in the mid-1980's. The tunnel was lined with a mechanically attached liner of high-density polyethylene (HDPE) to protect the walls from microbially induced corrosion. This system was completed and placed in service in 1988. In 1991, sections of the liner failed during several intense rainstorms. This caused sewer overflows, and subsequently led to removal of the liner from the tunnel. The exact failure mechanism is not fully agreed, although there are several apparent contributing factors and litigation continues. In August 1994, an inspection indicated that more than half of the tunnel was experiencing some degree of corrosion and some areas had significant accumulation of debris from concrete deterioration. The total debris volume was estimated at 3058 cu m (4000 cu yd).

This led to a study by the City of Houston Department of Public Works and Engineering - Greater Houston Wastewater Program. A comprehensive evaluation of the technical issues and economics associated with rehabilitation of the deteriorated sewer was performed by a team consisting of utility and consulting personnel, and a final report, "Evaluation of Alternative Cleaning and Rehabilitation Methods for the Northside Sewer Relief Tunnel," was issued in October, 1995. City of Houston (1995) This report provided significant useful material for the BWARI project because a large number of rehabilitation methods were screened in the Houston study, some of which were applicable to new construction.

The rehabilitation study identified 27 possible systems for rehabilitating the NRST, which can be grouped into coatings, liners, and slipliners. Of the 27 products, 14 coatings, 8 liners, and 5 slipliners were identified from extensive product surveys. The 27 methods were then evaluated by a set of minimum qualifying criteria that were developed based on the physical conditions existing in the NRST. These criteria were:

- Resistant to corrosion with pH $\geq$ 0.5 and H2S concentration $\leq$ 200 ppm
- Can be applied at 100% relative humidity
- Ability to withstand 210 kPa (30 psi) hydrostatic head
- Construction method allows evacuation of the tunnel within a 2-hour time period in response to a wet weather warning
- Cure time < 6 hours
- Suitable for rehabilitation of pipe diameters in the range of 1.8 to 3.66 m (72" to 144")
- Has a history of successful application under conditions similar to those existing in the NRST.

Screening of the 27 potential methods according to the above criteria resulted in elimination of 12 methods, with 15 methods remaining for final ranking and consideration. Included were 6 coating systems, 6 liner systems and 3 sliplining systems. Of these methods, coatings were considered the least desirable based on permanence (unknown design life and higher frequency of failure) and experience (less than 10 years). Liner systems were ranked second most desirable, with greater permanence (estimated at 25 to 50 years design life and moderate ease of repair) than coatings. They also exhibit longer experience records, have been more frequently installed than coatings, and were rated best in constructibility. Slipliner systems were ranked most desirable because of greatest permanence (study stated 50 year design life), greatest experience in terms of years and quanitity of materials installed. They were rated least constructible because of shape conformance problems, loss of flow capacity, need for jacking pits, etc.

Costs were estimated for each system, and a ranking based on cost and risk again separated the methods based on a combined evaluation into the following categories:

- Low Cost/Low Risk - This category included some of the liner systems.
- High Cost/Low Risk - This category included slip liner systems.
- Medium Cost/Medium Risk - This category included liner systems that had higher costs and less performance history than the Low Cost/Low Risk category.
- High Cost/High Risk - This category included coating systems. The coatings were in the high cost category because the analysis was a present worth analysis that included replacement provisions for the coatings given the uncertainties regarding their service life.

Systems recommended for potential application included two Low Cost/Low Risk liner systems and two slip lining systems for cases where slip lining would be more suitable. It was also recommended that the two lowest-risk coating systems be tested in a demonstration project in an accessible location where routine inspection can be performed to document performance. If the coatings perform better than expected, this could improve the risk rating and potentially reduce the present worth cost.

5.2 *Austin, Texas*

Some sections of the Onion Creek Interceptor in Austin, Texas were lined with a mechanically anchored HDPE lining system, installed by the manufacturer as a subcontractor to the tunnel contractor, in the mid-1980's. This 10-foot diameter tunnel was lined for 270° of its circumference; with HDPE liner anchored using a batten strip at the tunnel crown and a batten strip at each lower edge of the liner. Twenty-foot sections of liner were installed, with an overlap at the joints, where circumferential battens were installed. Personal communication with personnel from the Austin Water and Wastewater Utility indicates that there have not been any dramatic failures. A camera inspection of the shafts allowed a view of portions of the tunnel, and the liner was intact where it could be seen. No information is available regarding the condition of the concrete behind the liner, Vallejo (2001).

5.3 *Seattle Washington*

King County Wastewater Treatment Division has installed three sections of sewer with mechanically attached liners, the oldest having been installed in the early 1990's. All three either have been replaced or are scheduled for replacement. Being familiar with the failure of a similar system in Houston, they increased the number of longitudinal battens to reduce the spacing between them. Their experience has been that this lining system does not prevent corrosive gases from contact with the concrete and concrete deterioration has progressed anyway. They have not had any catastrophic failures, and the liners have stayed attached. By means of an active inspection program, they have observed progressive loss of concrete at the anchor locations, which would eventually permit a catastrophic failure and have proceeded with removal of these liners, Browne (2001).

5.4 *Singapore Deep Tunnel Sewerage System*

The government of Singapore has adopted an ambitious deep tunnel sewerage system (DTSS), which will be developed over the years ahead, with tunnels in soft ground and hard rock. The total planned length is 48 km (29.8 mi) of tunnel ranging in diameter from 3.3 to 6.0 m (10.8 to 19.7 ft). The first phase of the work will involve six major contracts to be executed using a design-build approach, the basic design having been performed by a joint venture of CH2M Hill and Parsons Brinckerhoff.

The specified design life of the facility is 100 years and to satisfy that requirement considering the high temperatures and humidity presented by Singapore's climate, a thick composite tunnel lining was specified. The primary tunnel support is a precast concrete segmental lining, inside which a dual CPL consisting of a PVC or HDPE anchored liner cast into a minimum 22.9 cm (9") thickness of unreinforced cast-in-place concrete is to be installed, over 330 to 350° of the circumference. The designers of the design-build teams will perform the detailed design of this system, Singapore (undated).

The General Conditions section of the Instructions to Design-Build Contractors were reviewed and some of the key requirements are listed below: Singapore (undated)

- No chemical dosing is permitted for corrosion protection
- H_2S concentration ≥ 100 ppm
- Very limited capability for future inspections and maintenance
- Design for full hydrostatic pressure, "applied as point sources at various points around the CPL."
- The bottom ends of the lining must be secured
- All manufacturers' recommendations to be met
- Use of trained personnel required with material supplier representative on-site
- Liner anchor to withstand a test pull of 19.6 kN/m (1340 lb/ft) at the liner crown perpendicular to the concrete surface for one minute without rupture or pullout of liner anchors
- Color choice to facilitate future video inspection (light color)
- Stationing marks to be applied
- Full welded joints between sections and of joints at intersecting structures
- Same lining system applies to shafts and other structures
- A 10 m (33 ft) long test section is required prior to production installation.

- Implement an inspection and test plan to address integrity of joints, anchorage of liner, voids, and tolerances. Necessary elements include drilling of a grout hole every 5 m (16.4 ft) for contact grouting (between precast segments and cast-inplace concrete liner) and probe holes to determine CPL system thickness.

One material supplier indicated that the concept of casting the anchored liner material directly onto the precast concrete segments might have been proposed as a value-engineering concept by one of the designers. This would require the specification to be altered, since it now clearly requires a dual CPL consisting of the polymeric liner with integral anchors cast into 22.9 cm (9") of cast-in-place sacrificial concrete.

This is a rigorous specification for a severe design condition and ambitious design life. A full dual CPL system is to be provided. It should be noted that the application of liner materials with integral anchors to cast-in-place concrete is not a simple task, and that the greatest success with these materials in the past has been when they were applied to precast concrete pipe in the casting yard. When applied to precast pipe, the forms are vertical and vibration can be properly performed to ensure flow of concrete and encapsulation of the anchor elements. This is much more of a challenge in a cast-in-place system where the liner material must be secured to a form in a confined annular space and access for vibration is much more difficult.

6 CPL SYSTEM DESIGN CRITERIA

It is useful to review the design criteria that would be employed in designing the CPL system for this project to aid in technical evaluation of the various systems. Highlights of the Singapore documents describing CPL system requirements were listed in Section 5.4 above. Most of those requirements are applicable to provision of a CPL for BWARI, either as is or with some modifications. These can be grouped into several categories, with fairly obvious results, the categories being:

- Criteria derived from the physical layout, sewage characteristics and planned operations of the sewer by the owner. These include H_2S concentration, difficulty of future access, and decision not to use chemical dosing for corrosion control.
- Practical design requirements applying to any system - e.g., color choice consistent with future video inspection, stationing marks, meeting of manufacturer requirements, use of trained personnel, and on-site presence of manufacturer's representative.
- Actual design requirements, either of loadings that must be resisted, or specific aspects of CPL system design, e.g., design for hydrostatic pressure, design anchor pull requirement for liner anchors, welding requirements at joints, application of same lining system to shafts and other structures, and the requirement to secure the bottom ends of the lining.
- Quality control issues - e.g., construction of test section prior to production installation, and requirement to implement an inspection and test plan to address all aspects of installation.

With regard to hydrostatic pressure, it is necessary for all systems except the mechanically anchored liner to resist full hydrostatic pressure, whereas the mechanically anchored system provides for dissipation of such pressure and drainage of any infiltrating water to the invert. The mechanically anchored system would not require liner pullout tests, but a quality control program for bolt installation would be required.

7 BWARI CPL SYSTEM SELECTION PROCESS

Selection of a CPL system involves integration of functionality factors, constructibility, the Owner's risk tolerance, and cost to develop an optimized CPL system. To prepare an integrated analysis of these factors, a list of potential systems for providing a CPL for the BWARI tunnel was developed. All but two of these were based on protecting a primary structural liner of precast concrete segments, some methods requiring over sizing the diameter to accommodate the CPL system. Methods evaluated included:

- Not providing a CPL and implementing a proactive inspection program to identify CPL need, while providing oversized diameter to accommodate space for retrofit liner
- Layers of sacrificial cast-in-place concrete or shotcrete
- Mechanically anchored HDPE sheet liner
- PVC or HDPE sheet lining with integral anchors, cast onto precast concrete segments, with welded joints.
- Non-structural HDPE membrane with integral anchors grouted in place inside segments
- Profile-wall HDPE pipe (formed by spiral welding of HDPE rectangular tubes) grouted in place inside segments. The section thickness proposed was about 17.8 cm (7").
- Placement of T-lock lined concrete pipe inside segment tunnel and grouting annulus.
- Casting of precast concrete segments using polymer concrete.

Evaluation of these alternatives involved a matrix process with the design team and City engineering and maintenance representatives and was documented at the 2001 Rapid Excavation and Tunneling Conference, Uhren, (2001). The cost premium added to the estimated cost per foot of the base tunnel ranged from 18 to 103 percent. The City was not willing to select a system with a very large premium, and most of these systems had other significant design problems as well.

Additional design attention focused on the more economical systems and involved provision of a detailed package of information to a number of material suppliers in order to obtain more accurate cost information for additional evaluation. Systems evaluated at the next level included:

System 1 - PVC or HDPE sheet lining with integral anchors, cast onto precast concrete segments, with welded joints.

System 2 - Mechanically anchored HDPE sheet liner.

System 3 - Non-structural PVC or HDPE membrane with integral anchors grouted in place inside segments.

System 4 - Profile-wall HDPE pipe (was retained in the study because the manufacturer could offer a more economical price by fabricating the pipe sections locally).

System 5 - Linabond Co-Lining System - Linabond, Inc. of Sylmar, California- This "chemically-attached" liner system had been reviewed previously but pricing appeared unattractive. It has been most often applied for rehabilitation projects. Clarification of the project details led to far more attractive pricing. Linabond was developed in view of the fact that concrete is never totally dry. The concrete surface is prepared for application of the primer and structural polymer by sandblasting or hydroblasting to remove laitance and salts and to open surface pores. The first system component is a hydrophyllic penetrating primer that also contains an agent that bonds to silica in the concrete aggregate. A two-part mix structural polymer is then applied to the concrete and while it is tacky, an extruded PVC sheet that has been treated on the back with an "activator" compound is applied to the structural polymer. The activator facilitates cross-linking of the mastic and liner, yielding a strong bond. Joints are made using a structural polymer joint material and overlapping the PVC sheet.

Results of the subsequent price comparison are shown in Table 1.

Table 1. CPL System Cost Comparison.

System #	CPL System	Cost Premium, % Increase Over Base Cost
1	Anchored PVC or HDPE sheet lining cast onto con-crete segments	20-22%
2	Mechanically anchored HDPE sheet liner	23%
3	Anchored PVC or HDPE sheet liner grouted in place inside segments	29%
4	Profile-wall HDPE pipe	68%
5	Linabond Co-Lining System	20%

System 2 was rejected because of case histories documenting problems with other applications and concerns about possible ongoing corrosion in spite of the liner. System 3 was rejected because of execution difficulties, because of concerns about performance under hydrostatic pressure potentially applied to the segment-grout interface, and because of potential erosion and loss of section at the invert. This could compromise the integrity of the grout ring, required for the stability of this liner system. System 4 was rejected because of the prohibitive cost in comparison with the other systems and the significant application details that would need to be addressed, since this system has not previously been used for such an application. This left System 1 and System 5, which are discussed further below.

7.1 *System 1 - Anchored PVC or HDPE Liner, Cast onto Segment Face*

Anchored PVC or HDPE liners have been in use for a few years in the case of HDPE, and over 50 years for PVC. They are finding increasing utility in both new construction and rehabilitation applications where protection of concrete against corrosion is required. Both materials are regularly cast onto precast elements, either pipe or panels, and the liner joints are welded following fabrication of the concrete elements. A schematic diagram of a liner joint detail is shown in Figure 1.

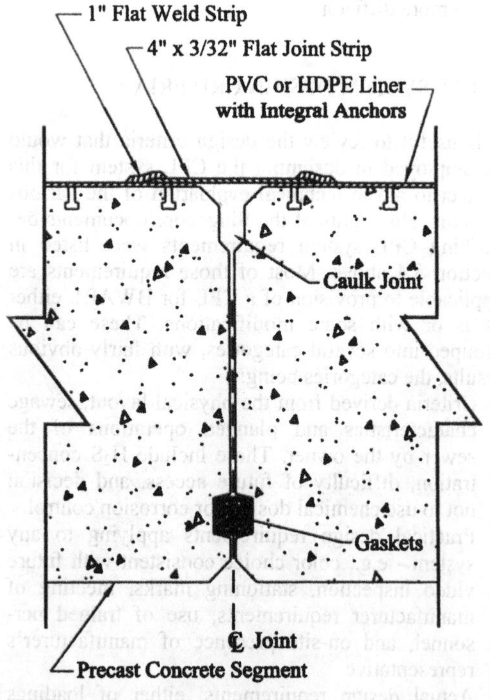

Figure 1. Anchored PVC or HDPE Liner Cast onto Segment Face - Welded Joint Detail.

These systems are designed to resist hydrostatic pressures up to levels nearly twice those expected on the BWARI project (413 kPa {60 psi} and higher versus 241 kPa {35 psi} expected). They are anchored to the concrete by means of closely spaced ribs or studs. The ribbed varieties can be oriented to permit any infiltrating water to drain to the tunnel invert, and the studded varieties allow this to happen in any orientation.

Welding of liner elements has progressed far beyond the high-production, quality-compromising procedures that became notorious during the peak of liner installation for landfill liners and caps. Welding of polymeric liners received a bad reputation in terms of quality control and durability of the welded seams. The welding procedures that have been developed by the part of the industry applying liners to structures have been codified in standards similar to those for welding metals, and when applied by properly certified installers, achieve high-quality results.

It is true that this system has not been applied in exactly the means proposed for this project, namely on precast segments used for tunnel support. However, all elements of the proposed alternative are proven, and the combination of these elements is not a major leap. The process of applying liner material to precast concrete segments during casting is straightforward and was considered feasible by manufacturers of tunnel segments. Sehulster (2001) The two disadvantages of this system are that it is in place and thereby subject to damage throughout construction and that it requires extensive welding in the tunnel. As noted in Section 4, 20.7 linear meters (68 ft) of joints exist per 1.22, (4 ft) tunnel advance and many welding techniques would require welding a cap strip on both sides of each joint, doubling the welding footage. Patches would also be requires at all bolt hole locations, adding significantly to the total footage of welding required.

7.2 *System 5 - Linabond Co-Lining System*

Linabond was developed primarily for rehabilitation applications, using the structural polymer layer to provide structural strength to deteriorated sections, and bonding to it a PVC liner to provide the necessary protection against sewage and corrosive gases. The system components are shown schematically in Figure 2. It has been use for about 10 years and has some enthusiastic supporters among wastewater utilities (Browne, 2001; Public Works, 1998). It appears to have been thoroughly tested and is well documented.

This system has a significant advantage in that it would be applied at the end of construction, and would not be subject to damage during construction. It is chemically welded, allowing much more rapid installation than the HDPE or PVC linings requiring extrusion or heat welding, both of which are slow

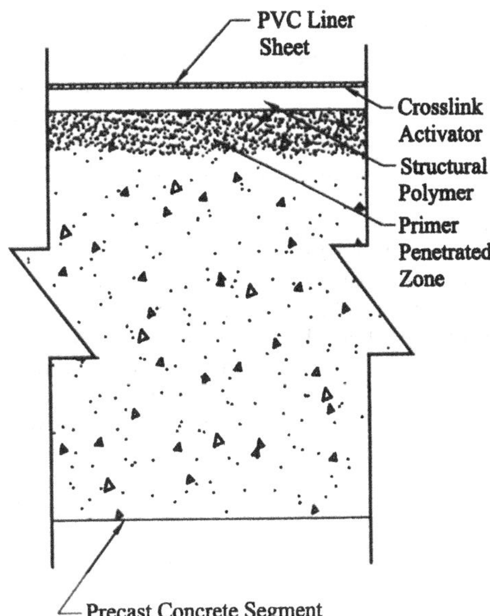

Figure 2. Schematic of components of Linabond Co-Lining System.

and labor-intensive, and subject to some percentage of defects which could become entry points for gases and initiation of corrosion. The Linabond system also provides great flexibility in design of connections at structures because it can be adapted to essentially any shape and the necessary connections made with the proprietary joint material.

8 TESTING OF LINABOND SYSTEM

8.1 *Rationale and Tests Performed*

Linabond presented impressive test results and demonstration tests with the co-lining system applied to relatively porous concrete blocks. This included pull-off strength and flexural strength testing of concrete beams with and without the Linabond co-lining system. The pull-off test typically resulted in a tensile failure in the concrete rather than at the interface and the beam with Linabond had substantially greater flexural strength and ductility. Results of hydrostatic testing using a Federal coating standard reported by the material supplier indicated that applied hydrostatic pressure typically resulted in tensile failures in the concrete rather than de-lamination of the lining system components. U. S. Government (1977).

There was concern on the part of the design team that application of this system to dense, relatively impermeable concrete segments might yield less successful results. A small trapezoidal key segment was obtained from a segment supplier so that Linabond could be applied to it and tests performed to gauge its effectiveness on this substrate. The test of greatest interest was direct pressure testing to the applied Linabond co-lining system. In addition, flexural strength testing was performed on beams cast of 41.4 MPa (6000 psi) concrete, and pull-out testing was performed on Linabond applied to flat concrete blanks.

The design team proposed that a hole be drilled entirely through the concrete segment so that pressure could be applied directly to the back of the liner. A 6.35 mm (0.25 inch) hole was drilled prior to installation of the Linabond to the face of the segment, which was trapezoidal, nine inches thick, and had maximum dimensions of about 1.0 and 1.2 m (39 and 47 in). The rear end of the hole was reamed to a diameter of 12.7 mm (0.5 inch) and a pipe nipple was cemented into the hole for attachment to a pump to apply pressure. A schematic of the test set-up is shown in Figure 3.

The face of the segment was sandblasted to remove any laitance or foreign substances and to roughen the profile of the segment face to facilitate bonding of the co-lining system. This can be done with either sandblasting or high-pressure water blasting. The primer was applied and allowed to cure. The structural polymer was then applied using a notched trowel to control the thickness to about 3.2 mm (125 mils). On a production scale, this material is sprayed on using a special metering system to mix the two-part system at the spray nozzle. Finally, the PVC liner sheet, which had previously been cut to shape and treated on the back with the cross-link activator was applied, pressed into place, and held to the curved segment face using a clamping system. In the field, this is done using a special form that is held in place using a system of jacks for the time required to achieve the initial set of the structural polymer.

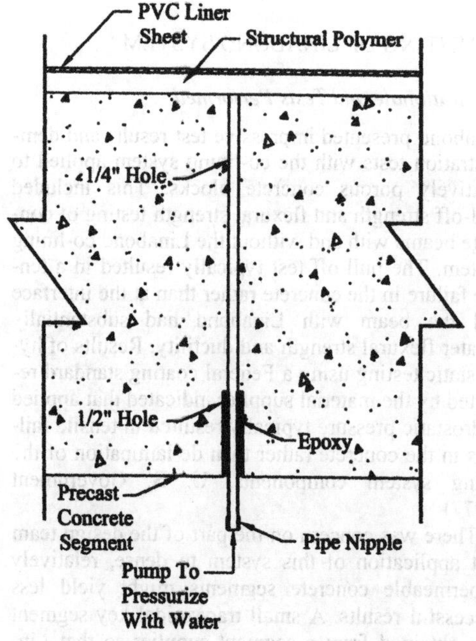

Figure 3. Test set-up for direct pressure test.

8.2 *Test Procedures and Results*

Pull-off testing involves coring through the lining system and into the concrete and cementing a steel loading fixture to the section of liner isolated by the coring, according to the procedures of A.S.T.M. D4541. After the cement is cured, a testing device is attached to the loading fixture, and is supported against the surface at three points. Tension is then applied to failure, and a dial gauge records the load, which can be converted to a value for tensile stress. The device used for this test was the Bond Test Instrument No. 12-2485 by Germann Instruments, Inc., having a loading fixture diameter of 75 mm (3"). Values of 1.86 MPa (270 psi) and 2.24 MPa (325 psi) were obtained for the two tests performed and in each instance the failure was entirely in the concrete. These values are not true tensile strengths, because the pulling force is not necessarily perfectly axial. This device is made for field-testing and quality control and primarily tests whether the lining material adheres such that the break is in the concrete, or if it de-laminates.

The flexural strength test was performed using 150 mm (6") square beams, two of which were plain and two of which were cut to 125 mm (5") thickness with one inch of structural polymer and the PVC liner sheet applied. This test was performed according to A.S.T.M. C78, the flexural strength test for beams using third-point loading. The rupture modulus values for the unreinforced beams were 4.9 and 5.0 MPa (710 and 725 psi), and those for the beams with Linabond applied were 11.0 and 10.3 MPa (1602 and 1500 psi). The latter beams did not fail by rupture at the extreme tensile fiber because of the Linabond, which adds considerable strength and ductility. At failure, the beams with Linabond exhibited diagonal shear cracks and signs of compression failure at the loading points, so the calculated rupture modulus cannot be directly compared. It is obvious that Linabond, besides providing corrosion protection, adds considerable mechanical strength to the section.

The direct pressure test was conducted by attaching a system of dial gauges and a pneumatically activated pump to pressurize water and supply it through a fitting to the hole that was drilled through the segment prior to application of the Linabond co-lining system. As indicated above, the design pressure is about 240 kPa (35 psi). Accordingly, it was decided to apply pressure increments of 414 kPa (60 psi) for 30 minutes, and 620 and 827 kPa (90 and

120 psi) for 15 minutes each. This was successfully accomplished, and then the loading was increased to failure, defined by water leakage from several locations in the concrete segment. The failure load of 9.7 MPa (1400 psi) was sustained for several minutes.

9 CONCLUSIONS

The high cost and operating problems caused by a need to rehabilitate deteriorated sanitary sewer pipes and structures warrant provision of a CPL system at initial construction for deep tunnel sanitary sewer conveyances. Bolted and gasketed precast concrete segments are an expedient choice for tunnel support in water-bearing glacial soils. The segmental nature of this system leads to some different challenges in designing a CPL system from those encountered with concrete pipe tunnel liners.

The CPL system selection and design process for the BWARI sewer project in Columbus, Ohio resulted in adaptation of a proven CPL system, i.e., System 1 - PVC or HDPE liner sheets with integral anchors cast onto the segment faces and welded joints. An additional system that had previously been used primarily in the rehabilitation market was also identified as having very strong potential, i.e., the Linabond co-lining system that is chemically attached. Final CPL selection may focus on one system, or the design may provide for application of either system. This decision will have been made and be available for reporting at the time of presentation of this paper at the NAT 2002 conference.

The testing program performed on the Linabond system showed encouraging results, particularly the direct pressure test, which far exceeded design requirements. The direct pressure test and the pull-off test confirmed the capability of Linabond to adhere strongly to the surface of high strength concrete used to manufacture precast concrete tunnel segments.

10 ACKNOWLEDGEMENTS

The author would like to acknowledge the support of Lachel & Associates, Inc., URS, and the City of Columbus, Ohio in preparation of this paper.

REFERENCES

Browne, Roger. 2001. Personal communication - Civil Engineering Supervisor - King County Wastewater Treatment Division Seattle, Washington.

City of Houston, Texas. 1995. Final Report - Evaluation of Alternative Cleaning and Rehabilitation Methods for the Northside Sewer Relief Tunnel. *Study Report for Greater Houston Wastewater Program*. October 1995. Houston, Texas.

U.S. Government. Federal Specification TT-P-1411A, *Paint, Copolymer Resin, Cementitious (for Waterproofing Concrete and Masonry Walls*. 1973 and 1977 amendment. June 28, 1977.

Frank, G.D. and Daniels, J., 2000. The Use of Borehole Ground Penetrating Radar in Determining the Risk Associated With Boulder Occurrence. *Proceedings, North American Tunneling 2000*. June 2000. Boston, Massachusetts.

Frank, G.D. and Chapman, D. R., 2001. Geotechnical Investigations for Tunneling in Glacial Soils. Proceedings, *Proceedings, Rapid Excavation and Tunneling Conference*. June 2001. San Diego, California.

Joyce, James, 1995. Odor and Corrosion Control in Collection Systems: A Growing Problem? *Proceedings, Water Environment Federation Specialty Conference entitled, "Sewers of the Future,"* September 1995, Houston, Texas.

Joyce, James, 1998. An Evaluation of Lining Technologies for Corrosion Protection and Rehabilitation of Large Diameter Sewer Tunnels and Interceptors. *Water Environment Federation WEFTech Asia Conference*. March 1998. Singapore.

Joyce, James, 2001. Personal communication.

Peggs, I.D. and Hammer, H.I., 1999. Cast-in Liners Gain Foothold. *Geotechnical Fabrics Report*, Vol. 17, No. 5, July, 1999, p 24-30. Roseville, Minnesota.

Public Works. 1998. Lining System Does the Job. *Public Works*, Vol 129, No. 2, February 1998, pp 46-48.

Sehulster, Joseph. 2001. Personal communication.

Singapore Wastewater Utility, undated. Design and Construction of Sewer Tunnels for the Deep Tunnel Sewerage System - Contract T-04 and Mo Kio Tunnel - Instructions to Design-Build Contractors.

Uhren, D.J. and Gilbert, G.W., 2001. Matrix Evaluation to Resolve Corrosion Risks for a Sewer Tunnel in Difficult Glacial Soils. *Proceedings, Rapid Excavation and Tunneling Conference*. June 2001. San Diego, California.

University of Houston. 1997. University of Houston Environmental Engineer Searches for Solution to Wastewater Pipe Corrosion. Press Release. August 15, 1997. http://www.uh.edu/admin/media/nr/archives97/0897/roberts.html

Vallejo, Yvonne, 2001. Personal communication - Engineering Associate - Austin Water and Wastewater Utility. Austin, Texas.

Session 3, Track 1

Contracting practices

North American Tunneling 2002, Ozdemir (ed.)
© 2002 Swets & Zeitlinger, Lisse, ISBN 90 5809 376 X

Should Firm-Fixed Price contracting to the low bidder be outlawed for underground construction?

Timothy P. Smirnoff
Parsons Brinckerhoff Quade and Douglas, Los Angeles, California

ABSTRACT: Clearly underground construction is a different animal and requires a different approach to contracting. Conventional bidding and tender practices are no longer satisfactory for the Owners of large underground construction projects, as have the later claims and changes becoming increasingly contentious and litigious. Alternative procurement methods have been used by several federal and local governmental agencies to obtain for their agencies a well-constructed and operational product and the Contractors have made a profit with a minimal number of after construction claims and litigation. This paper will discuss several of these unique procurement methods.

1 INTRODUCTION

Underground construction is clearly a different animal when it comes to the procurement, delivery, and eventual litigation for both Owners and Constructors. No other form of construction results in more disputes, larger claims, and judgments for differing site conditions and defective specifications. The use of the conventional sealed competitive bidding procedures, for most Public agencies requires the contract be awarded to the lowest responsive and responsible bidder. Most Owners have found that for underground construction, the Firm-Fixed Price (FFP) bid of the lowest bidder in the end, well after the last shove of dirt is moved, is often neither the lowest nor responsible. The time has come to change of bidding regulations and requirements to allow a way to ensure that the proposer has the resources, experience and has a selected, reasoned and fully developed means and methods which have a reasonable chance for success at a fair and equitable price. This FFP low-bid/award process does not usually allow a determination before the award is made and then only the most superficial verification of contractors experience and financial viability is performed. The reliance on a performance and completion bond has sadly left many owners with little or no recourse but to continue the folly and litigate a happy ending. Further more the increasing complexity of the projects attempted in marginal or poorer conditions, the need for specialty equipment requires that special consideration to ensure the basis for the development of contract documents that allocate the risks of construction and ensure a economical and timely delivery of the project.

1.1 *Contracting Objectives*

It is axiomatic that the type of project delivery method that an owner should select is that type which, is calculated to best achieve the owner's goals. In turn, the owner's goals can best be achieved by selecting the contract type that will most effectively motivate the contractor to the desired end. The most effective contract type will vary from one situation to another, depending on many factors including: the degree to which the plans and specifications or statement of work—are firm and final, the extent to which the owner is willing to assume some or all of the risk inherent in the work, the extent and ability of the owner to monitor or manage actual construction work, and the value the owner places each of his goals: most economical cost, specified quality and completion on schedule.

The owner's goals are rarely completely attained in any one contract because these goals are in part mutually exclusive. The most economical cost may compromise the specified quality and specified completion. The quality works against low cost and rapid completion. The most rapid possible completion works against low cost and high quality.

There is no single type of project delivery method that best fits these criteria in every possible situation. A set of goals has been identified as the principal criteria for selecting the appropriate project delivery method. In summary, these goals are:

- Compatibility with Owners policies and constraints
- Responsiveness and accountability to public
- Confidence in quality of the work
- Meet environmental requirements
- Minimize risk exposure and litigation
- Foster innovation and ingenuity to reduce costs
- Clear lines of responsibility and accountability
- Incentive for schedule improvement
- Flexibility to incorporate scope changes
- Continued involvement of designers.

From a purely practical and pragmatic point of view, other items have an effect on the general suitability of alternative delivery methods, such as design-build construction. For example, these include:

- Size of project – for smaller projects an agency's standard procurement method is usually preferred.
- Complexity of project – alternative delivery methods are most beneficial for more complex projects, especially those that might benefit from the use of proprietary systems.
- Definition of project scope – alternative delivery methods provide less time and opportunity for scope changes.
- Resolution of constraints – environmental constraints resolution and the acquisition of permits are best carried out by the owner or the owner's engineer; a design-build contractor might encounter costly delays.

2 PROJECT SPECIFIC FACTORS AFFECTING PROCUREMENT

Contracting for the construction of many of today's tunneling projects requires additional consideration and detail as discussed below.

- The tunneling for many projects is challenging, since many of the sites now chosen for development are 1) in older cities or more difficult ground conditions which may include contamination with hazardous or toxic materials, 2) on sites which include obstructions from long buried and undocumented foundations, utilities, rubble fills, 3) within areas of which provide special environmental or special remediation or mitigation concerns.
- While the elements of design are important, successful tunneling is more dependent on the details of construction technique and methods than any other form of heavy construction. Excavation, support, and maintenance of the tunnel face and excavated perimeter are crucial. The adverse effects of tunneling, i.e., surface subsidence and settlements, impacts on utilities and surrounding structures etc., are almost universally attributed to construction methods and techniques.
- While experience is gradually developing within the United States contracting community with pressure face tunneling (PFM) machines and with gasketed, bolted segmental, precast concrete linings, there are areas with this country with virtually no tunneling experience under the anticipated special geologic conditions mentioned earlier. The novelty of this work and the severity of tunneling conditions suggest that it would be unwise to leave such construction to the vagaries of the low bidder. Further, the contracting method must maximize the chances that the successful responsible bidder has the correct mix of experience, organization, and financial resources to execute the work.
- The unique and specialized nature of the work requires a Contractor with specialized experience and equipment, which is absolutely necessary for successful construction of the tunnel. Qualifications for prime contractor and subcontractor personnel must be provided and rigidly enforced, as must all contract requirements be so enforced.
- Qualified experts with experience in the supervision of such work must be an effective and integral part of the construction management group. They must have sufficient authority and empowerment to maintain control of the construction operations, to ensure the design intent and integral construction sequence, and to protect the owner's interests by issuing and enforcing direction to the contractor if required.
- Construction monitoring and enforcement of specification requirements for the tunneling process, alignment, liner installation and deformation, contact grouting and machine performance, are critical to successful construction of this project. Instrumentation and monitoring must be provided and used in a timely and accurate manner to provide the required feedback on ground behavior and the necessity for construction modifications. Construction management staff must have the required equipment, staff, experience, and authority to analyze the data and provide timely and competent assessments of conditions and to direct that modifications be made to construction means and methods, rates of production, use of supplemental grouting, probing and related construction expedients.
- It must be understood that the ground encountered belongs to the Owner. The Contractor may promise to deliver a tunnel in a competent and timely manner, but whatever happens, it is ultimately the Owner who pays for it. A contractor will load a hard money bid with contingency funds to the extent that he considers that there is either insufficient information or unwillingness on the part of the owner to stand behind the information and designs provided.

3 PROJECT DELIVERY ALTERNATIVES

Project delivery methods examined in this document include the following main methods:

- The conventional design-bid-build delivery method and several of its derivatives as they have been practiced by most agencies in the USA over the last several decades.
- The more innovative (for underground work) design-build method, in which the contractor is responsible for the final design; used in recent years for major rapid transit systems around the world.

3.1 *Conventional Design-Bid-Approach*

The procurement procedure typically used by the federal government, state and most other public agencies to promote full and open competition requires that it must advertise publicly for the receipt of sealed bids (FAR Part 6.100). This is similar to the regulations governing acquisition of Federal contacting (FAR Part 14.101) that may be submitted by any contractor who can obtain and provide a bid bond with his bid. Award may be made only to the lowest responsive bidder who has established that he is responsible (FAR Part 9.103). To determine the responsibility of contractor requires that bid be evaluated for adequacy prior to award of the contract (FAR Part 9.105). This generally entails the review of a Bidders General Information Submittal application of the Special Conditions and, once the contract is awarded, review and approval of Technical Provisions requiring descriptions of proposed methods for tunnel construction, experience of project manager, tunnel superintendent, methods and mean of tunnel support, etc. The bid must be to perform the work strictly as it is advertised and may not be qualified or conditioned concerning price, quantity, quality or time for performance. Such contracts are in the form of a Firm-Fixed-Price Proposal. In this type of contract, the parties agree that the contractor obtains full benefit in the form of profit or accepts full responsibility in the form of losses for all costs under or over the firm-fixed-price. This type of contract provides the owner with a presumably firm price for the work and passes the maximum risk to the contractor.

Normally there is little or no right of refusal or acceptance criteria for contractors with projects going to the low bidder. A bonding company guarantees completion of the project. In many instances low bidders who have obtained bonds have nevertheless, been rejected as not being responsible. Both federal, state and local contracting officers authority to reject bidders has been upheld (FAR Subpart 9.103(b)). The difficulty with the standard procedure is that determination of responsibility is made only after receipt of bids. There is a natural reluctance to declare a low bidder not responsible after he has incurred the expense of bidding and after he is declared the low bidder. Possible litigation, with attendant delay in award of contract is also possible. Rebidding or reprocurement of bids in instances of rejection of bids rarely result in the same advantage of the Owner and does not ensure a lower price or a more responsible and responsive contractor.

3.1.1 *Owner Furnished Equipment*

Using this conventional contracting approach, specification and contract documents for procurement of tunnel excavation and support are generally performance based. That is, these documents present a number of performance criteria for the execution of the work with the contractor to provide the means, methods, equipment, and procedures. The Contractor would decide on the details and designs of his own choosing regarding requirements for tunnel shield, geometry, tunnel initial/final support, and the machine's ability to handle the described ground and water conditions. Criteria for control of losses of ground and for allowable ground movements would also be set in the Contract Documents. These standards would be used to review and accept the tunneling machine and equipment that the contractor selects for the job as well as a means to ensure that ground losses are controlled and protection of the neighboring public and environment are maintained. This method requires the presentation of a minimum series of experience requirements of the contractor and listing of experience record of the contractor and his key personnel and key sub-consultants. In addition, the experience of tunnel shield manufacturers, segment suppliers and other specialty subcontractors must also be detailed. It is generally difficult, however, to hold contractors to these minimum experience and past performance requirements. Once a low bid is received, public agencies generally do not want to declare the low bidder unacceptable, especially when the owner may have realized large savings in the low bid. Similarly, once the bid is awarded and submittals of the equipment are reviewed and approved, contractors often argue (with success) that the owner thereby agreed the equipment was appropriate for the job and, since it isn't working, it is obvious that the contractor has a differing site condition.

The selection of details of means and methods of construction by the Owner, such as dictating the tunneling method or shield type (e.g., use of an EPB shield) in the traditional approach places added responsibility on the Owner and implies warranty to the contractor for the supplied equipment or materials, such as segmental liners. In this situation, the contractor asserts that the Owner warrants that these items will perform to complete the project successfully. Machine breakdowns, faulty parts, extra parts, delay etc., which arise during construction, therefore, become the responsibility of the Owner and not the contractor. Similarly, any financial ramification of these eventualities is asserted to be a liability of the Owner. The contractor claims no stake in the means and methods of the project and accepts responsibility only to provide hired help to perform the work. Often there is little or no way to induce the contractor to make the equipment work or perform under the conditions encountered. The contractor is not encouraged to use innovation or his construction experience to develop the means and methods of construction, which are talents he is generally expected neither to provide nor to perform the work of a higher quality.

In recent times, the use of owner provided equipment and materials has also demonstrated the vulnerability of the owner:

- At the St. Claire Tunnel project in Sarnia, Canada, the failure of the main bearing and seals that resulted in a need to sink a special shaft to rescue and repair the owner furnished machine and delayed construction for seven months became the responsibility of the owner. It is acknowledged, however, that the completed tunnel was quite satisfactory.
- Similarly, the discovery of boulders on the Sheppard Extension of the Toronto Subway would have been a tremendous liability to the owner had the tunnel construction not been delayed. During the delay period, construction of an open-cut station excavation revealed boulders, which necessitated the rebuilding of the tunneling machines to a fully breasted face with recessed disc cutters. This rebuild cost the Owner approximately $3.5 million and because of these construction delays required a $500,000 factory rebuild of the hydraulic seals and gaskets that had deteriorated during the storage of the machines.

Many of the large tunnel construction projects in the United States including subway projects in Washington, D.C., and Los Angeles, the deep tunnel projects in Milwaukee and Chicago, and the Superconducting Super Collider, had at one time or another considered such options. After careful deliberation and consideration of the risk to the Owner, none of these contracting entities chose to assume the risk associated with pre-purchase of equipment or materials.

In recent times, the use of owner provided equipment and materials has also demonstrated the vulnerability of the owner:

- At the St. Claire Tunnel project in Sarnia, Canada, the failure of the main bearing and seals that resulted in a need to sink a special shaft to rescue and repair the owner furnished machine and delayed construction for seven months became the responsibility of the owner. It is acknowledged, however, that the completed tunnel was quite satisfactory.
- Similarly, the discovery of boulders on the Sheppard Extension of the Toronto Subway would have been a tremendous liability to the owner had the tunnel construction not been delayed. During the delay period, construction of an open-cut station excavation revealed boulders, which necessitated the rebuilding of the tunneling machines to a fully breasted face with recessed disc cutters. This rebuild cost the Owner approximately $3.5 million and because of these construction delays required a $500,000 factory rebuild of the hydraulic seals and gaskets that had deteriorated during the storage of the machines.

Many of the large tunnel construction projects in the United States including subway projects in Washington, D.C., and Los Angeles, the deep tunnel projects in Milwaukee and Chicago, and the Superconducting Super Collider, had at one time or another considered such options. After careful deliberation and consideration of the risk to the Owner, none of these contracting entities chose to assume the risk associated with pre-purchase of equipment or materials.

The advantages of this conventional approach of contracting:

- Procurement procedures are simple and straightforward.
- Results are generally clear-cut.
- Means and methods generally are the responsibility of the contractor.
- The owner controls negotiations and permitting, etc., with third parties.

The disadvantages of such conventional procedures:

- The contracting officer (generally non-technical) makes the determination of contractor responsibility.
- The owner has little input regarding the methods of construction.
- The method generally favors staid and tried approaches with little inventiveness.

More over it must be understood, by the OWNER, that the cornerstone of modern underground contracting practice is that "the ground encountered belongs to the Owner". The Contractor may promise to deliver the tunnel in a timely manner, but whatever happens, it is ultimately the Owner who pays for it.

3.2 *Design–Build and Derivative Approaches*

3.2.1 *"Convention Design-Build"*

As the term design–build method is commonly used and understood in the underground construction industry, the project is awarded to the firm submitting the lowest bid price (low bidder) based on the preliminary engineering documents prepared by the Owner's Engineer. For complex and multi-disciplined projects, proposal documents are generally taken to a much further stage of design to assure all of the Owner's needs and requirements are fully detailed and explicitly defined in the tender package. The selected contractor becomes responsible for final design and construction. Prices can be adjusted based on an agreed formula once the contractor completes the final design and the actual work is better defined. It is claimed that the design–build process can shorten the delivery time by shortening the time required for final design and for procurement of major equipment. While records show that construction times can indeed be shortened, there is little evidence that lower costs result.

The design-build alternative now is finding interest and application within the underground construction industry. Design-build has been used on several toll road facilities and bridges in the United States and is often used for underground construction in Europe and Asia. In these countries most contractors maintain large in-house engineering forces and do almost all of the design for both the permanent works and construction-related design as well. To allow design of a Design-Build project the owner typically retains an engineer or other technical advisor to produce a project preliminary design to about a 30 to 35 percent level.

3.2.2 *Design-Bid-Design Hybrid*

A hybrid approach to this method combining aspects of the conventional design-bid-build delivery with the design-build method is the advanced design-bid award process. With the advanced bid, the owner's engineer carries the design well past the preliminary state, to a 70 or 80 percent level and then bids the project. The engineer and the contractor, typically the low bidder, then complete the design together while incorporating the contractor' input and ingenuity. Recently the Bureau of Engineering Services, City of Portland, Oregon, used a similar technique, whereby they selected the Contractor using a Qualifications based selection system, similar to that used to procure engineering service, and then carried that selected contractor into the final design effort with existing city engineering Consultant. Final construction costs are then negotiated once the design is finalized by the Team of owners engineer and the Contractor.

3.2.3 *Disadvantages of Design-Build*

The major disadvantage with the design-build approach and its deriviates is the overall reduction in control of the project ownership during the final design and construction, and the quality of the completed work, since the contractor and not the Owner or Owner's Engineer does so much of the work. Most domestic Owners are used to and demand a role in the review, determination of standards, and control of project components and materials. Controls of these things are lost in design- build unless clearly defined in the tender documents. Many owners have also found that the design-build team is solely concerned about the constructed product and that control of the effects of construction on adjacent structures, utilities, and environmental mitigations, and long-term operations and maintenance is of little concern. While Owners have generally tried to shift the total risk for construction to the contractor in design-build, the claims for adjudication for differing site conditions is difficult, unless all necessary geotechnical explorations are carried out and baselines properly interpreted before tendering. The design–build approach, however, has not been particularly successful for large "state-of-the-art" tunnel projects in complex regulatory milieus.

According to recorded experience, the design–build process has been found best to use when the work to be built is a routine structure such as a conventional building, and where interfaces are minimal or already fully settled. One-of-a-kind structures such as urban underground works with interaction with a large number of public and private agencies, which are all stakeholders in the project, are less suited for this design-build process. Design–build here often leads to delays (for example due to permitting issues) and claims due to design changes or differing site conditions.

Advantages of Design-Build Approach:

- Potential for earlier completion
- Contractor involvement in input to the design process occur early in the project
- Single contract with builder simplifies accountability and lines of responsibility
- May reduce owner's exposure to construction claims and change orders
- Earlier availability of project cost estimates
- Encourages contractor ingenuity and increased collaboration with engineer.

Disadvantages of the Design-Build Approach:

- Owner relinquishes control over design and will have less control over the quality of work
- Requires early and clear definition by owner of <u>all</u> project requirements; there is little flexibility for changes
- Potentially less competition-fewer teams may be competing for the work
- Environmental and permit issues, if not already resolved, could cause construction delays.

The use of design build is generally not allowed in most State and Local jurisdictions and requires legislative action to permit its use. Several toll road and revenue based projects in California and several other states have used design build, as has the Texas

Department of Transportation for several demonstration projects.

3.3 *Negotiate Compressed Procurement*

The Negotiated Compressed Process (NCP) requires the involvement of all parties (Owner, engineer, and contractor) in the selection and design of essential components of the project. If these are preordered and manufactured prior to the selection of the main tunnel contractor, the early action can reduce the overall project schedule.

The key elements of the negotiated compressed process for tunnels are as follows:

- Definition of long-lead-time items
- Prequalification of contractors and suppliers
- Formation of technical working groups
- Prequalification of suppliers of linings and TBMs, if accelerated procurement of these are desired
- Buy-in and consensus forming
- Risk management assessment
- Tendering, obtaining bid proposals from prequalified contractors and suppliers
- Evaluation, selection of contractor, negotiation of contract
- Contractor's final input to TBM design, now under fabrication
- Ownership of TBM and timing of conveyance.

Key to the NCP procedure is the assembly of a group of prequalified contractors and suppliers, who with early involvement in the design and development of the contracting packages provide early and timely input to the Owner.

To organize such participation, representatives of prequalified contractors and suppliers meet with the Owner and Owner's Engineer as the Technical Working Group. This group participates in deciding the parameters for key long-lead items and in effect, 'buys in' to these essential long-lead items before the bidding for the selection of the main tunnel contractor.

For the NCP delivery process to work well on most projects would necessitate the Technical Working Group to agree to the basic design and have say the Owner preorder the TBM and the segmental tunnel lining forms.

These steps in the process are further defined as follows:

- *Formation of the Technical Working Group* - The Technical Working Group is to include one representative from each of the prequalified contractors and suppliers, as well as the Owner and Owner's engineer. Each representative would bring the full expertise or his or her company to the process to assist in determining the most appropriate design for the TBM and the lining. Each contractor representative would have the full authority to act on behalf of his or her respective firms and to sign an acceptance of the TBM, the tunnel lining system, and any other aspect as may be required.
- *Technology Acceptance Agreement* - Another element of the NCP process is the Technology Acceptance Agreement. The purpose of the Technology Acceptance Agreement is to save the Owner harmless, and release and indemnify the Owner from all claims and demands attributable to the design, manufacture, delivery, assembly, and performance of the TBM. Only contractors who maintain their prequalification status and who execute the Technology Acceptance Agreement would be allowed to bid the project.
- *Function of the Technical Working Group* - This process is intended to simulate, as much as practical, the conditions of acquisition that would have applied had the contractor procured these items directly. It also provides the Owner with the advantages of competitive bidding, while simultaneously providing the contractor with an opportunity to participate in the design using the latest and most appropriate technology. The technical representatives are usually paid at a per diem rate for their participation in the regularly scheduled technical reviews. The Technical Working Group would be dissolved prior to the call for tenders for the tunnel contract.

Design details that the Technical Working Group determines can be delayed without prejudice to the schedule are left as much as possible and practical to the successful contractor to complete after award.

With long lead items such as TBM and lining options settled early in the process, their purchase can be made immediately after the main tunnel contractor is selected. The successful contractor would be required to approve the TBM and assume full responsibility and ownership for it before it is shipped from the manufacturer's plant. The contractors would be required to send their own operators, electricians, and mechanics to the TBM manufacturing facility for complete training on the TBM that would include operation simulation.

The advantages of the NCP are:

- Contractor input during design process
- Long lead equipment is accepted by prequalified contractors and may be pre-ordered
- Schedule advantage from preordering of long-lead equipment.

The disadvantages of negotiated compressed process:

- Limited experience within the United States
- Unfamiliarity with process by procurement officials and contractors

- Reliance on technology acceptance agreement
- Necessity to obtain authority to use this method from State Legislature
- Reliance on prepurchased equipment or owner furnished materials.

There is limited experience within the United States with this method in procuring public works contracts. The need for contractors to sign a Technology Acceptance Agreement and the need to obtain the needed statutory authority to use this method makes it unlikely as a method to procure work unless there is adequate time to draft and get approval of the needed legislation in most states. In project that require the long times required to build access shafts, approach roads, fabricate adequate supplies of segments, etc. there is often no need for a prepurchasing of the TBM. However, the use of some elements of the process including the Technical Working Group bears serious consideration.

4 PRE-QUALIFICATION

Over the last 25 years a number of important reference documents have been developed with the construction industry to address the rising costs of underground construction and to reverse these trends. The first of these reports was published in 1974 by the U.S. National Committee on Tunneling Technology (USNCTT), within the National Research Council of the National Academies of Science and Engineering. The report entitled *Better Contracting for Underground Construction* had a profound, positive influence on the tunneling industry. This document identified the fundamental need to improve the overall approach to contracting for underground construction projects. Pre-qualification of bidders was one such recommendation made. Subsequently the Underground Technology (UTRC), Technical Committee on Better Contracting Practices of the American Society of Civil Engineers (ASCE) published a booklet entitled *Avoiding and Resolving Disputes in Underground Construction* and an updated edition in 1991, *Avoiding and Resolving Disputes During Construction*. These documents have been generally credited as establishing the Geotechnical Baseline Report as a part of the risk sharing methods used by owners and contractors in underground construction, however the use of alternative bidding methods has been meet with less than universally acceptance.

Alternatively, the Owner may wish to consider several other approaches to contracting such unusual work to attempt to obtain: 1) a superior technical approach, 2) experienced personnel, and/or 3) an innovative and conscientious contractor. Two such approaches that are favored for this project which preserve the traditional design-bid-build approach are:

- Pre-qualification or pre-approval of specialty subcontractors or specialists
- Two Envelope Bid Approach.

4.1 *Pre-qualification or Pre-approval of Specialty Subcontractors or Specialists*

To avoid the post-qualification difficulties described above many underground projects require that prospective bidders be pre-qualified. The pre-qualification process, however may add to the overall duration of the delivery of the project. Pre-qualification is generally required for most alternative delivery approaches.

The owner in advance of the bid may require pre-qualification of the tunneling subcontractor, and pre-cast segment and machine manufacturer and similar components to ensure qualified bidders in advance. This method has been used in other countries and by private and public agencies throughout the United States.

Pre-qualification saves bidders the cost of preparing a futile bid and eliminates the problems created by rejection of a bidder as not being responsible. Substantial cost savings should also accrue to owners from performance by contractors who have been determined to be qualified before being invited to bid. The process for pre-qualification of bidders is well established in Europe, where most owners, public and private, have a policy of pre-qualifying bidders for invitation to tender bids.

Under this alternative, documentation of the contractor's and manufacturer's experience, and the applicability of past work is submitted prior to bidding for evaluation. The pre-qualification follows an established protocol for evaluation that includes such items as: approach, experience on similar projects, history of completion of similar projects, experience of the project team members, the solvency and financial viability of the firms involved, and the experience of the designated specialty subcontractors and design consultants if any. A committee (of the owner's staff) would review the pre-qualification packets from interested contractors or joint ventures. Those proposers submitting qualified packages would be allowed to bid the work; those unqualified or not pre-qualified would be disallowed. The low bidder from the pre-qualified bidders would be awarded the work.

The Massachusetts Highway Department, on its massive Central Artery/Tunnel Project, has recently used such a method to pre-qualify specialty tunneling contractors for the jacking of four extremely large jacked box tunnels under the South Station's active rail yard, which serves over 300,000 commuters daily, and for the bidding of a Mined Tunnel Alternative using the New Austrian Tunneling Method for construction of ramps and roadway beneath North Station and adjacent to the Spaulding Rehabilitation Hospital. The tunneling contractors bidding the South Bay Outfall Project in San Diego, the St. Claire Tunnel project in Ontario for the Canadian Nation Railway, and the H-3 tunnels in Hawaii were also pre-qualified in a similar manner. A review of Engineering New Record, the engineering trade publication, just over the last year revealed many other projects for which owners actively sought pre-qualification of tunneling contractors.

To pre-qualify contractors owners should obtain and evaluate the following information:

- The organizational structure of the company, including the experience, levels of responsibility, education and length of service in the company, or in other companies engaged in similar work, of the proposed project-management personnel.
- Recent history and personnel qualifications (within the last 5-7 years) of the lead people available within the company or venture, if jointly bid, for assignment to the project. Certain minimum requirements may be specified and must be met and generally the personnel named are actually required to be employed on the project.
- Recent history of the company or venture in similar construction, using similar means, methods, and procedures, particularly concerning timely completion and description, cost performance history, cause and effects of delays, claims and disputes. This documented performance should include bid value, value of extras, and value of disputed extras.
- Current audited financial statement of the company, and a minimum required quick asset capacity. This includes items such as the general organization and ownership or contractor, its financial health, net value, outstanding debt, ability to finance the project and other relevant matters. Some of this information can be supported or verified through Dunn and Bradstreet and credit rating firms.
- Other pertinent factors including safety record, availability of specialty equipment or significant other assets. Safety record should provide documented accident and lost time accident rates. Contractors should also present a description of the company standards for setting up safety organizations and safety programs, and lists of qualified safety personnel.

Generally the submittals are reviewed and evaluated according to a pre-established set of weighted criteria. In order to avoid conflicts and protests, most procurement regulations require a pre-established appeals procedure so that any prospective bidder who is considered to be unqualified on the basis of the information submitted may request a hearing to review the basis on which it is proposed to exclude him from bidding the project involved.

4.2 *Two Envelope Approach*

A somewhat similar process is used in other states where a two-envelope approach is taken. Using this method the qualification and adequacy of a Contractor and/or Contractor's means and methods are established prior to review of cost proposals.

In this process, contractors submit their bids in two envelopes on the bid date. The first envelope contains responses to the same type of questions tendered for the pre-qualification discussed above plus a detailed description of the technical approach to the work. The technical approach includes complete descriptions and discussions of contractor means, methods, equipment and procedures, including details of specialized equipment including shields, details of muck handling, segmental lining design and fabrication, design of false work and temporary construction, etc. This includes particular methods foreseen for executing the work and any other particular details as desired by the Owner. Discussions, assessment, and evaluation of principal risks associated with the project and suggested remedial activities to reduce or eliminate these risks. These responses are evaluated on technical terms and for evaluating the compatibility and adaptability of the systems chosen with the ground conditions to be encountered on the specific project. The proposal evaluation is against an established criterion, and is categorized as acceptable, acceptable with modification, and unacceptable. In the second envelope, the contractor submits his completed price proposal. The owner's technical review team reviews the first envelope or technical proposal, while the second envelope or cost proposal is reviewed by the contracting officer. Generally the low bidder from among those judged acceptable would be awarded the contract. The technical evaluation criteria, including the overall weight structure for the technical and cost/price evaluation, and award determination are divulged with the solicitation.

Variations on this selection technique allow the weighting of technical and cost proposals say 70% technical and 30% price, to enable the acceptance of a superior technical proposal even if it is not the lowest offer (FAR Part 9.103 (c)), or simple opening of the top rated technical proposal and negotiating with this single bidder. The State of Nevada, Colorado River Commission and several other agencies including the US Bureau of Reclamation have used variations of this technique very successfully on challenging tunnel projects.

Advantages of the two-step bidding approach:

- Provides evaluation of adequacy of qualifications and means, methods, equipment, procedures, and personnel for construction.

- Does not require a ranking of proposals.
- Evaluation of technical proposals is independent of cost proposals.
- More defensible determination of responsible contractor.
- Contact awarded on basis of low bid from responsible contractor.

Disadvantages of this method:
- Confidentiality of proprietary means and methods must be preserved.
- Possible protest from Proposors determined to be unqualified but submitting a low bid.

One other variation of the two-envelope procedure is for the owner to purchase the tunneling machines and segments, as owner furnished equipment. For the selection of machine manufacturers or segment fabricators, any of the allowable procurement procedures may be used. Private clients and developers often use a negotiated or competitive bidding process.

5 WHAT'S THE NEXT STEP

The industry as a whole must become active in the development of legislative processes in the States to obtain the necessary authority for the Departments of Transportation, Water and Sanitary Districts and Authorities, etc., to use alternative procurement techniques. This sounds easier than it sounds. Many of us have, for over the last quarter century, been involved in the development of a series of improvements for contracting practices, i.e., Baseline Reports, Disputes Review Boards, Escrow Bid Documents and the rest. While these are widely accepted by many Owners, the jury is still out for many others. The somewhat spotty history of some Owners with one or several of these innovations has been well publicized and is common knowledge to the industry. The perception of some owners to such bodies as Disputes Review Boards has found some Owners more resistant to their acceptance of these contracting practices. Public perception to changes from a firm-fixed price environment must also be garnered. The attempt must be made to educate the public to the advantages to the taxpayers and stakeholders of obtaining a fully qualified contractor doing work for a fair price to obtain a quality product which will be appreciated for the many years for which it is intended to service the needs of its Owners and that of the public as well.

North American Tunneling 2002, Ozdemir (ed.)
© 2002 Swets & Zeitlinger, Lisse, ISBN 90 5809 376 X

Proposed revisions to the Differing Site Conditions Clause

John M. Stolz
Associate, Jacobs Associates

ABSTRACT: Industry task forces have long endorsed the use of a differing site conditions clause in construction specifications. However, while the underground construction industry has focused on various methods for utilizing or triggering the clause, it has largely ignored the clause language itself. This paper reviews the Federal Differing Site Conditions Clause to identify potential conflicts with other contract language and wording that is itself contrary to the intent of the Differing Site Conditions Clause, and then proposes revised language.

INTRODUCTION

For many years, the underground construction industry has worked to provide viable triggers for the Differing Site Conditions (DSC) Clause, currently culminating in the use of two new contract documents, the Geotechnical Baseline Report (GBR) and the Geotechnical Data Report (GDR). However, the DSC Clause language itself, as exemplified by the standard Federal procurement language, has not received such attention, and has remained essentially unmodified since its inception.

This paper first examines the Federal Differing Site Conditions Clause with regard to language that may not only conflict with language elsewhere in the contract, but that also undermines the efficacy of the DSC Clause itself by encouraging the incorporation of bid contingencies. The paper then goes on to propose revised DSC Clause language.

Note that the discussion presented herein represents a business view and not a legal view for effecting changes to contract documents.

The Federal DSC Clauses[1] states:

(a) *The Contractor shall promptly, and before the conditions are disturbed, give a written notice to the Contracting Officer of (1) subsurface or latent physical conditions at the site which differ materially from those indicated in this contract, or (2) unknown physical conditions at the site, of an unusual nature, which differ materially from those ordinarily encountered and generally recognized as inhering in work of the character provided for in the contract.*

(b) *The Contracting Officer shall investigate the site conditions promptly after receiving the notice. If the conditions do materially so differ and cause an increase or decrease in the Contractor's cost of, or the time required for, performing any part of the work under this contract, whether or not changed as a result of the conditions, an equitable adjustment shall be made under this clause and the contract modified in writing accordingly.*

(c) *No request by the Contractor for an equitable adjustment to the contract under this clause shall be allowed, unless the Contractor has given the written notice required; provided, that the time prescribed in (a) above for giving written notice may be extended by the Contracting Officer.*

(d) *No request by the Contractor for an equitable adjustment to the contract for differing site conditions shall be allowed if made after final payment under this contract.*

1 MODIFY THE REQUIREMENT TO LEAVE CONDTTIONS UNDISTURBED

"The Contractor shall promptly, and before the conditions are disturbed, give a written notice..."

The intent of the requirement to leave conditions undisturbed is to give the owner an opportunity to verify the presence of a DSC and to mitigate its impact on the remaining work. Accordingly, this language is well suited to the more readily apparent

[1] 48 FR 42478, Sept. 19, 1983, as amended at 60 FR 34761, July 3, 1995.

Class 1-type DSC,[2] such as the encountering of an unknown utility or rock where soil is indicated. In such cases, direct observation justifies the existence of a Class 1 DSC and it is therefore appropriate that the owner verify the undisturbed condition.

Conversely, it is only through the act of disturbing the subsurface conditions that oftentimes alerts a contractor to the possibility of a Class 2 DSC; e.g., increased cutter wear when tunneling through limestone that "doesn't cut like typical limestone." In such cases, it is not reasonable to require that the existing conditions be left undisturbed prior to giving notice.

The foregoing suggests that one solution may be to simply require leaving the condition undisturbed for Class 1 DSCs only, because the innate inability to identify Class 2 DSCs without disturbing the conditions would neither help an owner identify the DSC, nor mitigate its impacts on the remaining work.

However, this reasoning relies heavily on the assumption that *all* Class 1 DSCs are readily discernable as such, and that *all* Class 2 DSCs are not — a dangerously explicit supposition to incorporate into language that is best left in terms that are more general.

Perhaps a better solution is to rewrite the DSC Clause to simply require giving the owner the opportunity to verify the presence of a DSC.

2 STRENGTHEN NOTICE REQUIREMENTS

"The Contractor shall promptly...give a written notice to the Contracting Officer..."

This notification process is troublesome because it does not require the contractor to provide any information about a DSC as it is discovered and as it relates to justifying a subsequent request for change order. On the other hand, the impact of a DSC and therefore the basis for quantum calculations of the request for change may not be immediately apparent. The language attempts to address this by imparting a subjective (and, some may therefore argue, useless) sense of urgency with the use of the word "promptly" that ties in with the preceding discussion, but that nevertheless does not *require* the contractor to provide any immediate, hard information about the DSC.

Some owners eschew tying the notification of a DSC to an immediate request for change when the extent and amount of impact is often not readily ascertainable. They reason that such an action starts a number of administrative clocks ticking which pressure the owner, among other things, to make a determination of the estimated change order amount, and to issue a timely response to the request for change.

[2] So named for the subparagraph number of the DSC clause under which it appears.

However, these administrative clocks are inserted in the contract language as a tool for an owner to manage the cost and schedule impacts of potential change orders and claims on the project budget and schedule. Therefore, the exclusion of the DSC notification process from the notice requirements of the Changes Clause serves only to distort the data upon which an owner relies to make critical project forecasts.

It therefore seems clear that these administrative pressures are much less onerous than allowing unknown or ill-defined DSCs to pile up at the end of the job for resolution.

3 AVOID SOLE RELIANCE ON THE GBR TO JUSTIFY A DSC

Many differing site condition tunnel claims are predicated on more or less of a particular condition, so some might argue that regardless of how much geotechnical investigation is performed, all tunnels are nothing more than one big DSC from portal to portal. Therefore, if an owner includes a Geotechnical Baseline Report (GBR) and Geotechnical Data Report (GDR) as contract documents, the DSC clause trigger is tied to the GBR baselines, regardless of where they are set, and data presented in the GDR. Since the GBR and GDR are not the only contract documents—and certainly not the only tools for determining the existence of a DSC—there is no compelling reason to specifically cite them; but to use them, their order of precedence must be established elsewhere in the contract.

A recent trend among owners has been to limit the contractor's ability to justify a DSC founded solely on the baselines presented in the GBR. However, this logic erroneously supposes that a GBR is a panacea for all DSCs, when in reality the GBR can only identify those DSCs for which a baseline has been established. Furthermore, if a GDR has been included as a contract document, its purpose is now muddled. Therefore, in a contract wherein DSCs are limited to those for which the GBR provides a baseline, bidders will include contingency for the downside risk that the GBR has not provided a baseline for a condition that may manifest in a DSC.

Clearly, such an approach serves only to increase bidder's contingencies—a tactic contrary to the purpose of the DSC clause itself.

4 ELIMINATE CREDITS FOR BETTER THAN ANTICIPATED CONDITIONS

"If the conditions do materially so differ and cause an increase or decrease in the Contractor's cost..."

There are many arguments for eliminating this language but perhaps the most compelling is that this language leads contractors to bid at the baselines established in the GBR, or if a GBR has not been prepared for the project, at the most unfavorable conditions indicated. It takes away a bidder's incentive to gamble on better-than-anticipated conditions because the bidder could be required to "return" the same savings that reduced the bid in the first place.

Bids on contracts containing this provision will therefore be generally high by a contingent amount that is anticipated to be returned; a result that is contrary to the intent of the DSC Clause.

However, to the extent that an owner is willing to accept higher bids in exchange for a potential credit, thc owner must make the following administrative contract modifications:

- The DSC Clause language must permit either party to notify the other of a better-than-anticipated differing site condition.
- The Changes Clause must be amended to require the Owner to submit a timely credit change order request, including basis of entitlement, schedule savings analysis, and change order cost proposal for the credit amount due to the better-than-anticipated conditions. Similarly, the Contractor's role must be expanded to provide a prompt response to the credit change order proposal.

5 TIE TIME IMPACT ANALYSES TO THE SCHEDULING SPECIFICATION

"If the conditions do...cause an increase or decrease in the...time required for, performing any part of the work under this contract, whether or not changed as a result of the conditions..."

This language seems to imply that the DSC need not affect the project critical path but must simply prolong a portion of the work. Certainly, were a contractor to request a time extension citing the provisions of this clause, an owner could simply declare that a zero day time extension would be the equitable adjustment. However, a far simpler solution—and one that encourages reading the contract documents in whole rather than in part—would be to reference the clause specifying the means for analyzing schedule impacts.

6 TIE COMPENSATION TO THE CHANGES CLAUSE

"...An equitable adjustment shall be made under this clause and the contract modified in writing accordingly."

The DSC Clause does not specify how an equitable adjustment is to be made and furthermore, a typical contract usually does not require that contract adjustments be equitable[3]. Instead, a typical changes clause sets forth some uniform procedure for calculating compensation for a change that may not recognize and reimburse all of a contractor's actual additional costs.

If the DSC language retains this "equitable adjustment" language, a contractor may successfully claim that any DSC is to be reimbursed by an equitable adjustment to the contract and not via the prescribed method set forth under the Changes Clause because (a) the DSC clause does *not* reference the Changes clause, and (b) the interpretation given in (a) above resolves the apparent conflict within the contract General Conditions.

Therefore, the language should simply reference the Changes Clause, which establishes how *all* work not in the contract is to be compensated.

7 ELIMINATE REPETITIVE NOTICE REQUIREMENTS

"No request by the Contractor for an equitable adjustment to the contract under this clause shall be allowed, unless the Contractor has given the written notice required; provided, that the time prescribed in (a) above for giving written notice may be extended by the Contracting Officer."

This language essentially duplicates paragraph (a) of the DSC Clause and repeats what should already be in the Changes Clause. It is therefore unnecessary.

8 ELIMINATE REPETITIVE FINAL PAYMENT LANGUAGE

"No request by the Contractor for an equitable adjustment to the contract for differing site conditions shall be allowed if made after final payment under this contract."

This language is also unnecessary. The General Conditions will already have language specifying the disposition of unfiled claims and requests for changes as they relate to making final payment.

9 PROPOSED REVISIONS TO LANGUAGE

Based on the foregoing, the following revisions to the standard Federal procurement language are proposed. Strikeout text represents existing text to be deleted, and underlined text represents new text to be inserted:

(a) The Contractor shall ~~promptly, and before the conditions are disturbed~~, give ~~a~~ written notice ~~to the Contracting Officer~~ of (1) subsurface or la-

[3] See 48 CFR 552, Equitable Adjustments Clause.

tent physical conditions at the site which duffer materially from those indicated in this contract, or (2) unknown physical conditions at the site, of an unusual nature, which differ materially from those ordinarily encountered and generally recognized as inhering in work of the character provided for in the contract.

(b) The Contractor's notice shall be made in accordance with the Changes clause and shall provide the Contracting Officer the opportunity to ~~The Contracting Officer shall promptly~~ investigate the ~~site~~ conditions giving rise to the differing site condition ~~after receiving the notice~~. If the conditions do materially so differ and cause an increase or decrease in the Contractor's cost of, or ~~the~~ time required for~~,~~ performing ~~any part of~~ the work under this contract,- ~~whether or not changed as a result of the conditions, an equitable adjustment shall be made under this cause and the contract modified in writing accordingly~~ an adjustment shall be made to the contract in accordance with the Changes clause.

~~(c) No request by the Contractor for an equitable adjustment to the contract under this clause shall be allowed, unless the Contractor has given the written notice required; provided, that the time prescribed in (a) above for giving written notice may be extended by the Contracting Officer.~~

~~(d) No request by the Contractor for an equitable adjustment to the contract for differing site conditions shall be allowed if made after final payment under this contrast.~~

North American Tunneling 2002, Ozdemir (ed.)
© 2002 Swets & Zeitlinger, Lisse, ISBN 90 5809 376 X

What every owner should know about Dispute Review Boards

D.M. Jurich
Jurich Consulting, Incorporated, Gold Canyon, Arizona, USA

ABSTRACT: Dispute Review Boards are used to resolve construction disputes between Owners and Contractors. This non-binding means of dispute resolution has been described as a cost-effective tool that creates a positive atmosphere for construction projects. While Dispute Review Boards have been used to successfully resolve disputes and avoid litigation, periodic critical review of the system should be made to determine if improvements are possible. Further, design engineers and construction managers have the responsibility of making Owners fully aware of the potential benefits and consequences of using the DRB process for underground construction projects.

1 INTRODUCTION

The Dispute Review Board (DRB) process has been used on 673 projects through the year 2000. Of 859 recommendations made by Dispute Review Boards, only 3% were not accepted by one or both of the parties and advanced to either arbitration or litigation (Rogers, 2001). This is a significant record of accomplishment. But is this the only measure of success? Can the process be improved and is the DRB process always the best option for an Owner? A comparison of the DRB, mediation, and arbitration rules and guidelines helps to identify the strengths and potential weaknesses of the DRB process. Design engineers and construction managers have the responsibility of working with Owners to identify and quantify the risks and financial implications of all aspects of potential disputes and use of the DRB process.

2 THE DRB PROCESS

A traditional Dispute Review Board (Board) for underground construction projects consists of a 3-person panel that is composed of recognized construction and geotechnical experts and is intended to act as "project standing neutrals". The Board is appointed early in the project with the Owner and Contractor each submitting a candidate for the other's approval and the two approved candidates selecting a third Board member to act as chairman. The Board meets periodically during construction, monitors progress, hears disputes, and makes recommendations to settle disputes. The findings and recommendations of the Board are non-binding but are commonly accepted by both parties.

A Board should fully understand: contract plans and specifications, the Contractor's bid documents, correspondence during construction, other documentation including as-built records and test results, and verbal and written presentations by both parties (Rogers, 2001). The Board can be directed by the parties to issue findings regarding 1) only the merit of a dispute or 2) the merit and monetary (and delay) value of the dispute.

3 THE DRB PROCESS, ARBITRATION, AND MEDIATION

3.1 *General*

The DRB process is less formal than arbitration and is used in lieu of mediation. The mediation and arbitration processes involve experts in dispute resolution who do not necessarily have technical expertise related to the dispute. Additionally, the mediation and arbitration experts do not typically have any knowledge of the parties or individuals involved in the dispute. A fundamental difference between the DRB process and arbitration or mediation is that the Board is not an intermediary between the parties of the dispute. The Board's function is not to "referee" a dispute but to merely provide a technical opinion that is intended to assist the parties in resolving the dispute.

3.2 *Board Members and Impartiality*

Members of the Board, recognized experts in underground construction and/or geotechnical engineering with many years of experience, are usually well known to the Contractor, and sometimes to a lesser extent the design engineer and construction manager. Given that the underground construction industry is relatively small, it is not uncommon for DRB members to have a personal relationship with individuals involved in the dispute through one or more professional societies, past associations, or technical committees.

Underground construction is not the normal business of many owners and often they have no prior knowledge of, or relationship with, the DRB candidates. Many owners will have one or only a few projects that involve large-scale underground construction. The fact that underground construction contractors and consultants work closely in a relatively small industry can cause an owner to pause and question the impartiality of a DRB candidate.

This issue is best addressed in two ways. First, the Owner should require full disclosure of prior interaction between DRB candidates and the contracted parties. The disclosure should not be limited to working (contractual) relationships but should include interaction related to professional associations, societies, and committees. An Owner should determine if DRB candidates have co-authored technical publications with employees or consultants of the Contractor. It would be concerning to be the defendant in a legal case in which the judge and prosecuting attorney had co-authored an article for a legal publication. If this situation were to occur in a legal proceeding, the judge would be expected to dismiss himself from the case.

The investigation of potential conflicts of interest before a Board is selected and a dispute arises can be difficult. The Owner's best sources of information on Contractor key personnel and consultants are the contract pre-qualification and bid submittal documents. Resumes included in these documents should be cross-referenced to the resumes of the DRB candidates to identify past associations.

Second, the Owner should structure the project construction contract so that DRB candidates are not put forward by either the Owner or Contractor. The Owner and Contractor should produce a "pool" of several mutually acceptable candidates from which the full DRB is selected at random. The selected Board members then vote to establish a Board chairman. The candidates would not be able to associate their selection and appointment to the Board with one of the contract parties.

3.3 *Procedures*

Participants in a DRB hearing are not required to swear to "tell the truth" or to sign an affidavit stating that their testimony is truthful. Everyone is expected to conduct him or herself in a professional manner. This lack of a basic legal procedure is perhaps the main weakness of the DRB. If the dispute advances to arbitration or litigation, DRB findings that may be based on inaccurate or misleading statements are typically upheld. DRB hearings would benefit from a formalized declaration of truthful testimony. Without some form of oath, the DRB's findings should not be allowed to go forward in the event of arbitration or formal courtroom hearings.

Additionally, there are no subpoenas, depositions, or discovery processes in the DRB system. Therefore the Board renders a decision based only on the "facts" that both parties choose to present. Clearly, this can also lead to findings that are based on an incomplete understanding of the dispute. Owners can mitigate this risk by having full time on-site representation during construction.

Complete detailed documentation of ground conditions, Contractor activities, and field changes recorded by experienced field personnel, inspectors, and testing services independent of the Contractor is critical. The Owner can not rely solely on the construction records kept by the Contractor and must be diligent in this effort. The Owner may benefit from periodic independent documentation review by one or more experts in underground construction and geotechnical engineering to confirm that the proper information is being recorded. Very often the value of detailed observations and measurements is only truly appreciated after a dispute has be brought by the Contractor and the Owner is required to reconstruct a specific sequence of events or set of conditions.

With the exception of the potential for reviewing escrowed bid documents, there is no opportunity for the parties to obtain claim-related materials from the opposition. Once again, the DRB process carries the risk of information critical to a hearing remaining undisclosed.

A comparison of the DRB, arbitration, and mediation rules and guidelines in Table 1 summarizes the strengths and potential weaknesses of the DRB process.

4 RISK MANAGEMENT FROM THE OWNER'S PERSPECTIVE

4.1 *Rules of Conduct*

Design engineers and consultants typically recommend that an Owner include a DRB in underground construction contracts as part of a risk management policy. It is important to identify risks and estimate costs associated with the full range of potential outcomes associated with each element of risk. The Owner can then make fully informed decisions with

Table 1. Comparison of DRB, Mediation, and Arbitration.

Aspect	DRB *	Mediation**	Arbitration**
Member Appointments	one by each party	one by AAA	one or more by AAA
Candidate Disclosure	no ownership interest	no financial interest	no interest, past or present
Legal Reps.	not permitted	ok for lawyers to represent parties	ok for lawyers to represent parties
Oathes	none req'd	none req'd	may be req'd
Evidence	parties offer relevant information	parties offer relevant information	determines admissibility and relevance
Subpoenas	none	none	arbitrator can subpoena witnesses
Info. exchange	parties prepare position papers	mediator can direct docu -ments be produced	arbitrator can direct docu -ments be produced
Stenographic Records	no	no	permitted
Findings:	not part of negotiations	directs negotiations	determines settlement
Binding	no	no	yes
Admissible	yes	no	no
Expenses	shared	shared	shared
Cost	medium	low	high

* Matyas et al., 1996
**AAA, 2001

regard to mitigation measures. The DRB process has been used successfully to minimize and even eliminate cost overruns associated with arbitration and litigation of claims of differing site conditions. However the DRB process is not without it's own set of risks.

The DRB hearing is not a legal proceeding and the guidelines for rules of conduct contained in DRB three-party contracts are intentionally general to permit flexibility in hearing disputes. The lack of sworn testimony and cross-examination at DRB hearings give experts hired by the parties great latitude in the statements they can make.

If a DRB hearing is not intended as a formal legal proceeding with the normal checks and balances of testimony, participation in the DRB hearings should be limited to persons directly involved in the project. This would include only those persons who have firsthand knowledge of the conditions and events concerning the dispute.

Participation of experts hired by the Owner and Contractor should be limited to data analysis. Having the Board composed of recognized experts eliminates the need for an outside expert to argue a party's case at the hearing.

4.2 *Geotechnical Baseline Reports and DRBs*

Most, if not all, disputes in underground construction are brought by the construction contractor and are usually related to a claim of a differing site condition (DSC). Current contracting practice is to include a Geotechnical Baseline Report (GBR) as part of the contract documents to establish a set of baseline conditions for bidding and for the evaluation of claims of differing site conditions. These documents are interpretive in nature and generally include the extrapolation of limited data over great distances between borings. One boring for several hundred feet of tunnel is common. Some tunnel projects have been constructed without a single exploratory boring.

Uncertainty is unavoidable in the interpolation and extrapolation of exploration data and estimation of ground behavior. The issue of "baseline" conditions is further complicated because it is also necessary for the Owner to make baseline statements regarding the behavior of the ground. Unless an Owner dictates the contractor's means and methods, baseline statements regarding ground behavior become problematic because means and methods greatly influence ground behavior. An Owner who dictates means and methods must be prepared for the consequences of potential contractor difficulties. GBR statements that address the behavior of the ground have proven to be fertile ground for contractors pursuing claims of a differing site condition.

This would suggest that there is always the possibility of differing site conditions for underground construction projects, particularly with regards to expected ground behavior. Is it in the Owner's best interest to make baseline statements regarding ground behavior? The Owner may best manage the risk of claims of a DSC by limiting GBR statements to descriptions of ground conditions and specifically stating that interpretation of ground behavior for Bidder-selected means and methods is the Bidder's responsibility.

A dispute involving a claim of a DSC puts the Owner in a defensive position with regard to the GBR. Vague and ambiguous statements should be avoided in a GBR. A DRB will give the Contractor the benefit of the doubt when it is possible to interpret the meaning of a GBR statement in more than one way. The Contractor is permitted to make the most optimistic, yet reasonable, interpretation of GBR statements.

4.3 *Owner Responsibilities*

Every Owner has the responsibility to settle a DSC claim when actual ground conditions and/or behavior are clearly adversely different from predicted

GBR baseline conditions. Claims that result from an obvious DSC can usually be negotiated by the parties working in good faith and should not be brought to a DRB. Most DSC claims heard by a DRB address the gray areas of even the most expertly written GBR baseline conditions. This is particularly true when the contractor has struggled and has not achieved their anticipated progress. Detailed documentation of ground conditions and records of construction activities are the owner's best defense of GBR statements.

Owners cannot expect to benefit financially from conditions that are materially better than predicted in a GDSR. That is, contractors are not expected to give a credit to an owner if less rock support is used or less groundwater is encountered than predicted in the GBR. Similarly, the only benefit the Owner can expect from project completion ahead of schedule is the potential added value of putting the tunnel into service early.

The Owner will be deemed as benefiting from the use of the completed project (Heuer, 1997) and can recoup cost overruns over the life of the project (20 to 50 years) whereas the construction contractor has only one opportunity to make a profit on the construction of a project. This conventional wisdom can (and does) favorably influence the DRB's attitude towards a Contractor claim of a DSC.

In the case of a dispute regarding a claimed DSC, the burden of proof of a DSC rests with the Contractor. The DRB findings and recommendations should be based on the information presented by the dispute parties (Matyas, et al, 1996). However, this procedure is not always followed and a DRB is free to consider more than just the documents and testimony presented by the parties. A DRB can, and in some cases is expected to, draw upon relevant experience, familiarity with the equipment and participants, and general impressions of the situation in reviewing a dispute. The DRB's freedom can lead to some unexpected results and the owner has to be prepared for a finding that is not based solely on the "facts" of the dispute.

4.4 *DRB Findings Carry Weight and Other Risks*

The recommendations of the DRB are admissible in arbitration and litigation. However, if a Mediator is used to settle a dispute, information exchanged and offers of settlement are not admissible in subsequent arbitration or litigation (AAA, 2001). By choosing the DRB process over mediation the Owner is exposed to the risk of having a potentially flawed DRB finding admitted as evidence if the dispute advances to arbitration or litigation. Recall that neither the parties nor experts are under oath and that there is no cross-examination in DRB hearings. The Owner would then be required at subsequent proceedings to present compelling arguments as to why the DRB findings were flawed.

DRB findings carry weight in arbitration and legal proceedings. One of the first questions an Arbitrator or Judge will ask in hearing a dispute is why did one of the parties not accept the DRB findings and settle the dispute? The Arbitrator or Judge can be expected to reason that Board members, as unbiased recognized experts, have the best credentials to understand the issues regarding the dispute and find accordingly. The Owner will be required to present the potentially complex technical issues of a dispute to an Arbitrator or Judge and demonstrate why the DRB findings were inappropriate.

The cost of a DRB is not insignificant and can range from 0.1% of the total cost for large projects to 0.3% of the cost of a small project. Regardless of the cost the owner may not realize any benefit from the DRB process. The DRB is considered by some as insurance against potential claims with the understanding that a board of technical experts would quickly dismiss faulty or spurious claims. While this may be true, the DRB should be considered as a line item cost of completing the project that may or may not be offset by a reduction in settlement costs (project overruns) associated with disputed claims.

Statistically, Dispute Review Boards typically find that there is some measure of merit to claims of a DSC brought by contractors. Boards commonly find that there is a shared responsibility for the costs and delays claimed in a DSC hearing (partially due to a DSC and partially due to the contractor's means and methods). Boards more often have recommended the majority of the claimed costs and delays attributable to a DSC be awarded to the contractor. The use of a DRB can not be expected to eliminate cost overruns. Instead the DRB process should be considered as one of many tools an owner employs to mitigate the impact of added costs and project completion delays related to claimed DSCs and resultant arbitration or litigation costs.

5 THE FUTURE OF DRBS AND A FEW SUGGESTIONS

Underground construction will always include an element of the unknown. It is unlikely that cost-effective exploratory methods will eliminate the risk of encountering differing site conditions for deep linear features such as tunnels. DRBs have proven themselves effective in the resolution of disputes and by all accounts are here to stay. Boards have heard and helped resolve numerous claims related to underground construction projects. But there is room for improvement. The following summary of suggestions put forth in this discussion should be considered by every Owner contemplating using a DRB:

1 Modify the DRB member selection process so that the members are chosen from a pool of qualified candidates agreed to by both parties. This will help make the Board a truly independent third party.
2 Require all participants in DRB hearings to take an oath to tell the truth. This will help reduce the temptation to embellish or make misleading statements.
3 Limit the participants at DRB hearings to only those persons directly involved in the project. Prohibiting the presence of experts from the hearing should help keep the presentations and discussion focused on the dispute.
4 Limit the scope of the Geotechnical Baseline Report to addressing ground conditions. Do not make baseline statements regarding ground behavior if means and methods are not specified. Require Bidders to submit statements of anticipated ground behavior with respect to their selected means and methods.
5 Write the specifications for the DRB to clearly state that the findings of the DRB do not go forward to arbitration or litigation if the parties cannot settle after a DRB hearing. The DRB may not have heard all of the relevant information and arguments before writing and opinion.
6 Shift the role of the DRB closer to the role of Mediator. The intent of the DRB process is for the Board to come to an understanding of the position of both parties regarding a claim and opine as to the merit and value of the claim. The Board might benefit from a chairman who is not necessarily from the industry but someone experienced in arbitration and mediation. This would bring more of a balance to the Board in terms of dispute resolution and would have the added benefit of giving the Owner confidence that the presiding member of the Board is not influenced by long-standing relationships.

Regardless of the future of the DRB process in underground construction projects, Owners and their consultants need to recognize the strengths and weaknesses of Dispute Review Boards and manage the process as they would any element of risk.

REFERENCES

American Arbitration Association, 2001, Construction Industry Arbitration and Mediation Rules (www.adr.org).

Gould, 1995, Geotechnology in Dispute Resoltion, Journal of Geotechnical Engineering, American Society of Civil Engineers.

Heuer, 1997, Discussion of "Geotechnology in Dispute Resolution (The Twenty-Sixth Terzaghi Lecture)", Journal of Geotechnical and Geoenvironmental Engineering.

Matyas, Mathews, Smith, and Sperry, 1996, Construction Dispute Review Board Manual, McGraw Hill.

North American Tunneling 2002, Ozdemir (ed.)
 ISBN 90 5809 376 X

Implementing a large tunnel project – The San Vicente Tunnel

Zachary Ahinga & Gary Stine
San Diego County Water Authority, San Diego, CA, USA

Gregg Sherry
Brierley Associates, LLC, Littleton, CO, USA

ABSTRACT: This paper will cover the factors an owner must consider when implementing a large tunnel project. The San Vicente Tunnel is proposed to be a twelve-mile long 102 inch inside diameter tunnel that will be a critical component of the San Diego county Water Authority's Emergency Storage Project. The San Diego County Water Authority anticipates it to be their largest capital improvement project to date. This paper will examine the process for implementing this project including environmental issues, public involvement, evaluation of contracting practices, contract delivery method (design bid build versus design build), project controls and consultant selection. This paper will conclude with lesson learned for consideration by owners who are considering a large tunnel project.

1 INTRODUCTION

The purpose of this paper is to present a model for implementing a large tunnel project by a public agency. This paper will discuss the project organization, consultant selection, environmental issues management, pubic involvement, evaluation of contracting practices, project controls and conclusions for the San Vicente Pipeline Project.

The San Diego County Water Authority (Authority) was organized on June 9, 1944. Its mission is to provide a safe and reliable supply of water to its 23 member agencies serving the San Diego region. The estimated population in San Diego County is 2.9 million people, 97 percent of which live within the Authority's service area. The service area lies within the foothills and coastal areas of the westerly third of San Diego County, encompassing approximately 1,400 square miles.

The Authority's existing supply system is vulnerable to natural hazards such as strong seismic activity, severe floods, and prolonged droughts. The Authority imports 90% of its water supply from the Metropolitan Water District of Southern California (MWD). The Authority's pipelines cross the Elsinore fault, and MWD's facilities cross both the San Andreas and San Jacinto faults in several locations. A major earthquake on these faults could render some or all of this delivery system inoperable for up to six months. The Authority's Emergency Storage Project (ESP) is designed to improve reliability of the region's existing water supply system by the addition of approximately 90,000 acre-feet of reservoir storage to supplement emergency water supplies available to the region. The project will also expand the existing transmission and distribution capabilities of the Authority system.

The San Vicente Pipeline will transmit water between San Vicente Reservoir and the Authority's Second Aqueduct. It is a key component of the Authority's ESP. Two routes were identified for this pipeline in the FEIR/FEIS. Route 4 would have primarily been a trench construction with several significant tunnel sections. Route 16B is the chosen alignment and would primarily be a tunnel construction with limited trench construction. This alignment is called 16B is shown on Figure 1.

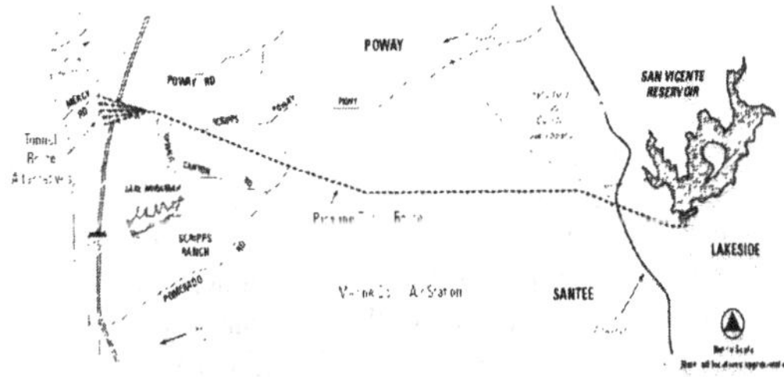

Figure 1. Project Location Map.

The east shaft for the proposed tunnel is anticipated to be located about one mile west-southwest of San Vicente Reservoir. The tunnel will extend westward about 12 miles, and will connect with the Second Aqueduct west of Interstate 15 near Mercy Road (as shown on Figure 1). The tunnel is expected to be a 102-inch (8.5-foot) finished diameter pipe within an 11- to 12-foot excavated diameter. The majority of the final lining is anticipated to be cast-in-place concrete. However, there may be approximately 10 to 20 percent of the tunnel that will require steel lining. The vertical alignment and the extent and type of the final lining has not been defined and will be determined during design. The estimate value of this project is approximately $200 million. The tunnel must be connected to the San Vicente Pump Station on the east end and the Mercy Road Pressure Control/Hydroelectric Facility on the west end. This project may also include construction of a required surge control facility on the east end near the San Vicente Pump Station.

2 PROJECT ORGANIZATION

As stated earlier the Authority was organized to bring safe and reliable water to it's twenty-three member agencies in the San Diego Region. The engineering department is the lead for major capital improvement projects like the San Vicente Pipeline. A principal engineer, project manager, staff engineers and consultants, as necessary, manage each capital improvement project. The organization for the San Vicente Pipeline is shown in Figure 2.

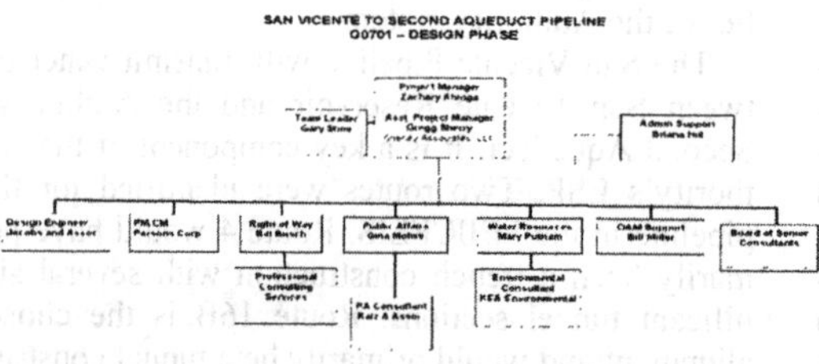

Figure 2. Project Organization Chart.

3 CONSULTANT SELECTION

Consultant support services were required to implement the project through the Environmental Documentation, Design, and Construction phases. The project requirements for consultant services were determined by reviewing the availability, past work experience, and expertise of Authority staff to implement a large tunnel project. After the needs were identified a schedule and action plan to retain the consultants was implemented. The components of the consultant selection plan and discussion of important considerations in the selection process follow.

Consultant Management: The lead responsibility for implementing the project is assigned to the Engineering Department. However, management responsibility for the consultant procurement process and consultant activities is assigned to the appropriate Authority operating Department. The Project Organization shown on Figure 2 depicts the formation of a multi-departmental Project Team. The Project Team consists of one staff person from each participating department. Each member and their respective Department Head are responsible for a project discipline. Public Affairs manages the public outreach consultant, Water Resources manages the environmental consultant, Right-Of-Way manages the Survey and Mapping consultant, and Engineering manages the project management support consultants, design consultant, and the construction management consultant.

Procurement Process: The Authority used two consultant procurement processes, 1) an on-call professional services list that is developed by prequalifying and ranking the consultant firms and 2) the Statement of Qualifications/Request For Proposal (SOQ/RFP) process. The type of work, anticipated contract value, time constraints, and compliance with the Public Contract Code are variables that were considered in selecting the procurement process.

Selection Panel: Where an on-call professional service list was used, no panel is required because the firm at the top of the list is selected for negotiation for a contract. Where the SOQ/RFP process was used, panel members were included that were familiar with the scope of work and services to be provided. Panel members included the contract manager, and staff within the discipline department, at least one member from other participating departments and one member form an outside agency. The City of San Diego, a water retail member agency of the Authority, will contribute a large share of the costs for the Emergency Storage Project. Therefore, City staff was invited to participate in the consultant selection for the project management support, construction management, and design consultant selection panels.

Public Outreach: The consultant will support Authority staff in a program to invite and include affected communities in the project implementation, design and construction process. Services include researching the public, planning of outreach strategies, and execution of the public outreach program. Katz and Associates was retained through the SOQ/RFP process and provides public outreach for the Authority's entire Emergency Storage Project.

Environmental Documentation: Consultant services are needed to assess the need for and prepare additional environmental documentation as necessary to comply with the California Environmental Quality Act and the National Environmental Protection Act. In November 2001, HDR Engineering was awarded a contract. HDR was selected from an on-call professional service list developed by the Water Resources Department.

Aerial Photography and Planimetric Mapping: The Authority as part of its efforts to have detailed and accurate records of its rights-of-way, had implemented a Geographical Information System (GIS) project that retained Merrick to provide aerial photography, 2-foot interval contour and surface feature mapping of its existing pipeline rights-of-way and the future ESP pipeline corridors. The alignment for the San Vicente Tunnel was already included in Merrick's scope of work. A contract was amended to provide the aerial photography, contour and planimetric mapping for the San Vicente Tunnel that will subsequently be provided to the Survey and Mapping consultant.

Survey and Base Mapping: The consultant selected is Project Design Consultants (PDC). The SOQ/RFP procurement process was utilized. PDC will conduct surveys to establish field control and reference for the project right-of-way, prepare record of surveys, right-of-way plats and legal descriptions for use in right-of-way acquisition, and prepare the base maps for the project Tunnel Design. PDC will provide aerial photography, 1-foot interval contours, and surface mapping for all access roads, portals and shaft areas. The mapping provided by Merrick will be integrated with PDC's surveys and mapping to produce the project base maps.

The Authority on past projects had assigned the survey and mapping scope to the prime pipeline design consultant. If that practice had been implemented on this project the prime consultant, the Tunnel Designer, would provide survey and mapping through a first tier subconsultant. The Authority determined that the Right-Of-Way Department would assign its licensed surveyor to manage the consultant in order to effectively manage the time critical mapping effort. This would also allow the Right-Of-Way staff to be involved in both the development of the right-of-way requirements and acquisition. The Authority believes that a result will be a smoother and timely acquisition process and cost effective management of the Authority's future rights-of-ways.

Tunnel Engineer: The Authority determined that it did not have sufficient expertise on staff to effectively manage the implementation of the large diameter tunnel components in the ESP projects. Brierley Associates was retained to provide assistance with project management and in-house staff consultation on tunnel engineering, project feasibility, consultant selection, project delivery methods, construction cost estimating and construction administration services for the overall ESP project. The SOQ/RFP procurement process was utilized.

For the San Vicente Tunnel project, Brierley assisted in developing the overall project plan, providing assistance with project management, developing the geotechnical exploration program, preparing proposals, developing interview evaluation criteria and preparing check estimates for consultant and construction costs. Through the design phase, Brierley will continue to provide project management assistance, design coordination and reviews, construction cost estimating, assist in the monitoring of the project budget and schedule, and provide bid evaluation assistance. It is anticipated that the Authority will retain Brierley during construction to provide construction review and administration support.

Project Management Support Services: The Authority retained Parsons Engineering Science to provide project management support and construction management (CM) services. An SOQ/RFP procurement process was utilized. The Engineering Department, as the lead for implementing the San Vicente Tunnel project had assigned a Principal Civil Engineer and Project Manager the responsibility to implement the project. It was determined that the department did not have adequate staff to effectively manage the project. The project management support includes providing one engineer through the design phase at the Authority's offices, and up to ten additional staff including engineers and technicians, as needed for short term assignments. The contract also provides services for preparation of a construction communication and construction management procedures manual, contract document reviews, cost estimate and schedule reviews, value engineer services, providing a Board of Senior Consultants for design review, partnering and bidding services.

Construction Management Services: As discussed above the Authority retained Parsons Engineering Science to provide construction management services. For this project, the Authority decided the CM consultant would be retained prior tp The Start of the design. The Authority's normal practice is to retain a CM consultant at the time the 90 percent design submittal is completed. The schedule called for the project construction to begin in mid 2004 and extend over a 48-month period to mid-2007. If standard practice was implemented, firm costs for CM services would not be available until January 2004. Estimates for CM services generally rage form 7 to 10 % of construction costs. This is a significant unknown.

The Authority determined that it could better manage the project budget through the design phase if all consultant contracts were executed and accounted for in the project budget prior to the start of the design phase. Then as design progressed budget impacts due to design changes and components could be accurately assessed and appropriate decisions on the use of contingency funds can be made. Then upon completion of the 100 percent design and construction cost estimate, with the construction management services budget already established, a timely evaluation could be completed to determine if additional funding authorization is required.

The construction management services covered multiple future years and would commence approximately 2.5 years into the future. It was vital that the work be well defined. The contract was based on a contractor implementing two TBM headings, two site offices, a well-defined staff and level of service, and an achievable construction schedule. Special contract provisions were included as follows:

1. Limits on the amount of annual increases of personnel direct labor rates to ensure that rate of increase matches closely to the cost of living index;
2. Specific definition of the timing of labor rate increases to ensure that the project can be completed within budget if there are no additions to scope of work;
3. Definition of the labor categories that are entitled to overtime compensation and requirements to obtain prior approval to work overtime;
4. Definition of the labor categories that require payment of prevailing wages;
5. Definition and provisions to compensate for other direct costs such as vehicles and outside inspection services;
6. A negotiated cap/limit on the maximum billing rate paid for the management and engineer's staffs. This is necessary to ensure that the compensation specified in the contract does not exceed competitive market rates at the time the work is performed;
7. Specific definition of the labor multiplier for home office and field office services that covers the direct and overhead costs and profit. Negotiations on the labor multiplier were based on external audit information provided by the consultant for the previous four years and future pricing audits provided by an agency of the federal government. An analysis of the projected state of the local economy during the contract period must be factored into the negotiations.

Tunnel/Pipeline Design Services: The Authority retained Jacobs Associates as the prime design consultant to design the San Vicente Pipeline. An SOQ/RFP procurement process was utilized. The design services includes preparation of construction documents, coordination and support for the environmental documentation effort, geotechnical exploration and well monitoring services, selection of the tunnel alignment and profile, hydraulic analysis and pipeline design, geotechnical engineering, tunnel engineering, surveying and access road design, surge control facility design, risk analysis and construction support. Jacobs will set up a project office in San Diego and has included thirteen subconsultants on its design team. Of the thirteen, seven are local San Diego firms.

The project scope of work clearly defines the services and deliverables expected. Design submittals are required at the 10, 30, 75, and 100 percent design completion milestones. Consideration was given to requiring a 50 and 90 percent submittal. The 75 percent submittal was required instead, because it was agreed that progress meetings held at the 50 and 90 percent progress milestone dates where progress drawings and specifications would be provided will provide a cost effective design review. Contract provisions similar to those that are indicated above for the construction management services contract was included in this contract.

4 ENVIRONMENTAL ISSUES

In 1997, the project Environmental Impact Report and Final Environmental Impact Statement (EIR/FEIS) had been completed. The EIR/FEIS approved two alternative alignments, 1) a cut-and-cover pipeline alignment that included two short tunnel segments, and 2) a tunnel alignment with short pipeline segments at the ends of the tunnel. The Authority in December 2001 selected the tunnel alignment for project implementation. In 1997, the tunnel alignment extended across mostly undeveloped rural areas. The environmental document may require amendment. The form of this amendment is not known but a supplemental EIR or a Negative Declaration of no Impact is being considered. The reason for this modification of the existing environmental document are the western third of the alignment has been significantly developed since the approved EIR/ FEIS were finalized, the middle third of the tunnel alignment now extends across land being purchased and planned for residential development and the Authority received a request from the Marine Corps Air Station Miramar that the pipeline be relocated outside of the base property for security reasons. Under consideration is a change in the method of tunnel muck disposal from on-site disposal to off-site export, which may also result in additional community impacts or effects. The additional environmental work will be performed concurrently with the 30 percent design.

5 PUBLIC INVOLVEMENT

Public involvement strategies are key to the implementation of the San Vicente Tunnel project. Effective public involvement strategies will allow project completion on schedule and within budget. In January 2002, the project began the tunnel design phase. The goal for the public involvement plan was to achieve two-way timely communication and to gain community support for the project. The key message to be delivered was: that the San Vicente Tunnel is a component of the ESP, vital to the regions water reliability, and that it would be designed to be constructed in the safest and least disruptive manner possible.

At that time, design challenges included obtaining permission to enter property to conduct geotechnical borings, obtain input and support for proposed construction access routes and communicating the associated noise, dust, visual, and traffic effects due to construction. Gaining input and support for proposed shaft/portal sites are a challenge because the shaft at San Vicente Reservoir at the east end could impact recreational uses at the reservoir. Work at one proposed intermediate shaft could affect recreation at the Goodan Ranch Regional Park, another intermediate shaft would be located within a planned residential development, and the shaft at the west end would be located within 100-feet of residences in an established community. An important project milestone is preparation of environmental documentation for the project that would be accepted by the communities and the Authority's Board of Directors.

The Authority planned to implement the following strategies at the start of the design phase to disseminate information and provide opportunities for community feedback:

Community Contact List – The list were researched and prepared to include school officials, community planning groups and civic organizations, interested residents, businesses, elected officials, media, individuals who called the project for information, and others anticipated to be affected by construction.

Presentations – Authority staff would attend civic group meetings, planning group meetings, homeowner association meetings, meetings with individuals as requested to provide information, resolve issues and obtain community input.

Project Information "hotline" – Project information line cards would be distributed within the community and to the contact list. The public would be able to call the 24-hour line to express concerns, request information, and ask questions. If the calls were received after business hours, return calls would be made during the next business day. The hotline would also provide a source on what project-related issues were creating community relations challenges.

Project Newsletter - As part of the overall communications plan for the ESP project, a project newsletter published quarterly, would be prepared to provide project status and inform the entire communities affected by the San Vicente Tunnel project, on project importance and benefits, and issue safety reminders and other timely information. The newsletter would also provide information on other component projects of the ESP.

Additional strategies that are to be implemented during the construction phase include:

Project Updates – Designed to keep community leaders, school officials and others requesting updates apprised of construction progress, traffic impacts and safety issues. Updates would be distributed twice monthly, as one-page, two-sided, single color publications with bullet-point format.

Special Project Advisories – Designed to contact specific groups to apprise residents of project construction changes that may affect them, safety alerts, or provide information of work under way. Some changes may include earlier work hours, weekend work, access route changes, and notifications should blasting be required. The advisories would be in a one-page format on colored paper so they will be noticed. Advisories will be distributed by direct mail or delivered door to door wrapped in a rubber band and placed on doorknobs.

The public involvement plan is a communication strategy designed to facilitate the design process by inviting the public to provide input on the design and construction, with the goal to build community support for the project. It will be evaluated for effectiveness and adjusted as necessary throughout the project.

6 EVALUATION OF CONTRACTING PRACTICES

The Authority evaluated the possibility of performing this project as a design/build project. A white paper was prepared to evaluate this alternative form of contract. The Authority typically contracts for design services to prepare contract documents illustrating complete design, accepts bids on these contract documents from all interested contractors; and awards the construction contract to the low responsive bidder.

The advantages of design/ build for this project were the integration and design and construction, schedule and a negotiated contract. The ability for direct and timely collaborative effort and communication between the owner, designer and contractor during the design and prior to contract formation provides a tremendous opportunity to enhance the design and construction process and to more clearly define reasonable and fair contract expectations. The integration of design and construction and a negotiated contract appear to be a tremendous plus for the design/ build method, but to date in the United States on underground public works projects the primary reason for using the design/build method has been the significant schedule reduction and the associated accelerated use or revenue stream.

The disadvantages to design/ build for this project were loss of control, quality control and the contract. These disadvantages make it hard for most public agencies to accept. These disadvantages also mean that any changes during the design will be more costly than with the traditional approach. In general to overcome these disadvantages the owner must spend significant time and effort defining the project and the contract clauses for the RFP.

Additionally a true cost comparison between the traditional approach and design/build is difficult to do. It is a general belief for public works underground projects in the United State that the cost is higher.

The use of the design/build on the San Vicente Pipeline was not implemented for the following reasons:

- A major advantage for design/build methods is the accelerated schedule. The San Vicente Pipeline construction is constrained not to start until Phase 1 of the Emergency Storage Project is complete. Phase 1 is not anticipated to be complete until late 2002 or early 2003. Given this constraint the design/build method will have no schedule advantage over the traditional design/ bid/ build method.
- The time and effort required by the Authority to develop documents for a design/ build contract could be quite extensive and costly.
- The use of design/build on public underground projects in the United States is relative new and the pros and cons are probably not fully understood. The use of this method on the Authority's biggest individual project is risky.

The Authority also plans to implement contract practices typical for tunnel construction. Tunnel construction is fundamentally different from construction of other types of civil projects and as a result, construction contracts are often tailored to meet the specific needs of tunneling. Included are issues relative to presentation of geotechnical data, ground conditions, and construction review. The Authority will evaluate these issues during design to decide on the final format for contract documents.

Other contract issues, which aren't necessarily used on all tunneling projects but may warrant consideration on this project, are:

- Incentives (schedule, disruption, or other issues)
- Liquidated damages
- Third party impacts minimization (such as traffic, local access to businesses, and utilities)
- Contractor-prepared contingency plans for broken utilities, excessive settlement, contaminated ground, and specific ground conditions such as faults, weak rock, and areas of thin rock cover or not full-face rock
- Instrumentation Monitoring
- Bid and contract allowances for potential negotiated items such as specific ground conditions and environmental encounters.

7 PROJECT CONTROLS

Project control on the San Vicente Pipeline is achieved by developing a detailed project plan and schedule for complete execution of the project from beginning to end, regular assessment of the plan and schedule regular monitoring of all components of the project budget. A project plan was developed and input into a Primavera P3 system along with manpower and financial resources assigned to each task. A more detailed project plan is developed for tasks and subtasks and the critical success factors are determined for the future 12 months. An Action Plan is developed which identifies these critical success factors as shown in Figure 3. This action plan is revaluated monthly and revised as necessary. If critical success factors are not accomplished executive management is informed and provides assistance to achieve success factors. The project schedule is also evaluated and updated in accordance the critical success factors developed. An example page of the schedule is shown on Figure 4. The design phase is scheduled to begin in January 2002 and extend eighteen months to mid-2003. Right-of- Way acquisition is scheduled to completed in early 2004.

The project control system is critical to achieving the project success. It requires the project manager and management to regularly evaluate the project and communicate to others. Other factors critical for project control are regular and structured communication between all team members, monitoring of services performed by consultants and communication with upper management. Communication is ac-

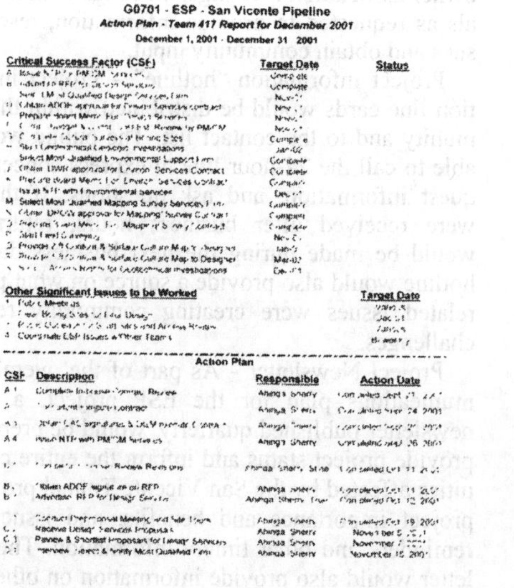

G0701 - ESP - San Vicente Pipeline
Action Plan - Team 417 Report for December 2001
December 1, 2001 - December 31 2001

Critical Success Factor (CSF) | **Target Date** | **Status**

Other Significant Issues to be Worked | **Target Date**

Action Plan

CSF | **Description** | **Responsible** | **Action Date**

Figure 3. Action Plan Example.

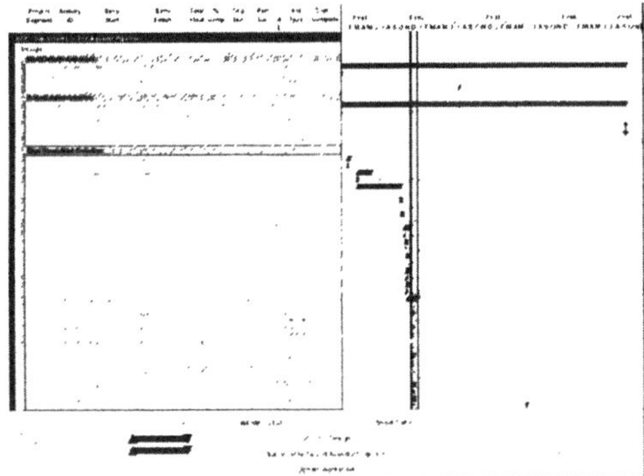

Figure 4. Project Schedule Example.

complished with regular meetings and subsequent conversation phone calls and emails.

8 CONCLUSIONS

The management of a large tunnel project requires a lot of work. It is important to establish a plan and management systems prior to project initiation. The conclusions concerning project management of the San Vicente Pipeline Project are:

- Establish a detailed project plan and assign responsibilities.
- Developed a project organization chart to establish lines of communication and consultant needs.
- Hire consultants to obtain required discipline knowledge and to augment organization.
- The project value and complexity should be considered when establishing management procedures.
- Select or develop appropriate procurement processes involving all stakeholders in the process.
- Consultant Requirements should be identified by evaluating in-house staff availability and expertise.
- Consider the benefits of retaining the Construction Management consultant prior to initiating the design in order to better manage the project budget through the design effort.
- Environmental documentation requirements should be developed based on at least the 30 percent design documents.
- Consider using Public Involvement programs as a tool to facilitate the design effort.
- Consider using Public Involvement programs build community support for the project to facilitate construction.
- Design/Build versus Design/Bid/Build should be evaluated based on the need for an accelerated schedule, the level of management and quality control required, and level of financial risk the agency can accept.
- Prepare a Project Team Action Plans indicating individuals responsible for completing tasks with completion dates.
- Establish a control system that requires detailed review and updating of the schedule and resources.

North American Tunneling 2002, Ozdemir (ed.)
© 2002 Swets & Zeitlinger, Lisse, ISBN 90 5809 376 X

Predicted and actual risks in construction of the Mercer Street Tunnel

L. Abramson
Hatch Mott MacDonald, New York, NY, USA

J. Cochran
King County Department of Natural Resources, Seattle, WA, USA

H. Handewith
Tunnel Consultant, Seattle, WA, USA

T. MacBriar
RoseWater Engineering, Seattle, WA, USA

ABSTRACT: The Denny Way/Lake Union CSO Control Project includes three short tunnels constructed under active railroad tracks and the Mercer Street Tunnel constructed in a densely populated urban setting. Tunneling was conducted in recent glacial and modern fill deposits.

During planning and design, tunneling risks were evaluated. These included groundwater inflow, ground settlement, running sands, boulders, methane, highly plastic silt, railroad ballast, and buried obstructions such as wooden piles, railroad trestles, shoreline sea walls, concrete, and other construction debris. Risk registers were prepared to evaluate the potential severities, ramifications, frequencies, mitigation measures, and expected costs. A Monte Carlo simulation addressed the probabilistic aspects of these risks. These studies led to key decisions incorporated into the project contract documents. During construction of the tunnels, some of the predicted risks were realized and managed. A comparison between the predicted and actual risks has been assembled and is the subject of this paper.

1 PROJECT PURPOSE AND OVERVIEW

The Denny Way/Lake Union Combined Sewer Overflow Control Project is a joint effort of King County and the City of Seattle. When complete in 2004, it will significantly reduce overflows of combined sewage and stormwater that occur in Lake Union and Elliott Bay. The "centerpiece" of the project is the Mercer Street Tunnel, which provides storage and conveyance of combined flows during storm events.

Frank Coluccio Construction Company commenced working on the $29.5 million Mercer Street Tunnel contract in June 2000. The contract includes four tunnels, all driven from a common shaft. This shaft is 150-feet long by 70-feet wide, and is 50-feet deep. It is also the principal excavation for the future Elliot West CSO facility.

Three short tunnels were driven west from the shaft and under multiple railroad tracks. The crown of the first and smallest tunnel excavated was just over five feet below the operating rail tracks. It had a 72" diameter and was 125 feet long. The deeper effluent and CSO tunnels were pipe jacked 50' below the rail grade. Both tunnels were about 350' in length and had a 96" inside diameter.

Mercer Street Tunnel (MST) is the largest of the four tunnels. It was excavated using a 16'-8" diameter Lovat earth pressure balanced (EPB) machine. The machine was lowered into the shaft in May 2001. In less than two months of single shift operation, nearly 600 feet of tunnel had been bored and lined. By then the TBM was operating in full production mode. The finished tunnel has a 14'8" inside diameter, and will be 6,212 feet long. It will generally follow under Mercer Street with a maximum ground cover of 172 feet. When completed, it will provide 7.2 million gallons of underground CSO storage.

Excavation of the Denny CSO Tunnels has proven to be a unique challenge. The contractor has worked five-day weeks consisting of a 10-hour working shift and an 8-hour maintenance shift. On an average production day, the Mercer Tunnel pro-

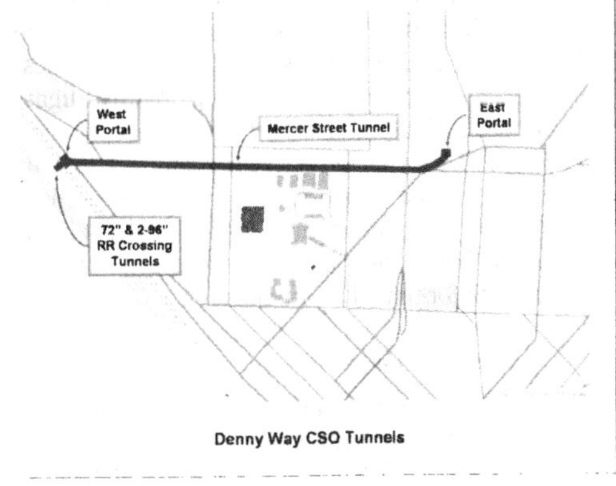

Figure 1. Plan of Denny Way CSO Tunnels.

duces nearly 1,300 cubic yards of muck. Operating a maintenance/haulage shift at night has allowed for minimum traffic disruptions when moving muck from the site.

2 RAILROAD-UNDER-CROSSING TUNNELS

2.1 *Background*

The three railroad crossing tunnels extend west of the West Tunnel Portal of the Mercer Street Tunnel. Two of the tunnels extend across 13 sets of railroad tracks, and the third extends across six tracks to the existing Elliott Bay Interceptor, a main sewer line leading to the West Point Treatment Plant. The two longer tunnels connect to future pipelines to be constructed along the waterfront.

Design considerations for the short tunnel were unique, in that its elevation was constrained by the need for gravity drainage into the existing Elliott Bay Interceptor. This tunnel will be used to drain the Mercer Street Tunnel and allow conveyance of the stored flows to the treatment plant. This dictated a shallow tunnel, with only six feet of cover over the 72-in. diameter pipeline.

The longer tunnels were not similarly constrained. The CSO pipeline will convey flows from the collection system to the Mercer Street Tunnel. The Effluent pipeline will convey treated CSO flows to the new outfalls located about a ½ mile south of the future CSO facility. The pipelines along the waterfront will be constructed under a future contract. Functionally, the two pipelines could have been installed within a range of depths varying from a minimum cover of approximately 37 feet to provide 8 feet of clearance below the existing EBI, to a maximum invert depth of 46 feet to allow gravity flow from the CSO pipeline to the Mercer Street Tunnel.

Open-cut construction would have been preferable for the shallow 72-in. pipeline, due to the low cover. However, in discussions with BNSF it was apparent that the railroad would not approve open-cut construction, because of the disruption to its operation on the mainline and spur tracks in this area. The designers were concerned that tunneling in this alignment would be challenging, given the sensitivity of the railroad tracks above, and the fill materials the tunnel would need to be constructed in. These concerns were foremost in conducting the geotechnical explorations discussed below and in developing the contract requirements for tunneling in this area.

2.2 *Geotechnical explorations*

The railroad crossing tunnel alignments are in a waterfront area that was formerly tidelands and had been repeatedly filled over the last 150 years. Review of historical data showed at least three episodes of fill occurred in this area. Near the turn of the century, the shoreline was east of the tracks, and two sets of tracks on pile-supported trestle were located on or near the present-day alignment of Burlington Northern Santa Fe Railroad (BNSF) tracks NP 26 and NP 27 (originally built by Northern Pacific Railroad). At some time prior to 1960, and possibly in more than one episode of filling, the shoreline was extended to a point west of the existing BNSF tracks and east of the Elliott Bay Interceptor.A timber pile bulkhead seawall with loose riprap protection was reportedly constructed along that shoreline. Today's shoreline is west of the grain terminal, and there is no evidence at the surface of the seawalls built in the past, and presumably abandoned in-place as the shoreline was extended.

Borings, test pits and horizontal directional drilling in the area confirmed the general locations of two abandoned seawalls, one under BNSF track D6 and a second under the grain terminal tracks. Tracks D6 and D7 are used by BNSF to make up trains for the grain terminal located just west of the tunnel construction site. King County and BNSF discussed the possibility of removing the abandoned seawalls from the surface to ensure that tunnel excavation or rail line operation would not be interrupted. BNSF felt that the work should be scheduled to avoid disrupting operation of the Cargill grain terminal, which is somewhat seasonal in nature.

As a result, King County elected to issue a separate construction contract for removal of the abandoned seawalls. The intent was to execute the removal work in seven days. Due to the need to take several tracks out of service to complete the work, it was necessary to restrict the contractor to a short timeframe and require the contractor to work all necessary hours to complete the work within the allowable number of days.

Work was authorized in late September 1999. The seawalls were located, removed, and the excavations were backfilled using controlled density fill (CDF). Work proceeded smoothly and was completed well within the seven-day timeframe. Removal was successful and eliminated the uncertainty of tunneling through the abandoned seawalls.

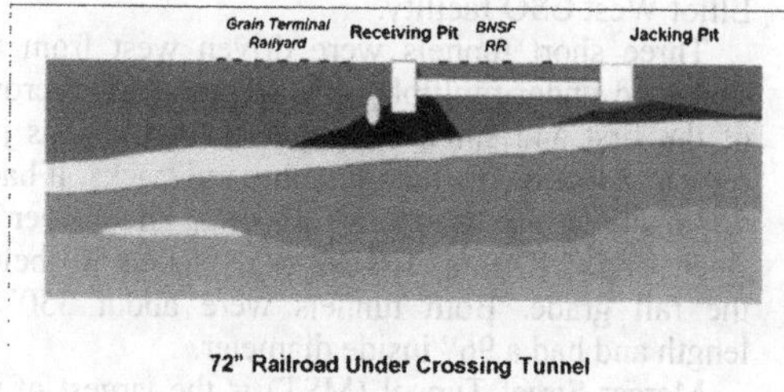

Figure 2. Profile of 72" RR under-crossing tunnel.

2.3 *Shallow 72" I.D. rail road under-crossing tunnel*

2.3.1 *Contract considerations- 72" tunnel*

With the decision to use tunnel construction to install the pipeline, the major risks were identified: 1) encountering obstructions that could not be removed through the TBM; 2) groundwater conditions that may interfere with obstruction removal; 3) heterogeneous soil conditions due to tunneling in fill materials; and 4) settlement or heaving of the railroad tracks. These risks and the actions taken to address them are discussed below.

The contract documents specified that the 72-in. tunnel be excavated using an open face digger shield and pipe jacking, with dewatering of the tunnel corridor. The pipe was specified as steel casing, with a 36-in. ductile iron drain to be installed inside the casing. The contract specified ground improvement (permeation grouting) in the tunnel corridor to limit ground loss into the face and control settlement of the railroad tracks to acceptable limits. Due to the high likelihood of encountering boulders, timber piles, and other debris in the tunnel zone, the specifications required that the shield have a digger arm/ excavator sufficient to excavate the ground and obstructions. Based on the results of the geotechnical investigations, dewatering was required along this alignment, with a requirement that the water level be drawn down at least 5 feet below the bottom of the tunnel.

In order to ensure minimal impact to railroad operation, the contractor was required to coordinate tunneling with the railroad's schedules, primarily to avoid tunneling under a track that had a train on it. In addition, extensive monitoring of settlement points on and adjacent to the tracks was required on a daily basis.

While installing grout pipes in the vicinity of BNSF tracks NP25 and NP26, the Contractor encountered an obstruction that appeared to be a sizeable piece of wood. The Construction Manager and the Contractor determined actions necessary to minimize any impact or excavation delay caused by the obstruction. The proposed work was coordinated with BNSF and Cargill. It was found that the two tracks were not as critical to BNSF's normal daily operation as the adjoining mainline tracks. The CM then contacted BNSF to see if obstruction removal could be accomplished from the surface. By removing the obstruction from the surface and backfilling with CDF, it would be possible to significantly reduce the risk of excavation under these tracks. Removing the obstruction in advance would have the advantage of getting it out of the way under controlled circumstances, rather than risking it as an emergency.

BNSF agreed to allow removal of the obstruction from the surface providing the work could be scheduled on a Sunday and be completed within 24 hours. It was also agreed that the contractor would excavate down to the tunnel invert under both sets of NP tracks, remove all obstructions in the tunnel zone, and backfill the excavated area with CDF.

The work began on a Sunday at 6 a.m. Excavation to the tunnel invert proceeded smoothly, and the wood obstruction—a large stump—was located and removed. It was observed that under the NP26 tracks, the timber piles were indeed still in place, but located on either side of the tunnel alignment. However, as the area under NP 25 tracks was excavated, three timber piles were encountered, centered on the tunnel alignment. They were removed and the entire excavation backfilled with CDF.

2.3.2 *Construction of shallow 72" tunnel*

Tunnel excavation began and the first 10-ft. section of steel pipe was installed on October 19, 2000 using a 74-in. diameter horizontal auger drill. Three obstacles were encountered during tunneling, including timber piles and a tree trunk (untreated). In each case, the auger was retracted to allow access to the face to break up the obstruction and remove it. Obstruction removal required delays of several hours up to one day. Tunneling and pipe jacking was completed on Nov. 8.

Groundwater was encountered but did not significantly affect the tunneling operation. No significant settling of the railroad tracks occurred. At the time when one of the obstructions was encountered, the tunneling operation caused a slight heaving of one mainline track. At completion of the tunneling, there was no significant change in the tracks. Coordination with railroad operations caused some brief delays in tunneling but was not a significant problem.

Figure 3. 72" steel pipe in jacking pit for shallow RR under-crossing tunnel.

2.4 *96" CSO and effluent railroad under-crossing tunnels*

The two 96" diameter tunnels cross below the BNSF railroad mainline and the right-of-way of the adjacent Port of Seattle Grain Terminal rail yard between Elliott Avenue West and Elliott Bay. The tunnel alignments pass under 13 rail tracks, resulting in

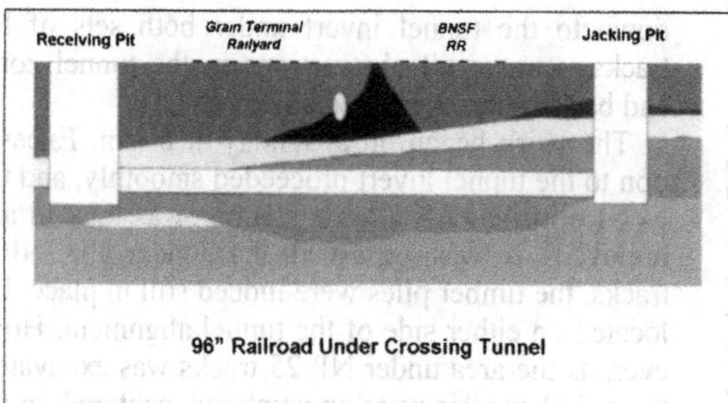

Figure 4. Profile of the 96" under-crossing CSO and effluent tunnels.

crossing lengths of approximately 350 feet. The invert of both the CSO and Effluent tunnels is 50-ft below the railroad bed.

2.4.1 *Contract conditions for 96" tunnels*

Alternatives considered for these tunnels focused on the vertical alignment of the tunnels. The tunnels needed to be deep enough to allow sufficient clearance below the existing EBI. Placement of the pipelines at the shallower depth would have allowed shallower shafts and permanent structures, but would have required tunneling through a mixed face of water-bearing beach deposits and overlying unclassified saturated fill materials. The deepest alignment considered was at an invert depth of 46 ft., which would allow gravity flow to the Mercer Street Tunnel. The crossings were aligned at the lower elevation to situate the pipelines primarily in hard silty clay glaciolacustrine deposits, which are relatively impermeable and particularly suitable for tunneling. As a result of the slope of the original bay bottom, the selection of the deeper tunnel alignment resulted in a mixed face of the silty clay glaciolacustrine deposits and the overlying water bearing beach deposits in a short reach at the western tunnel terminus. The beach deposits were expected to exhibit high permeability and low to moderate compressibility. Because these beach deposits lie below the ground water level, flowing ground conditions at the tunnel face were considered a risk. Consequently, the beach deposit layer to be penetrated by the tunnels was stabilized using grout injection techniques, which was accomplished from the surface prior to tunneling.

During the design phase, microtunneling, an Earth Pressure Balance (EPB) shield, and a Slurry Tunneling shield were all considered for these 96" diameter tunnels. All options were considered with pipe jacking as the means for installing the tunnel liner. The following factors led to the decision to specify an EPB machine for these tunnels:

- The silty clay glaciolacustrine deposit presents problems when using slurry based spoils handling. Both Microtunneling and a Slurry Tunnel shield handle spoils in this way. Tunneling through clayey deposits requires large volumes of slurry to cut and convey the spoils. Drilling fluids and polymers are often required to assist in spoil removal. Separation and removal of fine grain materials for disposal requires specialized equipment, which generates large volumes of waste slurry.

  The interface between the glaciolacustrine deposit and the overlying beach deposits presents a significant risk for encountering boulders or nested cobbles. Any such obstacles encountered by a Microtunneling machine, which could not be broken down by cone rock crushers mounted at the face, would present a potential impediment to progress without access from the surface or by back tunneling from the opposite direction. The inability to gain access to obstructions through the face of a Microtunneling machine made this option undesirable for the anticipated conditions.
- Microtunneling machines of a size necessary for mining a 96" diameter tunnel are at the high end of the size range of such equipment available internationally, but outside of the range of sizes available in North America at the time.
- EPB machines are generally used for tunnels larger than 144" diameter, but are available in sizes suitable for the 96" tunnels. The ability to gain access to obstructions through the shield face if required, and the use of a mechanical conveyor system rather than slurry based spoils handling system make the EPB the most appropriate choice.
- Because the EPB machine maintains pressure balance at the face using closeable panels at the cutter head, the potential "running" condition associated with the overlying beach deposits at the west end of the tunnels presented a concern. The panel closure reaction time may not be adequate to maintain the tunnel face under "running" conditions. Consequently, use of the EPB would require the stabilization of the beach deposits using grout injection. The potential for "running" conditions in this material where full grout takes may not be achieved is further reduced by the requirement to drawdown the groundwater table during tunneling.

Construction access constraints and spoils handling considerations dictated that the 96" tunnels be driven from east to west. This meant that the jacking pits for these railroad tunnels would occupy a portion of the same footprint as the Mercer Street Tunnel West Portal. Sequencing of the work was required to ensure that pipe jacking in a westerly direction below the railroad tracks could be accomplished with an adequate reaction block in place prior to full excavation of the West Portal shaft for launching of the large diameter TBM for the Mercer Street Tunnel. The recognition of the requirement for close coordination between these tunneling operations resulted in the combining of all large and small tunneling elements to be launched from the Elliott West site in a single construction contract.

2.4.2 *Construction of the 96" tunnels*

The contractor used the west portal as the starting point for installation of the two tunnels. A 120-inch diameter Lovat Model M-102, wheeled excavator-shield was used to bore both tunnels. The TBM cutterhead mounted four sliding flood control doors as standard equipment. It also had a muck ring and pressure-relieving gate that would allow it to work in EPB mode if required. This feature proved not to be necessary on the two 96-inch tunnels.

In order to create a jacking pit for the tunnels within the much larger pit for the Mercer Street Tunnel, a temporary sheet pile wall was installed across the west portal, a trench was excavated on the west side of the sheet piles and backfilled with concrete. The west side of the concrete wall was then excavated to create a jacking pit about 12 ft. deep. Beyond the concrete thrust reaction wall, the remainder of the west portal was left un-excavated until after completion of the 96-in. tunnels. A pipe-jacking frame was placed on the work slab and braced against the thrust reaction block. The jacked pipe had a 9" wall, was twelve-foot long, and weighed about 18 tons. When tunneling in the glacial soils the pipe thrusting forces seldom exceeded the dead weight of the pipe. Tunneling for the CSO tunnel was begun on March 3, 2001, and completed on March 12. The contractor worked around the clock except for Sundays. Two boulders were encountered but did not cause significant delays to the tunneling machine. The grouted beach sands were encountered just east of the receiving shaft and were tunneled through without incident.

Tunneling for the Effluent tunnel began on March 19 and was finished on April 3, 2001. Mechanical delays occurred on several days. One boulder was encountered, and became stuck in the floodgate. It was removed without significant delay.

3 MERCER STREET TUNNEL

3.1 *Mercer Tunnel geotechnical conditions*

The glacial soils encountered in explorations for the Mercer Street Tunnel included hard, clays and silts from a large pro-glacial lake, which were subsequently overridden, outwash sands and gravels, basal lodgement till, and several ice contact and near-contact drift deposits. Local advances and retreats of glacial ice and location of the project site at the leeward end of a pre-glacial hill, has created a complex sequence of glacial soils. Except for the uppermost deposit of recessional outwash, these glacial soils are highly over-consolidated and are very dense or hard. The outwash sands and gravels are generally saturated, with permeabilities occasionally in excess of 0.1 cm/sec; however, the till, drifts, and clays generally have permeabilities less than 10^{-4} cm/sec. In general, the relatively dense glacial soils in the project area present predictable and mostly favorable tunneling conditions. Conditions in the fills at the eastern and western ends of the project are much less predictable and much less favorable for tunneling.

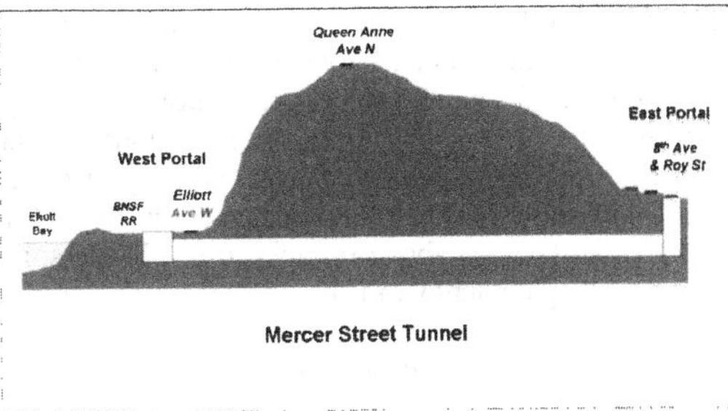

Figure 5. Profile of the Mercer Street Tunnel.

Tunneling conditions predicted during design can be summarized into five reaches, each reach representing approximately one fifth or 20 percent of the alignment. From the west portal, the tunnel is situated beneath landslide deposits and through disturbed lacustrine clays. The upper part of the heading was expected to be close to saturated beach sands, potentially causing a flowing condition in the crown. There was also concern during the design that the initial 50 ft (15.3 m) to 100 ft (30.5 m) of tunnel excavation would encounter buried logs along the old shoreline and concentrations of boulders eroded out of the glacial soils that comprised the shoreline bluff. Water levels were anticipated to be 20-ft (6.1 m) above tunnel crown, resulting in a risk of inflows concentrated along fracture planes. Actual geologic conditions and tunneling behavior encountered during the first 1,200 ft of tunneling are discussed below.

The second fifth of the tunnel was expected to pass out of the landslide-disturbed clays and into relatively intact lacustrine clays, with localized fractures and blocks. Seepage was expected to be negligible, except for minor seeps along sand seams or lenses.

The third fifth of the tunnel was expected to pass into mixed face conditions with clay in the upper face and glaciomarine drift and outwash sand moving progressively upward into the tunnel from the invert. Boulders were expected to be concentrated along the weathered erosion surface between the overlying glaciolacustrine clays and underlying clayey to sandy glaciomarine drift.

The fourth reach of the tunnel was expected to continue in mixed face conditions; however, the upper 10 ft (3 m) to 20 ft (6.1 m) of the outwash sand was expected to be very fine-grained and silty, making it "livery" in tunneling terminology, and very difficult to drain or otherwise improve with grouting techniques. The upper portion of the tunnel was expected to be capped in glaciomarine drift and glaciolacustrine soils, resulting in a hard, resistant arch and upper face and a soft lower face and invert.

The final fifth of the tunnel was expected to be advanced from under the overlying layer of glaciomarine drift into a variable outwash/sandy till-like material. Boulders were expected to be concentrated along the contact between the overlying glaciomarine drift and the underlying outwash. Along the extreme eastern end of the tunnel along Broad Street, the face was expected to be entirely in clean sand outwash with the groundwater level 20-ft (6.1 m) to 30-ft (9.1 m) above tunnel crown. These soils have a strong tendency to flow.

3.2 *Mercer Tunnel contract considerations*

Of most concern during design was the potential presence of running saturated granular soils and the possibility of encountering hard boulders. Types of tunneling methods considered for the project included:

- Digger shield with dewatering and two-pass lining system
- Digger shield with dewatering and one-pass lining system
- Earth pressure balance tunnel boring machine with two-pass lining system
- Earth pressure balance tunnel boring machine with one-pass lining system.

To evaluate these four potential tunneling and lining methods, a risk analysis was performed to compare the potential risks and impacts of the various tunneling methods. The potential risks considered included:

- Encountering boulders
- Encountering running sands
- Excess groundwater disposal
- Damage to existing utilities
- Ground settlement
- Tunnel misalignment.

The number and potential impacts of these occurrences on the four methods considered were evaluated, including cost and schedule considerations. From these evaluations, it was concluded that the project should proceed with the requirements for an earth pressure balance tunnel boring machine and a one-pass lining. Although the open face machine would likely have a lower bid price than an EPB machine, the high likelihood of boulders pushed the "final" cost of the open-face option above that of the EPB option.

The analysis also led to development of contract provisions that would mitigate and administrate these potential risks, including:

- Differing site conditions clause
- Escrow bid documents
- Dispute review board
- Geotechnical Baseline Report
- Ground improvement bid items
- Obstruction removal bid items.

3.3 *Construction of the Mercer Street Tunnel*

TBM excavation of the Mercer Street Tunnel commenced in May 2001. At the time this paper was written (August 2001) nearly 1,300 feet of tunnel had been excavated and lined. All excavation was in the first tunnel reach, consisting of highly overconsolidated glacial soils that are very dense and hard. However, from the launch shaft to the Elliott Avenue under-crossing, the tunnel encountered landslide deposits and disturbed lacustrine clays. Surface monitoring indicated a slight settlement both under Elliott Avenue and in the overbearing land-

Table 1. Features of the Lovat EPB machine used to excavate the Mercer Street Tunnel.

The Earth Pressure Balanced Machine (EPBM)

Manufactured by: Lovat Inc., Toronto Canada
Model: RMP200SE Series 15900
Year Built: 1996
Operating Mode(s): EPB, Semi-Closed, or Open
Weight: 305 tons
Total Length (with trailing gear):259-Feet

Cutterhead features:

- Mixed Face
- Flood control doors
- Face injection ports
- Probe drill ports
- Air Lock (for access to pressurized EPB cutterhead chamber)
- Variable speed drive (0 to 4.8 rpm)
- 1,050 horsepower
- 3,200 tons of thrust (maximum)
- 2.63×10^6 ft lb Maximum torque

Special Features:

- Segment Erector
- Class I Division II Electrical Systems
- Ground Conditioning System capable of Injecting:
  - Foam
  - Polymer
  - Water
- TACS guidance system* (from Germany)
- Programmable Logic System* (PLC)

*Capable of recording 32 data channels of the EPBM's operating parameters and gas levels at the tunnel heading (up to 10 samples per second). With computer displays onboard the machine and at both the project managers and the construction managers offices.

Operation of the tunnel boring machine (TBM) is in the Earth Pressure Balanced Mode, using the screw conveyor and a pressurized cutterhead (pressurized to approximately one atmosphere or higher) over the tunnel length.

slide deposit. As the tunnel advanced under increasing gained ground cover no further evidence of settlement was noted.

The tunnel was lined with precast concrete ring segments. Each ring lined four feet of tunnel and consisted of five segments and a key. All segments had seals and gaskets. Grout was placed behind each ring to reduce chances of ground settlement and to minimize any water inflows.

Progress of the TBM is outlined in Figure 7. This graph reflects a standard TBM start-up curve for coordinating the operating people and equipment.

Figure 6. Cutterhead for the Mercer Street/Lovat tunnel boring maching.

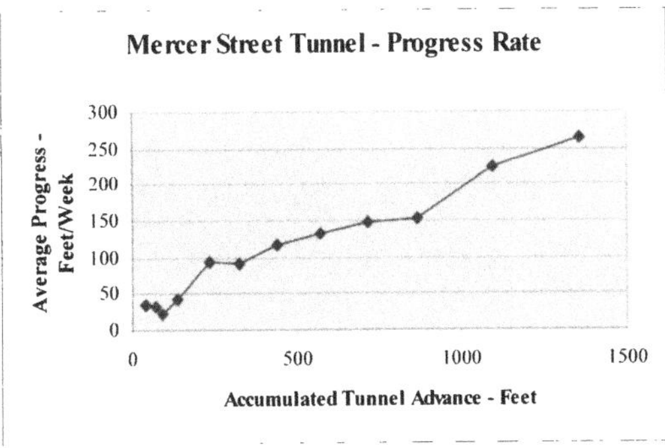

Figure 7. Mercer Street Tunnel progress rate from May through August 2001.

4 LESSONS LEARNED AND CONCLUSIONS

4.1 *Lessons learned*

Excavation of the three small railroad undercrossing tunnels is complete. The successful construction of these tunnels reflected the efforts made during the design and exploration process to understand the likely ground conditions and take active steps to address problems prior to beginning the tunnel excavation phase. These steps included removal of the abandoned seawalls through a separate construction contract; locating the 96-in. tunnels at a deeper elevation than functionally necessary in order to ensure tunneling through native ground; and removing known and unknown obstructions under the railroad tracks prior to beginning tunneling. Although these tunnels are a relatively small part of the project, their successful completion was not taken for granted. Rather, the design and construction team actively considered the risks that might affect completion of the tunnels and addressed each risk factor as appropriate in the process. The result was that the tunneling operation itself was uneventful and did not cause any interruption to rail service.

The Mercer Street Tunnel is something over 20% complete. Many lessons have yet to be learned. Most of the risks considered in the design appear to have been successfully addressed in the project documents and execution. For example, the likelihood of encountering boulders was a serious concern with regard to the requirement for an EPB machine. However, the boulders encountered to date have been broken up by the machine and have not caused any delays. Although surface settlement occurred in the first few reach of tunnel, in a zone of shallow ground cover, there is no significant damage to utilities. Extensive geotechnical monitoring during construction provides the team with current data on the impacts of tunneling so that the information can be used as appropriate to modify procedures. The laser guidance system and TBM steering capabilities have resulted in tunnel line and grade within the design tolerances at all times.

5 CONCLUSIONS

On this project Lovat Tunneling has provided the contractor with the capability of recording 32 data channels of real-time TBM physical operating and performance data. The contractor has, by contract, made these data available to the owner. The TBM designer and the geotechnical engineer now have the capability of understanding qualitative TBM performance in relationship to the actual geologic conditions encountered.

The future of underground development in North America depends on our ability to accurately forecast underground tunneling conditions, estimate costs, and complete tunnels within those estimates. The contractor's collecting and sharing these performance data with equipment suppliers, and tunnel owners should assist in minimizing misunderstandings relating to the development future of underground infrastructure development.

North American Tunneling 2002, Ozdemir (ed.)
© 2002 Swets & Zeitlinger, Lisse, ISBN 90 5809 376 X

Managing uncertainty and risk – The exploration program for Seattle's proposed Light Rail Tunnels

D.C. Ward, R.A. Robinson, & T.W. Hopkins
Shannon & Wilson Incorporated, Seattle, Wash., USA

ABSTRACT: The risk of changed condition claims on tunnel projects has been directly related to the thoroughness of the geotechnical exploration program by several authors. This paper assesses the benefits of a detailed investigation of very complex geotechnical conditions along a 4.5-mile-long tunnel alignment with 4 stations for the Sound Transit Light Rail Tunnels (Link) in Seattle utilizing a phased geotechnical exploration program. The exploration program used a range of exploration techniques ranging from standard soil borings and sampling procedures to state-of-the-art sonic drilling and statistical assessments of sample data. The goal of every exploration program is to balance the costs associated with explorations and testing against the risks associated with not having enough data to adequately characterize the subsurface conditions. The phasing of a geotechnical exploration program maximizes the benefits obtained from borings as the design and alignment selection proceeds from conceptual, to preliminary, to final design and helps to balance cost, schedule, and risk. Based on our experience with the Link project, an adequate level of geotechnical investigation may be greater than generally accepted guidelines on exploration depth and spacing. Further, this level of geotechnical investigation by the owner may be appropriate regardless of whether or not the project is Design/Build or Design/Bid/Build.

1 INTRODUCTION

This paper presents a general discussion of the general geology of the Seattle area, the risks and uncertainties associated with the geology, the exploration and testing program for the Sound Transit Link Light Rail project (Link), and a discussion of managing uncertainty and risk using a phased exploration program.

The proposed Link project currently includes approximately 5.5 miles of twin-bore, 21-foot diameter tunnels. The alignment for the North Corridor Tunnels is about 4.5 miles long and stretches between Downtown Seattle and the University District. The alignment for the 1-mile-long Beacon Hill Tunnels crosses the long-axis of Beacon Hill. This paper will focus on the North Corridor Tunnels.

The original schedule for the North Corridor Tunnels called for the project to go from the 30% design phase to operation in 7 years. The project includes 3 mined stations, 1 mined cross-over and 1 cut-and-cover station, and 1 cut-and-cover crossover. The relatively short project schedule prompted Sound Transit to select a Design/Build format for awarding the contract for this portion of the project.

2 REGIONAL GEOLOGY AND ASSOCIATED RISKS AND UNCERTAINTY

Seattle is located in the central portion of the Puget Lowland, an elongated topographic and structural depression bordered by the Cascade Mountains to the east and the Olympic Mountains to the west. Broad outwash plains, low-rolling relief, some deeply cut ravines, and broad valleys characterize the lowland. Geologists generally agree that the Puget Sound area was subjected to six or more major glaciations during the Pleistocene Epoch (2 million years ago to about 10,000 years ago), which filled the Puget Lowlands to depths in excess of 3,500 feet with glacial and nonglacial sediments (Galster and Laprade, 1991).

A review of case histories of tunneling and other large civil facilities projects in the Seattle area provided abundant information regarding construction difficulties related to the ground conditions associated with alternating glacial and interglacial episodes. These difficulties included cobbles and boulders, highly abrasive sands and gravels, methane, fractured and slickensided clays, abrupt changes in geologic contacts, multiple perched groundwater levels, flowing and running ground, and potential faulting (Shannon & Wilson, 1999). In order to try to quantify the uncertainty and risks associated with these construction difficulties, the subsurface exploration and testing programs for underground projects in Seattle must be designed appropriately.

Local experiences with explorations for tunnel projects in Seattle indicate that borings must be relatively closely spaced to detect variations in soil properties and abrupt changes in soil contacts. As an example, borings along the 3.3-mile-long Lake City Sewer Tunnel were spaced 800 to 2,000 feet apart. The exploration program missed two deep sand-filled valleys in a predominately clayey tunnel horizon. These omissions resulted in significant claims for this tunnel project. The borehole spacing along the Downtown Seattle Bus Tunnels (Robinson et al., 1991) was about 50 to 200 feet along the 1.2-mile-long alignment. This relatively close spacing proved sufficient to accurately extrapolate critical soil layer contacts at tunnel depth to with 2-vertically. Likewise along the 1.2-mile-long Denny Way CSO borings spaced about 250 to 350 feet apart have proven capable of locating critical soil contacts within 4 feet when extrapolating between borings. In soil tunneling, this level of predictive accuracy is necessary for designing dewatering and grouting and scheduling construction maintenance operations such as pick and cutter replacements.

3 SUBSURFACE EXPLORATION AND TESTING PROGRAM

The field exploration program for the North Corridor Tunnels was divided into three separate phases, conceptual engineering (CE), preliminary engineering (PE), and tender design. The tender design phase was undertaken for Sound Transit's use in the procurement process for a Design/Build team. The geotechnical scope of work for the tender design phase was similar to the effort for final design on most Design/Bid/Build projects.

3.1 *Borings*

The phased geotechnical borings ranged in depth from 65 feet to 311 feet and were primarily drilled using mud rotary drilling techniques. Mud rotary was selected because of the depth of the borings, the relative consistency and density of the glacially overconsolidated soils, the presence of boulders and cobbles, and the desire to obtain relatively undisturbed soil samples in hard silts and clays.

The planned depths of the borings were determined by adding one tunnel or station tube diameter to the depth of excavation invert. The extra depth allowed for some flexibility with regard to potential changes in the vertical alignment and engineering considerations related to the potential for excessive hydrostatic pressure below the invert of the tunnels, stations, cross-overs, and cross-passages.

The CE phase consisted of 11 geotechnical borings, including two overwater borings. These borings had an average spacing of over 2,000 feet with a few borings more closely spaced at critical locations along the proposed alignment. The borings were located where subsurface conditions were anticipated to have a greater potential impact on the vertical alignment.

The PE phase consisted of an additional 31 geotechnical borings, including four overwater borings. During this phase, boring spacing averaged less than 800 feet. The boring locations were selected to enhance the understanding of subsurface conditions along the alignment to assist in the preliminary level civil design.

The third (tender design) phase of explorations consisted of 71 additional geotechnical borings, including one overwater boring. These borings were spaced approximately 300 feet apart. The boring locations were selected to fill in the gaps between the borings drilled in the CE and PE phases and to further reduce the level of uncertainty in the interpreted geologic contacts and piezometric levels.

In addition to the standard geotechnical borings, four sonic core borings were also performed. Sonic drilling uses high-frequency vibration applied to the top of the drill column, along with down pressure, and on occasion rotation, to obtain a nearly continuous, approximately 3.75-inch-diameter soil core. The costs associated with sonic drilling were nearly twice that of the mud rotary geotechnical borings, but the volume of soil collected and the level of detail for the complex subsurface conditions revealed in the samples were significantly greater. Sonic core drilling was able to obtain continuous samples of glacial till, which has similar strength and behavior properties to moderately weak rock or lean-mix concrete. In addition, cobbles were drilled and sampled and at least one boulder was pulverized, which enabled us to confirm the presence of boulders as opposed to inferring their presence from drilling action, as is typically done. Although the sonic core borings provide a much clearer picture of subsurface conditions because the core is nearly continuous, the samples should not be used for strength testing because of the degree of sample disturbance associate with sonic drilling (Robinson and Ward, 2001).

3.2 *Soil sampling*

Except for sonic core drilling, the sampling program for the geotechnical borings consisted of a 5-foot spacing to within one tunnel diameter above the crown of the tunnel, and a reduced spacing of 2.5-feet down to one tunnel diameter below the proposed tunnel invert. Both disturbed and relatively undisturbed samples were collected in the geotechnical borings.

The majority of the disturbed samples from the geotechnical borings were obtained from the Standard Penetration Test or a 3-inch diameter splitspoon driven by a 300-pound hammer. The 3-inch sampler was used to try to obtain a larger volume of soil in very dense and gravelly soil.

The relatively undisturbed samples were obtained from Osterberg, Shelby, or Pitcher barrel samplers depending on the relative consistency of the soil. The Osterberg sampler was used to collect very soft and soft cohesive soils encountered in vicinity of the Portage Bay under-crossing.

Shelby tubes were used to collect samples of medium stiff to very stiff and occasionally hard glacial silts and clays. More typically a Pitcher barrel sampler was used to obtain samples of hard silts and clays.

Sonic core samples, which were very helpful for characterizing the subsurface conditions, could not be used for laboratory strength testing because of the degree of disturbance. The area ratio, the ratio of the cross sectional area of the sampler to cross sectional area of the sample, is in excess of 115% which exceeds the recommended 10% limit for a relatively undisturbed sample (Peck et al., 1974). In part because of the need to characterize the physical properties of the geologic units through laboratory strength testing, in part because of the relative newness of the drilling method, and in part because of the cost, the use of the sonic core drilling was limited to one hole per station.

3.3 *Groundwater*

Multiple groundwater measuring devices were installed during the three phases of explorations to assess the complex hydrogeologic conditions along the alignment. The devices included: 7 observation wells and 5 vibrating wire piezometers (VWPs) during the CE phase; 17 observation wells and 12 VWPs during the PE phase; and 42 observation wells and 47 VWPs during the tender design phase. At selected locations, multiple VWPs were installed in the same borehole as the observation well. This installation method was useful for evaluating potential groundwater gradients and differing responses to aquifer testing.

The data from the groundwater measuring devices revealed that between one to three distinctly different area-wide piezometric surface often exists in the complex stratigraphic sections along the alignment.

A total of 237 individual slug tests, both rising head and falling head tests were conducted in the observation wells installed during the three phases of the project. These tests helped provide information on the hydraulic conductivity of the more granular geologic units encountered.

Aquifer pumping tests were performed at four locations during the tender design phase. Three tests were performed in the vicinity of proposed stations and one test in an anomalous crosscutting sand filled -trough on Capitol Hill. The aquifer testing provided information on hydraulic conductivity, storativity, and transmissivity in the vicinity of the tests.

3.4 *Gas measurements*

During previous tunneling projects in the Seattle area, the presence of methane gas has been identified (Shannon & Wilson, 1999). Because of the potential hazard to tunneling operations, the testing program included checking for the presence of methane.

Methane gas was detected in soil samples obtained from Pitcher tubes and in headspace and groundwater samples obtained from observation wells. At selected locations, the screened portions of the observation wells were positioned near the tops of granular units, immediately below low permeability clayey units in order to increase the likelihood of detecting methane. In addition to methane gas, hydrogen sulfide gas and oxygen deficient conditions were also detected. Based on the results of the testing, we classified the tunnel as potentially gassy. Testing of soil samples and screening of well headspaces were undertaken both during the PE and tender design phase. Testing of water samples was performed during the tender design phase.

3.5 *Boulders*

Cobbles and boulders, which can be a hazard to tunneling, are very common in the glacial soils encountered in the Seattle area. The boulders are found in varying amounts in all of the glacially deposited soil. Greater concentrations of nested and grouped boulders may occur along contacts between nonglacial and glacial deposits.

During the PE phase, a statistical assessment based on data collected during the field exploration phase was undertaken to quantify the potential number of cobbles and boulders that could be encountered during tunneling. The assessment was based on the frequency and size of cobbles and boulders inferred during the drilling of the borings. The model was calibrated based on boulder quantities determined from two local case histories. The model was updated and refined during the tender design phase.

3.6 *Laboratory testing*

The combined laboratory testing program for the three phases incorporated over 1,300 physical property tests (including grain-size distribution, Atterberg limits), and approximately 250 strength tests (including triaxial compression tests and direct shear tests).

In general, the laboratory testing for each of the phases was proportional to the number of samples collected during that phase. The complexity of the tests increased slightly as the project advanced (e.g. the creep tests were performed only during the tender design phase).

3.7 *In situ tests*

A total of 42 pressuremeter tests were performed during the three phases of the exploration program; 5 in the CE phase, 15 in the PE phase, and 22 in the tender design phase. The pressuremeter testing in the CE and PE phases was performed to provide input parameters for numerical modeling completed by the preliminary civil design team. The additional testing during the tender design phase was to provide the potential Design/Build contractors with additional data to help characterize the properties of the different geologic units and spatial distribution of these properties along the alignment for their use in the final design of the project.

Vane shear tests were performed at four locations during the CE phase to characterize some of the softer sediments encountered in the vicinity of Portage Bay under-crossing. The alignment was subsequently lowered to avoid these softer sediments and therefore no additional vane shear tests were performed during the PE or tender design phases.

Eight downhole seismic tests were performed during the tender design phase. These tests were performed in non-slotted, 2-inch diameter PVC casings grouted into boreholes to provide input parameters for numerical modeling to be completed by the selected Design/Build team for their final design. While the need for non-slotted casing precluded the ability to install an observation well in the boring, VWPs could be installed adjacent to the casing. Although the VWPs could provide groundwater level information, they could not be used to screen for methane and other gases or for slug testing. In addition, whereas VWPs are in general very reliable, an observation well can provide additional assurance concerning the absence of groundwater.

4 PHASING AND SPACING OF EXPLORATIONS

The phasing, and the associated reduction in spacing between borings, of an exploration program helps balance the costs associated with explorations against the risks associated with not having timely and sufficient data to adequately characterize the subsurface conditions.

4.1 *Phasing*

The phasing of a geotechnical exploration program decreases the amount of money spent on borings not suitable for the final design of the project. For example during the CE phase, the borings were drilled at critical locations where conditions could significantly effect cost, alignment, and viability of the project. One such location was the crossing under Portage Bay. Prior to performing the explorations, the depth to glacial soils was assumed to be relatively shallow. However, during drilling of the second boring it was discovered that the softer sediments extended to depths as much as 64 feet below mudline. The discovery of this trough of softer sediments forced the whole vertical alignment of the project deeper. The two stations north of Portage Bay, which were originally planned as cut-and-cover stations, were forced deep enough that it was deemed less expensive to mine twin binocular caverns 500-feet long and 35 feet high and sink two 60-foot diameter shafts at each of the proposed station locations. The overwater borings, which were originally considered to be too costly for inclusion in a CE exploration program, helped prevent subsequent PE phase borings north of Portage Bay from being too shallow for the design of a deepened alignment.

4.2 *Spacing*

Although there are no industry standards for levels of exploration for tunnels, most tunnel jobs have between 0.25 to 1.25 feet of boring per foot of tunnel (USNCTT, 1984). The report recommended 1.5 feet of boring per foot of tunnel based on 87 tunnel jobs and was a function of contractor claims made on the projects. One major problem with this using this analysis to design an exploration program is that as the tunnels get deeper, the spacing of the borings would increase. Based on our experience on the Denny Way CSO project, currently under construction in Seattle, a standard based on an average spacing of about 300 feet is better suited for running tunnels in glacial soils. In addition, our experience developing subsurface profiles on the Link project, indicates an even tighter spacing, perhaps as close as 150 feet, in the vicinity of critical structures, such as mined stations, is also appropriate. The following sections provide a discussion of how our understanding of the complexity of the geology of Seattle evolved as the phased exploration program moved through the various exploration phases and the spacing between the borings decreased.

4.2.1 *Conceptual geologic model*

The previous understanding of the geology underlying the hills of Seattle was developed from relatively shallow explorations and excavations along the tops and sides of the hills. The conceptual geologic model of Capitol Hill at the onset of the project was essentially that of relatively horizontal layers connecting similar deposits observed on the flanks of the ridge. Some draping or capping of sediments from the most recent glaciation down over the sides of the ridge was anticipated. The geologic model consisted

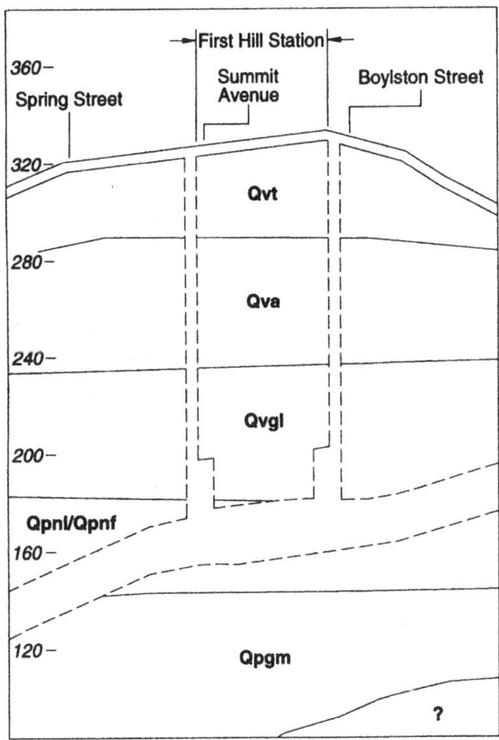

Figure 1. First Hill Station Conceptual Geologic Model.

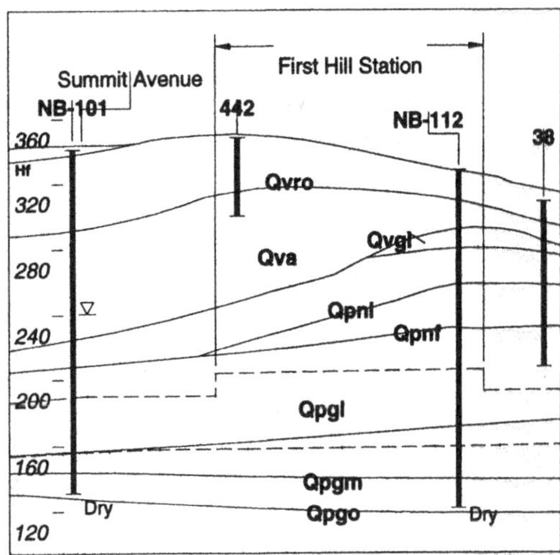

Figure 2. First Hill Station Conceptual Engineering Profile.

of a surficial veneer of till and post-glacial deposits, underlain by outwash sand and gravel underlain by glaciolacustrine silt and clay. In the vicinity of the First Hill station at the south end of Capitol Hill, these Vashon-age sediments were thought to comprise the uppermost 200 feet of the would be underlain by nonglacial sediments and glacial (glaciomarine) sediments from the previous interglacial and glacial periods, respectively (Figure 1).

4.2.2 *Conceptual engineering*
During the CE phase of the exploration program, it was soon apparent that the geology beneath Capitol Hill was significantly different from the conceptual geologic model. Except on the flanks of the ridge, pre-Vashon deposits were encountered at depths less than about 50 feet below the ground surface. These pre-Vashon deposits consist of a complex sequence of glacial and nonglacial sediments, some with potentially significant impacts on the design of the proposed tunnels. Further, the geology encountered in widely spaced borings was commonly widely divergent between borings and stratigraphic extrapolation was difficult and sometimes little more than guess work. The CE phase explorations, which were spaced on average over 2,000 feet apart, largely provided a basic understanding about the gross stratigraphy, but little understanding of the detailed stratigraphy.

In the vicinity of the First Hill station, which lies at the top of the west flank of Capitol Hill, the two deep CE phase borings contrast with the shallower pre-existing borings as shown on Figure 2. This figure also shows the pre-Vashon deposits encountered. However, the geometry of these deposits can be only crudely inferred, and stratigraphic detail of the deposits is lacking. The CE borings also revealed a perched piezometric surface within the glacial outwash (Qva) and nonglacial fluvial sand (Qpnf and nonglacial lacustrine silt (Qpnl) above the relatively impervious pre-Vashon lacustrine silt and clay (Qpgl) and glaciomarine drift (Qpgm). Surprisingly, the observation wells located in the underlying pre-Vashon glacial outwash sand (Qpgo) were dry.

It should be noted that originally only one boring, NB-101, was scheduled to be drilled for the First Hill station during the CE phase. However, boring NB-112 was added to the exploration program after all of the other borings were completed to provide a borehole for pressuremeter testing. The spacing between these borings drilled for the First Hill station was approximately 650 feet.

4.2.3 *Preliminary engineering*
The PE phase of the exploration program provided an increased understanding of the stratigraphy and allowed for more reasonable geologic correlation between borings. A general decrease in the spacing of the borings, to less than 800 feet, improved the overall stratigraphic interpretation, and borings specifically located in poorly understood areas better defined enigmatic stratigraphic structures and groundwater conditions. Substantial laboratory testing was also accomplished during the PE phase, so that the geotechnical properties of the various geologic units and the likely tunneling conditions were better known. As the result of the increased under-

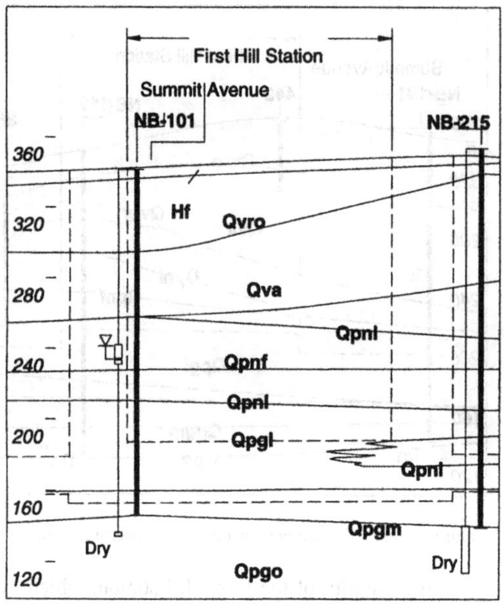

Figure 3. First Hill Station Preliminary Engineering Profile.

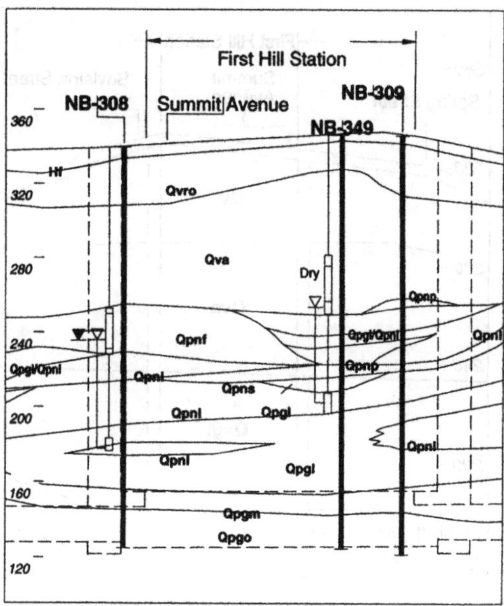

Figure 4. First Hill Station Tender Design Profile.

standing in stratigraphy, soil properties, and groundwater levels, both horizontal and vertical changes to the proposed alignment were made. At the First Hill station (Figure 3), an improved resolution of the geometry of the deposits and the recognition of problematic wet silts (Qpnl) within the thick layer of silty clay (Qpgl) resulted from the additional boring drilled in the vicinity of the station during the PE phase. The average spacing between the borings in the vicinity of the First Hill station at the completion of this phase was 325 feet.

4.2.4 *Final/Tender design*

The final geotechnical phase of the exploration program was largely directed at providing an increased understanding of the local geologic conditions around critical structures, such as stations and crossovers. Specifically, these borings were performed to better define the locations of geologic units that could have significant adverse impacts on mined structures. As a result of the final phase of explorations, changes were made to the location of the critical structures, both laterally and vertically, in an attempt to optimize structure locations relative to the geology. Figure 4 shows significantly more complex geologic conditions at the First Hill station than previous geologic interpretations. The final phase of explorations at the First Hill station were drilled to fine-tune the proposed location of the structure to avoid flowing silts (Qpnl) in the crown of the excavation and to keep sand and gravel (Qpgo), prone to raveling, below the midline of the station excavation.

These borings helped confirm that the relatively impermeable silts and clays encountered at a depth of 120 feet in the vicinity of First Hill act as an aquitard separating the upper sand and gravel from the lower sands and gravel. The surface water infiltrates through the upper sand until it encounters the silt and clay layer where it forms a slight mound before moving laterally. A portion of the water continues to move downward through the silt and clay and along fractures and joints until it reaches the lower sand. The difference in hydraulic conductivity between these two units, which is at least three orders of magnitude, is large enough that at least the upper 60 feet of the lower sand (Qpgo) is dry.

The average spacing between the borings at the completion of this final phase in the vicinity of the First Hill Station was about 150 feet.

5 CONCLUSION

The goal of every exploration program is to balance the costs associated with explorations and testing against the risks associated with not having enough data to adequately characterize the subsurface conditions. A phased exploration program with a final borehole spacing of about 300 feet has generally been shown to be sufficient to assess: 1) variations in soil properties; 2) locations of stratigraphic contacts to with about 5 feet vertically and 50 feet horizontally; and 3) the presence of multiple perched groundwater levels. In addition, although there is no apparent industry standard for sampling interval, a tighter, near continuous spacing some distance above to some distance below the proposed structure makes sense. The exact distance should be a function of the likelihood that the vertical alignment will change, the proposed method of tunneling, and potential impact of hydrostatic forces below the invert on the tunnel construction method.

Because the owner is often seen as being ultimately responsible for damage to buildings due to excessive settlements caused by unanticipated ground conditions or contractor costs associated with differing site conditions, it is in the owner's best interest to see that subsurface conditions are adequately characterized prior to design and construction. On the Design/Build Tren Urbano project, for example, which had an average boring spacing of 450 feet when the contract was awarded, the tunnel contractor recently filed a $45-million claim over differing site conditions (ENR, 2001).

Providing adequate geotechnical information helps level the bidding process, provides for an acceleration in schedule, and provides the owner with a level of assurance that an appropriate amount of geotechnical data will be collected. Although the North Corridor of the Sound Transit Link Light Rail project was to be a Design/Build project, over 90% of the needed explorations has been drilled prior to the bidding phase. Therefore, regardless of whether or not the project is Design/Build or Design/Bid/Build, we contend that the owner should perform the same level of geotechnical investigation.

REFERENCES

Engineering News Record (ENR). 2001. Rising Costs, Slipping Schedule Have Tren Urbano in Hot Water. September 17. p. 15.

Galster, R.W. & Laprade, W.T. 1991. Geology of Seattle, Washington, United States of America, *Bulletin of the Association of Engineering Geologists,* Vol. 28, No. 3: 235-302.

Perk, R.B., Hanson, W.E., & Thornburn, T.H. 1974. *Foundation Engineering, Second Edition.* New York: John Wiley & Sons.

Robinson, R.A., Kucker, M.S., & Parker, H.P. 1991. Ground Behavior in Glacial Soils for the Seattle Transit Tunnels. In sxxx (ed.), Proceedings of the Rapid Excavation and Tunneling Conference, Seattle, Washington, June 16-20. Baltimore: Port City Press.

Robinson, R.A. & Ward, D.C. 2001. Vibrating Your Way Through Glacial Soils, *AEG News, 44/5:* 19-20.

Shannon & Wilson, Inc. 1999. Geotechnical Characterization Report, prepared by Shannon & Wilson, Inc., Seattle, WA, for Sound Transit.

U.S. National Committee on Tunneling Technology (USNCTT). 1984. *Geotechnical Site Investigations for Underground Projects.* National Academy of Science.

Session 3b, Track 2

Policy and decisions about underground space

North American Tunneling 2002, Ozdemir (ed.)
© 2002 Swets & Zeitlinger, Lisse, ISBN 90 5809 376 X

Waller Creek Tunnel Project cost and funding

D. French, P.E., R.P.L.S.
Halliburton KBR, Inc., Houston, Texas, USA

D. Ivor-Smith, P.E., F.I.C.E.
Halliburton KBR, Inc., Houston, Texas, USA

W. Espey, Ph.D., P.E.
Espey Consultants, Inc. Austin, Texas, USA

G. Oswald, P.E.
City of Austin, Watershed Protection and Development Review Department, Austin, Texas, USA

ABSTRACT: Waller Creek, located in Austin, Texas, is a normally dry creek that runs through the University of Texas campus and several blocks of under-developed downtown property before discharging into Town Lake on the Colorado River. The Waller Creek Tunnel Project was conceived to intercept floods downstream of the university campus and convey a 100-year event through a 5000-ft. long tunnel to discharge directly into Town Lake. During the conduct of the preliminary design, it was determined that the original goals for the project would incur project costs significantly beyond the $25 million approved in the May 1998 bond election. The project costs for the recommended alternative are now estimated at over $50 million to meet the original 100-year flood control goals. Based on these findings, the City authorized the design team to undertake additional analysis to develop alternative flood protection goals and design recommendations to reduce the project cost and to evaluate the benefit-cost ratio for the City that could be derived from implementing the project.

1 INTRODUCTION

1.1 *History*

The City of Austin has long been interested in improving flood control and providing water quality enhancements to the lower Waller Creek watershed. The City has conducted several studies that indicate that a stormwater bypass tunnel with surface level inlet and outlet structures and a recirculation system would meet the City's needs. Based on earlier studies, Austin voters, in May 1998, approved the proposed tunnel project and its financing. The project would be financed in conjunction with the expansion of the City's Convention Center through the imposition of an increase in the hotel occupancy tax rates to support bond issuance debt. A total of $25 million was authorized for the flood control project. In May 1999, the Brown & Root / Espey Padden Joint Venture was contracted for the preliminary engineering of the Waller Creek Tunnel Project and subsequently was authorized to perform a "Scope Reduction and Benefit-Cost Analyses".

1.2 *Project description*

The Waller Creek watershed encompasses the area between N. Lamar Street and I-35, including a significant portion of the University of Texas campus, as shown Figure 1. The creek outfalls into Town Lake between Trinity Street and Red River Street. The original project goals generally require a 22 feet diameter tunnel with a length of over 5000 feet, an in-channel diversion structure, inlet and outlet structures, and a recirculation pump system. Stormwater is to be intercepted at Waterloo Park below 15th Street with discharge to Town Lake near the existing Waller Creek outfall.

Figure 2 illustrates the conceptual profile of the tunnel and general location of the facilities. If the project were implemented, the 100-year flood event would be contained within the creek channel downstream of the inlet structure. Approximately 42 commercial and residential structures and 12 roadway crossings currently subject to flooding will be afforded flood protection.

Further reductions in flood levels in the lower reach of Waller Creek (downstream of 12th Street) may also be achieved by capturing additional stormwater flows downstream of Twelfth Street through diversion of floodwaters from intervening storm sewer areas into the tunnel. This additional stormwater interception, while not being part of the proposed project, was examined for its feasibility and approximate cost.

2 RECOMMENDATIONS AND COSTS OF DIVERTING 100-YEAR FLOOD

The recommended facilities and their associated costs for accomplishing the original preliminary engineering phase objectives are summarized below. Twelve combinations of alternative inlet and outlet sites and tunnel alignments were developed and evaluated. Each alternative was analyzed with respect to its construction cost, construction schedule, reliability, right-of-way requirements, impact to the park environment, required operation and maintenance costs, and resulting total annual costs. The selected alternatives which best meet the entire range of criteria are outlined below.

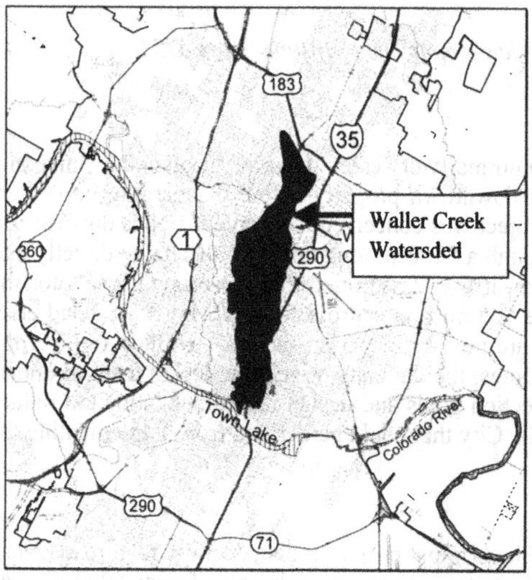

Figure 1. Austin Vicinity and Waller Creek Watershed.

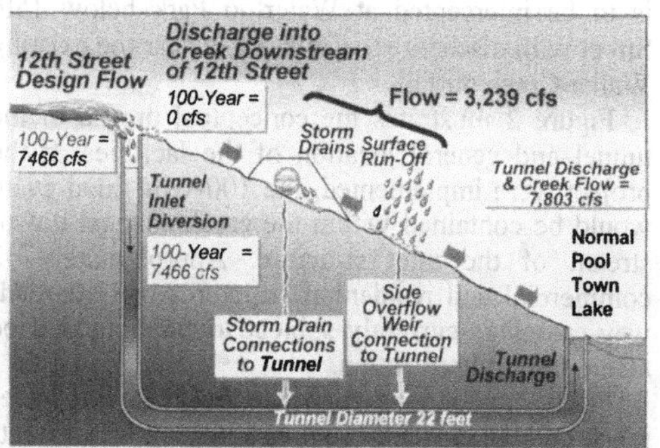

Figure 2. Conceptual Profile of Proposed Waller Creek Tunnel.

2.1 *Inlet*

The recommended inlet design is the "Morning Glory" alternative located immediately north of 12th Street in Waterloo Park, as Figure 3 illustrates. Major components include an excavated inlet pond formed by a downstream dam across Waller Creek, a circular-shaped weir at the tunnel inlet shaft, and an octagonal screen structure with mechanical cleaning. Adjacent to and immediately to the west of the dam is a pump station/control building structure housing recirculation pumps and additional screening equipment. These facilities provide the recirculation water supply for enhancing the water quality of Waller Creek downstream of the inlet. A bridge provides access for operation and maintenance. Architecturally, the pond will be sculpted to allow retention of large trees within the pond area, thus creating peninsulas surrounded by large native limestone boulders. The dam will also be covered by large boulders. The pump station will be veneered in native limestone and will be designed to make it visually compatible with Waterloo Park. Landscaping within the park will also be enhanced to achieve compatibility with this facility.

Figure 3. Perspective View of Recommended Inlet Facilities.

2.2 *Outlet*

The proposed outlet uses a "lagoon" concept that will be created immediately west of the Waller Creek confluence with Town Lake. Major components of this alternative include a tunnel outlet shaft, outlet weir, recirculation screens, a tunnel dewatering pump station, and a control building. Because the site is heavily used as a recreational area, the opportunity exists for an outdoor amphitheater to be developed in conjunction with the outlet facilities.

Figure 4 illustrates the siting and general layout of the outlet structure. Land north of the lagoon will be terraced with limestone on the front edge of each tier forming bench seating for the amphitheater. The area behind the limestone seating will be grassed with ADA access provided. A floating stage is proposed to be wrapped around the dewatering pump station access shaft. In addition there will be an upper walkway bridge connected into the existing jogging trail. The existing boathouse near this site

Figure 4. Perspective View of Recommended Outlet Facilities.

would need to be relocated for the outlet structure construction. The relocated boathouse is proposed to have a second story to be used as the control room for the outlet pumps and instrumentation. The boathouse will continue to be a limestone veneered building and the existing boat docks will be relocated a short distance further west from their present location. An additional boat launch site can be conveniently incorporated into the outlet training walls.

2.3 *Tunnel*

A total of 12 different alignments were compared for the base case of the 22-ft diameter tunnel, on four principal corridors as shown in Figure 5. The principal corridors considered were Trinity Street, Neches Street, Red River Street, and Sabine Street. Each of these options had between one and four variations depending on the assumed locations of the inlet and outlet structures.

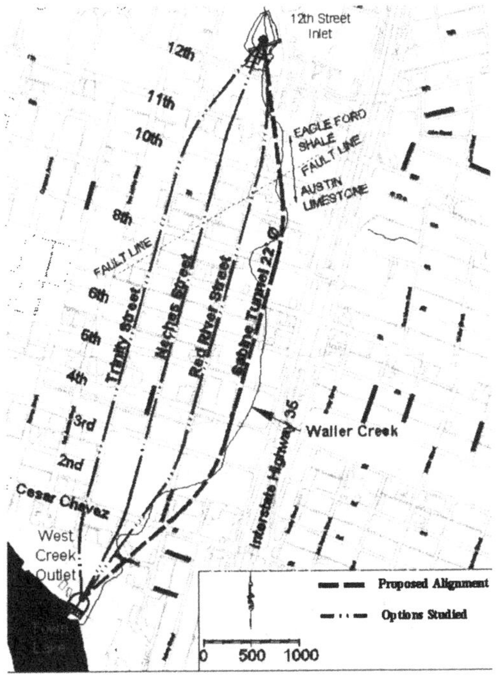

Figure 5. Tunnel Alignments Studied.

Geotechnical conditions along each of the four major corridors were examined, based on existing data to the extent possible, and supplemented by a few additional boreholes in key areas. It became apparent that the tunnel on any of the alignments would be predominantly in Austin Limestone, but that the northern and upstream portion of the tunnel would necessarily be situated in Eagle Ford Shale Figure 6. Due to a diagonal fault running south-west to north-east across this part of the City, the further east the tunnel is situated, the shorter would be the length of tunnel in this more difficult soft rock formation.

Construction schedules were developed for the various tunnel alignments based on historical data, assuming a variety of tunnel construction techniques, namely TBM, roadheader, and drill-and-blast, both with and without an intermediate shaft. This shaft, intended for ease of future maintenance, offers an advantage for tunnel construction particularly for roadheader and drill-and-blast methods, by offering a second working face. The liner options included cast in place for all tunnel construction methods, and segmental concrete for the TBM. Tunneling was assumed to be double shifted, for 10X10 hour shifts per week. Average advance rates were based on historical averages for each of the conditions prevailing.

The outlet shaft was assumed to be used as a launching shaft for a TBM, on the premise that no tunnel construction could be carried out from Waterloo Park, due to the proximity of Brackenridge Hospital, and to reduce the construction impact on the Park. A temporary TBM receiving shaft would be required at 12th Street to avoid delaying the construction of the inlet works. For drill-and-blast, and roadheader construction, schedules were based on single face and double face options, both with and without an intermediate shaft.

Not surprisingly, the tunnel operations are on the critical path, delaying the construction of the outlet works until such time as the tunneling operations no longer require the use of the outlet works site. The construction periods for several options were found to be very similar, with total contract durations ranging from 194 weeks (based on the tunnel constructed by roadheader or drill-and-blast on two faces, with intermediate shaft) to 252 weeks for TBM drive with cast in place liner, and no intermediate shaft. The schedule for a TBM drive with segmental concrete liner and the use of an intermediate shaft assumed a conservative average TBM advance rate of 180 ft per week, and resulted in an overall construction contract duration of 196 weeks. This latter method was assumed for cost estimating purposes.

For a number of reasons the preferred inlet structure would be located in Waterloo Park just north of 12th Street, on the west side of Red River Street. The proposed outlet structure would be located on the north shore of Town Lake, just west of Waller Creek. The shortest tunnel length between these points (5,004 feet) would be on Neches Street. However, on this alignment the tunnel has to pass under existing buildings at the north end, and beneath the Convention Center, where underpinning of the foundations would be required. In addition, the length of tunnel in Eagle Ford Shale is significantly more than on the more easterly alignments.

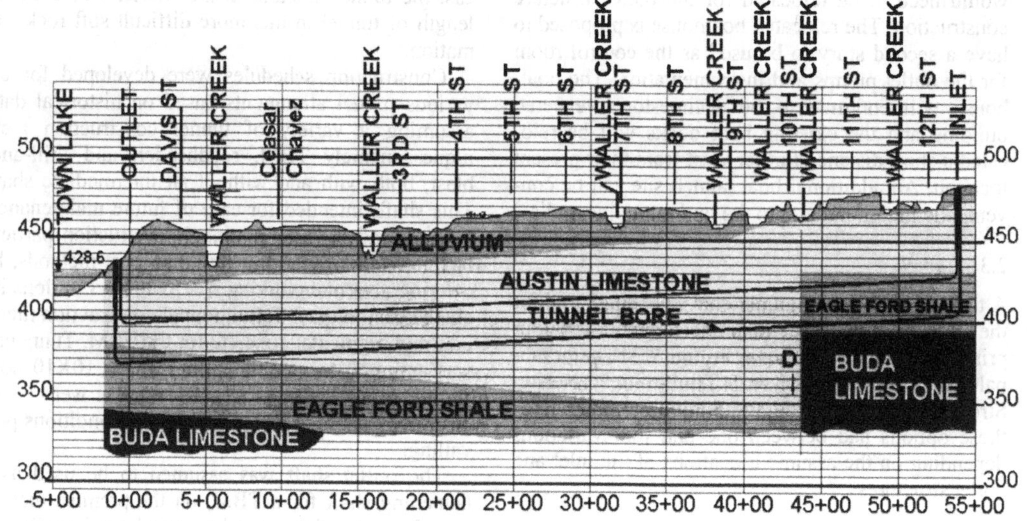

Figure 6. Sabine Geological Profile West Creek Outlet to 12th Street Inlet.

The lowest cost for the tunnel is for the alignment following Red River Road. However, if the intermediate storm drains connections are included in the costs, then the Sabine alignment becomes marginally more economical due to the shorter connections of the storm drains to the tunnel. This results in a tunnel length of 5,139 feet for the Red River Street alignment or 5,376 feet for the Sabine Street alignment.

For the recommended inlet and outlet locations, a tunnel alignment was selected which generally locates the tunnel under Sabine Street as illustrated in Figure 3. In profile, the tunnel is basically 'U' shaped with vertical shafts at its inlet and outlet. The tunnel elevation has been selected to minimize the construction cost by locating the tunnel in geological formations that provide better tunneling conditions along its length. See Figure 6. The crown of the tunnel is below Town Lake water level and therefore under normal operating conditions, the tunnel will be continuously underwater. The finished tunnel diameter is 22 feet. Capital and Operating Costs.

Table 1 summarizes the costs for the recommended facilities based on January 2000 costs. Annual costs were developed, assuming financing of capital at 6% and 4% inflation for O&M. The $52.8 million project, of which $40.8 million is the construction cost, meets the original project goal of keeping a 100-year flood within the existing channel of the lower Waller Creek.

3 SCOPE REDUCTION COST ANALYSIS

The total project costs are expected to exceed $50 million without capture of intermediate stormwater

Table 1. Summary of Costs - Waller Creek Tunnel Project - Sabine Alignment.

TUNNEL PROJECT		($ in millions)
Inlet –	Morning Glory with Tunnel Recirculation and Water Feature	14.31
Tunnel –	22ft Diameter with Intermediate Maintenance Shaft	18.74
Outlet –	West Creek – with lagoon and floating amphitheater	7.78
	Total Construction Cost	40.83
	Right-of-way (Tunnel)	0.46
	Engineering, Testing, Construction Management	11.51
	PROJECT COST	52.80
	Annual O&M Cost	1.40

flows, and would exceed $60 million if this element is included. The cost of the project therefore exceeds previous estimates and the City's authorized bond limits of $25 million. Accordingly, the City authorized additional investigations with the goal of identifying a tunnel and inlet/outlet alternative with a project cost closer to the $25 million budget. More specifically, the goal of this additional study included:

- Evaluating various alternatives using reduced diversion-capacity to determine the impact on cost and floodplain reduction in the lower watershed.
- Evaluating opportunities to minimize cost by reducing features for addressing operational and aesthetic issues associated with the project.

3.1 *Alternatives using reduced capacity*

The tunnel facilities for the base case provide for the full diversion of the 100-year design flood upstream of 12th Street. However, it is possible to provide a similar level of flood protection downstream of 12th

Street by diverting a smaller design flood, albeit with corresponding increase in the channel flow.

Using a 100-year flood as the diversion criteria will reduce the total floodplain area by approximately 1.243 million square feet. Using 77% and 55% diversions of the 100 year flood as diversion criteria upstream of 12th Street will result in floodplain reductions of approximately 1.243 million square feet (same as 100-year) and 1.225 million square feet, respectively. While the 100 year diversion will eliminate flooding of all existing road crossings (10 total) downstream of 12th Street, a 77% and 55% diversion of the 100-year flood will result in the flooding of one road crossing (9th Street) for the same reach.

Various elements of the project can be reduced in size if the flow capacity-reduction scenarios are acceptable. For example, tunnel diameters of 17.5 feet and 15.5 feet will handle 77% and 55% diversion of the 100-year flood events, respectively, rather than 22 feet required for the full 100-year diversion.

3.2 *Alternatives using reduced features*

Various alternatives were evaluated to further reduce the construction costs of the inlet, outlet and tunnel. In general, certain operational or appearance-related features can be eliminated, or various construction items can be staged or scheduled for a future phase, in order to minimize initial capital cost.

3.2.1 *Inlet*

For the inlet, all cost reduction alternatives are based on the "morning glory" arrangement with various high capital cost facilities removed or deferred. Features, which could be eliminated or reduced, include: mechanical cleaning of the intake screens; certain water features and associated pumps and screens; recirculation screens, and the pump station structure. Recirculation pumping at a reduced level must be retained; although, a simplified lift station can be supplied with chopper-type pumps to handle debris. In general, the facilities that can be eliminated directly impact the ease of operation of the inlet and require increased operator attention and a greater compliment of heavy construction-type equipment for debris removal and maintenance of the facilities. In other words, the resulting decrease in capital costs will be offset by increased operational cost and reduced reliability. The appearance of the inlet will be greatly impacted by the deletion of the water features and landscape treatment. Due to the unreliable nature of this "bare bones" inlet alternative, it is only considered as an initial phase to the other two alternatives, and should not be relied upon as a permanent solution.

3.2.2 *Outlet*

Outlet features, which could be eliminated, include re-circulation screens, outlet weir, dewatering pumps, boathouse renovation, floating stage, and amphitheater. Landscaping and walkway access across the lagoon structure must be retained. In general, the facilities that might be eliminated directly impact the ease of operation of the overall recirculation system and tunnel dewatering operations. With the elimination of the outlet weir, the yearly dewatering for cleaning and inspection of the tunnel would require the purchase and temporary installation of stop-logs to isolate Town Lake from the tunnel. Likewise, elimination of permanent dewatering pumps would require the purchase or rental of portable dewatering pumps. Use of a full construction crew would also be required for all dewatering operations because of this change. Again, due to the unreliable nature of this "bare bones" outlet alternative, it could only be used as an initial phase and not relied upon as a permanent solution.

3.2.3 *Tunnel*

For the tunnel, cost reduction alternatives involve eliminating the intermediate access shaft and reducing the tunnel diameter to accommodate reduced-size flood diversions. A 20 foot diameter tunnel is still capable of handling the 100-year flood diversion, with a maximum flow velocity of 24 feet per second; however, a tunnel diameter of less than 22-feet cannot hydraulically satisfy the connection of lower watershed storm sewers to the tunnel. A two-phased approach was examined, wherein a 15.5 feet diameter tunnel is initially constructed along with 100-year inlet and outlet structures, followed by the future construction of second parallel 15.5 feet diameter tunnel. At the completion of the second tunnel, the entire system would be able to handle a 100-year flood event. The ultimate cost of the two tunnels would be siginicanlty higher than for a single larger tunnel.

3.3 *Cost comparison*

The project cost for the least-cost scenarios for 55% diversion of the 100-year flood are summarized in Table 2. Due to the unreliable nature of the "bare bones" alternative, it should only be considered-as an initial phase for the other two alternatives, and not relied upon as a permanent solution. Cost comparisons of all the various cost-reduction alternatives considered were made, but only the least cost scenario is presented here for brevity.

4 BENEFIT-COST ANALYSIS

After completion of the Scope Reduction Analysis a further evaluation the potential benefits which might accrue to the City as a result of the Waller Creek Tunnel Project was performed, specifically to assess:

Table 2. Project Cost - 55% Diversion of 100-Year Flood - Recommended 15.5 Feet Diameter Tunnel Alternative1*.

DESCRIPTION	Alternative 1 "Operationally Friendly & Aesthetically Pleasing"	Alternative 2 "Operationally Friendly"	Alternative 3 "Bare Bones" – *Recommended Only as Initial Phase For Alternatives 1 & 2*
Inlet – 12th Street Morning Glory	$14,310,000	$12,830,000	$5,460,000
15.5 Feet Dia. Tunnel – Red River	$12,740,000	$12,740,000	$12,740,000
Outlet – West Creek Lagoon	$7,780,000	$6,040,000	$3,990,000
Total Construction Cost	$34,830,000	$31,610,000	$22,190,000
Right-of-Way	$40,000	$40,000	$40,000
Engineering, Testing, & CM, Inspection, Small Bid Pkgs.	$11,030,000	$10,680,000	$9,710,000
PROJECT COST	$45,900,000	$42,330,000	$31,940,000

*Costs are in January 2000 Dollars

- the expected annual damage that could be avoided by reducing the floodplain after the tunnel is complete.
- the future tax revenues that could be produced by new development in the resulting reclaimed floodplain area.
- the economic benefits of the tunnel diversion project as compared to its estimated costs.

4.1 *Expected annual damages*

The benefit-cost analysis requires estimation of the Expected Annual Damage (EAD) for the existing and proposed conditions. The difference between the existing and the proposed condition EAD represents the net benefit created by the project in terms of potential avoided damages. For this analysis, the proposed condition included a 55% diversion of the 100-year peak flow into the proposed tunnel. This 55% diversion is approximately equal to the 10-year peak flow rate.

Computed water surface elevations along Waller Creek for the existing and proposed conditions were determined using HEC-RAS for the 2-, 10-, 25-, 50-, 100- and 500-year frequency design storms. The resulting water surface elevations along Waller Creek were used to compute corresponding flood damage cost through application of the HAZUS program and the HEC-EAD Program.

The HEC-EAD program (March 1989 version) was used to estimate the EAD from the HAZUS damage for different flood levels. The program is based on the principle that flood damage to an individual structure, group of structures, or floodplain reach can be estimated by determining the dollar value of flood damage for different magnitudes of flooding, and by estimating the percent chance of exceedance of each flood magnitude. To compute the damage which can be expected in any year, the damage corresponding to each magnitude of flooding is weighted by the percent chance of each being exceeded (damage caused by rare events is thus weighted less). In general, the sum of the weighted damage values for each storm represents the expected annual flood damage.

Using the HAZUS program, damage estimates were developed by the City of Austin staff for the various flood frequencies as shown in Table 3 for both existing conditions and the proposed 55% diversion of the 100-year flood.. The computed value for expected annual damages is $490,000 for the existing conditions and $40,000 for the 55% diversion condition. The difference is therefore approximately $450,000 in EAD savings through construction of the proposed tunnel project.

Table 3. HAZUS Damage Estimates for Various Frequency Flood Events for diversion of 55% of 100-year flood.

Flood Frequency (Years)	Existing Conditions Damage Estimate (Year 2000 $)	Proposed Conditions Damage Estimate (Year 2000 $)
2	$500	$0
10	$716,000	$0
25	$2,421,000	$0
50	$7,540,000	$500
100	$9,419,000	$1,800
500	$13,793,000	$7,454,000

4.2 *Developable land area*

The total land area capable of being developed within the 100-year floodplain along lower Waller Creek was estimated under existing and alternative diversion scenarios and is summarized in Table 4.

The total developable land area is shown by block in Figure 7 for the most severe flood-prone areas down stream of 12th Street. For the same area, Figure 8 illustrates the increased developable area based on a proposed 55% diversion of the 100-year flood. However, changes in developable land area alone are not indicative of the potential for increased tax revenue to the community because the changes do not reflect the enhancement provided by contiguous land and its increased viability for development after removal of the parcel from the 100-year floodplain.

Table 4. Alternative Diversion Scenarios - Gross Land (Surface) Area Recovered for Development.

Condition	Gross Land (Surface) Area Available for Development
Existing Condition (No Action)	116,000* square feet
55% of 100-year Flood Diversion	1,225,000 square feet
77% of 100-year Flood Diversion	1,243,000 square feet
100% of 100-year Flood Diversion	1,243,000 square feet

* Estimated 50% reduction due to required mitigation.

A better indication of the potential tax revenue to the City can be found in considering the "building area" changes created by the tunnel project, as further explained below.

4.3 *Development assumptions*

It was assumed that two previously relocated historic structures could be relocated to another suitable site. All other parklands and historic lands and structures were excluded from the gross building area density projections. Land value gains and improvements to the property were also excluded from the tax revenue scenarios projections for many of the city blocks which had historic improvements, state or municipal facilities, existing hotels and other already existing improvements. After removing all the proprety exceptions described above eight remaining blocks were used to estimate the total GBA on a block-by-block basis along the Waller Creek corridor for the properties immediately adjacent to the creek.

Existing data on tax values, land area and building restrictions were used to develop future scenarios for land usage and development density. The improvement scenario assumptions were jointly arrived at by the project team engineers, architect and appraiser in a series of brain storming meetings. The project which would apply to each property. Improvement scenarios to property consisted of office, retail, hotel and residential development and were based on absorption rates and other assumptions. Absorption ratcs for the improvement scenarios were estimated using historical data. The estimated absorption rate for office is 600,000 square feet per year, for retail 150,00 square feet per year, for residential 400 units per year, and for residential 600 units per year. The development density assumptions for each were established for office, retail, hotel and residential development types. These assumptions for the development types were made by the project team using the development limitations stated above and existing data on the development densities in the Austin MSA and the Austin Central Business District. The value of all property and improvements are

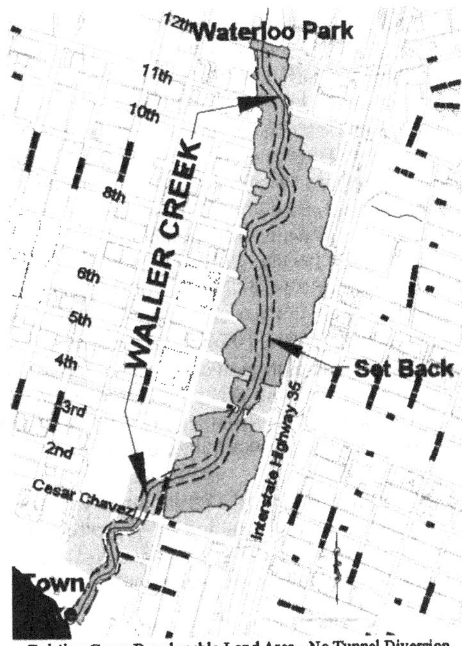

Figure 7. Existing Gross Developable Land Area With No Tunnel.

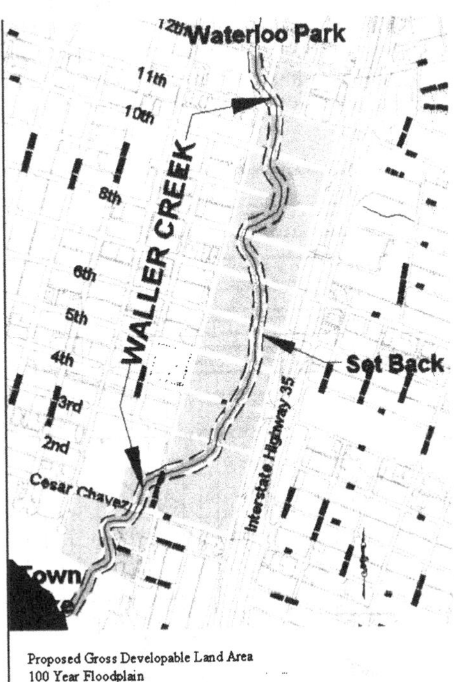

Figure 8. Increased Gross Developable Land Area Resulting From 55% Tunnel Diversion of 100-Year Flood.

inflated at 5% per year. The net value of improvements only includes those estimated values above what presently exists along the watershed.

The revenue scenarios include increases in property value and associated tax, the improvements' construction sales tax, retail sales tax, hotel tax, sales tax on utilities, and professional services (janitorial and pest control) sales tax. Tax revenue estimates were made for Austin, which include all city tax revenue sources. Local tax revenue estimates include property taxes for Austin, Travis County, the Austin Independent School District and the Austin Community College District. Total taxes revenues include all local taxes plus Texas' sales tax. Again, only increases above the existing tax revenue stream were considered. All revenue cash flow streams were brought back to present value using a 6% discount rate.

The present worth cost of the tunnel was developed using a 6% bond rate and a 5% percent inflation rate on operation and maintenance cost. The tax revenues for the following scenarios are shown in Figure 9.

4.3.1 *Scenario 1 – existing conditions - 50% of maximum density*

Scenario 1 illustrates the estimated tax revenues produced without the tunnel, assuming 50% of maximum development density in the area between the 25 year and 100 year existing floodplain under current City of Austin building restrictions. Under this scenario, a potential developer must demonstrate that the proposed improvements would not result in identifiable adverse flooding to other properties.

4.3.2 *Scenario 2 – existing conditions - 50% and 20% of maximum density*

Scenario 2 illustrates estimated tax revenues produced under current building restrictions without the tunnel, assuming 50% of maximum development density in the area between the 25 year and 100 year existing floodplains, plus 20% of maximum development density in the area between the 25-year existing floodplain and a 60-foot creek centerline setback. Under this scenario, the developer must demonstrate that the proposed improvements would not result in identifiable adverse flooding to other properties. In addition, variances would have to be given to allow development in the 25-year floodplain up to the 60-foot creek setback. Any variance granted by the City to allow building construction within the limits of the 25-year floodplain must demonstrate that the proposed improvements would not result in any increase in floodplain elevation to be in compliance with minimum FEMA National Flood Insurance Program requirements. This scenario therefore represents a maximum potential development under existing (no tunnel) conditions and is useful as a conservative estimate for baseline conditions for the evaluation of minimum benefits.

4.3.3 *Scenario 3 – 50% maximum development density in reclaimed flood plain*

Scenario 3 estimates the tax revenues produced with the tunnel in place assuming 50% of maximum development density in the reclaimed floodplain area under current building restrictions. The developer would not have to obtain variances for any improvements in the reclaimed floodplain area.

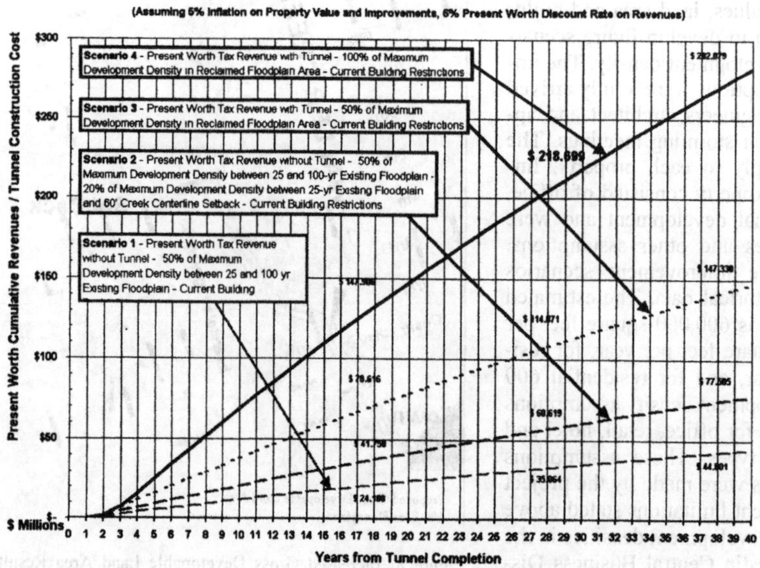

Figure 9. Comparison of Present Worth Cumulative Austin Tax Revenue Scenarios.

Table 5. Project Cost "Break-Even" Point (Years).

Cumulative Tax Revenue Sources	Tunnel Construction Cost [1]				Tunnel Net Present Worth [2]			
	Tax Revenue Scenario 3[8]		Tax Revenue Scenario 4[9]		Tax Revenue Scenario 3[8]		Tax Revenue Scenario 4[9]	
	15.5 ft Dia.	22 ft Dia.	15,5 ft Dia.	22 ft Dia.	15.5 ft Dia.	22 ft Dia.	15.5 ft Dia.	22 ft Dia.
	Scenarios 3 & 4							
Austin [3]	11.75	15.75	7.25	9.10	14.00	18.00	8.50	10.75
Local [4]	5.75	7.20	4.20	5.00	4.10	5.10	3.40	3.90
Total [5]	3.40	4.20	3.10	3.50	4.10	5.00	3.40	3.90
	Scenarios 3 & 4 Reduce by Tax Revenue Scenario 1 [6]							
Austin [3]	16.75	22.75	8.50	10.60	19.30	25.25	10.00	12.4
Local [4]	8.10	10.40	4.90	5.75	9.600	12.00	5.50	6.40
Total [5]	4.75	6.00	3.60	4.10	5.75	7.10	4.00	4.50
	Scenarios 3 & 4 Reduce by Tax Revenue Scenario 2 [7]							
Austin [3]	25.00	34.50	9.50	12.25	27.00	35.50	11.20	14
Local [4]	12.25	16.00	15.50	6.25	14.20	18.20	6.10	7.10
Total [5]	6.25	8.10	4.00	4.50	7.75	9.50	4.30	4.90

(dark shading) = Best Case "Break-Even" Point; (light shading) = Worst Case "Break-Even" Point

[1] Tunnel Construction Cost = January 2000 Construction and Engineering

[2] Tunnel Net Present Worth = Present Worth January 2000 Construction and Engineering, Bond and Operation and Maintenance Cost Minus the Present Worth for Elimination of 9 Bridge Replacements Minus Present Worth Cumulative Annual Flood Damage Reduction Flood Damage

[3] Austin Tax Revenues = Austin Property, Sales & Hotel Taxes

[4] Local Tax Revenues = Austin Property, Sales & Hotel Taxes Plus Property Tax for Austin ISD, Travis County and Austin Community College

[5] Total Tax Revenues = Austin Property, Sales & Hotel Taxes Plus Property Tax for Austin ISD, Travis County and Austin Community College Plus Texas State Sales Tax

[6] Scenario 1 = Present Worth Tax Revenue without Tunnel - 50% of Maximum Development Density between 25 and 100 yr. Existing Floodplain - Current Building Restrictions

[7] Scenario 2 = Present Worth Tax Revenue without Tunnel - 50% of Maximum Development Density between 25 and 100-yr Existing Floodplain - 20% of Maximum Development Density between 25-yr Existing Floodplain and 60' Creek Centerline Setback - Current Building Restrictions

[8] Scenario 3 = Present Worth Tax Revenue with Tunnel - 50% of Maximum Development Density in Reclaimed Floodplain Area - Current Building Restrictions

[9] Scenario 4 = Present Worth Tax Revenue with Tunnel - 100% of Maximum Development Density in Reclaimed Floodplain Area - Current Building Restrictions

4.3.4 *Scenario 4 – 100% maximum development density in reclaimed flood plain*

Scenario 4 estimates the tax revenues produced with the tunnel in place assuming 100% of maximum de under current building restrictions. Again, the developer would not have to obtain variances for any improvement in the reclaimed flood plain area.

4.3.5 *Benefit-Cost comparisons*

Table 5 below summarizes the "break-even" point for the various tax revenue scenarios. The tunnel construction cost and net present worth cost including financing, bond, operating and maintenance cost are all included. The expected annual damages estimates prior to tunnel construction, developed in Section 4.1, are then subtracted from the tunnel present worth cost as a credit for not having occurred if the tunnel is in place. Without the tunnel in place, several bridges would also have to be re-built. Consequently, the tunnel cost is further reduced by the annualized present worth cost for those bridges as shown in Figure 10 and Figure 11. These two figures illustrate the present worth tunnel cost versus an adjusted cumulative present worth bridge replacement credit with annual flood damage. Examples of the comparison curves of the tunnel present worth costs versus the cumulative tax revenues for thebest case and worst case scenarios are shown in Figure 12 and Figure 13, respectively. For most of the 18 scenarios developed the "break-even" point occurs very quickly at the intersection of the tunnel cost curves with the revenue curves for Scenarios 3 and 4 as illustrated in Table 5. The worst case revenue scenario with a 35.5 year "break-even" point is shown in Figure 13. This scenario is very unlikely. The best case "break-even" point of 3.10 years is shown in Figure 12.

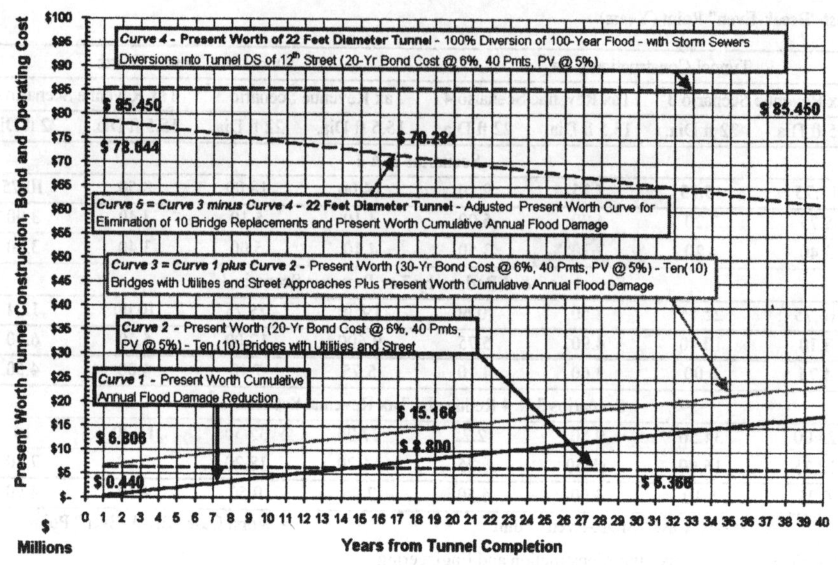

Figure 10. Present Worth 22 Feet Tunnel Construction, Bond, Operation Cost and Cumulative Present Worth Annual Flood Damages.

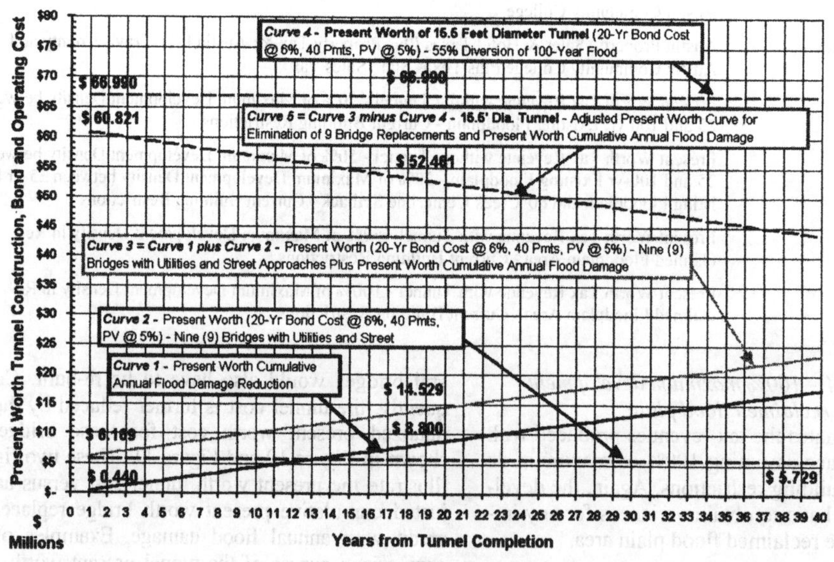

Figure 11. Present Worth 15.5 Feet Tunnel Construction, Bond, Operation Cost and Cumulative Present Worth Annual Flood Damages.

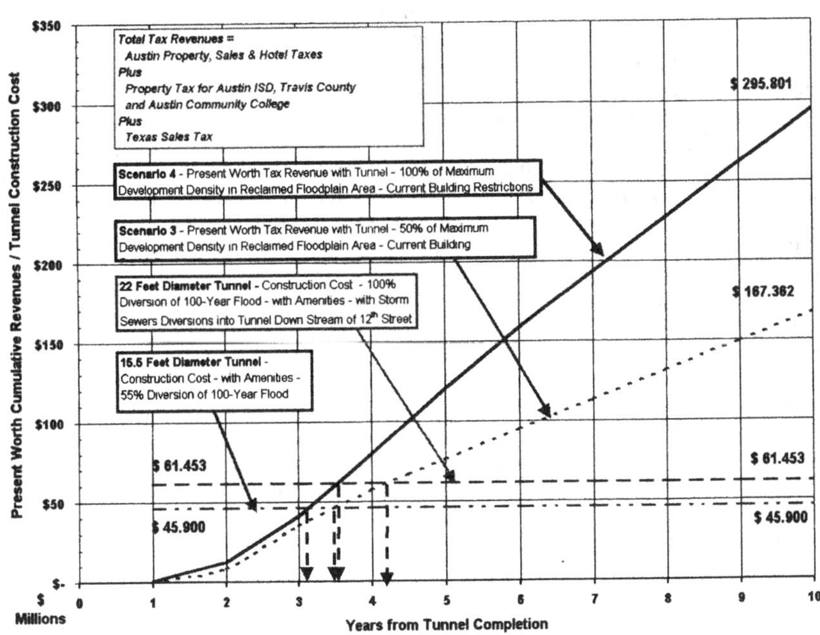

Figure 12. Present Worth Tunnel Construction, Bond, Operation Cost and Cumulative Present Worth Annual Flood Damages.

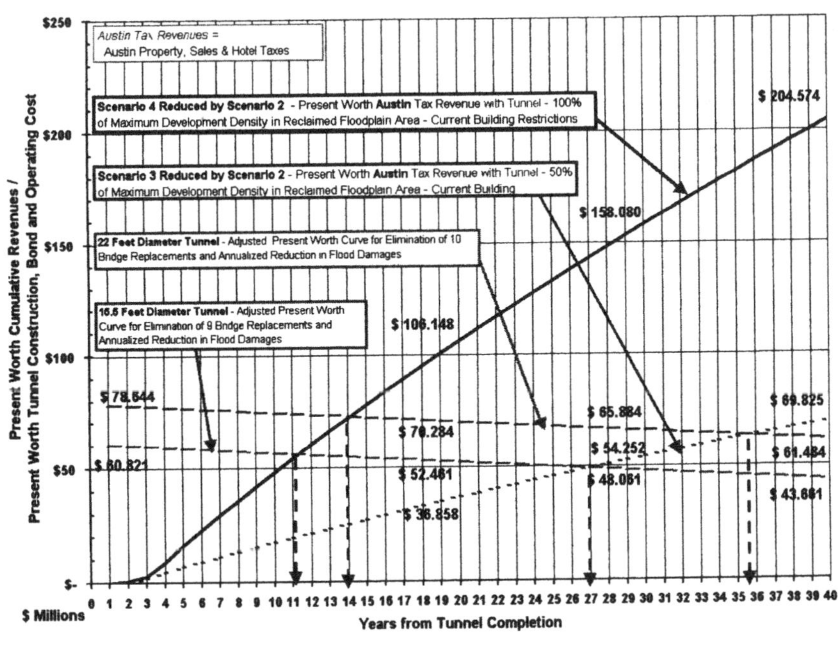

Figure 13. Comparison of Tunnel Net Present Worth Cost with Austin Tax Revenue Scenarios 3 and 4 Reduced by Scenario 2.

5 CONCLUSION

This project, like many others before it, has been well studied. The major difference in the estimated cost of the project from earlier studies is in the mechanical screening of debris that the current studies have included. The stormwater diversion tunnels that protect the San Antonio River Walk also depend on mechanical screening for their effectiveness. These screens were retrofitted by the San Antonio River Authority following unsatisfactory experience without them. The Waller Creek diversion system would also be very difficult to operate and maintain without similar mechanical screening. The debris would otherwise accumulate on the screens or in the tunnel, seriously compromising the effectiveness of the diversion system.

The cost reduction studies for the Waller Creek Tunnel did not result in any significant viable savings, but several items, if omitted in the initial construction, could potentially be deferred until some time in the future. Retrofitting mechanical screens, while offering the greatest apparent cost cutting measure, does not appear to be a wise option.

The greatest hope for the project's implementation appears to be in the recognition of direct and indirect benefits that the project would bring to the City by eliminating the somewhat blighted area of downtown Austin that currently exists along much of the creek. Several imaginative concepts have been developed through private initiatives, that would effectively create a longitudinal park in the Waller Creek channel, with cycle paths and pedestrian access through a landscaped environment, thereby encouraging compatible development in the many undeveloped lots currently within the flood plain. The only way such a development could be successfully implemented appears to be through the interception of damaging storm flows.

The City's goal is to continue to seek adequate funding for this project, recognizing that the current financial climate within the City and the region has necessarily further diminished the immediate hope of its implementation. However, we believe its time will come.

6 ACKNOWLEDGEMENTS

The authors would like to thank Al Simmons, AIA and Jihad Zebdaoui of the Austin architectural firm Graeber Simmons & Cowan, AIA, for preparation of the inlet and outlet renderings and their contribution to the land development scenarios within the Waller Creek neighborhood. We are grateful to Mr. Vance Powell, III, MAI, SRPA, SRA, a local commercial real estate appraiser experienced in downtown Austin development who assisted in developing the land build-out rates and tax revenue scenarios. Our appreciation is also due to Brian K. Reis, P.E., of Espey Consultants for his assistance in the floodplain delineation and the HAZUS analysis and to Prakash Donde, P.E, Jenney Engineering, for his to the tunnel design alternatives. We would like to express our gratitude to John Lacy of Halliburton KBR for his tireless efforts in the preparation of the graphics for this report. A special thanks to Gary Kosut, P.E., Project Manager for the City of Austin whose review and input into the original reports prepared on his subject were invaluable.

North American Tunneling 2002, Ozdemir (ed.)
© 2002 Swets & Zeitlinger, Lisse, ISBN 90 5809 376 X

Evaluating the use of trenchless technology on the Big Walnut Sanitary Trunk Sewer extension

D.A. Day & H.L. Marsh
DLZ Ohio, Inc., Columbus, Ohio, USA

T. Arsh
City of Columbus, Columbus, Ohio, USA

T.P. Kwiatkowski
Jenny Engineering Corporation, Springfield, New Jersey, USA

ABSTRACT: The extension of the Big Walnut Sanitary Trunk Sewer in Columbus, Ohio has been designed for underground construction. The decision to design and construct the sewer extension using trenchless technology was made following numerous economical, engineering, environmental, and social evaluations.

Several issues led to the need to evaluate tunneling methods for the construction of the extension. The project area includes the Hoover Reservoir, a major source of drinking water for the residents of Columbus. The project area also includes numerous park sites, a golf course and environmentally sensitive areas. The majority of residents in the project corridor rely on private well systems, which required careful evaluation during design. Topographical features of the extension area dictated that open-trench construction could not be completed without incorporating pump stations, aerial sewers, or inverted siphons. In addition to the issues specific to the project area, past experiences with similar sewer projects has caused the City and the Design Team to carefully evaluate impacts that the project may have on the public.

This paper describes the extensive evaluations that were undertaken during the preliminary engineering phase of the project and how the evaluations were incorporated into the Detailed Design of the sewer extension.

1 INTRODUCTION

The City of Columbus constructed its first sanitary sewer in 1841 and by 1900 approximately 150 miles of sewers were completed.

Today the City's sewer system consists of over 413 miles of sewer that serves not only the City of Columbus but also 22 suburban communities.

Figure 1. Early Sewer Construction Crew.

The central Ohio area has experienced and continues to experience rapid population growth. The City of New Albany and the City of Westerville, two suburban communities in northeastern Franklin County, have undergone extensive commercial and residential development. The area is the home of the Limited's corporate headquarters, the Easton Towne Center and the Polaris Fashion Mall. Areas within northeastern Franklin County and western Licking County are arguably the most desirable undeveloped real-estate within the City of Columbus's Sanitary Sewer Facilities Planning Area.

The City of Columbus, Department of Public Utilities, Division of Sewerage and Drainage, has responded to the growth in the northeastern Franklin County area by extending several major trunk sewers and interceptors to the New Albany area. The extension of the Big Walnut sewer represents the final effort that will be required to ensure that the area receives adequate sanitary sewer service.

2 KEY ISSUES

Numerous economical, engineering, environmental, and social implications were realized during the evaluation and design of the Big Walnut Sanitary

Trunk Sewer Extension. The impacts of these implications led to the selection of an entirely underground structure. The following subsections describe the key issues of the project, which led to the decision to design the sewer using trenchless technology.

2.1 *Topography*

Much of the area near Hoover Reservoir is heavily wooded and moderately to heavily undulating. The topography gently levels and becomes rolling further east of the reservoir. For approximately 3,500 feet, from the terminus of the existing 84" Big Walnut Sanitary Trunk Sewer to where Lee Road dead ends, the terrain is moderately to heavily undulating and moderately to heavily wooded. The east bank of Hoover Reservoir contains numerous ravines through which surface and drainage water flow to Hoover Reservoir. Some of the ravines are wide and extensive in size while others are narrow and small in comparison. Four of the larger ravines extend eastward beyond Lee Road making it necessary to utilize pump stations, inverted siphons, or aerial sewers for a shallow open-trench design.

Surface elevations in the project area vary within 190 feet. The ground surface elevation climbs from about 820 feet MSL at the terminus of the existing sewer to approximately 1010 feet MSL at the Franklin/Delaware County line near Harlem Road.

2.2 *Geology*

The stratigraphy in the area consists mainly of a thin layer of overburden (except in the bedrock valley areas) and the Berea Sandstone, the Bedford Shale and the Ohio Shale. The Berea Sandstone is only present in the upper elevations of the project area and is approximately 50 feet thick. The Bedford Shale underlies the Berea Sandstone and is 20 to 100 feet thick. The Ohio Shale lies beneath the Bedford Shale and is a relatively competent shale with iron pyrite concretions. Methane gas has been detected in the Ohio Shale.

Open trench excavations would encounter substantial amounts of bedrock requiring drilling and blasting operations during construction. This type of excavation would also impose greater costs due to the depth and size of the pipe.

A bedrock valley exists in the eastern section of the project area, between Lee Road and Harlem Road. This valley appears to be filled with sand and gravel glacial outwash and is a major carrier of groundwater. Excavation through this portion of the project would require impermeable structures due to the nature of the aquifer.

2.3 *Access*

The location of the existing trunk sewer terminus is approximately 100 feet east of Big Walnut Creek in a heavily wooded area behind a residential subdivision. Access for construction and maintenance of the sewer extension would impact nearby residents and the wooded area surrounding the terminus. Access must be provided for the construction of the connection structure, excavation for the trunk sewer, pipe placement, backfill, restoration, and maintenance. Gaining and maintaining temporary and permanent access to the existing terminus would be costly and disruptive to the residents and the natural beauty of the area. Several alternatives for access to this area were discussed including a bridge over Big Walnut Creek, an access road across the golf course, an existing easement (of only about 50 feet wide) which ran between two residential homes, or an access road through the wooded area which would parallel the residential area.

Access to the initial portion of the alignment would also be difficult because most of the Phase 1 portion of the project could not take advantage of existing right-of-way and many tracks of land would be crossed with the sewer alignment.

2.4 *Maintenance and operation*

Open trench construction of the trunk sewer would require one or more pump stations, aerial sewers, or inverted sewer siphons. All of these structures would require future access for maintenance and operations. The structure required to cross the first major ravine would be particularly difficult for future maintenance due to the extreme topographical changes and the dense woodlands.

2.5 *Hydraulic design*

The hydraulic design of the sewer posed a challenge due to the potential growth patterns of the project area. The design required the sewer to convey relatively insignificant initial flows to an ultimate peak design flow of 92 cfs for the 18,730 acre tributary area. Greater slopes or a low flow channel would be required for initial flows in an all gravity sewer and complicated siphons and pump stations would be required for a shallow sewer.

2.6 *Well protection*

Most of the residents in the area utilize residential water wells as a potable water source. These wells are constructed in the overburden within the bedrock valleys and within the Berea Sandstone formation. The underlying lower shale formation does not carry an adequate (quantity or quality) water source to supply a residential household. Therefore, if water wells were to be dewatered during construction another supply of water must be made available to the residents, as drilling a deeper well may not be an option. Tunneling within the Ohio Shale will minimize impacts to residential wells.

2.7 *Nature preserves*

The project area has two known nature preserves which would be crossed, as well as a park on the east side of Hoover Reservoir. The impacts to these natural areas have been an issue for public concern on other projects in the area. Therefore, the design team is very conscience of the need to mitigate impacts to the preserves.

Figure 2. Nature Preserve in Project Area.

2.8 *Environmental*

Several environmental issues have previously been discussed. However, several issues remain including the amount of land clearing, natural beauty of the area, riverfront property, exposed sewer pipe, blasting, and construction water runoff to Hoover Reservoir/Big Walnut Creek. All of these issues created a challenge for the Design Team.

2.9 *Public acceptance*

The City of Columbus has nominated this project for a Water Pollution Control Loan Fund, which requires that the City solicit public input and acceptance of the project. As noted previously, the public is protective of the nature areas within the project area. In addition, the protection of residential wells in the area will be a concern of the public. A similar project was completed in the close proximity, which was perceived to have resulted in the dewatering of residential wells and impacts to the water quality.

2.10 *Economics/construction cost*

The cost of construction plays an important role in any publicly funded program. A cost-effective approach is required for the project to be successful.

The economic study for this project went beyond the typical present worth analysis. The design team was also challenged to develop a plan to ensure that the sewer alignment, profile, and connection structures would be the lowest cost alternative and located to ensure the ease of development.

3 ALTERNATIVES

Four corridors were developed within the project area (See Fifgure 3.).

The first corridor developed, Corridor A, represented a route that minimized the length of the trunk extension. Almost the entire length of Corridor A is characterized by numerous topography changes and moderately to heavily wooded terrain. This corridor is relatively close to Hoover Reservoir, a drinking water storage site for the City of Columbus. In addition, a large portion of the area along the reservoir has been designated for a park or nature reserve.

Corridor B was developed to utilize Cubbage Road and Red Bank Road right-of-ways. The areas along the roadways represent improved topography and terrain for construction. However, the sections not along the roadways are, like Corridor A, characterized by numerous topography changes and moderately to heavily wooded terrain.

Corridor C utilizes Lee Road and Red Bank Road right-of-ways for the majority of its route. Only the initial portion of this corridor includes difficult topography and terrain.

Corridor D was developed to mitigate the impacts of the two major drainage ways that flow to Big Walnut Creek/Hoover Reservoir. The initial portion of this corridor, from the existing sewer terminus to Walnut Street, is essentially an east/west alignment. At Walnut Street the route turns north and extends to Duncan Run Creek.

After developing the corridors, several factors led to the decision to no longer consider corridors A and B as viable alternatives. As noted previously both of these corridors included difficult terrain that would make access for construction difficult. Open trench sewer construction would require siphoning, bridging or embanking at numerous locations. In addition, open trench construction would require excessive clearing of the wooded terrain. Although a deep tunnel profile sewer was possible, access to intermediate shaft/construction sites would be difficult, and if tunnel excavation equipment required repair, access would be nearly impossible. Corridors A and B were not considered to be constructable alternatives.

Numerous profiles were developed for Corridors C and D. The project team wanted to evaluate each corridor relative to shallow profile and deep profile sewer construction.

Open-trench, shallow profile construction within Corridor C would require the construction of aerial sewers and/or pump stations at several places along the alignment. The special structures were required to cross ravines encountered in several locations along the corridor. Corridor D was located further from Hoover Reservoir, which reduced the impacts of the ravines. Therefore, only one aerial sewer was needed for an open-trench alternative for corridor D.

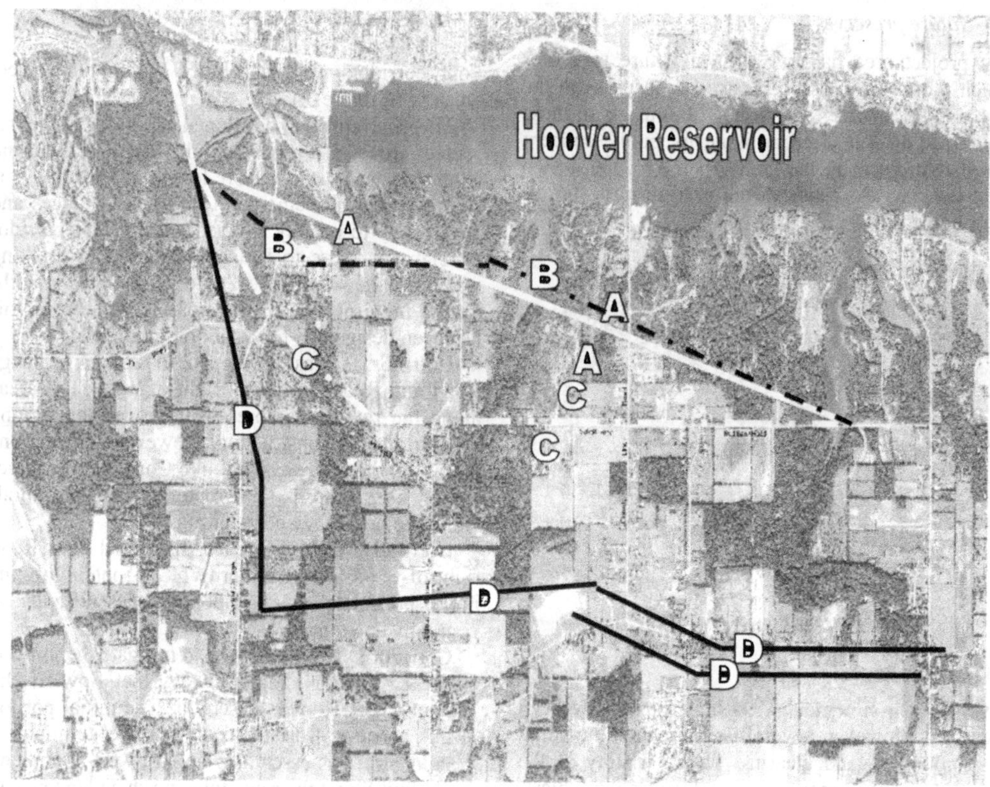

Figure 3. Corridors Evaluated for Sewer Construction.

Several deep/shallow profile alignments were developed to reduce the number of aerial sewers or pump stations required. The initial sections of the alignments were designed for deep construction to pass below ravines, once the sewer extended beyond the ravine; the profile was raised for the remainder of the alignment. This process was repeated for each corridor based on the number of ravines that required crossing.

The tunnel profile alternatives were developed at different elevations based on the shale formations. The shales are anticipated to be water tight and good tunneling mediums.

4 EVALUATION OF ALTERNATIVES

Each of the suitable alignment and profile alternatives were evaluated through a process addressing each key issue noted previously.

One of the major economic factors reviewed for the project was the cost to replace or protect the residential wells within the area. These costs were anticipated to be high for the shallow profile alternatives but relatively low for the deep tunnel alternatives.

Shallow sewer construction also requires the use of either aerial sewers and/or pump stations. While these items are accepted methods for the conveyance of sewage, they typically require more maintenance and the operation costs are higher than gravity sewers. The shorter life of these structures and the possibility of requiring upgrades with increased population also added to the life cycle cost of these structures. In addition, the construction of these items has many impacts associated with them that tunnel construction does not.

Because the project could not be constructed entirely in public right-of-way, the impacts associated with the project to residents, especially those that would be requested to provide land, was considered.

The evaluation process also included an independent Value Engineering Study. The Value Engineering Team reviewed the Final Design Report developed by the Design Team and made recommendations for the improvement of the design. The value Engineering Team decreased the impact to residents and the environment by recommending a connection point approximately 1,200 feet downstream of the existing terminus. This enabled the issue of construction and maintenance access to disappear as the sewer could be constructed within existing right-of-

Figure 4. Selected Sewer Alignment.

ways at the connection structure and the initial 2,130 feet.

After the V.E. Study, the Design Team further refined the recommended alternative by changing the routing of the Phase 3 portion of the project.

5 ALTERNATIVE BEING DESIGNED

The alternative selected for design is a combination of alternatives developed in Corridors C and D as a result of the extensive evaluation conducted by the Design Team and the Value Engineering Study. Photo 4 depicts the selected alignment for the sewer.

The sewer will be 72-inch diameter and constructed by trenchless technology utilizing both conventional tunneling and pipe jacking/microtunneling methods. Shaft sites abut existing right-of-way allowing for easy access from roads for construction and maintenance.

The contract will be designed and constructed in three phases. Phase 1 consists of approximately 625 feet of pipe jacking and 9,230 feet of tunnel. This phase will have one connection structure and four shafts. It is anticipated that the tunnel will be bored downslope.

Phase 2 consists of approximately 5,720 feet of tunnel within the Ohio and Bedford shales. Three shafts are anticipated for this phase, one of which will be a connection to the Phase 1 project.

Phase 3 consists of approximately4,500 feet of tunneled sewer with two shafts anticipated.

Odor control facilities are also expected at several locations along the sewer alignment.

Underground construction of the Big Walnut Sanitary Trunk Sewer Extension will benefit Northeast Franklin County by providing the most cost effective and least disruptive sanitary sewer service.

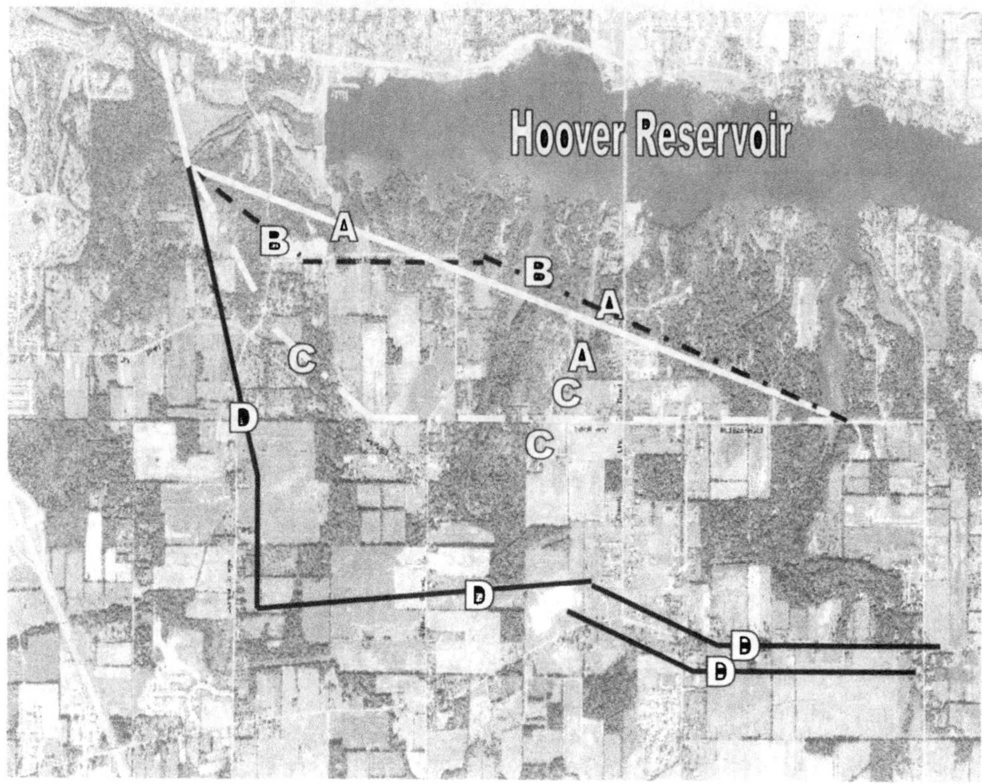

Figure 3. Corridors Evaluated for Sewer Construction.

Several deep/shallow profile alignments were developed to reduce the number of aerial sewers or pump stations required. The initial sections of the alignments were designed for deep construction to pass below ravines, once the sewer extended beyond the ravine; the profile was raised for the remainder of the alignment. This process was repeated for each corridor based on the number of ravines that required crossing.

The tunnel profile alternatives were developed at different elevations based on the shale formations. The shales are anticipated to be water tight and good tunneling mediums.

4 EVALUATION OF ALTERNATIVES

Each of the suitable alignment and profile alternatives were evaluated through a process addressing each key issue noted previously.

One of the major economic factors reviewed for the project was the cost to replace or protect the residential wells within the area. These costs were anticipated to be high for the shallow profile alternatives but relatively low for the deep tunnel alternatives.

Shallow sewer construction also requires the use of either aerial sewers and/or pump stations. While these items are accepted methods for the conveyance of sewage, they typically require more maintenance and the operation costs are higher than gravity sewers. The shorter life of these structures and the possibility of requiring upgrades with increased population also added to the life cycle cost of these structures. In addition, the construction of these items has many impacts associated with them that tunnel construction does not.

Because the project could not be constructed entirely in public right-of-way, the impacts associated with the project to residents, especially those that would be requested to provide land, was considered.

The evaluation process also included an independent Value Engineering Study. The Value Engineering Team reviewed the Final Design Report developed by the Design Team and made recommendations for the improvement of the design. The value Engineering Team decreased the impact to residents and the environment by recommending a connection point approximately 1,200 feet downstream of the existing terminus. This enabled the issue of construction and maintenance access to disappear as the sewer could be constructed within existing right-of-

North American Tunneling 2002, Ozdemir (ed.)
© 2002 Swets & Zeitlinger, Lisse, ISBN 90 5809 376 X

Conceptual design of underground space on the Central Artery/Tunnel Project

Brian Brenner
Parsons Brinckerhoff

William Lindemulder
Wallace Floyd Design Group

ABSTRACT: The Central Artery/Tunnel Project in Boston has designed and constructed kilometers of underground highways and transit structures. A large part of the project replaces an overhead viaduct expressway, the Central Artery, with a cut-and-cover tunnel. The viaduct is scheduled for demolition in 2004, exposing 30 acres of prime, developable land parcels in downtown Boston. The future development of these parcels is a dramatic demonstration of the benefits associated with tunneling, and this above-ground space has been the subject of much discussion and debate.

Less heralded, but perhaps of equal importance, is the availability of new underground space along the artery corridor. The underground space has become available due to the geometry and construction requirements of the Central Artery tunnels. This paper will discuss:

- Central Artery tunnel alignments which have led to opportunities for underground space development
- Design of the South Boston Piers Transitway, a new subway line which takes advantage of the Central Artery construction
- Conceptual plans for different uses of underground space along the corridor above the highway tunnels
- Planning for underground uses to support above ground parcel development.

1 INTRODUCTION

The Central Artery/Tunnel Project in Boston has designed and constructed kilometers of underground highways and transit structures. A large part of the project replaces an overhead viaduct expressway, the Central Artery, with a cut-and-cover tunnel. The viaduct is scheduled for demolition in 2004, exposing 30 acres of prime, developable land parcels in downtown Boston. The future development of these parcels is a dramatic demonstration of the benefits associated with tunneling, and this above-ground space has been the subject of much discussion and debate.

Less heralded, but perhaps of equal importance, is the availability of new underground space along the artery corridor. The underground space has become available due to the geometry and construction requirements of the Central Artery tunnels. This paper will discuss:

- Central Artery tunnel alignments which have led to opportunities for underground space development
- Design of the South Boston Piers Transitway, a new subway line which takes advantage of the Central Artery construction
- Conceptual plans for different uses of underground space along the corridor above the highway tunnels
- Planning for underground uses to support above ground parcel development.

2 VERTICAL ALIGNMENT

Figure 1 shows a vertical profile and Figure 2 shows a plan of the cut-and-cover I-93 Central Artery tunnel currently under construction in downtown Boston. This 8–10 lane wide tunnel will replace the existing 6 lane overhead viaduct expressway in a phased traffic sequence planned to begin in November, 2002. By the mid to late 2004, the existing viaduct will be demolished, and all traffic will travel in the tunnels. In addition to improving traffic, depression of the existing Central Artery in a tunnel creates a great opportunity to improve the urban downtown landscape. What was an ugly, rusting highway hulk will be replaced by 30 acres of land parcels available for parks and other uses. The thirty acres measures the surface land area. Additional opportunities are provided in the space below grade and above the tunnels.

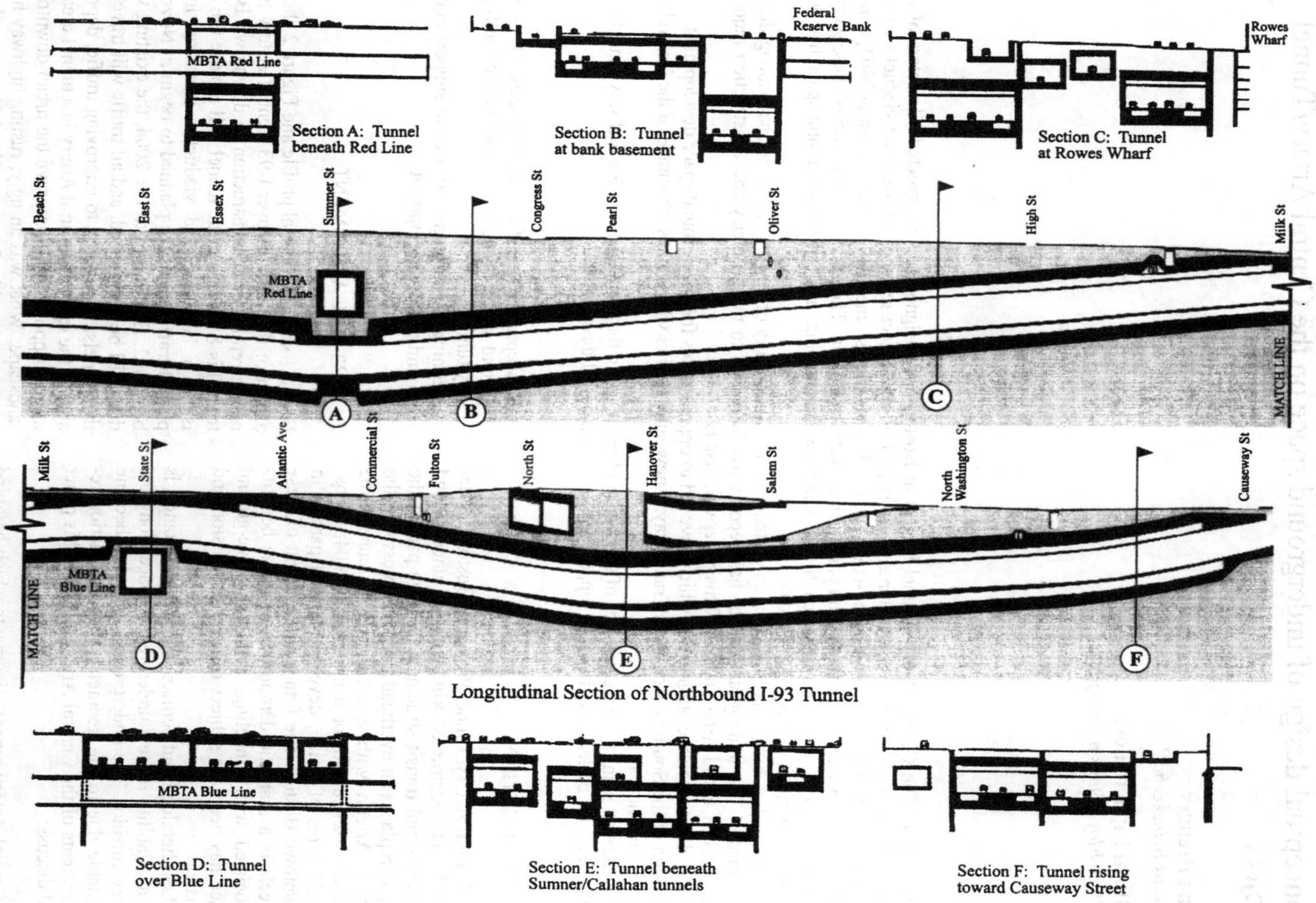

Section A: Tunnel beneath Red Line

MBTA Red Line

Section B: Tunnel at bank basement

Federal Reserve Bank

Section C: Tunnel at Rowes Wharf

Rowes Wharf

Beach St
East St
Essex St
Summer St
Congress St
Pearl St
Oliver St
High St
Milk St

MBTA Red Line

A
B
C

MATCH LINE

Milk St
State St
Atlantic Ave
Commercial St
Fulton St
North St
Hanover St
Salem St
North Washington St
Causeway St

MBTA Blue Line

MATCH LINE

D
E
F

Longitudinal Section of Northbound I-93 Tunnel

MBTA Blue Line

Section D: Tunnel over Blue Line

Section E: Tunnel beneath Sumner/Callahan tunnels

Section F: Tunnel rising toward Causeway Street

Figure 1.

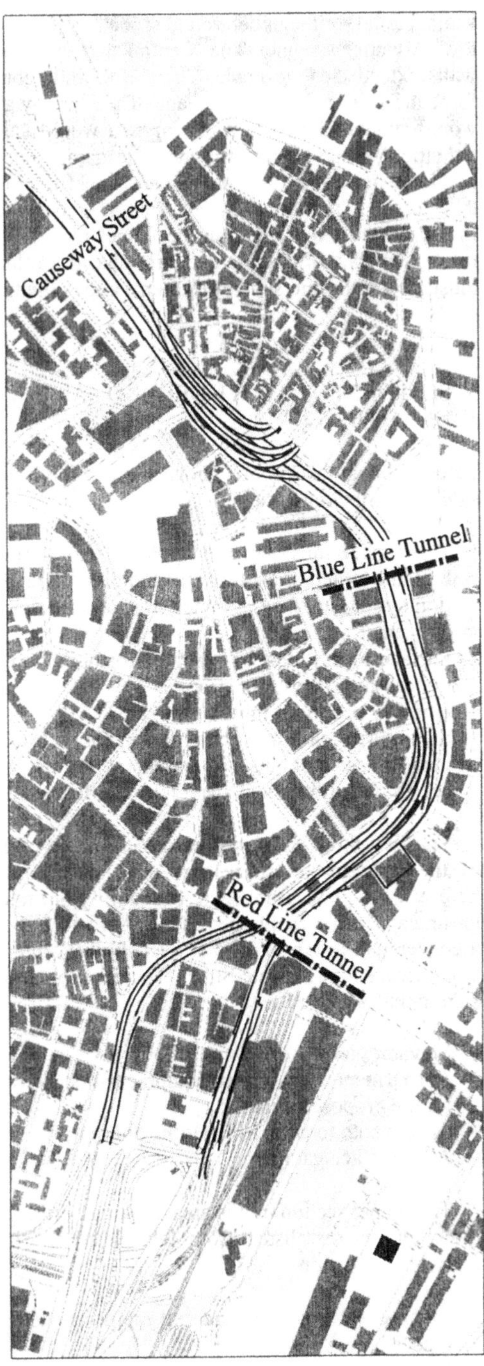

Figure 2.

The I-93 vertical profile design can be summarized as follows:

- The NB tunnel needed to avoid the existing Red Line subway tunnel. A decision needed to be made whether the highway tunnel should pass under or over the existing subway. At the time, the Massachusetts Bay Transportation Authority planned a fifth mass transit subway line for the city, the "Silver Line." The Central Artery highway tunnel was thus routed below the existing Red Line subway. The underground space above provided room for the new Silver Line subway.
- At State Street, the NB and SB highway tunnels needed to cross the path of the existing Blue Line subway tunnel. This profile places the new highway tunnels on top the existing subway tunnel. As part of the Central Artery project, the MBTA's plan to lengthen and modernize the Aquarium subway station was included in a joint construction contract. The rebuilt Aquarium Station will have access and headhouses on both the east and west sides of the new highway tunnel. Subway passengers walk down to the lengthened and rebuilt station platform, which passes below the new highway tunnel.
- Further to the north, the southbound highway tunnel has underground connections to the existing Sumner and Callahan Tunnels, which cross beneath Boston Harbor towards East Boston. The underground highway ramps are designed to pass above the mainline tunnel, pushing the mainline tunnel profile even deeper.
- Utility corridor crossings are placed at various spots all along the alignment.

These vertical design control points have had the effect of providing "nooks and crannies" of potentially usable underground space above the new tunnels, but below finished grade. How this space is or may be used is further described below.

3 SOUTH BOSTON PIERS TRANSITWAY

At the southern part of the I-93 NB tunnel alignment, the new tunnel passes below the Red Line at South Station. The underground space above this area is available for the "Silver Line," the MBTA's fifth subway line, also designated as the South Boston Piers Transitway. The Silver Line, first phase, will extend from South Boston to the South Station area. The western part of the new tunnel alignment is placed in the underground space above the new Central Artery NB tunnel.

The Central Artery tunnel was constructed with slurry walls used for both excavation support and as part of the final tunnel structure. The slurry walls also were used to support temporary traffic decking for Atlantic Avenue. The Transitway alignment was designed to take advantage of the Central Artery vertical profile. From the south the alignment starts in a turn around loop above the highway tunnel. It proceeds north, above the tunnel and through Dewey

Square, continuing north to a point where it turns east off the Central Artery alignment. This turning point was originally even further to the north than the final design alignment. Originally, the Transitway continued above the Central Artery Tunnel and turned sharply to the east, through the site of Central Artery Ventilation Building #3. During design review, the MBTA determined that the curve to the east was too sharp, and a more gradual curve was designed. The resulting Transitway alignment places the subway tunnel box beneath the existing Russia Wharf Building, to the southeast corner of Ventilation Building #3, and through an immersed tube tunnel below the Fort Point Channel to South Boston.

Overall, the Transitway design followed the artery tunnel alignment in plan with one exception: the turn around loop at the southern end was wider than the highway tunnel, and required modifications to slurry walls and sequencing. At Dewey Square, where the highway tunnel passed below the existing Red Line Subway, the new transit tunnel was short enough to fit in a space above the subway tunnel. The existing Red Line subway station was modified to include a revised mezzanine structure and provisions for transfer between the old Red Line and the new Silver Line. At this intersection, the underground space will have four levels of uses: the MBTA entrance mezzanine, the new Silver Line, the existing Red Line, and, at the bottom, the new I-93 northbound highway tunnel.

Use of underground space above the highway tunnel led to cost savings, when compared to building the two projects separately. The highway and transit project were combined in one construction contract, and were able to share costs of excavation support walls, soil removal, cross lot bracing, utility staging, traffic staging and decking, construction support costs, and other costs.

4 CONCEPTUAL PLANS FOR DIFFERENT USES OF UNDERGROUND SPACE

The I-93 Central Artery vertical alignment has several areas of available vertical clearance above the tunnel and below final grade. These "nooks and crannies" are potentially available to support other below grade uses. The CA/T project has conducted several conceptual studies of possible uses for different spaces. To date, none of these specific studies has progressed to final design and development. However, they are illustrative of potential uses of additional underground space.

Figure 3 shows a plan and section of one concept for an underground parking garage. The concept was developed for the below grade space above the I-93 NB tunnel and entrance Ramps C/D, just to south of the turn around Transitway loop. The conceptual design includes two levels of parking, taking full advantage of all of the underground space available below Atlantic Avenue. The Central Artery slurry walls extend close to grade. The slurry walls could form the side walls of the garage. The slurry walls would support the roof, which in turn would carry Atlantic Avenue traffic at grade. The garage space would be mostly air, in comparison to backfill which the Central Artery tunnel roof section is designed for. To avoid extra loads on the tunnel roof, structural elements of the garage are placed to land mostly on walls and support elements above the tunnel below.

5 SUPPORT FOR ABOVE GROUND PARCEL DEVELOPMENT

Later building construction above a tunnel must consider the following issues:

- The tunnel must be strong enough to support building structure loads, or the building's foundation system must be designed to avoid the tunnel altogether.
- Surface access must be provided for building entrances and support functions.
- Utility connections must be provided.
- Subsurface space must be provided for the building, for elevator pits, for basement storage and access, for utility rooms, and for other functions.

The first three issues do not directly involve use of the underground space, and are not discussed here. It is this last issue where availability of underground space comes into play. Figures 4 and 5 show a conceptual design of a hotel over the I-93 tunnel, at a parcel south of State Street. The conceptual design includes space for:

- Elevator pits
- A basement for support equipment and storage for the proposed building
- Access path to enter and exit the building
- A subsurface garage.

The tunnel section in this area, looking north, features four boxes, listed from left (west) to right (east) in the section:

- Ramp CS-P, a southbound exit ramp that rises to grade;
- Mainline I-93 southbound, which in this area dips lower below grade so that ramp CS-P can rise on top of the mainline box at a point further south;
- Combined ramps A-CN and R-T, both northbound on ramps that have dipped down at this section to join the adjacent northbound box to the east;

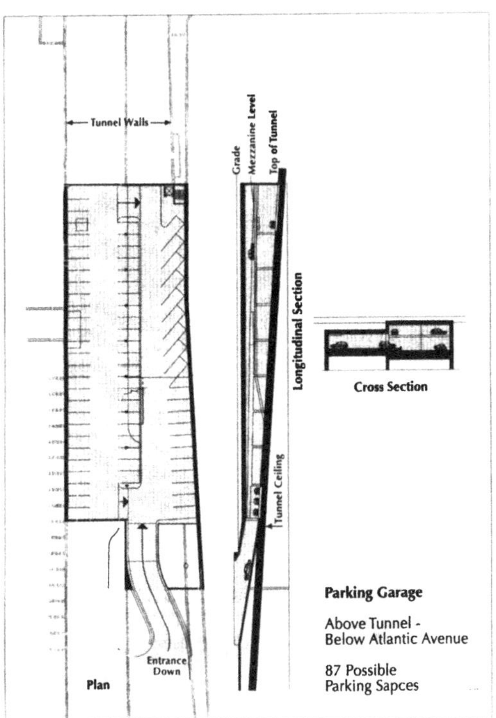

Figure 3.

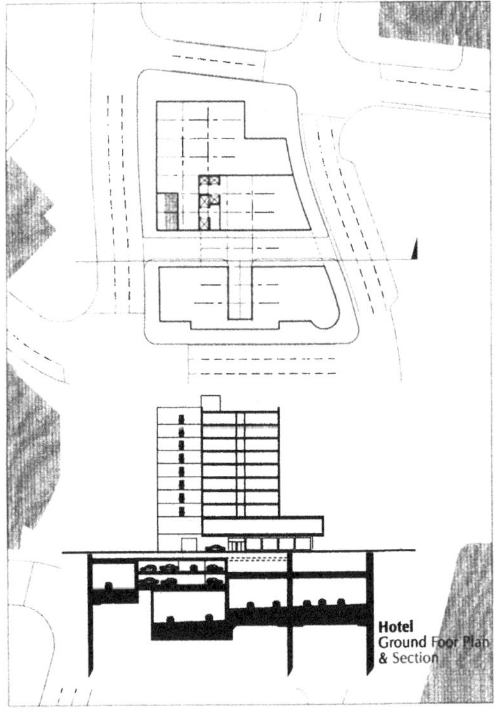

Figure 4.

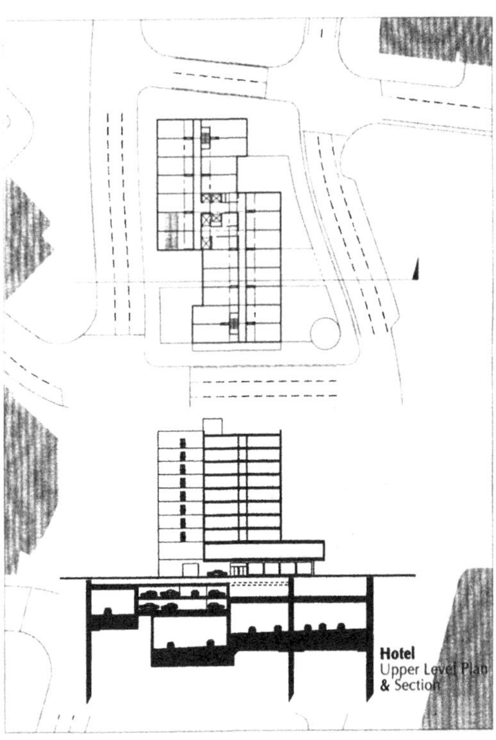

Figure 5.

– Mainline I-93 northbound, which in this area rises from south to north, to cross the Blue Line subway a few hundred feet further north. At this point, the top of the box is still 15 to 20 feet below grade.

Based on the tunnel section geometry, all four boxes provide areas for the use of underground space. The conceptual hotel design takes full advantage of the space. Figures 6 and 7 show a plan for subgrade parking, utility rooms and elevator pits. Direct access is provided from the garage to the hotel above. Central Artery/Tunnel slurry walls are used as side walls of the underground space, and for support of above-grade building structural elements.

Figures 8 and 9 illustrate another conceptual design for the same parcel. This design is for a museum. The figures show how the below grade space can be utilized. Basement access is provided by a truck ramp on top the I-93 northbound tunnel as shown in Figure 7. Likewise, space is provided for elevator pits and utility connections.

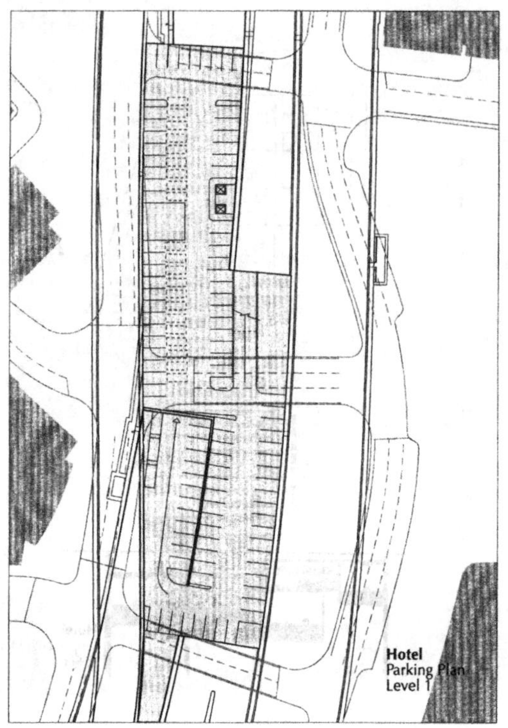

Figure 6.

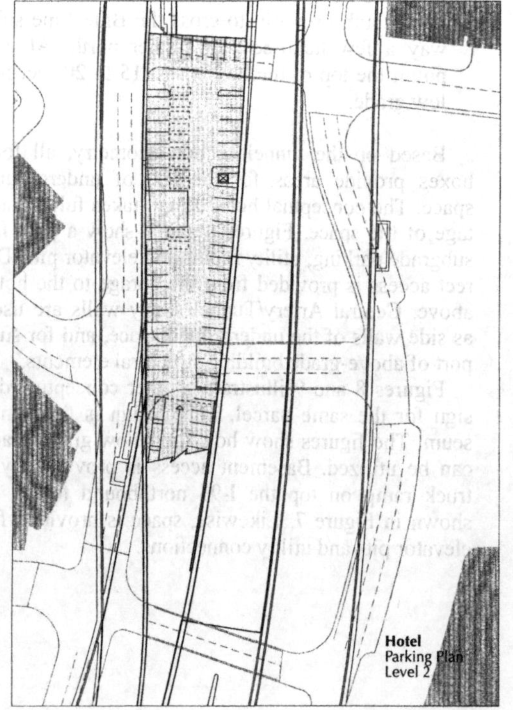

Figure 7.

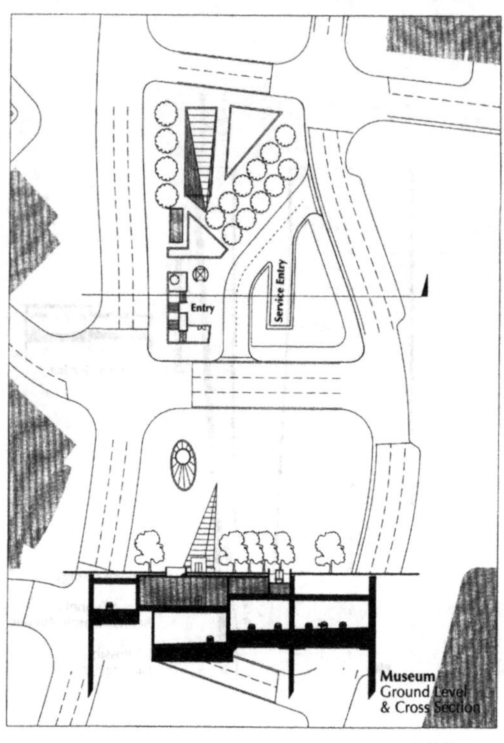

Figure 8.

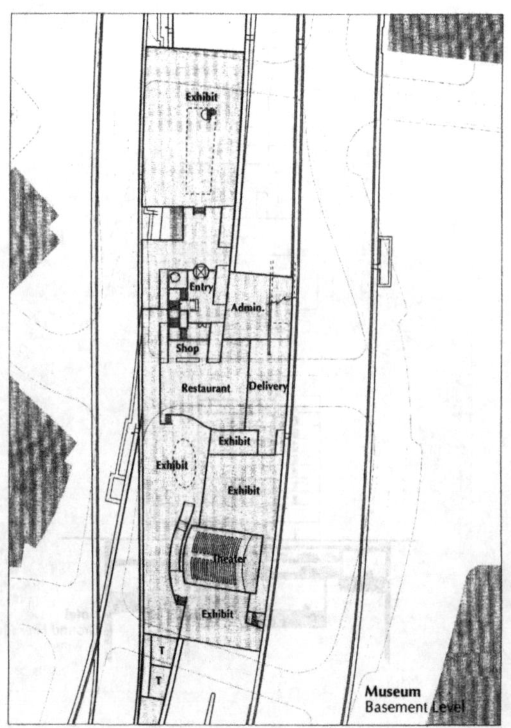

Figure 9.

6 SUMMARY AND CONCLUSIONS

The Central Artery/Tunnel project's primary goal is the extension of I-90 to Logan Airport, and the reconstruction and depression of the I-93 Central Artery. Tunnel construction alignment provides peripheral benefits by making available additional underground space not directly used for the CA/T Project highway structures. Use of some of this underground space has been programmed now as part of the project, through, for example, construction of a large segment of the city's fifth subway line. Other spaces have been evaluated in support of potential future underground uses, and underground uses to support above grade construction.

Evaluating every "nook and cranny" of underground space in a tunnel project provides additional benefits. Overall, it increases the efficiency and effectiveness of designing underground facilities.

REFERENCES

Brian Brenner, Dan Wood and Robert Valenti, "Future Air Rights Development Above Cut-and-Cover Tunnels" in *Proceedings, 1998 Transportation Research Board.*

Session 3, Track 3

Ground treatment and grouting techniques

North American Tunneling 2002, Ozdemir (ed.)
© 2002 Swets & Zeitlinger, Lisse, ISBN 90 5809 376 X

Rock joint sealing experiments using an ultra fine cement grout

Rajinder Bhasin
Norwegian Geotechnical Institute, Oslo

Per Magnus Johansen
Norconsult AS, Oslo

Nick Barton
Nick Barton and Associates, Oslo

Axel Makurat
Shell International Exploration and Production, The Netherlands

ABSTRACT: Rock joint sealing experiments have been conducted in a laboratory using a bi-axial coupled shear-flow test apparatus (CSFT). The laboratory test programme was designed to investigate the penetrative potential of a grout using different water/cement ratios on joints having different joint roughness (JRC) and joint conducting aperture in different stress conditions (total normal stress, joint water pressure and grouting pressure). The results indicate that joints with a conducting aperture (e) as small as approximately 25 microns can be grouted using a stable mixture of superfine cement, water and a super plasticizer (dispersing agent). The penetration capacity of a specific cement grout depends, in addition to the joint's characteristics, on the maximum grain size, the water/cement ratio and the injection pressure used. The tests reveal that the minimum physical aperture (E) that can be grouted corresponds to approximately four times the cement's maximum grain size.

1 INTRODUCTION

In many cases water leakages are governed by flow along the joints. An understanding of how the groundwater moves in rocks is one of the most important factors in the solution of rock engineering problems. This is especially true with regards to the planning and design of tunnels, storage caverns and underground waste disposals. Concerning nuclear waste repository safety a key aspect is the confidence of being able to successfully seal underground excavations and demonstrate methods of reducing the permeability of adjacent rock by sealing joints and fissures.

This paper describes the laboratory sealing experiments conducted on rock joints using a unique testing equipment designed by NGI. The equipment, called coupled shear flow temperature testing (CSFT) apparatus, has basically been used to derive the experimental data needed to quantify the effect of joint deformation on joint conductivity (Fig. 1). With the CSFT apparatus, joints can be closed, sheared and dilated under controlled normal stress conditions and at the same time cold or hot fluids can be flushed through the joint. Deformations, flow rate and stresses are recorded simultaneously. The CSFT test is designed to simulate as closely as possible the in situ state of critical joints and its modification by increases or decreases in normal and/or shear stress. In the present series of tests cement grout mixture was injected in the joint samples with increasing injection pressures. The rate of grout flow and the injection pressure versus time were recorded simultaneously to study the penetrability of a grout.

2 PENETRATION POTENTIAL OF GROUT MIXES

From a rheological point of view, a grout mix corresponds to a Bingham body exhibiting both cohesion and viscosity. A stable grout mix is defined as a mix having virtually no sedimentation e.g. less than 5% sedimentation in 2 hours (Lombardi, 1985). Water on the other hand follows Newton's law and is therefore a Newtonian body due to its viscosity and its lack of cohesion. A Newtonian fluid is represented by the following equation:

$$\tau = \eta \; dv/dx \qquad (1)$$

where τ = shear stress (Pa); η = kinematic viscosity (Pa·sec), dv/dx = strain rate (sec^{-1})

A Bingham body or a stable mix is represented by the following equation:

$$\tau = c + \eta \; dv/dx \qquad (2)$$

where c = cohesion (Pa)

Unstable mixes of cement and water will show intermediate rheological properties compared to the two cases mentioned above.

By decreasing the water content in the mix, either by decreasing the initial water-cement ratio or by loss of excess water during grouting, the mix may be so dry that it will have an internal friction angle and the rheological law of such a body will be (Lombardi, 1985):

$$\tau = c + \eta dv/dx + p \cdot \tan\varphi \qquad (3)$$

where τ = shear stress (Pa); c= cohesion (Pa); η = kinematic viscosity (Pa·s), dv/dx= strain rate (1/s), p= internal pressure (Pa), φ= internal friction angle (°)

Even if the internal friction angle is assumed to be small in such a mix the potential travel distance will be only a fraction of that of a Bingham body for a given pressure gradient and joint aperture. This mix will form a "plug" in a joint at a very short distance from the grout hole (or even in the grout pipes or hole itself).

This form of "plugging" or "pressure filtration" may occur even with a thin, unstable grout. Grouts with high water content and low viscosities are thought to have higher penetrative potential than the thicker mixes - the limit being the grain size of the grout compared to the joint opening, and the setting ability. When the larger grains become stuck in the joints, the gradient and pressure drop across this plug increases and the excess water will be pressed out of suspension. Penetration tests on natural or artificial joints have been performed by several authors using different cement and/or bentonite based suspensions (see e.g. Ran and Daemen, 1992, Widman, 1991 and Håkansson et al. 1992). It is a general opinion that cement grout suspensions may penetrate rock joints with an aperture varying from about of 0.1 to 0.5mm.

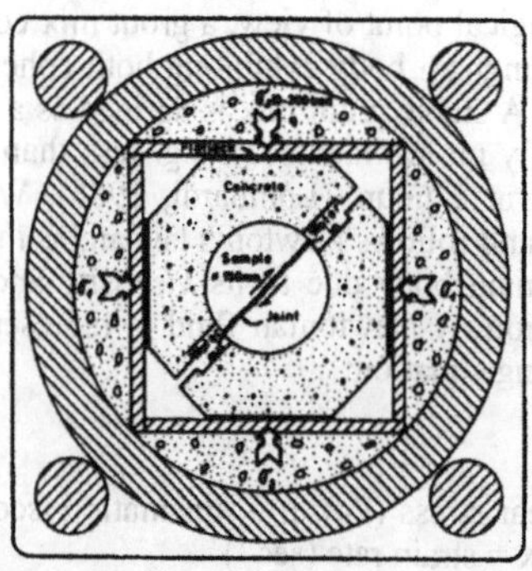

Figure 1. NGI's biaxial apparatus for CSFT testing of joints.

3 PROPERTIES OF THE GROUT USED

In the present rock joint sealing experiments an ultra fine cement grout, in which 98% of the material was finer than 12 microns, has been used to study the penetrative potential of grout mixes with different water cement ratios. The grout mixture comprised of cement (Spinor A), tap water and dispersing agent (Mighty 150). The dispersing agent (corresponding to 2% of the weight of cement) was added to the water and mixed for 1 minute. Then the water and cement was carefully mixed by hand before the grout was thoroughly mixed for approximately 10 minutes using an ULTRA-TURRAX T25 mixer with a 17mm diameter rotor. The rotor speed was set at 8000 rpm and gradually increased to 24000 rpm. The injection test was started immediately after the mixture was prepared in order to avoid any hardening of the grout.

A Marsh funnel calibrated to an outflow of one quart (946 cm^3) of fresh water at a temperature of approximately 21°C per 26 seconds was used to measure the viscosity of the grout at different water/cement ratios. Table 1 shows the results of the tests at water/cement ratios of 0.6, 0.8 and 1.0; with and without a dispersing agent.

A sample of each mixture was cured under cover. The apparent bleed of the grout mixture with the dispersing agent was considered as negligible and no cracking was observed in the cured samples.

Table 1. Results of Marsh Funnel Tests on Grout Mixtures.

Water/Cement	Time, Seconds	
Ratio	Dispersing agent	No dispersing agent
0.6	38.9	∞
0.8	31.5	46.3
1.0	28.9	36.8
Water	26	

4 DESCRIPTION OF ROCK JOINT SAMPLES

Three different rock joint samples were used for injection testing in the CSFT apparatus. The joint samples were procured from the field in mated condition and were prepared such that the horizontal length of the joint plane was around 85 mm. All the samples were characterised and profiled using the method described by Barton and Choubey, 1977. Table 2 shows the joint characterisation of the individual samples. Figure 2 shows a rock joint sample (sample 2) used for the injection tests. In one of the parts of the sample a hole of approximately 15 mm diameter is bored through the sample to facilitate injection of the grout mixture. The re-assembled joint sample is then cast in a reinforced epoxy block (see Figure 1).

All the three joint samples were tested in NGI's CSFT apparatus. In this apparatus by applying the same oil pressure to all four flatjacks, only normal stress is applied over the joint. From the two compartments adjacent to the sample in the epoxy block a back pressure is applied to simulate insitu ground water conditions The volume of material injected, total normal stresses, back pressure, injection pressure and joint conductive aperture were measured for all tests.

Table 2. Rock joint samples used for the injection tests.

Sample No.	Rock Type	JRC	Joint Description
1	Sandstone	5-7	Slightly stepped joint surface
2	Welded tuff	3-6	Rough planar with <1mm calcite infill, rough uneven
3	Welded tuff	8-12	Rough uneven with haematite coating, rough planar

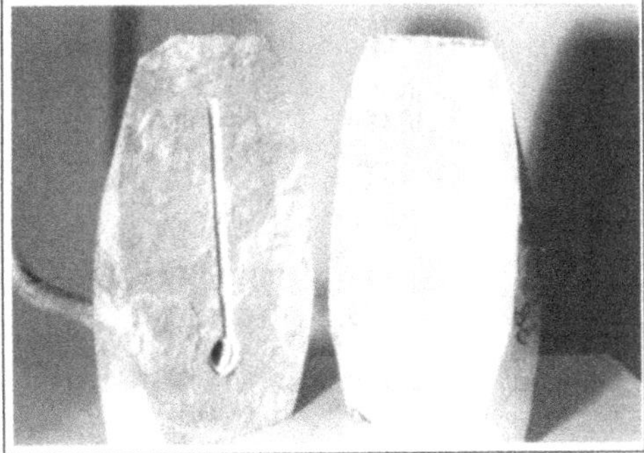

Figure 2. Photo of the tested joint (sample 2).

The following sequential order was adopted for the injection tests.

1) Saturation of the sample in water for at least 24 hours before testing.

2) Performance of water injection tests for measuring joint conductive aperture at different normal stresses (not carried out for sample 2).

3) Measurement of the joint conductive aperture at the normal stress selected for the injection test by measuring the amount of fluid that passes through the joint starting with a low injection pressure.

4) Injection of the cement grout mixture in the joint with increasing injection pressures.

5) Flushing of the cement grout mixture with water.

Repetition of Steps 2 to 5 with increasing normal stress.

The above procedure has been carried out for different water/cement ratios.

For all the tests joint conducting apertures were calculated from the flow through the joint assuming laminar flow between parallel plates using the following equation:

$$e = \sqrt[3]{\frac{Q \cdot 12 \cdot \upsilon}{g \cdot w \cdot i}} \qquad (4)$$

where w= width of flow path (m); e= conducting aperture assuming parallel-plate flow(m); ν= kinematic viscosity (m^2/s); Q= flow rate (m^3/s); i= hydraulic gradient; g= gravity acceleration m/s^2.

The results of water injection tests for measuring joint conductive aperture at different normal stresses for the samples 1 and 3 are shown in Fig. 3. This figure indicates that in sample 1 (sandstone), the joint conducting aperture is reduced from approximately 45 μm to 25 μm as a result of an increase in effective normal stress of approximately 35 bars. In sample 3 (welded tuff), the joint conducting aperture is reduced from approximately 220 μm to approximately 100 μm as a result of an increase in effective normal stress of approximately 60 bars. As stated earlier, the hydraulic conducting aperture is calculated assuming laminar flow between parallel plates.

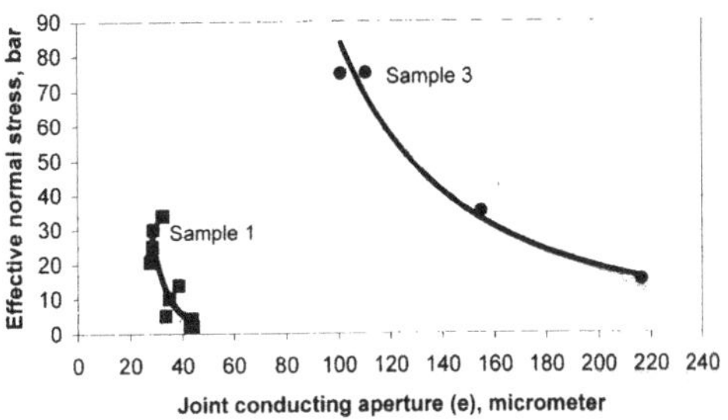

Figure 3. Conducting apertures as a function of effective normal stress for samples 1 and 3.

5 INJECTION TEST RESULTS

The injection test results for the three samples are summarised in Table 3. This table shows the joint conducting apertures obtained for the water injection tests for different normal stresses. The results from grout injection tests are also summarised in this table. During the tests, the rate of grout flow and the injection pressure versus time were recorded automatically. Typical results of these recordings for sample 1 are shown in Figures 4, 5 and 6. Figure 4 indicates that an injection pressure of 5.05 bars, which is slightly above the back pressure of 5 bars, was needed for penetration of the grout into the joint (water/cement ratio = 0.6). In this case the joint conducting aperture was 47 microns at a total normal stress of 10 bars (see Table 3). Figure 5 shows that, when the normal stress was increased to 20 bars, no

Table 3. Summary of water and grout injection tests results for the rock joint samples 1, 2 and 3.

Sample No.	Test No.		Total Normal Stress (bar)	Back Pressure (bar)	Injection Test Pressure (bar)	Injection Test Penetration	Joint Cond. Aperture (μm)
1 JRC=5-7	1:	a) Water Injection b) Grouting Test 1 (W/C=0.6)	10	5	a) 5.06 b) 5.05	- GOOD	47
	2:	a) Water Injection b) Grouting Test (W/C=0.6)	20	5	a) 5.50 b) 15.0	- NO	25
	3:	a) Water Injection b) Grouting Test (W/C=1.0)	20	5	a) 5.56 b) 15.0	- GOOD	25
2 JRC=3-6	1:	a) Water Injection b) Grouting Test (W/C=0.6)	10	5	a) 5.50 b) 12.65	- GOOD	66
	2:	a) Water Injection b) Grouting Test (W/C=0.8)	10	5	a) 5.50 b) 5.50	- GOOD	61
	3:	a) Water Injection b) Grouting Test (W/C=0.8)	20	5	a) 5.50 b) 15.0	- MINOR	50
	4:	a) Water Injection b) Grouting Test (W/C=1.0)	20	5	a) 5.50 b) 15.0	- MINOR	45
	5:	a) Water Injection b) Grouting Test (W/C=1.0)	40	5	a) 5.50 b) 20.0	- MINOR	43
3 JRC=8-12	1:	a) Water Injection b) Grouting Test (W/C=0.6)	20	5	a) 5.10 b) 5.05	- GOOD	137
	2:	a) Water Injection b) Grouting Test (W/C=0.6)	40	5	a) 5.09 b) 5.05	- GOOD	111
	3:	a) Water Injection b) Grouting Test	80	5	a) 5.14 b) 5.05	- GOOD	90
		(W/C=0.6)	75	0	c) 0.43	-	22
		c) Water Injection after Grouting	75	0	c) 0.47	-	15

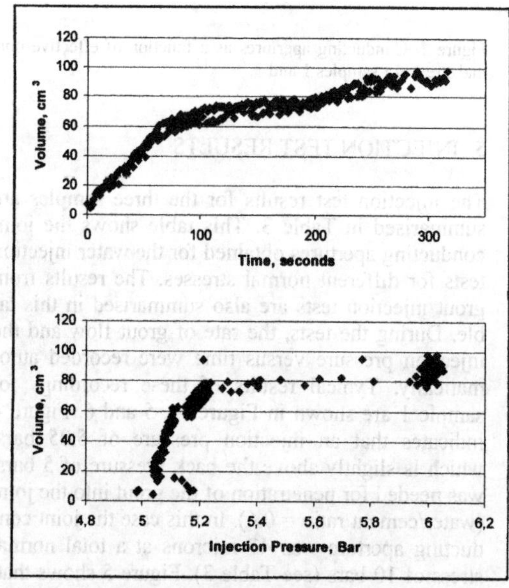

Figure 4. Injection test for Sample 1 with water/cement ratio= 0.6, total stress = 10 bar.

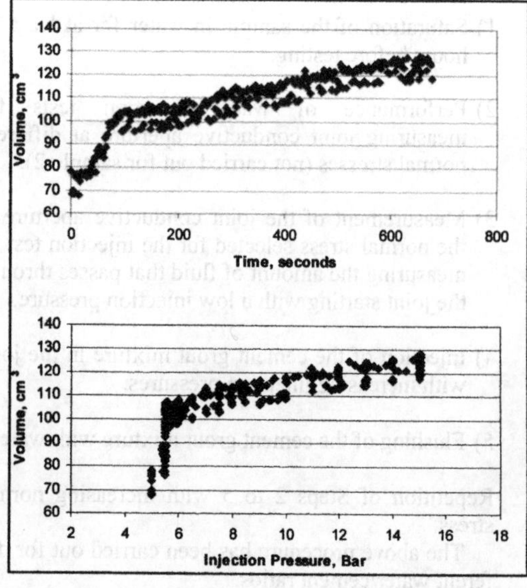

Figure 5. Injection test for sample 1 with water/cement ratio=0.6, total stress=20 bar.

grout penetrated the joint, even when the injection pressure was increased to 15 bars. This is presumably due to the decrease in joint conducting aperture (25 microns) as a result of increased normal pressure. However, as Figure 6 shows, once the water/cement ratio was increased from 0.6 to 1.0, the grout was very slowly penetrating the joint at an aperture of 25 microns. It is clearly evident from the figures that the injection pressure required for penetrating the grout depends on the water cement ratio and the normal stress across the joint. The results from samples 2 and 3 are summarised in Table 3 with similar trends.

As mentioned earlier, the cement used in the tests has a grain size distribution curve that indicates that 98% of the particles are finer than 12 microns. It may be seen from Table 3 that the minimum conducting aperture for sample 1 is approximately 25 microns and for sample 2 it is between 50 and 60 microns. For sample 3, no limiting aperture could be registered. These theoretical smooth wall apertures (e) can be converted to the real mechanical apertures (E) between the irregular joint walls. Generally, E is larger than e, implying that a rough-wall joint requires a larger aperture than a smooth-wall joint for the same water capacity. Wall friction and a tortuous path are considered responsible for flow losses.

It may be noted that the equivalent smooth-wall aperture (e) and the physical aperture (E) are related to the joint roughness coefficient (JRC) in the following manner (Barton, 1982):

$$e = \frac{JRC^{2.5}}{(E/e)^2} \tag{5}$$

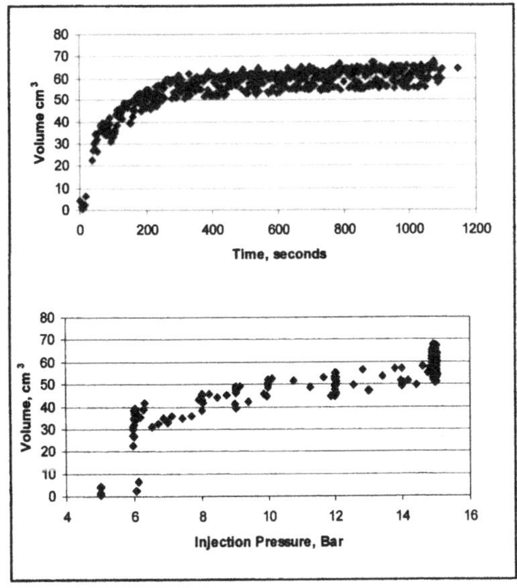

Figure 6. Injection test for sample 1 with water/cement ratio=1.0, total stress=20 bar.

The above empirical relationship is illustrated in Fig. 7.

Using equation 5, the physical aperture (E) for sample 1 (JRC=6, e=25 μm) is calculated as 47 μm. For sample 2 (JRC=4.5, e= 50 μm) the physical aperture (E) is close to the limit value of E=e, i.e. 50 μm (see Fig. 7). The above values correspond roughly to 4 times the maximum grain size of the cement.

The tests support the surprising notion that a joint with a lower Joint Roughness (JRC) will be less easily penetrated than a rougher joint, the hydraulic joint aperture being equal. At a physical aperture of 47 microns, Sample 1 is easily grouted using a water/cement ratio of 0.6 and an excessive grouting

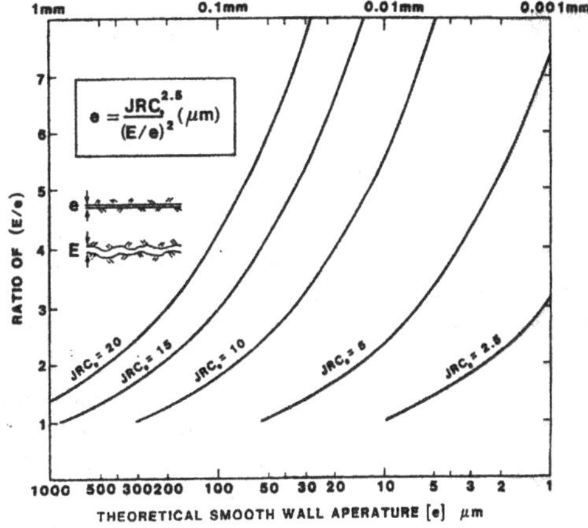

Figure 7. Empirical relationship between the mechanical and equivalent conducting flow aperture based on the JRC-value (Barton, 1982).

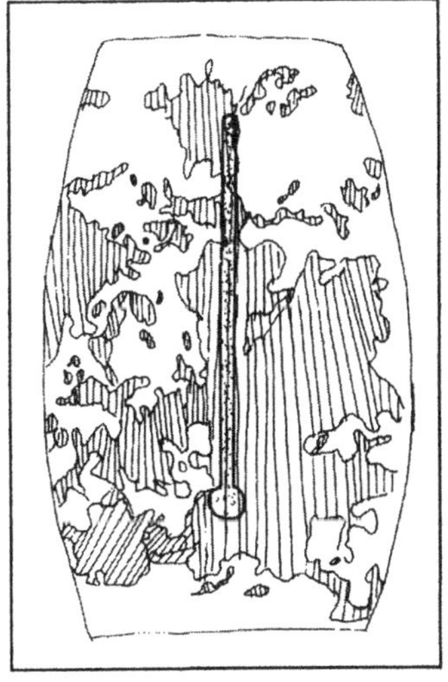

Figure 8. Grouted area of sample 3, after test 3.

pressure of 0.05 bar while Sample 2, having a smoother surface and a 50 micron physical aperture, is not groutable using a water/cement ratio of 1.0 and a grouting pressure of 10-15 bars. The tests support the predictions of Barton et al. 1985 and Barton and Quadros (1997) and support findings of Widmann (1991) and others that the grouting efficiency seems to be increased by an increase in the grouting pressure. There is of course the possibility of hydraulic jacking here.

Figure 8 shows the result of the grouting for Sample 3, Test 3. Approximately 50% of the area is covered by grout material. This indicates that even a 80-100 micron joint is not 100% effectively grouted using the super fine (Spinor) cement. The effective hydraulic aperture has been reduced to about one fourth. If one assumes parallel plate flow (laminar), the hydraulic conductivity has been reduced 64 times.

6 CONCLUSIONS

The laboratory tests on rock joint sealing have concentrated on the penetration capacities of a superfine cement grout with respect to different conducting apertures. It has been shown that grouts with a maximum particle size of 12 microns have the capacity of penetrating joints with a corresponding joint conducting aperture of 25 microns. Although the penetration capacity of a cement grout depends on the maximum grain size, the water/cement ratio and the injection pressure used, the tests do not conclusively support the thesis that a joint with a lower joint roughness will be more easily penetrated than a rougher joint, the hydraulic joint apertures being equal. For nuclear waste repository safety, such tests, which can be site specific, provide not only guidance for conducting in-situ test programmes but give recommendations for selecting appropriate grout mixtures.

REFERENCES

Barton, N. 1982. Modelling rock joint behaviour from in situ block tests – implications for nuclear waste repository design. Office of Nuclear Waste Isolation, Columbus, OH, ONWJ-308, p.96.

Barton, N., Bandis, S. And Bakhtar, K. 1985 "Strength, Deformation and Conductivity Coupling of Rock Joints", Int. J. Rock Mech. & Min. Sci. & Geomech. Abstr. Vol. 22, No. 3, pp. 121-140.

Barton, N. & Quadros, E.F. 1997. Joint aperture and roughness in the prediction of flow and groutability of rock masses. Proc. 36th U.S. Rock Mechanics Symposium and ISRM Symposium, Vol. 2 pp. 907-916.

Håkansson, U., Hassler, L. & Stille, H. 1992. Rheological Properties of microfine grouts with additives, ASCE-conference, New Orleans: Grouting, soil improvement and geosynthetics, Vol. 1, pp.551-563, New York.

Lombardi, G. 1985. The role of cohesion in cement grouting of rock. ICOLD, Q.58, R13, Lausanne.

Ran, C. & Daemen, J.J. K. 1992, Fracture grouting with Bentonite slurries, ASCE-conference, New Orleans: Grouting, soil improvement and geosynthetics, Vol. 1, pp. 360-371, New York.

Widman, R. 1991. Injeksjonen i Fels und Beton – Zwischenbericht einer arbeitsgruppe des Verbundkonzerns. Felsbau 9, pp. 147-154, Wien.

North American Tunneling 2002, Ozdemir (ed.)
© 2002 Swets & Zeitlinger, Lisse, ISBN 90 5809 376 X

Methods used in the excavation of shafts supported by artificial ground freezing

J.A. Sopko & G.F. Aluce
Layne Christensen Company, Pewaukee, Wisconsin

M. Vliegenthart
J.F. Shea Company, Boston, Massachusetts

ABSTRACT: Artificial Ground Freezing has been used as a method of providing temporary earth support and ground water control for shaft construction for well over 100 years. Design analyses of the thermal and structural methods for evaluating the freezing method and systems have been presented extensively in the literature during the past 25 years, while the excavation methods and procedures are often neglected. This paper evaluates the methods of shaft excavation used on several shafts in North America. Consideration is given to the initiation of excavation related to surface water control and initial shaft alignment. Excavation techniques such as drilling and blasting, raise boring and road header operations are discussed.

Artificial Ground Freezing is a technique used to provide temporary earth support and groundwater control during subsurface construction of shafts and deep excavations. Ground freezing should not be considered as a substitute for conventional support methods such as steel sheeting, soldier piles and lagging, and sunken caissons due largely to the higher installation and maintenance costs. In comparing ground freezing to slurry wall construction, the slurry wall method is typically the cost effective solution when the slurry wall can be an integral component of the finished shaft or structure. As shaft depths increase beyond 30 meters however, there is a marked decrease in the unit price of freezing as the shaft depths increase. With the economic factors considered, ground freezing is typically the choice of contractors when a proposed project has one or more of the following characteristics:

1 Soft or loose unconsolidated water-bearing soils;
2 Sandy soils with too many fines to implement a dewatering system;
3 Organic soils;
4 Irregular contact surface between overburden soils and underlying bedrock;
5 Fractured, water-bearing bedrock;
6 Hydraulic bottom instability requiring a cut-off system to depths greater than 1.5 times the invert elevation;
7 Presence of cobbles or boulders that preclude the driving of steel sheets or soldier piles; and
8 Contaminated groundwater that would require excessive treatment if pumped and discharged.

Several different methods can be used in designing frozen shafts. In implementing any of these design procedures the engineer or contractor should keep in mind that, unlike steel or concrete design and construction, ground freezing is a "process" that is on-going and subject to changes in ground conditions and system operation. Ground freezing should never be considered a combination of components such as drilled pipes, coolant manifold and refrigeration plants configured from a rigid design selected from a text book or published article.

The "process" of ground freezing should always begin with a thorough understanding of the subsurface conditions. Quite often however, there is not adequate information or material properties (frozen and unfrozen) readily available and the designer must use all available resources to establish estimated parameters for design. The major components of a ground freezing system include the depth and number of freeze pipes, coolant temperature and required freezing time. As ground freezing is commercially available in today's industry, it is usually implemented by a specialty company working as a subcontractor to a project owner or general contractor that quite often has the design capability required for a successful project. The subcontractor will install the system, freeze the ground, then turn it over to the general contractor to excavate, providing a quality assurance program is implemented. This "process" of ground freezing is then culminated with the excavation and completion of the final subsurface structure. The excavation process is the focus of this paper.

1 PRE-EXCAVATION

1.1 *Preliminary considerations*

The first step of shaft excavation typically begins, before the freeze pipe drilling and installation, with the construction of a temporary or permanent shaft collar at the ground surface. This collar serves as both an excavation and freeze pipe template and as a method to prevent the inflow of surface water. A simple method of constructing the collar is illustrated in Figure 1.

Figure 1.

On this particular project, the concrete collar was formed and poured prior to the initiation of the drilling program for the freeze pipes. The collar illustrated above was used for the excavation template, however the completed shaft was of a considerably smaller diameter, therefore the collar was not integral in the final construction and was discarded at project completion.

Figure 2.

Figure 2 illustrates a collar that not only provided an excavation template, but was also used for shaft excavation support in the unsaturated overburden soils above the water table. In several ground freezing projects completed by the author, the static ground water table has been located at depths approaching ten meters. In each of these cases, the unsaturated soil was a coarse-grained cohesive soil, requiring a conventional excavation support system such as ribs and lagging boards. Other shafts have been successfully completed using a steel rib with liner plates to support unsaturated loose soils. In the case of methods, successful projects were completed when the initial sections were constructed prior to excavation, thus ensuring a uniform, circular template. While maintaining a uniform circle is critical to any shaft excavation, added importance is emphasized on ground freezing projects to ensure that the excavation is symmetric with the circular shape of the freeze pipes at the ground surface.

In the case of the concrete collar, or the selected method for initial excavation support, placement should be made to ensure that surface water will not drain directly into the shaft. In the cases of flooding, heavy rain, or rapidly thawing snow, ground water flowing toward an excavated shaft tends to develop a preferred path, which can lead to rapid thawing and "cutting" into the frozen earth wall. A thawed zone can rapidly develop downward as the flows continue, which can lead to rapid instability of the structural frozen earth wall. If a larger or non-circular excavation is required and a collar is not feasible, care should be taken to ensure that surface water drainage is directed or pumped away from the shaft. In such a case, excavation control points and procedures must be developed to maintain verticality of the excavation face.

1.2 *Diameter of freeze pipe plan*

While the shaft collar or template is sized to the excavation diameter, the diameter of the freeze pipe circle is based on the required thickness of the frozen earth wall and the excavation procedures planned by the contractor. Generally, the deeper the shaft and/or lower the strength of the frozen soils, the wider the frozen earth zone must be. Once the frozen wall thickness is determined in the preliminary design, the diameter of the circle of freeze pipes is established based on the concept that the freeze wall will develop both medially and laterally from the circle of freeze pipes. Half the wall is internal to the circle and the other half external. As freezing time progresses, the wall will continue to thicken internally and externally. Since the inside thickness is truncated once the excavation begins, the frozen zone on the exterior will continue to grow. Additionally has freezing time progresses, the frozen earth temperature decreases resulting in an increase in frozen soils strength. Both the exterior thickness and final frozen earth temperature should be considered when designing the ground freezing system.

The encroachment of the frozen earth towards the center of the shaft is also considered when determining the diameter of freeze pipe installation. The magnitude of frozen soil encroachment is primarily a function of soil type. Coarse-grained cohesionless soils will freeze at a faster rate than fine-grained cohesive soils. In a deep shaft where there are several strata of different soil types, the encroachment of the frozen soils may vary as the excavation progresses.

The amount of frozen soil located inside the excavation can be varied somewhat based on the diameter of the freeze pipe plan and the required freezing time. A frozen earth wall having sufficient structural thickness with minimal frost encroachment can be formed with an increase in diameter of the freeze pipe plan. The amount of frost encroachment can be considered a design variable, which can be suited to the excavation method selected by the contractor. There are three different approaches to the handling of frost encroachment which are often considered:

1 Freezing Across the Entire Excavation—Freezing a shaft solid is often used when raise boring excavation methods are used and in cases where freeze time is not a critical element and the excavation contractor would rather dispose of frozen cuttings instead of soft materials with excessive free water.
2 Freezing Close to the Excavation Line with Minimal Encroachment- In the last ten years several seven meter diameter shafts have been constructed on coal mine projects where freezing has been completed with no encroachment beyond the excavation line in an effort to minimize frozen soil excavation. The advantage of this method is that frozen soils can be much more difficult to excavate than unfrozen soils. A uniform sandy overburden soil permitted such a technique which incorporated a freeze pipe plan diameter about one meter larger than required to satisfy structural requirements. When using this technique, the contractor must make every effort to complete the excavation quickly before extended freezing times will result in frozen ground encroachment which is more difficult to excavate and defeats the purpose of installing additional freeze pipes to get the larger plan diameter. When using this method, sometimes the contractor will encounter unfrozen soils at the face of the excavation which may slough off back to the frozen zone and result in an uneven excavation face.
3 Freezing with the Intention to Excavate Encroached Frozen Soil – The process of preparing to excavate the shaft with the intention of mining encroached frozen soil is the most popular method employed by contractors. If the proper equipment and personnel are available to excavate the frozen soil that has encroached into the excavation, the mining of frozen earth can be readily accomplished with two significant advantages. First, proper mining of the frozen earth will result in a circular excavation with an even face suitable for single forms for placing concrete. A single form can be placed in the excavation permitting concrete to be poured directly against the insulated frozen earth. The time saved in this forming technique almost always compensates for any additional time required to mine the frozen earth. The second, though less obvious advantage of mining frozen earth is the psychological effect of the miners encountering the very hard, rock-like frozen soil. Several projects that the authors have constructed are the first frozen earth shafts encountered by the mining crews. New miners are typically skeptical of the ground freezing technology and it is not until they encounter the hard frozen material that they become aware of the strength and safety of a frozen earth support system.

2 INITIATION OF EXCAVATION

If the excavation template is installed after the installation of the freeze pipes, as usually done with the ribs and lagging boards or liner plates, it is critical that the template be placed concentric with the circle formed by the freeze pipes as shown in Figure 3.

Figure 3.

A slight offset towards freeze pipes will result in a frozen earth wall thinner than planned and could result in structural deficiencies. A combination of the previously mentioned irregular circle form of the template, combined with imprecise installation can result in significant structural consequences during excavation.

Depending on the diameter of the excavation, at least four plumb-bobs placed 90-degrees apart should be installed on the template. These plumb-bobs are used to ensure verticality of the excavation and used in all shafts, whether freezing is incorporated or not.

During the initiation of excavation, the contractor should be aware of any freeze pipes that have deviated into the interior of the excavation. During drilling, it is not uncommon for freeze pipes to deviate from the vertical in several directions. In some cases however, freeze pipes may deviate into the excavation zone. If the deviation is too great, an additional pipe may have to be re-drilled to compensate for the gap in the frozen zone created by the deviation. In some cases, a pipe deviating towards the center of the shaft is used as a temperature-monitoring pipe as it provides an excellent opportunity to measure the temperature of a section across the entire frozen zone. Freeze or temperature pipe deviation is measured using gyroscopic surveying equipment during the drilling phase of the project.

3 EXCAVATION TECHNIQUES

Several methods have been used to excavate the frozen shafts. The authors have presented a description of the methods they have directly been involved with.

3.1 *Raise boring*

Raise Boring is undoubtedly the most expedient method for excavation of a frozen shaft. A pilot hole is first drilled at the center of the shaft into the underlying tunnel or mine excavation. Once the pilot hole is drilled, the raise equipment is installed inside the tunnel or mine. As the bore is elevated from the surface, the frozen material is deposited into the tunnel or mine and removed. Most raise bore equipment is designed primarily for rock boring and some modification of the equipment may be required. After completion of the boring, a permanent steel or concrete liner is installed. The major advantage of this method is the speed at which the excavation can be completed. Additionally, the resulting excavation is extremely smooth for rapid installation of the final liner.

Even though this method requires less time than other methods, there are several drawbacks. The equipment required for raise boring is often expensive and not readily available. Another problem associated with raise boring is the difficulty of drilling a vertical pilot hole. Any deviation in the pilot hole will result in a unsymmetrical frozen zone which can lead to structural weakness. It should be also noted that few projects are designed where ground freezing is completed over an existing tunnel, mine or underground works that even permits consideration of raise boring.

3.2 *Road header*

A road header as illustrated in Figure 4 is the most effective method for trimming the frozen earth and forming a vertical wall in a circular excavation. Several manufacturers have provided equipment that has been very successful in frozen ground excavation. Contractors contemplating the use of the road header should consult the appropriate manufacturer to select the proper header teeth that will work with frozen soils. A major disadvantage of the road header is that tooth life is quite short in sand or gravel overburden. Additionally, when cobbles or boulders are encountered, it may be necessary to abandon the road header temporarily and resort to hand mining methods.

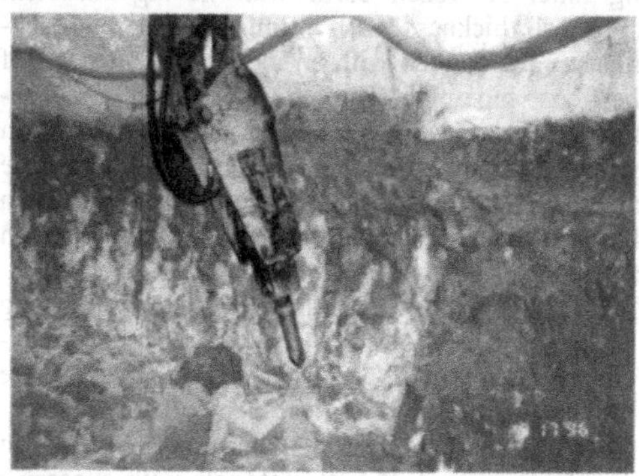

Figure 4.

3.3 *Hoe rams and spades*

Excavation with a conventional hoe ram with hand operated pneumatic spades is the most common technique on ground freezing projects, due largely to the availability and expense related to the equipment. This technique is the most labor intensive, but with experienced crews production rates are often as high as with a road header. Interesting projects have occurred in California or the Southern United States when crews unaccustomed to working in cold weather are exposed to ambient air temperatures of 20° to 30° and then lowered into shafts where the temperature is between 0° and – 10°C.

4 INSULATION

Frozen earth walls exposed to warm ambient air temperatures must be insulated continuously during the excavation process. Insulation assists in maintaining a lower internal frozen earth temperature, resulting in higher frozen soil strengths and increased structural capability. On several projects, shaft excavation has occurred during the winter months when ambient temperatures remain below freezing and insulation has not been used. Care should be exercised however in ensuring that direct sunlight does not induce any melting, even though temperatures appear to be cold.

As is the case on any underground projects, plans do not always develop as originally intended. The authors are familiar with several projects where the excavation process has taken longer than planned due to process inefficiency or increases in excavated depth. Springtime weather then resulted in sloughing or melting of the frozen earth wall. It should also be noted that insulation does reduce the amount of sublimation of ice in frozen soils even though the temperatures remain below freezing. A drying frozen earth wall will exhibit signs of individual soils particles falling off the facing of the excavation in almost powder-like form.

Figure 5.

The most effective insulation method is spraying a polyurethane foam against the frozen surface. Prior to the foam application however, a small gauge wire mesh must be in place for the foam to adhere (see Figure 5). This wire mesh is typically installed from the ground surface. Once anchored at the ground surface, the rolls of mesh are lowered as the excavation progresses. In some cases the mesh must be fastened directly to the face of the frozen earth wall. This can be accomplished by drilling a 1 to 2 centimeter hole in the frozen earth with a hand drill and then driving a steel rod or nail into the opening. If done quickly, the slightly thawed soil in the small opening will re-freeze around the rod and secure it firmly in the frozen earth. This technique provides a relatively strong anchor for the wire mesh.

Insulation need not be applied any thicker than ten centimeters. Care should be taken in selecting a product that meets fire code requirements and that will also form and expand when sprayed against a cold surface. Experience has shown that this insulation can be applied as the excavation proceeds in 2 to 4 meter lifts. In almost all cases, and depending on ambient air temperature, frozen earth should not be exposed to warm temperatures for time periods greater than 48 hours.

4.1 *Lining while excavating*

One of the safest and most reliable methods of placing the final concrete in the shaft is to install the forms and pour the concrete in 3 to 5 meter lifts as the excavation progresses. While frozen earth shafts

Figure 6.

can be excavated to the invert depth without any additional support, the lining with excavation progress has several significant advantages:

1 Insulation may not be required if the forming and pouring process is completed in short time frames;
2 Creep deformation in cohesive soils will not begin due to the short time period of being subjected to compression stresses;
3 Should a breach occur in the frozen earth wall resulting in flooding, minimal thawing or damage occurs to the excavated and lined portion of the shaft; and
4 There is a less stringent need for back-up electrical power or refrigeration equipment in the event there are interruptions or failures.

In designing the lining system, the friction between the frozen earth and concrete is the greatest single component adding to the forces supporting the shaft as the excavation proceeds below each installed lift. For this reason, progressive lining is not recommended for shafts raise bored, as the smooth face minimizes the frictional component. Additional forces adding to supporting the lining can be added by incorporating a larger contact area if a concrete collar is installed. In this case the collar will act as a footing, distributing the load of the hanging liner at the ground surface. If the surface area of the collar is too large, sleeves will have to be installed to facilitate the drilling of the freeze pipes after collar placement. If this approach is used, consideration should be given to the reduction in strength of the frozen soil once the system is terminated and thawing begins.

5 CONCLUSION

Several techniques have been presented to provide assistance in the excavation of shafts in frozen soils. As with the design and freezing of the shafts, the excavation of frozen earth shafts should be treated as a "process" requiring constant observation and monitoring of the temperature regime within the frozen earth. Exterior ground water levels should be monitored to ensure there are no abnormal flow patterns occurring during excavation. As previously stated, this process of freezing and excavation is based largely on experienced contractors with an understanding of the mechanics of frozen ground. The authors do not recommend the techniques presented in this paper to be implemented without the assistance from experienced work crews.

North American Tunneling 2002, Ozdemir (ed.)
© 2002 Swets & Zeitlinger, Lisse, ISBN 90 5809 376 X

Innovative long distance pumping of backfill material and contact grout for the South Mountain Reach 3B Tunnel Project

Raymond W. Henn
Haley & Aldrich, Inc., Denver, CO, USA

Patrick J. Stephens
Pacific International Grout, CO, Bellingham, WA, USA

William D. Leech
Haley & Aldrich, Inc., Phoenix, AZ, USA

Mark R. Rybak
Affholder IncoYporated, Chesterfield, MO, USA

ABSTRACT: A 8 foot diameter, 6,060 foot long TBM tunnel was excavated beneath the southwest flank of South Mountain in Phoenix, Arizona. A 48 inch diameter prestressed concrete cylinder pipe (PCCP) water transmission pipeline was installed within the tunnel. The annulus between the PCCP and the excavated tunnel was backfilled with cellular concrete. The contract documents also required contact grouting of the tunnel crown to fill any remaining voids after completion of backfilling. The City of Phoenix and their pipeline design engineer would not allow penetrations in the PCCP, therefore, cast-in grout ports were not allowed in the pipe. The contract documents required a maximum discharge distance of 500 feet between backfill injection points. Since no penetrations were allowed in the PCCP the 500 foot maximum distance between injection points requirement would require bulkheads to be constructed throughout the tunnel. Bulkheading was not acceptable to the tunnel contractor based on cost and scheduling concerns. The methods developed involved injecting the cellular concrete backfill from the southeast and northwest tunnel portals via long delivery pipes installed within the annulus between the tunnel and the PCCP.

1 INTRODUCTION

The 6,060 foot (1,848 m) long, Reach 3B Tunnel under the southwest flank of South Mountain in Phoenix, Arizona was completed in December 2001. The Reach 3B Tunnel is a portion of the overall South Mountain Water Facilities Project. The tunnel was excavated using a refurbished 8 foot (2.4 m) diameter Robbins Double Shielded Tunnel Boring Machine (TBM). Grippers allowed the TBM to thrust forward in the rock and cemented soil portions of the tunnel, while a thrust ring was used to push off of the steel rib and wood lagging temporary support system in the loose soil portions of the tunnel. Once tunnel excavation was completed the tunnel contractor installed the 48 inch (120 cm) inside diameter, 20 foot (6.1 m) long sections of prestressed concrete cylinder pipes (PCCP). The PCCP pipeline is designed to carry potable water at high pressure. The annulus between the excavated tunnel and the PCCP was designed to be backfilled with cellular concrete. After backfilling was completed the contract document required the crown area of the tunnel to be contact grouted to fill any remaining voids. The installation of grout ports in carrier pipes to allow for backfilling and contact grouting is a standard industry practice for tunnel projects. Therefore, the tunnel contractor had planned to inject both the backfill cellular concrete and contact grout through grout ports which would have been installed in the PCCP during the pipe fabrication. However, the owner of the South Mountain Reach 3B Tunnel Project did not want any "penetrations" (grout ports) in their water pipe. This "no penetrations in the pipe" requirement was clearly stated in the contract documents.

Since no penetrations in the pipeline were allowed, an alternate method to inject the backfill cellular concrete and contact grout had to be designed. The contract documents also required a maximum spacing of 500 foot (152 m) between injection points for the backfill. The first method considered required the installation within the tunnel of 500 feet (152 m) of PCCP at which point pipe installation would be stopped and a bulkhead built. The 500 feet (152 m) length of PCCP would be backfilled with cellular concrete and then contact grouted through the bulkhead. After contact grouting was completed the process of PCCP installation, bulkhead building, backfilling and contact grouting would be repeated in lengths of 500 feet (152 m) at a time for the remainder of the tunnel. This method was rejected by the tunnel contractor because it required starting and stopping of the pipe installation operation which would have had an unacceptable cost and schedule impacts to the project. As an alternative, a system of long backfill delivery pipes, installed within the annulus between the pipe and the excavated tunnel, was developed. Each backfill delivery pipe discharged at approximately 500 foot (152 m) intervals of increasing lengths from either of the portal bulkheads. A contact grout delivery pipe, the full length of the tunnel, from portal bulkhead to portal bulkhead was also installed within the annulus at the tunnel crown. These alternative methods of backfill cellular concrete and contact grout injections are the subject of the paper.

1.1 *Project Team*

The project owner is the City of Phoenix, Arizona. Stanley Consultants, Inc. served as the general engineering consultant and provided construction inspection services. Haley & Aldrich, Inc. was the tunnel designer and provided onsite tunnel advisory and construction management services during construction. HDR provided overall construction management for the South Mountain Water Facilities Project of which the Reach 3B is a part. Affholder, Incorporated was the tunnel contractor. Specialty subcontractor Pacific International Grouting Co. performed the cellular concrete backfilling and contact grouting.

1.2 *Geologic Conditions*

The geology for approximately 85 percent of the 6,060 foot (1,848 m) long tunnel was hard rock comprised of gneiss and granite. The remainder of the tunnel was comprised of alluvium and cemented alluvium (caliche). There was one section in the tunnel which had open rock fractures which might cause the backfill or the contact grout to be lost to the surrounding ground during injection. The fractures and several open exploratory boreholes were sealed off before installing the PCCP. The groundwater level was below the tunnel invert. The amount of groundwater entering the tunnel was nominal and not a factor during the backfilling and contact grouting operations.

1.3 *Backfilling*

The tunnel slopes uphill from the southeast portal to the northwest portal. The southern 3,880 feet (1,183 m) of the tunnel is at a 0.25 percent grade and the northern 2,180 feet (665 m) of the tunnel is at a 2.05 percent grade. The total difference in elevation from the southeast portal to the northwest portal is 54.4 feet (16.6 m).

The project team agreed to start backfilling from the southeast portal. The plan was to inject about one half of the estimated volume of cellular concrete backfill from the southeast portal, with the other half of the backfill being injected from the northwest portal. A backfill delivery piping layout was developed as shown in Figures 1 and 2. Table 1 gives distances from the portal bulkheads to the discharge point of the delivery pipes. Figure 3 shows a picture of the backfill and contact grout delivery pipes installed in the tunnel. As shown in Figure 3 the backfill delivery pipes were all installed on both the left and the right tunnel ribs below springline, however each delivery pipe discharged at the tunnel crown. As a delivery pipe reached its discharge point elbows and short pieces of pipe were used to bring the delivery pipe's discharge to the tunnel crown. There was also a short, 2 foot (0.6 m), backfill delivery pipe at each portal bulkhead (pipes SP and NP). A masonary block bulkhead was built at each portal. A third bulkhead was also built at approximately the midpoint of the tunnel, Figure 4. Figure 5 shows the southeast portal bulkhead with eight backfill delivery pipes (SP thru S7) installed. The contact grout delivery pipe is pictured in the tunnel crown. There are also three electrical conduits pictured on the left side of the portal. Four 1½ inch (38 mm) PVC vent pipes were installed through the bulkhead at 0, 90, 180 and 270 degrees to vent air during backfill injection and to verify the level of backfill at the bulkhead.

The delivery pipes used were 3 inch diameter schedule 20 steel with Victraulic couplings. At first, PVC pipes were considered, but the idea was rejected because it was believed the pipes might partially melt during backfill injection because of the elevated temperature within the annulus generated by the hydration of the cellular concrete.

Originally the backfill was planned to be injected from the southeast portal starting with the shortest delivery pipe than hooking up to the next longest delivery pipe to the next longest delivery pipe and so on. For example, starting with delivery pipe SP then hooking up to delivery pipes S1, S2, S3 through delivery pipe S7. Using this injection sequence would have meant that the cellular concrete being injected would have had to travel through ever increasing lengths of delivery pipes. These delivery pipes would have already been embedded in the previously injected cellular concrete. It was believed the heat generated by the hydration of the previously injected cellular concrete surrounding the delivery pipes could cause the cellular concrete being injected to flash set inside the delivery pipe.

Table 1. Backfill and Contact Grout Pipe Discharge Distances from the Portal Bulkheads.

Pipe No.	Distance From Portal FT (meters)	Pipe No.	Distance From Portal FT (meters)
SP	2 (0.6 m)	NP	2 (0.6 m)
S1	460 (140 m)	N1	495 (151 m)
S2	995 (303 m)	N2	1005 (306 m)
S3	1480 (451 m)	N3	1495 (456 m)
S4	2020 (615 m)	N4	2030 (619 m)
S5	2395 (730 m)	N5	2395 (730 m)
S6	2775 (846 m)	N6	2795 (852 m)
S7	3130 (954 m)		

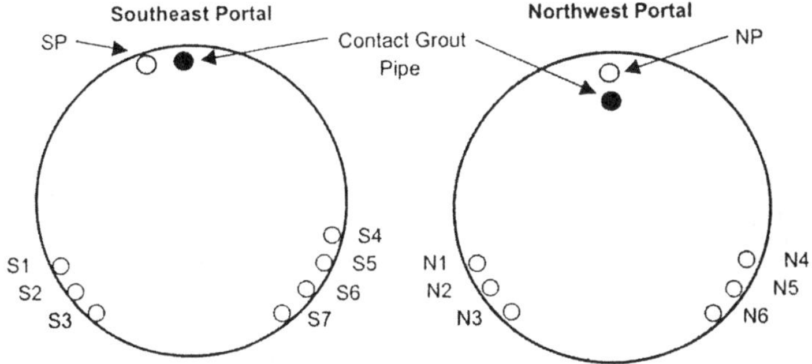

Figure 1. Backfill and Contact Grout Delivery Pipe Layouts at Portal Bulkheads.

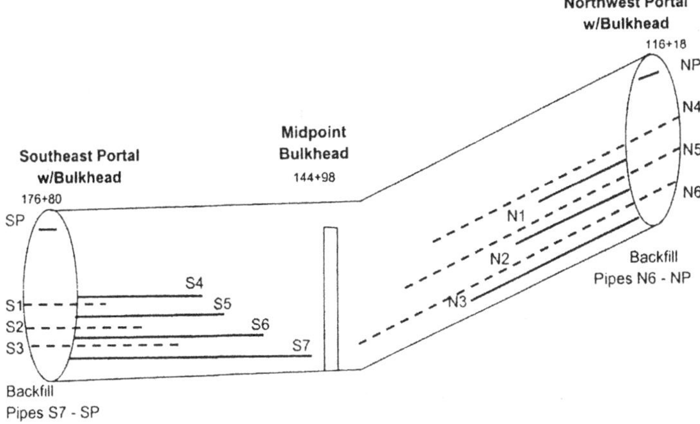

Figure 2. Backfill Pipe Layouts within the Tunnel.

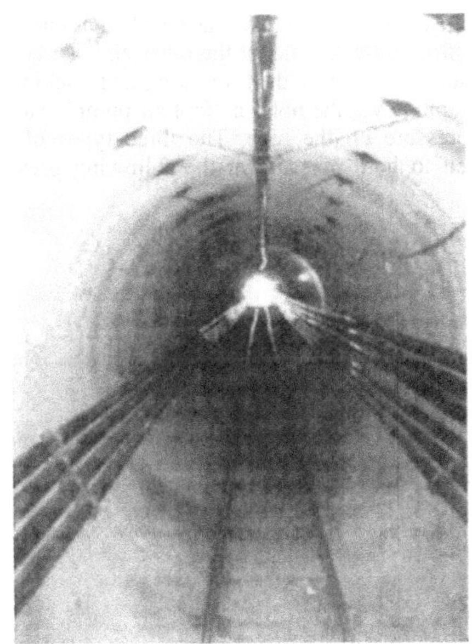

Figure 3. Looking Towards the Southeast Portal, Delivery Pipes S4 thru S7 are on the Left and Delivery Pipes S1-S3 are on the Right, with the Contact Grout Delivery Pipe at Crown.

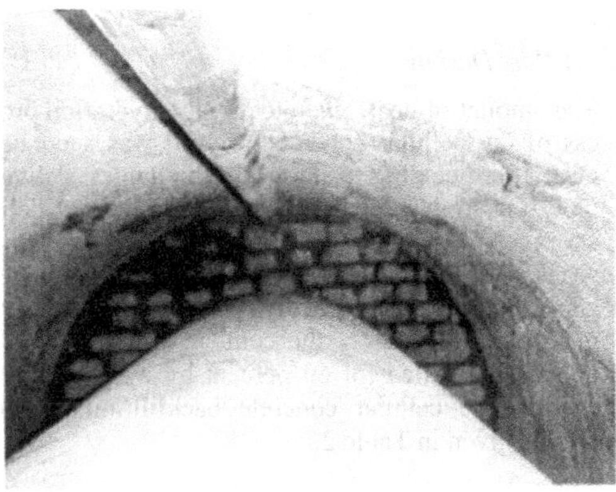

Figure 4. Bulkhead at the Approximate Midpoint in Tunnel with the Contact Grout Delivery Pipe at the Tunnel Crown.

Because of the unacceptable risk of flash set it was decided to start backfill at the southeast portal using the longest delivery pipe (S7) first and working towards the portal using ever shorter lengths of delivery pipes. For example, starting with delivery pipe S7 then hooking up to delivery pipes S6, S5, S4 through delivery pipe SP.

Figure 5. Southeast Portal Bulkhead with the Cellular Concrete Pump Discharge Hose Hooked Up to Backfill Pipe Delivery Pipe SP.

The backfill for northern half of the tunnel was injected from the northwest portal starting with delivery pipe N6, the longest pipe, through delivery pipe NP, the one which discharged 2 feet (0.6 m) inside the tunnel from the portal bulkhead.

1.4 *Mix Design*

The amount of heat generated by the hydration process of the cellular concrete backfill was a concern even after deciding to start with the longest delivery pipe first from the southeast portal. The steel delivery pipes would still get very hot within the confined space of the annulus. In an attempt to reduce the heat generated, the amount of flyash used in the mix was maximized. Therefore the cellular concrete backfill mix design called for 60 percent by weight Type F flyash. The cellular concrete backfill mix design used is given in Table 2.

Table 2. Cellular Concrete Mix Design.

Material	Weight Lbs	Volume CF	Unit Weight PCF
Cement Type II	427.3	2.17	
Flyash Type F	709.1	4.96	
Water	647.8	10.38	
Subtotals	1784.20	17.51	101.88
Foaming Agent	23.7	9.49	
Totals	1807.9	27.00	66.96

1.5 *Contact Grouting*

The contract documents required contact grouting of the crown area of the tunnel to be performed after backfilling was completed to fill any remaining voids which might exist. A 3 inch (75 mm) diameter schedule 20 steel contact grout delivery pipe with Victraulic couplings was installed along the tunnel centerline at the crown of the tunnel, Figure 3. The contact grout delivery pipe had two ½ inch (13 mm) diameter, shop drilled, discharge holes located every 10 feet (3.1 m) along the pipe. The two holes at each location were orientated; one hole at the top of the pipe and the second hole spaced 90 degree to the left or right of the tophole, putting the second hole on the pipe's horizontal centerline. The holes located on the horizontal centerline alternated leftright, left-right every 10 feet (3.1 m) for the entire length of the tunnel. Figure 6 shows this contact grout discharge hole layout.

The ½ inch (13 mm) diameter discharge holes were covered over with duct tape before the contact grout delivery pipe was installed in the tunnel. Taping over the discharge holes was done to help prevent any backfill cellular concrete from entering the contact grout delivery pipe during backfill injection.

During the planning stages for the contact grouting, field tests were conducted to evaluate the best type of tape to use to cover the discharge holes in the contact grout delivery pipe. To conduct the tests a short section of 3 inch (75 mm) diameter steel pipe with two ½ inch (13 mm) diameter discharge holes drilled 90 degrees apart was used. The pipe was capped at both ends by welding plates over the ends of pipe. A pressure gauge was installed at one end of the pipe and a air inlet at the other end. Three types of tape were tested; duct, masking and packing. After taping over the holes a "foot air pump" was used to pressure up the pipe. The three types of tapes failed to hold pressure at the following pressures:

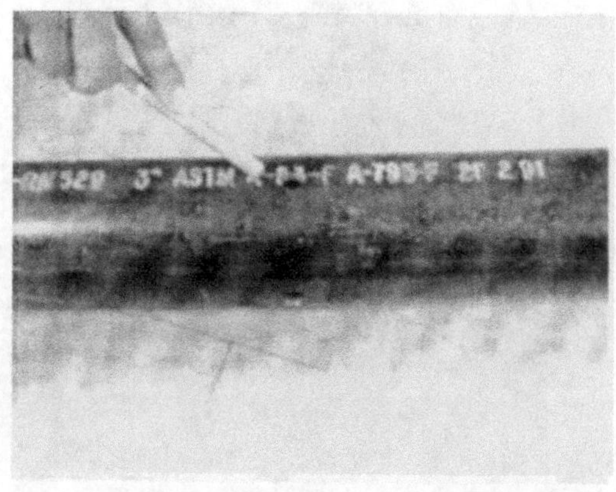

Figure 6. A Section of 3 inch (75 mm) Diameter Steel Pipe Showing a Set of ½ inch (50 mm) Diameter Shop Drilled Contact Grout Discharge Holes.

Duct tape 14 psi (0.95 bars); Masking tape 5 psi (0.34 bars); and Packing tape 12 psi (0.82 bars).

Eight tape tests were conducted. In all but one of these tests the tape failure was at the bond between the tape and the pipe. In each case the air escaped under the tape away from the drilled holes. Only during one test did the duct tape repture over the hole. Based on the results of these simple tests it was decided to use duct tape to cover the contact grout delivery pipe discharge holes. Figure 7 shows tape testing equipment.

The contact grout mix design used is given in Table 3.

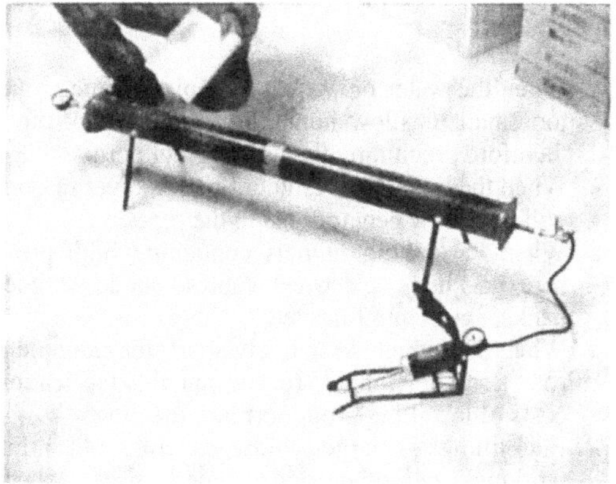

Figure 7. Testing Equipment Used to Test Tape which Covered Contact Grout Discharge Holes.

Table 3. Contact Grout Mix Design.

Material	Weight Lbs	Volume CF	Unit Weight PCF
Cement Type I/II	1578	8.03	
Water	1184	18.97	
Subtotals	2762	27.00	102.30
Foaming Agent	78	26.64	
Totals	2840	21.64	55.00

1.6 *Batching, Mixing and Pumping*

The cellular concrete/contact grout batch plant consisted of two material storage silos, two mixing augers, an electrical generator, a high shear mixer, two foam concentrate storage tanks, a foam generator and a positive displacement cellular concrete delivery pump. The cement and flyash were delivered to the project site in tanker trucks. The materials were transferred from the tankers to the silos. There was one silo for cement and one silo for flyash. The cement and flyash were metered out of the silos using a rotary valve. The measured material discharged into a mixing (blending) screw auger where the correct amount of water was introduced. Each of the materials, cement and flyash, had its own screw auger.

As the cement and water, and the flyash and water moved along the screw auger, which was approximately 6 foot (2 m) long, the water and the materials were partially mixed to form a slurry.

The cement slurry and flyash slurry were discharged from the two screw augers into a high shear colloidal mixer. Figure 8 shows the two mixing screw augers and the colloidal mixer. The mixer discharge was connected directly to a progressing helical cavity (Moyno) delivery pump.

Foam concentrate was delivered to the project site in SS gallon drums. The concentrate was transferred from the drums to holding tanks, which were located in close proximity to the batching operation. From the holding tanks the concentrate was feed into a foam generator where it was mixed with air. The foam was introduced to the cement/flyash slurry at the intake of the delivery (injection) pump.

The delivery pump discharge was connected to the backfill delivery pipes and later to the contact grout delivery pipe located at the tunnel portals via a flexible hose. Figure 9 shows the batching, mixing and pumping set up.

Figure 8. Two Mixing Screw Augers and Colloidal Mixer.

1.7 *Field Testing*

The unit weight of the cement slurry, flyash slurry, cement/flyash slurry and the cellular concrete were tested every 30 minutes using a "mud balance". The unit weight of the foam was tested every 30 minutes using a balance scale to weigh a standard concrete unit weight test pot which had been filled with foam. Approximately every 4 hours four 3 inch by 6 inch (75 mm by 150 mm) cellular concrete cylinders were cast in a Styrofoam mold for compressive strength testing at 7 and 28 days.

The average 7 day compressive strength test results for the cellular concrete backfill was 226 psi (1.6 kPa). The average 28 day compressive strength test results was 519 psi (3.7 kPa). The average dry unit weight of the cellular concrete was 58.6 PCF (0.99 g/cm^3).

Figure 9. Cellular Concrete Contact Grout Batch Plant.

1.8 *Conclusion*

It is believed, based on the actual volume of material injected, that the backfilling and contact grouting of the South Mountain Reach 3B Tunnel Project was completed successfully. However, there were no physical means to verify the actual condition of the in situ backfill and contact grout materials or if any voids remained unfilled after completion of contact grouting.

It must be emphasized that the preferred industry wide method of injecting backfill materials and contact grouts into the annulus behind pipes installed in tunnels is through pre-installed grout ports fabricated into the pipe during manufacture. If the tunnel is located at a relatively shallow depth and surface access is available another acceptable backfilling method is to inject the backfill through drop holes from the surface. In this method the backfill material is discharged into a hole(s) drilled from the surface into the tunnel crown. However, utilizing the long length delivery pipe(s) backfilling and contact grouting methods discussed in the paper maybe required in the following backfilling and contact grouting applications:

- When the water or waste water pipe diameter are too small to allow human entry inside the pipe, therefore precluding the use of grout ports.
- When the owner or designer, for whatever reason, will not allow penetrations in the pipe.
- When backfilling tunnels containing high pressure gas lines, electrical cables/ conduits/ duct banks, petroleum lines, etc.
- When the tunnel is relatively short, for example a road crossing tunnel, to warrant the fabrication costs of installing grout port into the pipe.
- Backfilling of large voids, caverns, old mine workings and other underground spaces where there is no large diameter carrier pipe.

The main purpose of the paper is to give the reader some idea of the methods, pumping distances, delivery pipe layouts which have been used successfully on the South Mountain Reach 3B Tunnel Project to inject cellular concrete backfill and contact grout relatively long distances within the tunnel annulus.

Session 4, Track 1

Tunnel case histories

North American Tunneling 2002, Ozdemir (ed.)
 ISBN 90 5809 376 X

Integrated Management System for tunneling projects

A. Moergeli
moergeli + moergeli consulting engineering, CH-8716 Schmerikon, SG, Switzerland (http://www.moergeli.com)

ABSTRACT: An Integrated Management System (IMS), based on ISO 9001:2000, including occupational safety and health, risk, environmental and project specific management helps you to achieve the owner's targets within expected time, quality and costs effectively and efficiently.

1 HOW DOES A MANAGEMENT SYSTEM HELP YOU?

A management system (MS) must increase your productivity – from your customer's perspective as well – or it is useless ....

Any MS is a tool to back up your leadership's responsibility. The MS provides you with a minimum set of principles, rules and guidelines. It is always based on your clear and unambiguous commitment to reach the outlined targets.

A management system establishes for you and your customer

- a transparent and comprehensible project order sequence
- an unequivocal organization of interfaces and information flows
- a common language.

It is not a coincidence that on bids for complex projects the professional owner often requests a proof of application of a quality management system (QMS).

The international norms ISO 9000 are today worldwide the best known and accepted standards in all questions of quality management (QM).

Up to the end of December 2000, at least 400'000 ISO 9000 certificates had been awarded in more than 150 countries worldwide; an increase of more than 60'000 certificates over the end of 1999 (ISO Survey, tenth cycle).

The international norm ISO 9001:2000 "promotes the adoption of a process approach when developing, implementing and improving the effectiveness of a quality management system, to enhance customer satisfaction by meeting customer's requirements." "..." "An activity using resources, and managed in order to enable the transformation of inputs into outputs, can be considered as a process." "..." "The application of a system within an organization, together with the identification and interactions of these processes, and their management, can be referred to as the "process approach" (ISO 9001:2000, 0.2 Process approach).

Your quality management system ensures that all your core processes are

- directed (managed)
- controlled
- transparent
- comprehensible
- effective and efficient
- documented.

Systematic documentation is a very important factor in the equation of success. However, in everyday operations it is always your own best execution that turns out to be the crucial factor. As we all

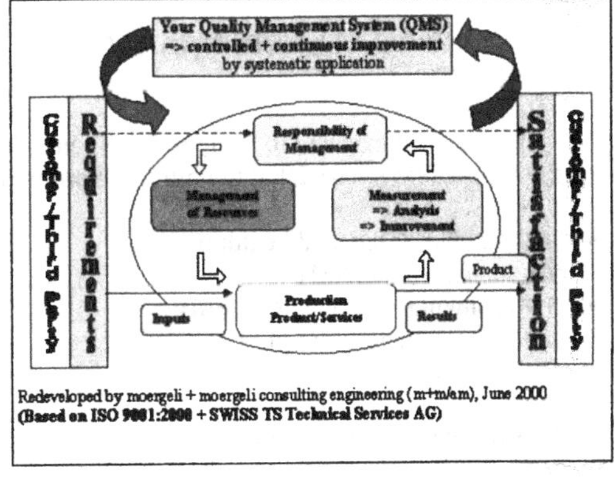

Figure 1. A possible QMS model.

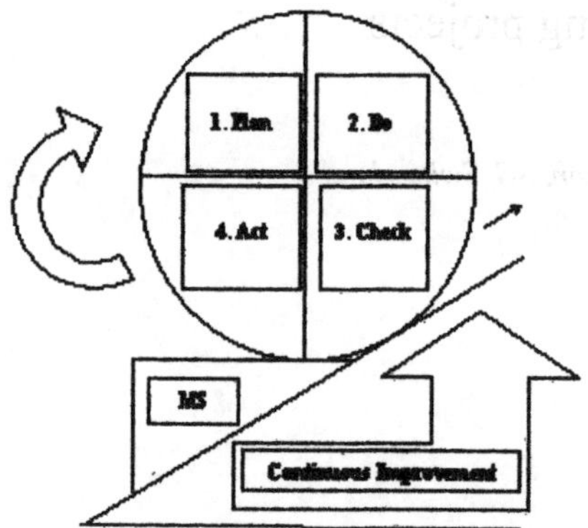

Figure 2. Continuous improvement process.

know from our own experience, even the best documentation will never compensate for our effective performance in real life. But without an effective documentation even your best performance will never be good enough.

As a rule of thumb, documentation contributes with about 30% to success, but 70% of a QMS are daily operations and continuous control of contradictions between the two.

"In addition, the methodology known as "Plan-Do-Check-Act" (PDCA) can be applied to all processes. PDCA can be briefly described as follows.

Plan: establish the objectives and processes necessary to deliver results in accordance with customer requirements and the organization's policies.

Do: implement the processes.

Check: monitor and measure processes and product against policies, objectives and requirements for the product and report the results.

Act: take actions to continually improve process performance" (ISO 9001:2000, 0.2 Process approach).

Please make sure that planning and controlling have nothing to do with checking and in no case are used as synonyms.

2 WHAT IS AN INTEGRATED MANAGEMENT SYSTEM?

An Integrated Management System (IMS) enables on the basis of ISO 9001:2000 the full integration of

- occupational safety and health (OSH)
- risk management (RM)
- environmental management (EM)
- and other systems on request
  into your already existing MS.

An IMS therefore helps you to

- increase your efficiency
- control your daily business effectively
- improve your processes continuously versus Business Excellence (BE), Total Quality Management (TQM).

3 WHY DO WE USE AN INTEGRATED MANAGEMENT SYSTEM IN TUNNELING?

Any underground construction is considered to be a high risk environment. This is for several good reasons:

- rock/ground always remains unpredictable
- unforseen water in big quantities can always be a big, crucial factor anywhere anytime
- available space is very limited
- heavy weight, high energy transport activities
- darkness presumes, lights are rare
- high construction noise
- high temperatures, high moisture
- dealing with explosives, high voltage
- fresh air is very limited, etc.

In addition any underground activity normally brings with it a:

- high public profile
- high capital investments
- work schedules around the clock.

All these puzzles bring about owners requests for at least some sort of proven quality management (quality assurance/control or whatever the preferred terminology will be).

Usually your various project partners are controlling the quality in their own sphere of venture – with or without a MS (qualified by ISO 9001).

So far so good. But how are you covering the overlapping areas between the different enterprises? How do you ensure the achievement of your targets within time- and budget-frame? How can you guarantee standard guidelines and a common language?

There is only one answer for you. You build up a project specific management (PQM):

- define processes
- state responsibilities
- appoint resources.

The document specifying these interactions is normally referred to as a (project-specific quality) management plan (PQM-Plan).

To activate your PQM-Plan you have to:

- train all concerned
- implement the system
- apply the system in your workday
- maintain it periodically as needed.

A PQM is the only available tool to provide you with the utmost control of all interfaces between the different involved parties, companies and persons.

You don't have to reinvent the wheel. The Swiss Society of Engineers and Architects (sia) provides you with Instruction-Sheet 2007 „Quality Management in Constructions" (Version 1997; currently under revision and shortly reissued in German as well as in English) making the fundamental theories available to you.

Different large projects (in Switzerland e.g. AlpTransit, Rail 2000 etc) are some of the well-known, successful applications.

4 HOW DO YOU PROFIT FROM AN INTEGRATED MANAGEMENT SYSTEM IN TUNNELING?

Integrate occupational safety and health into your management system:

1. start with a clear, unambiguous commitment
2. define responsibilities (management and crew)
3. define/integrate the documentation into your MS
4. analyze the risks of your business/tasks (e.g. by the Swiss suva method)
5. set up an OSH-Plan including all legal and statutory requirements
6. prepare action plans for a) emergencies, b) special operations, c) maintenance and d) your normal operations
7. train your crew
8. implement and deploy your plans
9. check and assess your system's effectiveness by Internal Audits (measure => analyze => improve)
10. review and improve your system periodically.

Experience shows that measuring, analyzing and improving the system are for most of us the most difficult, but valuable tasks. They turn out to be the source of real added value if performed systematically.

Failures often result due to an apparent lack of quantitative goals (visions/commitments broken down in figures). Without clearly defined, "hands-on" process targets at the very beginning, you will never end up with any measurable benchmarks at the end of the day. And to end up with measurable results, you have to define your measuring tools first.

Once you have mastered these tasks successfully, there are a lot of directly measurable profits such as:

- less risks
- any remaining risks transparently controlled
- higher safety and security
- less accidents
- less " Monday Blues"
- higher staff's satisfaction
- a smaller fluctuation rate
- less Total Cost of Ownership (TCO)
- less construction costs
- faster construction process
- less environmental impact
- higher owner's satisfaction
- a better public image
- reduction of your (product/third party) liability.

5 WHERE HAS AN INTEGRATED MANAGEMENT SYSTEM IN TUNNELING ALREADY BEEN USED?

An random selection of several (larger) tunneling projects in Switzerland (CH) in which the author is currently involved, may illustrate the benefits of an IMS approach for controlling the construction processes and their inherent risks.

5.1 *Alptransit, Gotthard Base Tunnel/Switzerland*

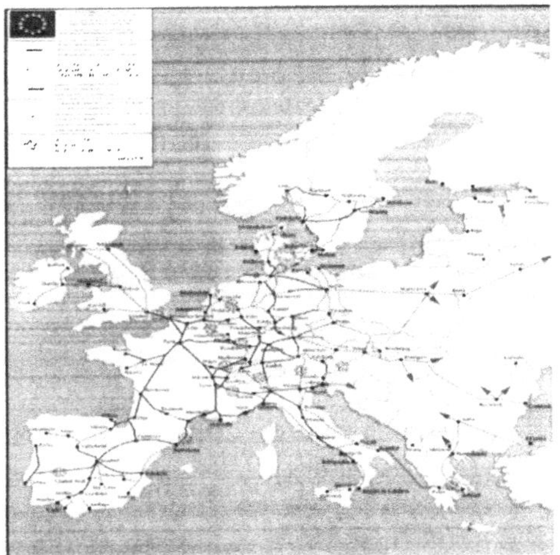

(Picture courtesy of AIB)

Figure 3. European Highspeed Railway Links.

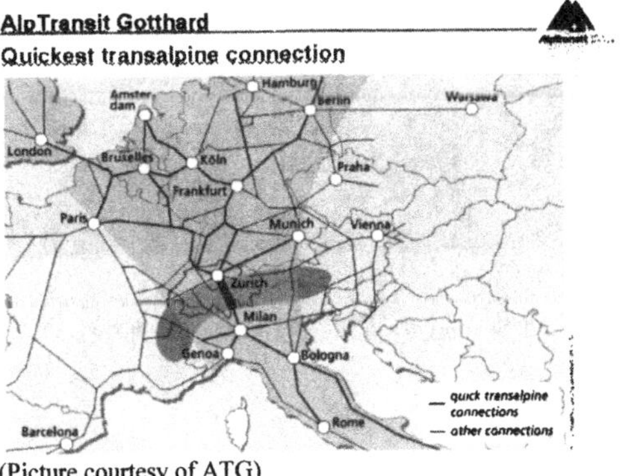

(Picture courtesy of ATG)

Figure 4. Fastest European transalpine lines cross Switzerland.

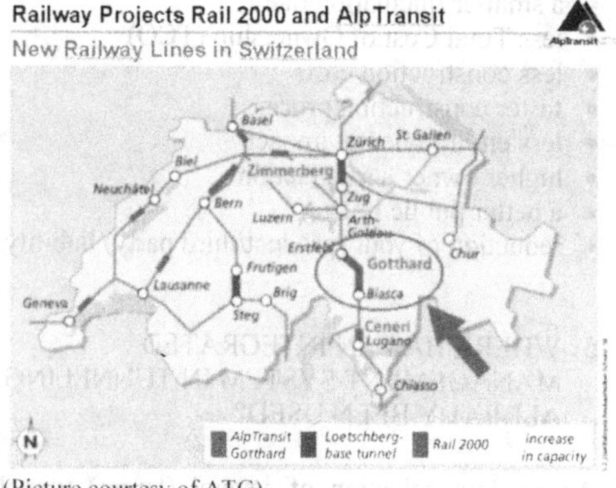

(Picture courtesy of ATG)

Figure 5. New railway lines in Switzerland.

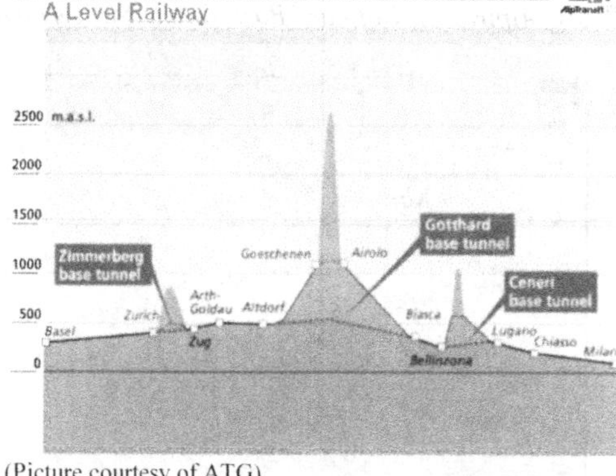

(Picture courtesy of ATG)

Figure 6. Gotthard Base Tunnel: A level railway system.

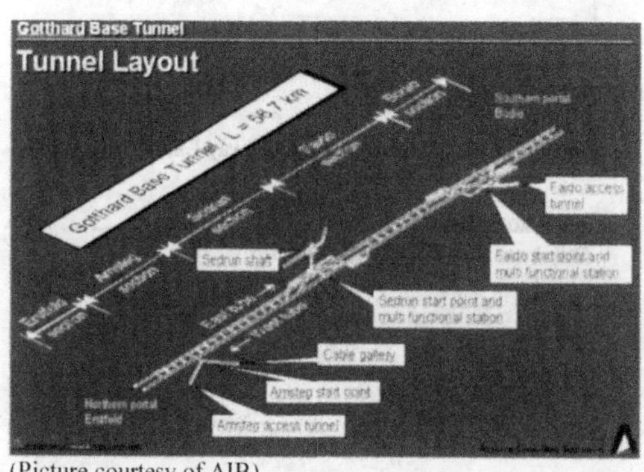

(Picture courtesy of AIB)

Figure 7. Gotthard Base Tunnel: Overview tunnel layout.

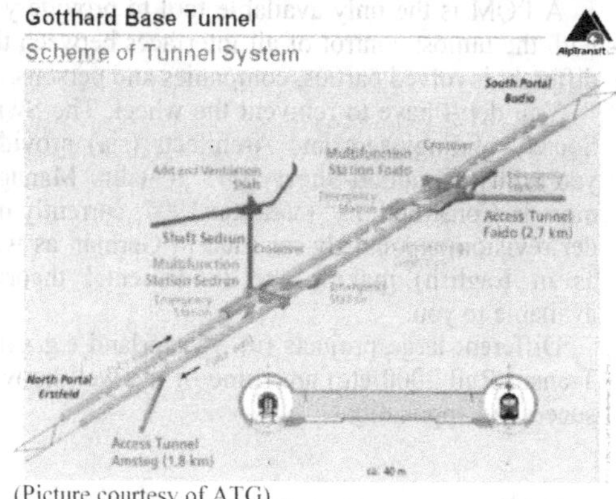

(Picture courtesy of ATG)

Figure 8. Gotthard Base Tunnel: Tunnel scheme.

Gotthard Base Tunnel Safety Concept
Emergency Station

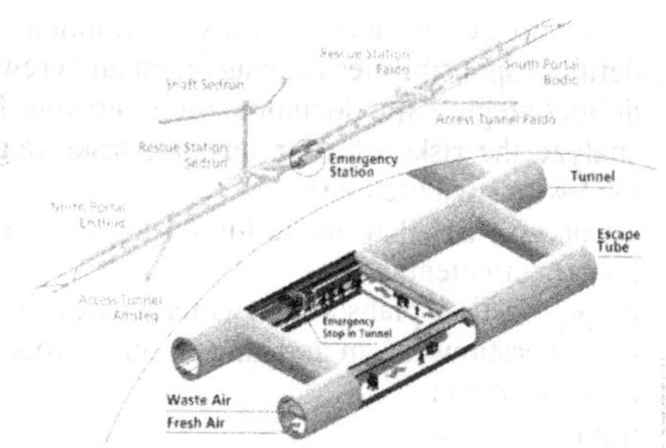

(Picture courtesy of ATG)

Figure 9. Gotthard Base Tunnel: Emergency exit system.

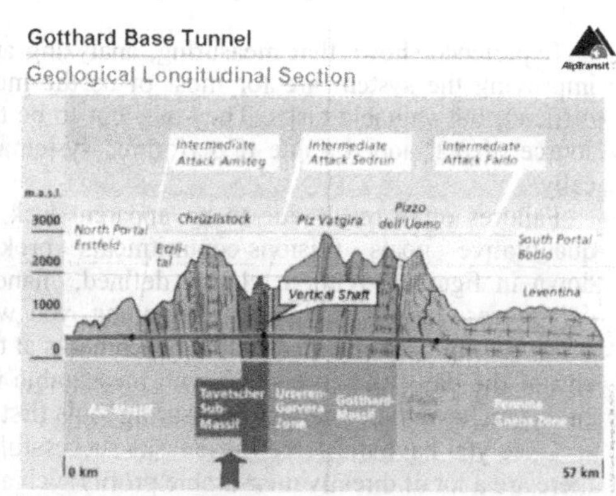

(Picture courtesy of ATG)

Figure 10. Gotthard Base Tunnel: Geological profile.

Standard Cross Section

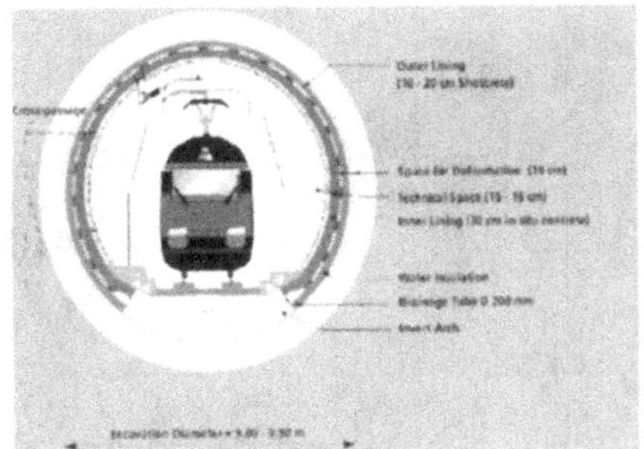

(Picture courtesy of ATG)

Figure 11. Gotthard Base Tunnel: Standard cross section.

5.2 *Gotthard Base Tunnel, lot 360, Tunnel Sedrun*

Table 12. Selected project data Lot 360, Tunnel Sedrun.

Owner	AlpTransit Gotthard AG
Project	Gotthard Base Tunnel
Location	Sedrun, Canton of Grisons/ Switzerland
Lot	360, Tunnel Sedrun
Designer & Client Representative	IG GBTS
Contractor	...?*
Tunnel length	Ca 14'000 m
Cross sections	Ca 60 – 140 m^2
Construction method	Drill & blast
Tunnel construction costs	Not (yet) disclosed*
Construction time frame	06/2002 – Ca 06/2012
Project status (11/2001)	Award expected for 12/2001*
Author's mandate	Assistance for one bidding joint venture in negotiations with the owner

* An update on the current project status can be offered through a presentation at the AUA Conference.

For more information please log on to the owner's website http://www.alptransit.ch – thank you.

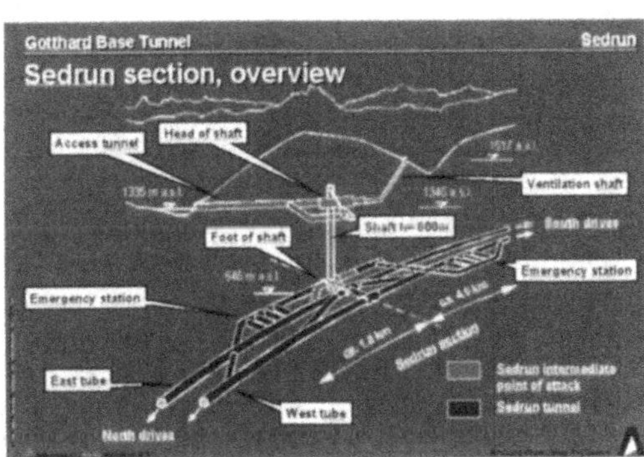

(Picture courtesy of AIB)

Figure 13. Gotthard Base Tunnel: Sedrun Overview.

(Picture courtesy of MBT)

Figure 14. Sedrun: Underground installation for shaft sinking.

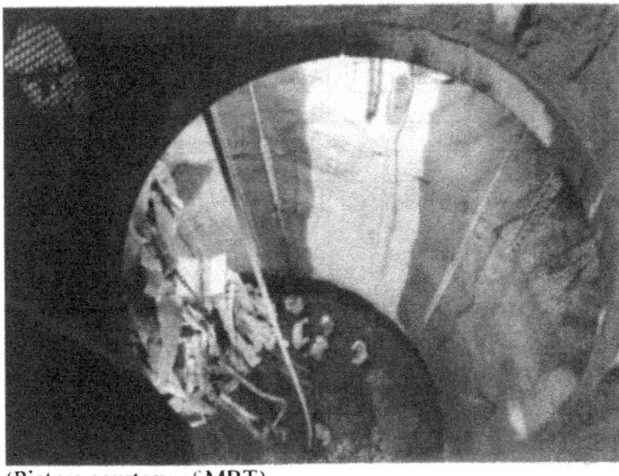

(Picture courtesy of MBT)

Figure 15. Shaft sinking Sedrun.

(Picture courtesy of MBT)

Figure 16. Sedrun: Shaft bottom.

(Picture courtesy of MBT)

Figure 17. Sedrun: Shaft bottom.

5.3 *Gotthard Base Tunnel, lot 251, Amsteg Access*

Table 18. Selected project data Amsteg Access.

Owner	AlpTransit Gotthard AG
Project	Gotthard Base Tunnel
Location	Amsteg, Canton Uri/Switzerland
Lot	251, Amsteg Access
Designer & Client Representative	ING GBTN
Contractor	ARGE Zugangsstollen Amsteg (AZA)
Tunnel length	Ca 1'800 m
Cross section	Ca 60 m^2
Construction method	Drill & blast
Tunnel construction costs	Ca CHF 25 Mio*
Construction time frame	10/1999 – 06/2001
Project status (11/2001)	Finished**
Author's mandate	Contractor's support in PQM, Internal Auditing, OSH, Price Increase Calculation with Object Index-Procedure (OIV) by sia 118/121

* USD = Ca CHF 0.60 (11/2001)
** For more information please log on to the owner's website http://www.alptransit.ch – thank you.

(Picture courtesy of AZA)

Figure 19. Amsteg Access: Boring with computer assisted jumbo.

(Picture courtesy of AZA)

Figure 20. Amsteg Access: Boring at the face.

(Picture courtesy of AZA)

Figure 21. Amsteg Access: Charging liquid explosives.

(Picture courtesy of AZA)

Figure 22. Amsteg Access: Charging the face.

5.4 *Engelberg Tunnel*

Table 23. Selected project data Engelberg Tunnel.

Owner	Luzern Stans Engelbergbahn LSE
Project	Steilrampe Tunnel Engelberg
Location	Grafenort (near Luzern)/ Switzerland
Lot	North + South
Designer & Client Representative	Bucher + Dillier Ingenieu-runternehmung AG + IG LSE
Contractor	ARGE Tunnel Engelberg (ATE)
Tunnel length	Ca 4'040 m
Cross section	Ca 30 m²
Construction method	Drill & blast
Tunnel construction costs	Ca CHF 42 Mio*
Construction time frame	05/2001 – Ca 07/2004
Project status (11/2001)	Under Construction**
Author's mandate	Contractor's support for OSH by suva's Integral Safety Plan

* USD = Ca CHF 0.60 (11/2001)

** An update on the current project status can be offered through a presentation at the AUA Conference.

For more information please log on to the contractor's website http://www.ast-holzmann.at – thank you.

(Picture courtesy of ATE)

Figure 24. Engelberg Tunnel: Site installation at south portal.

(Picture courtesy of ATE)

Figure 25. Engelberg Tunnel: Acces bridge to tunnel south portal area.

(Picture courtesy of ATE)

Figure 26. Engelberg Tunnel: South Portal.

(Picture courtesy of ATE)

Figure 27. Engelberg Tunnel: North tunnel.

6 CONCLUSIONS

6.1 *What is quality management in operation?*

The well known ISO 9001:2000 defines the world-wide accepted basis of quality management.

Quality management stands for continual improvement through controlled processes.

A complete documentation trail is a fundamental key to success. However in everyday's practice the crucial factor to an effective and efficient improvement of an organization's performance is full, unambiguous commitment, dedication and the executive's ability to live up to the standards of a quality ma-nagement system (QMS). It consists of about 30% documentation, 70% operation and continuous control of contradictions between the two.

6.2 *What is an Integrated Management System?*

An Integrated Management System (IMS) controls the latest state of the art for

- all relevant core processes of any involved and interested parties by ISO 9001:2000
- all relevant statutory and regulatory requirements as well as technical standards
- occupational safety and health
- a full risk management
- an environmental management
- all interfaces between all parties, including the owner by a project-specific MS (PQM)

6.3 *Why do we use an Integrated Management System in tunneling?*

The owner often requests a highly developed QMS by his contractors. However, there is plain evidence that it is in the very interest of all parties involved, especially contractors, to control their production processes by an IMS (or at least integrate their own MS into a PQM).

6.4 *What are your profits from Integrated Management Systems in tunneling?*

Measurable profits are:

- less risks
- any remaining risks transparently controlled
- higher safety and security
- less accidents
- less “Monday Blues”
- higher staff’s satisfaction
- a smaller fluctuation rate
- less Total Cost of Ownership (TCO)
- less construction costs
- faster construction process
- less environmental impact
- higher owner’s satisfaction
- a better public image
- reduce your (product/third party) liability.

7 ACKNOWLEDGMENTS

Figure 28. AST-Holzmann logo.

Figure 29. AlpTransit Gotthard AG (ATG) logo.

Figure 30. MBT logo.

The author thanks and acknowledges the very competent support and kind provision of plans, schemes and pictures by:

- AST-Holzmann Baugesellschaft m. b. H. (http://www.ast-holzmann.at), Mr H. Zelenka, Mr R. Steinscherer, Mr F. Kapfinger and many others and their efficient crews on site
- AlpTransit (ATG), Mr K. Aerni, Mr P. Unterschuetz with their team
- Client Engineer & Representative IG GBTS
- MBT International Underground Construction Group.

Without their big help this document would not have been possible.

The biggest thanks go to all crews on site safely coping with the unforeseeable as their daily routine. Every day they move into places where no human being has ever been before. Always just one small step for a man, but a giant leap for mankind ...

The author’s apologies go to the readers for any inconvenience dealing with small print, reduced tables and pictures.

The original paper will be available for download on http://www.moergeli.com/archive.htm after the AUA Conference 2002, May 18–22, The Westin, Seattle, WA/USA.

Table 31. Abbreviations.

#	Abbr	Definition/Explanation
1.	#	Number, No
2.	$	US dollar, $
3.	≈	about, circa
4.	(...)	option or issue date
5.	"..."	Quotation (within quotation marks)
6.	/	or
7.	[...]	dimension and/or directions/instructions
8.	+	and
9.	=>	[goal/target]
10.	abbr	abbreviation
11.	am	Alfred MOERGELI
12.	ARGE	joint venture (German: Arbeitsgemeinschaft)
13.	ATE	Joint venture Tunnel Engelberg (German: ARGE Tunnel Engelberg): Achermann AG - AST)
14.	AUA	American Underground-Construction Association
15.	AZA	Joint venture Amsteg Access (German: ARGE Zugangsstollen Amsteg): Wüest AG - AST)
16.	BE	Business Excellence (TQM)
17.	ca	circa, about
18.	CH	Switzerland
19.	CHF	Swiss Franc, always without VAT if no other mention
20.	CL	checklist
21.	D & B	Drill & blast
22.	Doc(u)	document
23.	e.g.	for example (Latin: example gratia)
24.	EM	Environmental management
25.	engl	English
26.	GBTN/ GBTS	Gotthard Base Tunnel North/South (German: Gotthard Basis Tunnel Nord/Süd)
27.	M	Management (leadership)
28.	m(+)m	moergeli + moergeli consulting engineering
29.	Mio	Million
30.	MS	Management system
31.	OIV	Object Index-Procedure (German: Objekt-Index-Verfahren)
32.	OSH	Occupational Safety and Health
33.	P(Q)M	Project-specific (Quality) Management
34.	PQM	project-specific Quality Management
35.	(P)Q(M) Plan	(project-specific) Quality (Management) Plan
36.	Q-	Quality-...
37.	QM	Quality Management
38.	QMDoc	Quality Management Documentation
39.	QME	Quality Management Executive
40.	QMS	Quality Management System
41.	RM	Risk management
42.	S	staff
43.	sia	Swiss society of engineers and architects (German: Schweizerischer Ingenieur- und Architekten-Verein)
44.	suva	Swiss National Accident Insurance Fund
45.	TQM	Total Quality Management (BE)
46.	USD	US dollar, $

8 ABBREVIATIONS

All abbreviations/terms can be used in singular or plural, with or without capital letters and with or without a period when abbreviated. All abbreviations/terms are explained in the text upon their first use.

REFERENCES

AIB, Amberg Ingenieurbuero/Amberg Consulting Engineers Ltd., Trockenloostr. 21, CH-8105 Regensdorf-Watt/Switzerland (http://www.amberg.ch).

AlpTransit Gotthard AG (ATG), Zentralstrasse 5, CH-6003 Luzern (http://www.alptransit.ch).

ARGE Tunnel Engelberg (ATE), CH-6388 Grafenort, (c/o AST).

ARGE Zugangsstollen Amsteg (AZA), Grund 51, CH-6474 Amsteg, (c/o AST).

AST-Holzmann Baugesellschaft m. b. H., Eduard-Ast-Str. 1, A-8073 Feldkirchen/Graz (http://www.ast-holzmann.at).

ISO, ISO 9001:2000 (http://www.iso.ch).

ISO, The ISO Survey of ISO 9000 and ISO 14000 Certificates, Tenth cycle: up to and including 31 December 2000 (http://www.iso.ch).

MBT, MBT International Underground Construction Group, Vulkanstrasse 110, CH-8048 Zuerich/Switzerland (http://www.ugc.mbt.com).

North American Tunneling 2002, Ozdemir (ed.)
© 2002 Swets & Zeitlinger, Lisse, ISBN 90 5809 376 X

Mitigating tunneling delay claims via CPM scheduling techniques

J.L. Ottesen, PE, Senior Associate
The Nielsen-Wurster Group, Inc., Seattle, Washington, USA

ABSTRACT: Owners and Contractors agree that tunneling is risky business. Still, Owners and Contractors contractually bind themselves to completing tunneling projects within predetermined limits of both cost and time, perhaps the two most uncertain variables related to tunneling work. Claims often result if planned costs or times are breached. This paper focuses on mitigating the unknown time variable in tunneling projects by looking at causes for time-related delays as well as delay mitigation actions taken on several tunneling construction projects. These factors formulate the basis for CPM scheduling recommendations intended to more accurately estimate the amount of time required to successfully complete tunneling work, and subsequently, reduce the filing of delay-related claims, including delay claims related to differing site conditions.

Owners and Contractors agree that tunneling is risky business. Tunneling is risky because of cost and time factors related to removing building material that cannot be chosen in advance, but is revealed continuously by advancing excavation (Kovari, May 26, 1997). Despite exploratory methods to alleviate subsurface uncertainty, such methods disclose only a fraction of the material to be excavated (Bickel, 1996) and thus, actual materials excavated proceed largely unknown. Still, Owners and Contractors contractually bind themselves to completing tunneling projects within predetermined limits of both cost and time, perhaps the two most uncertain variables related to tunneling work. Claims often result if planned costs or times are breached. This dichotomy has been understood for decades, consequently, risk management techniques emerged and are continuously being updated to minimize the likelihood of claims.

This paper focuses on mitigating the unknown time variable in tunneling projects by looking at causes for time-related delays experienced as well as successfully executed delay mitigation actions taken on several tunneling construction projects. These factors formulate the basis for CPM scheduling recommendations intended to more accurately estimate the amount of time required to successfully complete tunneling work, and subsequently, reduce the filing of delay-related claims. This paper is presented in four sections. First, a basic overview of differing site condition claims is presented. This type of claim is the most common between an Owner and Contractor when excavation or tunneling is involved. Next, factors that impact planned tunneling productivity rates are presented, followed by delay mitigating factors that were identified as effective means to completing tunneling projects on time. Information in these two sections was taken from actual tunneling completion. Information in these two sections was taken from actual tunneling construction projects. Finally, recommendations related to CPM scheduling techniques are presented. These recommendations intend to more accurately depict the possible delay causing factors that are likely to occur in the schedule, thereby reducing the likelihood of late completion and filing of delay-related claims.

1 DIFFERING SITE CONDITION CLAIMS

The age-old cliché "Time is money" applies to claims filed on tunneling projects. If a Contractor fails to complete work on time, it incurs additional costs. When profitability is breached, claims often follow. An encountered differing site condition can significantly prolong expected time to complete work, and therefore, the cost of the project. Because of this cost impact, differing site conditions is considered the most frequently occurring Owner - Contractor dispute in the construction industry (Richter, 1982).

According to federal government contracts related to differing site conditions claims, (Federal Acquisitions Regulations) a Contractor may be entitled to an equitable adjustment and/or contract modification if the Contractor encountered either a Type I or Type II condition (Kirsch), (Jensen). Likewise, several conditions must be proven under each claim type before equitable adjustment is made. Although defined in a federal government document, similar proof is often applied for private industry as well:

Type I: Subsurface or latent physical condition(s) at the site, which differ materially from those indicated in the present contract.

Type I Proofs:

- Subsurface conditions are represented in the contract
- Contractor reasonably interpreted the subsurface representations in the contract
- Contractor reasonably relied upon the subsurface representations in the contract
- Condition encountered differed materially from the representations in the contract
- The encountered condition was unforeseeable
- Added costs to the Contractor are due solely to the encountered differing condition

Type II: Unknown physical condition(s) of an unusual nature at the site, which differ materially from those ordinarily encountered and generally recognized as inherent in the character of work performed under the present contract.

Type II Proofs:

- Subsurface condition was unknown
- Subsurface condition was unusual and could not be reasonably anticipated based on review of the contract documents or site inspection
- Condition encountered differed materially from those ordinarily encountered and generally expected for the type of work performed

For either type, the underlying proof relies upon availability, use of and accuracy of information prior to excavating.

Type I cases occur more frequently than Type II. For Type I cases, research of 101 cases found that the majority of disputes occurred during the bidding phase and that the two elements of proof most frequently occurring included whether the Contractor acted in a responsibly prudent manner when interpreting the contract and whether the contract contained indications of material to be encountered (Jensen). These arguments look back to the early stages of the project. There can be many other factors, however, that occur during construction that may also contribute to delays that are unrelated to a differing site condition, but perhaps are wrapped into such claims.

Emphasizing the timing of differing site condition disputes, it is typically during the bidding phase that the project's planned duration is set and the Contractor's proposed baseline schedule is presented. Following award, the proposed baseline schedule may become the official baseline schedule by which contractual deadlines are established and by which liquidated damages are measured. However, it is not until the differing conditions are encountered, usually months later, that the contractor realizes that the originally planned durations and budgets cannot be met the field. Herein begins the Contractor's battle of proving that it is entitled to equitable adjustment according to the appropriate proofs required.

2 OTHER DELAY CAUSATION FACTORS ON TUNNELING PROJECTS

Although differing site condition disputes are the most common between an Owner and Contractor, a review of several different tunneling projects from around the world revealed that many other factors contribute to the breach of planned tunneling durations and subsequent cost overruns. Many of these factors can be identified during the bid phase and should be considered when the Contractor submits its proposed baseline schedule and budget. A summary of these factors is presented below, with brief explanations of projects that experienced these factors.

- Insufficient Data
- Construction Means and Methods
  1. Drill & Blast (over-excavation due to inadequate blasting methods) vs. Tunnel Boring Machine
  2. Removal of Spoils (resources, equipment)
  3. Lining type (mix design, application methods)
  4. Peripheral treatments (slope erosion protection at portals, adits)
  5. Rework due to heaving, water intrusion, other post-tunneling conditions
- Financing Difficulties
- Communication between Design and Construction Teams
- Noise Restrictions (Sodra Lanken Tunnel Construction, Stockholm, Sweden)
- Safety & Material Deformation Behavior

From these factors, a list of CPM scheduling recommendations is made relative to establishing scheduling tunneling work.

2.1 *Provo Canyon, Utah*

Beginning in the 1980's the Utah Department of Transportation (UDOT) began looking at alternative alignments for future plans to upgrade US Highway 189 through Provo Canyon from a two-lane highway to a four-lane, divided highway. One alternative presented in the supplemental to the Environmental Impact Statement included a twin tunnel section in lieu of a bridge crossing over the Provo River. Because of ongoing legal disputes and threats by environmentalist groups, UDOT was restricted from taking equipment into environmentally sensitive areas to take core samples at the required intervals for tunnel construction and the large cut sections leading to the portals. Facing possible loss of approved federal funding for further delay, UDOT found itself in a 'use it or lose it' situation. During construction (1998), the contractor discovered that material believed to be solid rock actually contained fissures with significant amounts of water intrusion. The contractor's chosen drill and blast method proved to remove more material than planned for both large cut sections. Landslides and other cave-ins occurred resulting in significant time and cost overruns. The contractor's claim was settled in mediation.

2.1.1 *Scheduling Recommendations (as applicable)*
Tunneling excavation activities should include predecessor activities that:
- Show availability of funds;
- Give the contractor authorization to proceed;
- Provide completed subsurface geotech reports as defined by contract and standard engineering practices

2.2 *Casecnan, Philippines*

The construction of this hydrodam and the tunnels related thereto experienced more than 13 months delay. Tunneling delays were attributable to several factors:
- The contractor's chosen methodology for assembling its Tunnel Boring Machine (TBM) was different than originally planned and took more time to assemble;
- Unavailability of spare parts for the TBM impeded progress;
- Low production rates in tunneling occurred due to insufficient manpower, too few trucks hauling spoils and too little equipment;
- Unqualified laborers improperly installed grouted rock bolts thereby requiring rework;
- Out-of-spec shotcrete was applied, performed poorly and consequently required rework;
- Failure to instigate a slope stabilization program at adits during rainy months caused erosion and landslides, which impeded movement of equipment at the tunnel's entrance

Any alleged differing site conditions for the delays in lieu of the many other delaying factors quickly dismissed such a claim.

2.2.1 *Scheduling Recommendations (as applicable)*
Tunneling excavation activity planned durations should consider:
- Planned productivity rates for the means and method of excavation utilized;
- Time required to procure spare parts in event of repair;
- Quantity of labor and equipment resources available, which is tied in part to the financial strength of the contractor and actual conditions surrounding the project location;
- Availability of skilled laborers relative to the chosen means of excavation; and
- Factors external to actual excavation that could impact actual durations (such as slope protection at tunnel entrance points)

2.3 *City Link Project, Australia (article from web site)*

Opening of this 3.4 km Burnley tunnel was scheduled for September 2000, but was delayed into mid-2001. Allegations that the contractor failed to conduct a thorough geological investigation of ground conditions under the Yarra River cast doubt on the contractor's claim of a differing site condition. It was suspected that the contractor sought cost-saving measures, which caused insufficient subsurface investigation prior to the contractor starting construction.

2.3.1 *Schedule Recommendation*
Include predecessor activities that provide completed geotech investigation reports prior to start of tunneling.

3 DELAY MITIGATING FACTORS

Research of other tunneling projects provided insights to other recommended schedule modifications.

3.1 *Washington Area Metropolitan Subway System, Washington, DC (American underground-construction association's featured project)*

Touted as the "Most Distinguished United States Underground Project" (from 1975 to 2000), the subway (underground) portion of the system is 51 miles (82 km) long with 47 stations. Success of the project stemmed from several factors, some of which included:
- Full disclosure of geotechnical information through geotechnical data and design reports;
- A quantified approach to predicting settlement was included for all types of underground conditions encountered (from hard rock to soft water-bearing material);
- Challenges posed by the geology were correctly defined up-front on a segment-by-segment basis, and solutions were implemented incrementally; and
- All tunneling designs underwent a peer review process for quality and safety purposes before being constructed

3.1.1 *Schedule Recommendation*
Include an incremental (segmental) approach of planned tunneling construction into the CPM schedule. Standard practice indicates maximum length of a construction activity to be about 15 work days in duration.

3.2 *North Cap Sub-Sea Road Tunnel, Norway (Advantages of permanent wet mix sprayed concrete tunnel linings)*

This tunnel is 6.8 km long and was excavated by drill and blast techniques. Poor sandstone and shale was encountered with approximately 20% tunneling remaining. At this point the project was already eight months behind schedule. In order to increase production of the concrete lining, the contractor changed its concrete pouring methodology to a wet mix sprayed concrete approach, which more than doubled its productivity. Use of a sprayed concrete lining methodology requires a competent, skilled workforce (Dimmock).

3.2.1 *Schedule Recommendation*
Recognizing differences in cost of finished product to the useable life of the facility is another factor to consider when establishing planned tunneling durations.

3.3 *Swiss Jura Highway Tunnel T8, Switzerland (Field instrumentation in tunnelling [sic] as a practical design aid)*

Research at the Federal Institute of Technology, Zurich Switzerland, focused on measuring material behavior in tunneling projects with aim "*to obtain adequate safety for a minimum of cost expenditure, whereby the manifold influence of the construction time is also included in the costs.*" Using various measuring devices, the deformation properties of an excavated area can be monitored to determine the material's reactive characteristics to various excavation methods. Tunnel T8 experienced rock swelling at the bottom of the tunnel. Swelling is largely due to water adsorption of clayey materials and impairs the serviceability and stability of underground structures (Anagnostou, 1995). Remedies to account for the swelling were presented, each of which lead to design and construction enhancements.

3.3.1 *Schedule Recommendation*
Because swelling or other deformation may occur in some portions of a tunneled section, but not in others, include an activity related to monitoring movement of previously excavated sections in order to adjust planned future tunneling work. Whereas in tunneling projects excavation work is typically on the critical path, (Hafer, May 15, 2000) allowing extensions to tunneling durations without affecting the planned completion date requires existence of sufficient float or an appropriate extension of time. Building this type of contingency into the schedule is perhaps best done during the bid stage, rather than during construction when so many other factors can undermine a differing site condition.

4 SCHEDULING FOR UNFORESEEABLE CONDITIONS

Scheduling for unforeseeable conditions can be difficult. Moreover, Contractors generally show reluctance to include such unknowns into its schedule, particularly in a competitive bidding process. When competing for a tunneling project and depending on the type of contract followed, the Contractor faces the dilemma of selling itself to the Owner on its capabilities to provide the best solution for an acceptable (often lowest) price. Whereas total time spent on site is directly related to the total project cost, providing the most optimistic schedule (i.e., the schedule with the earliest feasible completion date) theoretically increases the chance of the Contractor winning the job.

Under an earlier section herein, it was presented that for differing site condition claims one of the most argued points was whether the Contractor acted in a responsibly prudent manner when interpreting the contract. By incorporating several of the schedule recommendations into its CPM schedule in event of a differing site condition, a Contractor's position that it did act in a responsibly prudent manner when interpreting the contract is strengthened.

The following 'Before and After' example is made relative to a typical string of activities (assumed to lie on the critical path) taken from a baseline CPM schedule. The 'Before' condition contains three sequential tunneling activities (shown as bars) each with 20 days duration. The tunneling work is divided into equal 100-meter lengths. This level of detail is commonly observed in tunneling baseline CPM schedules. At time of the bid, the Contractor estimates the durations of the planned tunneling activities according to its planned excavation methodology relative to the geotechnical information provided. For example, if a worst-case drill and blast methodology was expected to generate 5 m/day productivity, a 100-meter tunnel section duration would equate to 20 working days planned duration in the CPM schedule. Depending upon the Contractor's experience, an additional portion of time may be added to account for unexpected difficulties. This additional time is often referred to as 'contingency' or 'activity float'. Similar procedure is followed for remaining tunneling activities. Depending upon the contractually required completion date, the Contractor then adjusts the planned durations to 'fit' the completion date, or may compress the schedule to show an early completion date, particularly if a bonus clause rewards early completion. Thus, the proposed baseline schedule is created, wherein the Contractor has effectively accepted a given level of uncertainty (risk) given the time and cost included in its bid (see Figure 1).

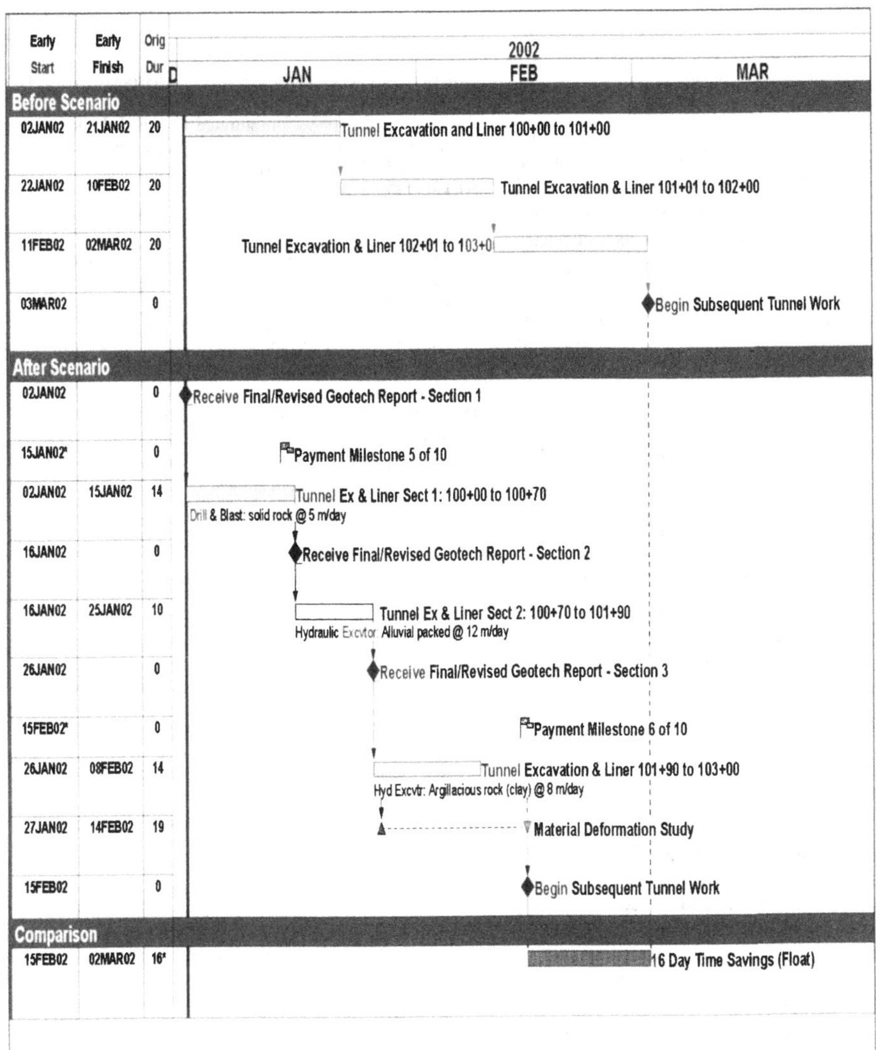

Figure 1.

The 'After' condition included many of scheduling recommendations made earlier. Changes made to the schedule are based upon the undesirable task of proving a differing site condition and explaining time delay, if encountered. Rationale and/or benefits for the following modifications made to the simple baseline schedule are given below:

- Receipt of final revised geotech reports precedes start of each tunneling activity. These milestones call out information needed by the Contractor to meet its planned durations as shown. Whereas each section requires revised geotech information, the most recent information is used with hopes of mitigating any unforeseeable ground conditions for that section. Addition of these milestones also heightens awareness of the responsible party for providing the information.
- The planned tunneling methodology by soil type was used to estimate the daily excavation rates wherein the planned tunneling durations were each generated individually. The methodology, understood soil type and anticipated excavation rates are listed for each section. Additionally, sections are defined by soil conditions, not simply by evenly distributed lengths. The benefit of this change is increased accuracy in establishing planned tunneling durations. In this example, 16 days less time resulted in comparing the 'Before' and 'After' conditions. Less time on site can be beneficial to both the Owner and Contractor. If more time were found, the result may alert the

Contractor of additional risk such that adjustments could be made to the chosen cost and methodology proposed at the bid phase.

- For tunneled sections through rock containing clay, an activity (shown as a dotted line) called "Material Deformation Study" has been included. This study is intended to complete prior to start of subsequent tunneling work with belief that future excavation efforts may be adjusted as necessary according to knowledge acquired from the study. Adding this activity shows that the Contractor is taking appropriate steps with information provided in the contract, coupled with actual material behavior in the field, to reduce uncertainty for future excavation.
- Payment milestones have been added to the schedule to heighten visibility and remind the Owner of necessary cash flow to keep the project moving forward.

Although overly simplified, this example shows that by inserting information from the contract documents into the CPM schedule, proof of the Contractor having acted in a responsibly prudent manner when interpreting the contract is substantially strengthened. Additionally, the revised schedule allows for easier identification of delay causation in event of a delay claim.

5 CONCLUSION

Differing site condition claims frequently occur on tunneling projects. Because proof of such claims falls back on information used and relied upon during the bid phase as contained within the contract documents, adding this information into the baseline CPM schedule strengthens proof of such claims if encountered. A secondary benefit of making these scheduling modifications is in identifying delay responsibility.

REFERENCES

American Underground-Construction Association's Featured Project; http:\\www.auca.org

Anagnostou, G., 1995. Unsaturated flow and deformation pattern in tunneling through swelling rock. Swiss Federal Institute of Technology, Zurich, Switzerland

Bickel, J.O., T.R. Kuesel, & E.H. King, 1996. Tunnel Engineering Handbook (Second Edition), Chapman & Hall, 544 pages

Dimmock, R.H., Practical Solutions for Permanent Sprayed Concrete Tunnel Linings, MBT Underground Construction Group, UK

Federal Acquisitions Regulations, Title 48, Section 52.236-2

Garshol, K.F., T.A. Melbye, Advantages of Permanent Wet Mix Sprayed Concrete Tunnel Linings – Project Experiences, MBT International Underground Construction Group

Hafer, R.F., Dixie Contractor, May 15, 2000, Risk Management and Dispute Avoidance in Underground Construction, (http://www.kilstock.com/site/print/detail/Article_Id=836)

http:\\www.theage.com.au/news/200000527/A20777, article dated May 26, 2000, regarding City Link Project

Jensen, D., Analysis of a Type I Differing Site Condition Claim: An Empirical Study to Determine Which Proof Element is Most Frequently Disputed and Which Party Interest Most Often Prevails, at the ASC Proceedings of the 37th Annual conference, University of Denver, Denver, Colorado, April 4 – 7, 2001, pp 87 – 94

Jensen, D., A Type I Differing Site Conditions Claim: Analysis of the Reasonable Reliance Element, at the ASC Proceedings of the 37th Annual Conference, University of Denver, Denver, Colorado, April 4 – 7, 2001, pp 199-212

Kirsch, J.T., Esq., proofs summarized at Jenkens & Gilchrist law firm, Washington, D.C. office; published at www.constructionrisk.com

Kovari, K., & Amstand, C.H., Field Instrumentation in Tunnelling [sic.] as a Practical Design Aid, Federal Institute of Technology Zurich

Kovari, K., May 26, 1997, Safety against Cave-ins and Rock Fall in Tunneling, Institute of Geotechnical Engineering

Richter, Irv & R.S. Mitchell, 1982, Differing Site Conditions. Handbook of construction low and claims, Reston Publishing Company, Inc., Reston, Virginia, cited by Donald Jensen, A Type I Differing Site Conditions Claim: Analysis of the Reasonable Reliance Element, at the ASC Proceedings of the 37th Annual Conference, University of Denver, Denver, Colorado, April 4 – 7, 2001, pp 199-212

North American Tunneling 2002, Ozdemir (ed.)
© 2002 Swets & Zeitlinger, Lisse, ISBN 90 5809 376 X

Monitoring the performance of earth pressure balance tunneling in Toronto

S.J. Boone & S. McGaghran
Golder Associates, Ltd., Mississauga, Ontario, Canada

G. Bouwer
Hatch Mott MacDonald, Mississauga, Ontario, Canada

T. Leinala
University of Toronto, Toronto, Ontario, Canada

ABSTRACT: Tunneling for a new subway in Toronto was monitored to assess construction performance. Ground conditions included over-consolidated lacustrine silt, sand, and clay deposits below groundwater levels and hard glacial till interspersed with boulders. Two earth pressure balance tunnel boring machines (TBMs) were used for mining. Control of ground movements and their subsequent effect on nearby facilities depended on appropriate machine operation. The TBMs included automated data acquisition systems to monitor operating parameters. Additional data was manually recorded. Precise leveling of over 1,200 settlement points was carried out during tunneling. Data from the TBMs and instruments were simultaneously transmitted to the engineer and contractor. Semi-automated database systems were used by the site staff to plot and interpret the data on a continuous basis. This paper presents example results of the monitoring program with emphasis on the program design and purpose, organization of responsibilities, data management, and lessons learned related to the performance of this tunneling project.

1 INTRODUCTION

The Rapid Transit Expansion Program (RTEP) originally consisted of five major projects including two new subway lines. The Eglinton West Subway was to have been about 4 km long with five stations, one river crossing, and twin bored tunnels. Several hundred structures were within the zone of influence of tunneling. The Eglinton West Subway, which was to be the first of the major projects, was halted early in construction due to budget constraints. The 6.4 km long Sheppard Subway, now near completion, includes five stations, 4 km of twin bored tunnels, tunnelled cross-passages, and a river crossing. More than 80 residential and commercial properties and 160 major utilities were within the tunneling zone of influence. An extension to the existing Spadina Subway line had also been planned, but this work was also postponed due to budget limitations. For technical and management efficiency, the owner retained professional services for surveying, geotechnical instrumentation, construction management and inspection, and geotechnical consultation during construction of all RTEP projects to assist in minimizing construction risk to adjacent facilities and for project quality assurance. Though the organization of engineering and management services was developed for the overall RTEP project, this paper examines the geotechnical monitoring program and tunneling performance for the Sheppard Subway twin tunnels project. The performance of cut-and-cover stations and related excavations is presented by Boone et al. (1998, 1999), Busbridge et al. (1998), and Westland et al. (1999).

1.1 *Ground conditions and tunnel construction*

Construction of the tunnels was carried out through dense and uniform lacustrine sands, over-consolidated lacustrine clay and silt, and hard/dense glacial till. The hard/dense glacial till was encountered near the surface with the older, layered deposits of lacustrine sand, silt and clay below. The layered glacial stratiraphy also created multiple aquifers and aquitards along the alignment. Tunneling conditions varied from a full face of hard and relatively dry glacial till with boulders to a full face of fine uniform sand below groundwater levels.

1.2 *Tunneling machines*

Two 5.9 m diameter Lovat earth pressure balance (EPB) machines were used for tunnel construction and were fitted with rippers, disc cutters, and soft-ground teeth to excavate the bouldery ground (Boone et al., 1998). Muck was extracted from the forward chamber through a screw conveyor mounted near the chamber invert. Ports were provided at the face, within the front chamber, and within the screw conveyor to add water, foam, polymers, bentonite or other additives to condition the muck for consistency, EPB pressure control, and to help limit machine wear (Busbridge et al., 1998, Leinala et al., 1999). The one-pass tunnel lining consisted of bolted pre-cast concrete segments with gaskets (Garrod et al., 1996). Each seven-segment liner ring was about 1.4 m long, 225 mm thick, and 5.2 m inside diameter. Grout ports were included within the TBM tail shield to allow nearly continuous extrusion of grout around the lining. The lining segments were also provided with grout ports. Within the contract, grouting was permitted through either of the two available systems. It was also considered that injection of foams would assist in reducing machine wear during tunneling (e.g. Peron and Marcheselli 1994). Therefore, it was specified that a minimum foam volume equal to 10% of the tunnel cut volume be injected during tunneling. The foam was specified to consist of air:liquid at a 6:1 volumetric ratio and a water:powdered foaming agent ratio of 100:1 (by weight) composing the liquid part. Since foam injection could have other operational benefits including developing the paste-like spoil consistency for EPB tunneling (soil conditioning), stabilizing EPB pressures, and reducing mechanical loads on the TBMs (e.g. Maidl et al., 1994, Kanayasu et al. 1995) the selection and use of foams for achieving such benefits was left to the contractor.

1.3 *Level 1 and Level 2 assessments, and construction performance requirements*

Early in the RTEP work it was recognised that damage to nearby facilities could not be entirely eliminated and would depend on construction methods and performance. Therefore, a two-step risk assessment process (Level 1 and Level 2) was developed for evaluating potential building and utility damage in relation to construction techniques. Details of these assessments are provided by Boone et al. (1998). The combined results of the evaluations were then used to develop a "most likely" settlement profile. If settlements were maintained to this profile, it was expected all structures within the zone of influence would likely suffer only "slight" damage or less. Therefore, settlement limitations based on the "most-likely" deformation profile were chosen as the prime method of stipulating the acceptable performance. A ground movement instrumentation and monitoring program was then designed to measure the ground responses to tunneling activities.

For each instrument in the performance monitoring program (detailed below) "review" and "alert" levels (readings) were established. The "review" level defined a measurement, typically 80% of the "alert" level, at which the contractor and engineer were to discuss causes of movement, possible trends, and to identify measures that would be taken to prevent the movement from exceeding the "alert" level. "Alert" levels, from the ML case, generally reflected the ground movement at which unacceptable damage for particular facilities could initiate. If "alert" levels were reached, the engineer could stop the work and have the contractor make the site safe and secure until alternative protection measures could be adopted.

It was also recognised that many factors under the contractor's control could affect ground deformations and tunneling performance including EPB pressures, thrust, soil ingress rate, soil extraction rate, soil conditioning, and annular tail void grouting. Providing detailed limits for each of these operating parameters, however, would have unreasonably restricted the contractor's means and methods to complete the work. To assist in assuring that construction performance goals were achieved a range of suitable operating EPB pressures was provided in the project's Geotechnical Baseline Report (GBR) and a detailed construction inspection program was also implemented by the owner. It was considered that through experienced review of all survey, ground instrumentation, TBM, guidance, grouting, and construction operations data, performance trends and workmanship issues could be rapidly addressed, thus protecting the owner's interests in obtaining a high-quality project with minimum impact on neighbouring properties and services.

2 INSTRUMENTATION AND MONITORING

For the RTEP projects, the owner undertook a number of responsibilities for monitoring and instrumentation to limit the potential for conflicts of purposes and interests in evaluating the performance of the underground project and to improve the efficiency of the overall work. Table 1 outlines the general responsibilities for the instrumentation and monitoring program.

Table 1. General responsibilities for instrumentation and monitoring.

Item	Owner/Engineer	Contractor
Materials	geotechnical instruments, gauge reading units	grout, surface covers, sand backfill, tie-back load cells
Installation & up-keep	inspection, review/refinement of field layout	drilling, grouting, welding, traffic control, field layout from plans, repairs
Reading	surveying, inclinometers, piezometers, extensometers	strain gauges, convergence gauges

For the Sheppard twin tunnel project, the ground monitoring program primarily consisted of settlement points installed about 2 m below the ground surface at 10 m intervals along the center-lines of each tunnel. Each settlement point consisted of a sleeved steel rod with the bottom 0.3 m grouted into a borehole. These settlement points were surveyed as the tunnel face approached and passed the point location (see Table 2). Other instruments including deep settlement points (installed every 100 m) and probe extensometers were installed at other locations along the alignment as secondary measures for judging overall patterns of ground response and construction workmanship. Arrays of settlement points, positioned transverse to the tunnel centerlines every 200 m, were used to evaluate the shape and magnitude of the settlement troughs caused by tunneling. Utilities were monitored by strain gauges or settlement points at critical locations. All buildings within the zone of influence were also regularly monitored for settlement. Vibrating wire and open standpipe piezometers were installed at critical points around braced excavations or tunnelled cross-passages to observe the effectiveness of dewatering systems within the multiple aquifers present at each site. Vibrating wire strain gauges were installed on selected struts within braced excavations to monitor strut loads.

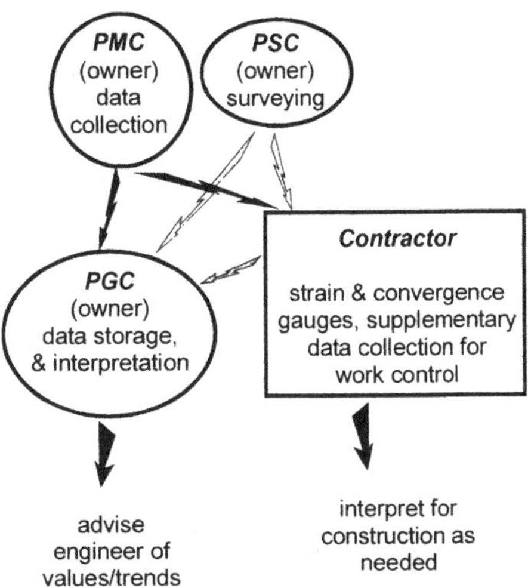

Figure 1. Data colection and transimision for monitoring instruments.

Survey data was captured by a digital levelling system and checked data was forwarded to the PGC (see Fig. 1) for database input, plotting, review, and evaluation. Survey data was simultaneously transmitted to the contractor to allow rapid responses if necessary. A similar approach was implemented by the PMC for inclinometers, probe extensometers, and piezometers. All data was normally transmitted electronically via an "e-mail" system to allow rapid

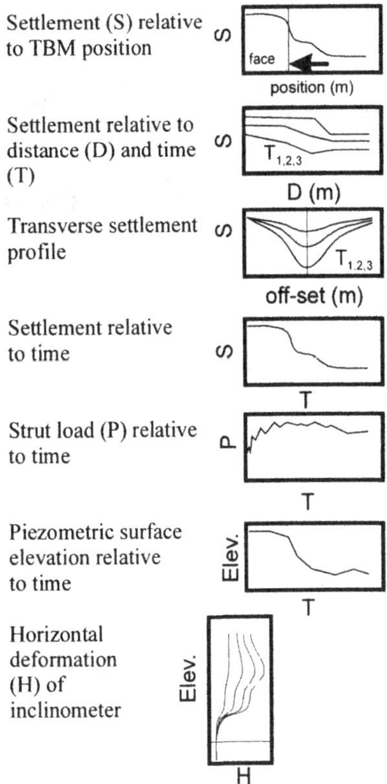

Figure 2. Examples of semi-automated routine graphs generated by generated by monitoring data base system.

sharing of data from many sources and to provide a record of data transfers (Fig. 1). The geotechnical database consisted of a series of linked spreadsheets and embedded commands to semi-automate the data storage process. A number of automated templates were also included within the database to allow rapid generation of routine graphical data plots (see Fig. 2). The automated plots allowed rapid evaluation of performance issues such as control of ground losses at the face, control of grouting, and dewatering effectiveness.

Within each TBM, five EPB pressure sensors were located on the forward bulkhead, at the 2, 5, 7, 10, and 12 o'clock positions. Two EPB pressure sensors were located near each end of the screw conveyor. Each TBM was provided with an automated data acquisition system to record EPB data as well as screw rotation rates, head rotation rates, thrust loads, and torque as well as other machine functions such as oil pressures and temperatures, bearing temperatures, and flood or guillotine door status. Foam injection rates (air, water, and foaming additive) were also recorded by the same system. A laser guidance system (ZED) was used for vertical and horizontal control of tunneling. Data was simultaneously transmitted from each TBM and ZED guid-

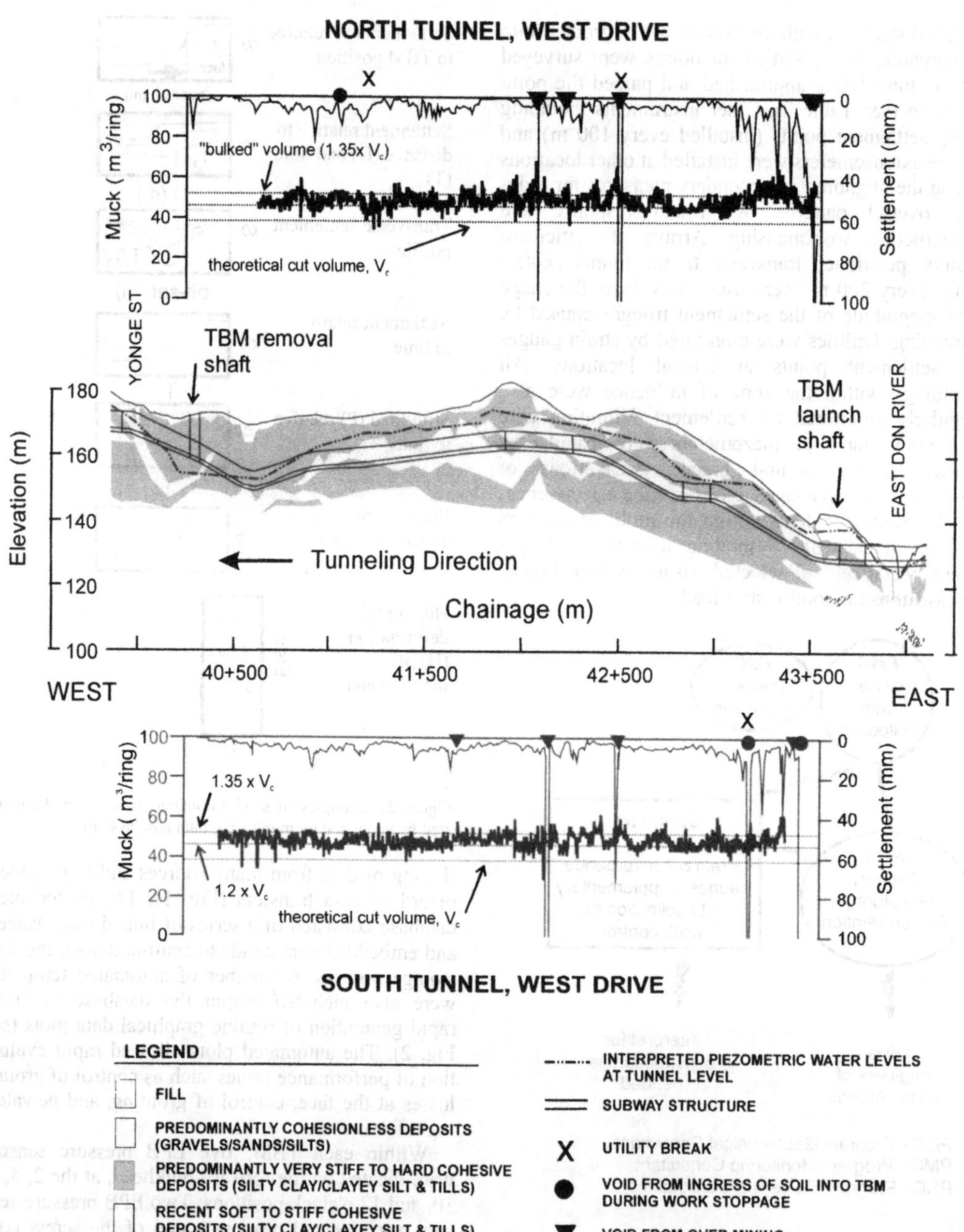

Figure 3. Profile of simplified subsurface conditions, settlement, and muck volumes from tunneling along west drives.

ance system on a "real-time" basis to remote storage and reporting systems within the contractor's and engineer's site offices.

To supplement the TBM data systems, the shift engineers manually recorded: muck car volumes (to the nearest 1/4 car or 2 m^3), conditioning agent usage, grout takes, machine operator, and other qualitative information on tunnel construction. Samples of the muck were also taken to assist in making comparisons to the GBR conditions.

3 MONITORING RESULTS

The majority of the settlement was less than the specified review and alert levels (Figs. 3 and 4), in-

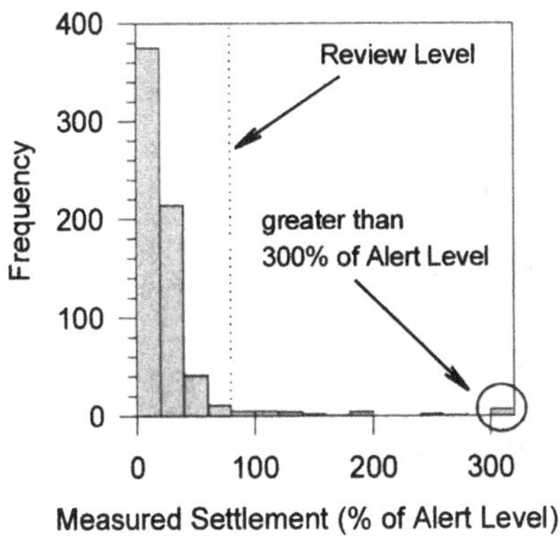

Figure 4. Tunnel settlement performance for center-line settlement points above west tunnel drive.

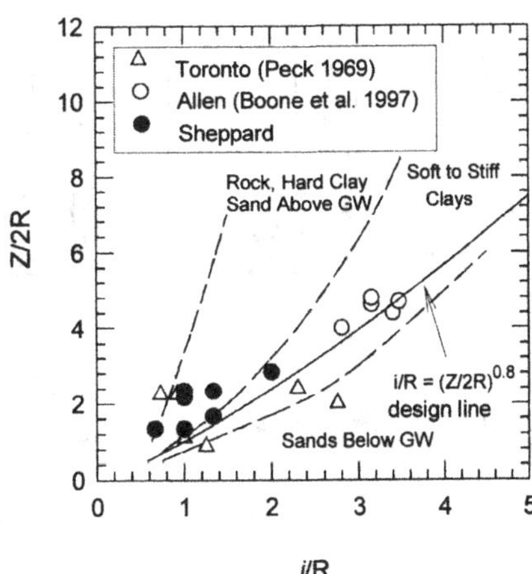

Figure 5. Relationship between off-set distance to settlement trough inflection point (i), depth to axis (Z), and tunnel radius (R), after Peck (1969).

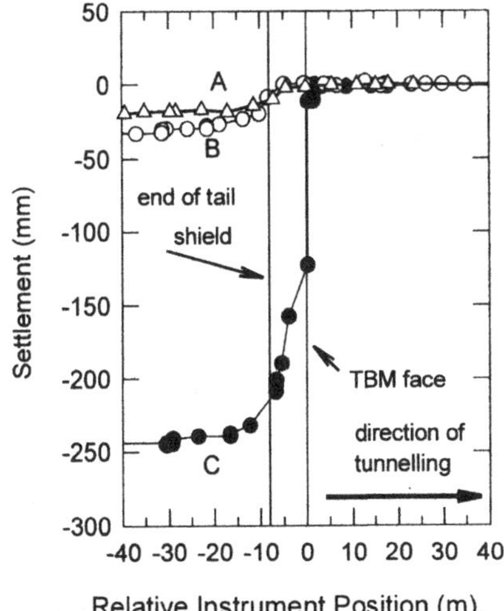

Figure 6. Surface settlement relative to TBM position: A) good grouting and mining; B) fair to poor grouting; C) ground losses due to ingress of soil into TBM during maintenance shut-down.

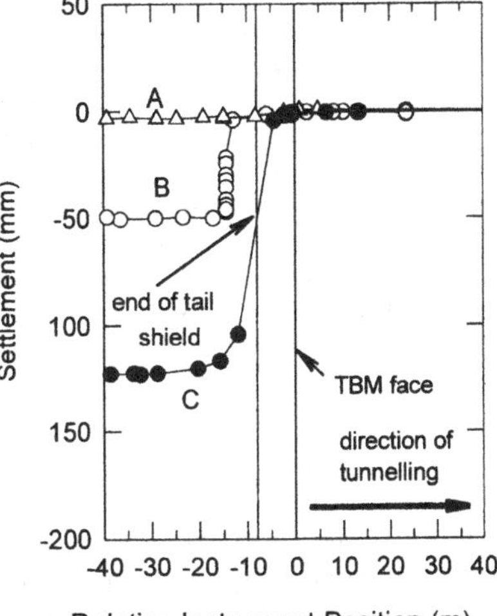

Figure 7. Deep settlement (1 m above crown) relative to TBM position: A) good grouting and mining; B) fair grouting and good mining; C) poor grouting.

dicating good overall performance. In general, the settlement trough width was less than assumed during design (Fig. 5), but generally fell within a zone reflective of the mixed saturated sand and hard clay overburden. Excessive settlement occurred at a number of locations as illustrated in Fig. 3.

During tunneling, one of the more important routine plots was that showing settlement relative to TBM position. Examples are provided in Figs. 6 and 7. From such plots it could be determined whether the largest proportion of the settlement occurred at or near the face, over the TBM skin (cut to skin diameter gap), or over the lining beyond the TBM where grouting was to take place. This plot was especially relevant for the deep settlement points. Although surface settlement was delayed somewhat, the patterns of settlement clearly indicated tunneling behavior and grouting performance.

TBM and conditioning agent data from construction of the north tunnel is illustrated in Fig. 8. Foam was used as the primary spoil conditioning agent during tunneling and the foam was composed of a surface active agent in a water:agent ratio of 160:1, and air at air:liquid ratios typically ranging between

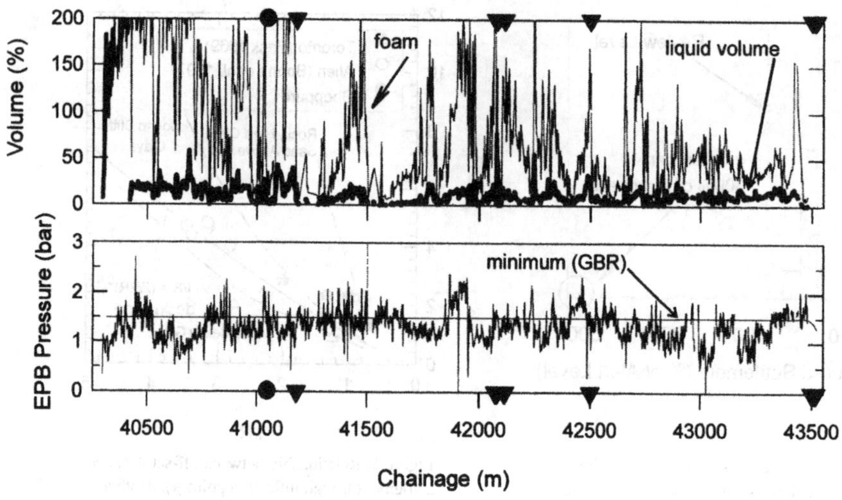

Figure 8. Foam use EPB pressure along north tunnel, west tunnel drive.

4:1 and 10:1. Foam injection rates (percent of excavated volume) were highly variable during construction (Fig. 8). EPB pressures (taken as the average from the bottom two sensors within the front chamber) were on average about 85% to 90% of the minimum value provided in the GBR (Fig. 8). TBM data and inspection reports indicated that the operators maintained the screw rotation at a nearly constant 10:1 ratio of the head rotation (Fig. 9). EPB pressures and muck discharges were largely controlled using the guillotine door on the screw conveyor.

4 DISCUSSION OF PERFORMANCE

The general settlement profiles shown in Fig. 3 clearly illustrate the ground response to the "learning curve" during the start of tunneling. As expected from earlier tunneling (see Boone et al., 1997), settlement at the surface was smaller and deep movements were attenuated where there was a cover of hard clay or cohesive glacial till above the crown. Large settlements occurred in a number of areas and appeared to propagate in a "chimney" fashion simi lar to the effect of loss of support at the face illustrated by Chambon and Corte (1994). Settlement beneath concrete road pavements was only observed because the settlement measurement points "disappeared" from view. The road surface often appeared unaffected. Examination of the muck records clearly indicated that over-excavation occurred at these locations. Where muck records indicated that overmining might have occurred, boreholes were advanced to explore for any other ground losses undetected by settlement points. Ground losses between settlement points were found in a number of areas and it has been assessed that where the volume excavated (from muck records) exceeded approximately 1.35 to 1.4 times the theoretical volume, voids occurred in areas where the overburden consisted mainly of sand (Figs. 3 and 9). Without the combination of instrumentation and muck volume records, potential voids beneath pavement areas might have otherwise gone undetected.

For areas with the largest ground losses, mining times were significantly greater than in other areas. As a result, foam injection volumes and total screw and head rotations were also greater (Fig. 9). Although head revolutions per minute (RPM) were greater in this area, screw RPM were within the range of fluctuation for other tunneling areas. In this area, EPB pressures were set to exceed the ground water pressures and were near the sum of effective active earth pressure and hydrostatic pressure. These EPB pressures should have been adequate under typical mining conditions. However, it is suspected that the dynamic drawing-in of soil from continued head rotation combined with screw operation within typical RPM ranges, in spite of the slow advance rate, allowed initial ground losses and fully active earth pressure conditions to develop. Mining times in this area were likely increased by both cutter wear and encountering the hard clay in the tunnel invert. These observations emphasize the problems involved with machine tunneling in mixed-face conditions where hard clays are en countered in the invert and relatively uniform sand is present from about tunnel axis to the ground surface, e.g. Clough and Leca (1993), see also Fig. 3. Though total screw revolutions per ring were only generally related to muck volumes (Fig. 10) as a result of guillotine door control, muck extraction rates must be monitored closely with TBM advance rates (rather than head RPM or other parameters) to avoid over-mining.

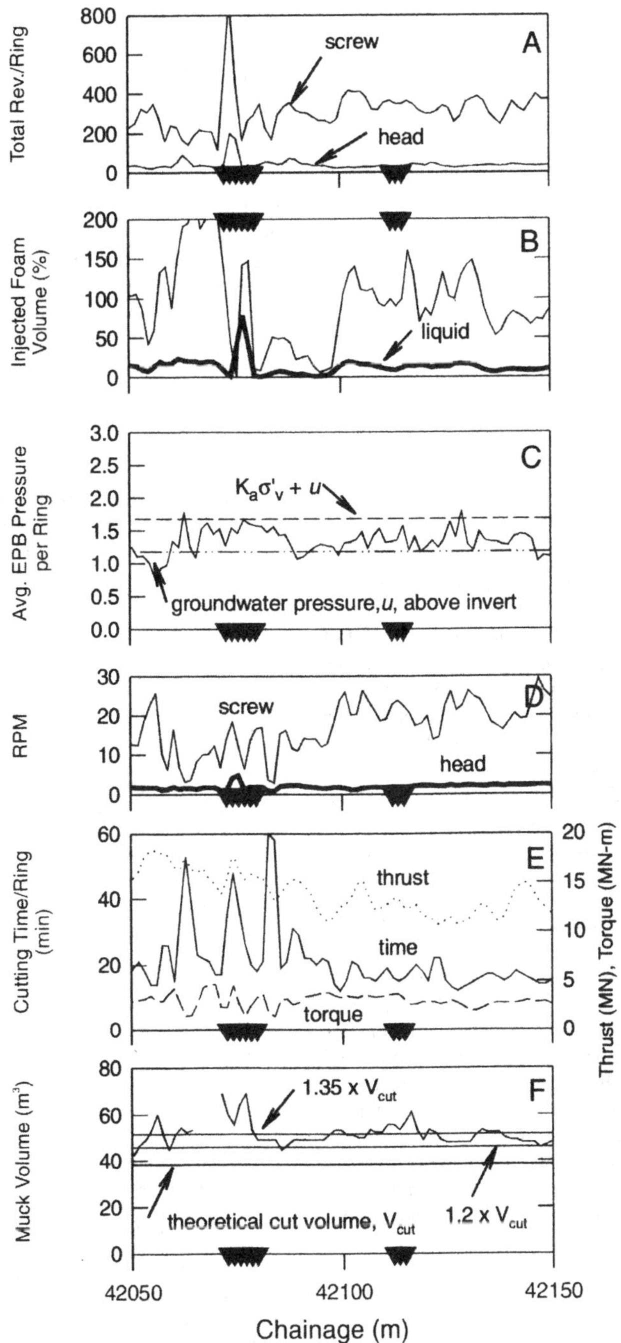

Figure 9. TBM mining and data in areas ground losses. A) total number of revoliutions per liner ring; B) foam injected as % of cut volume (1 atm.); C) average EPB pressure per liner ring; D) average revoliutions per minute of screw conveyor while mining (per ring); E) cut time, average thrust, and average torque per ring; F) approximate muck volume per ring. Ground loss areas shown as inverted triangles.

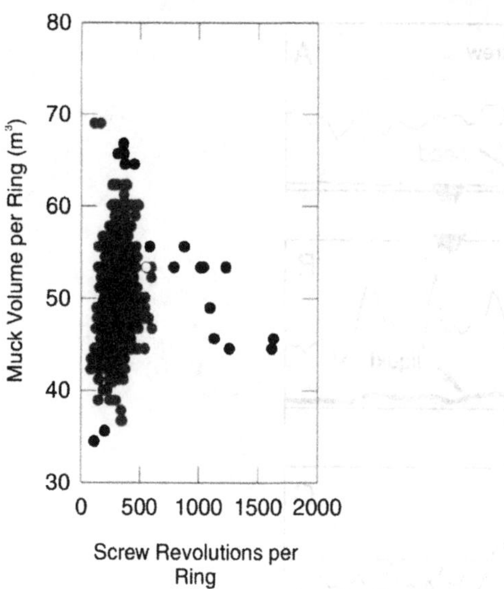

Figure 10. Muck volume relative to total number of screw

Where excavation was controlled, as over most of the alignment, settlement was generally governed by grouting performance. The contractor generally chose to limit grout volumes to the theoretical gap volume and grout pressure measurements were not taken. Grout was also injected through the liner ports rather than through the tail shield ports. Therefore, gap closure between the tail and lining was not always well controlled and subsequent probing above the liner indicated full closure in a number of locations.

Detailed statistical analyses of the data following construction have also permitted examination of the tunneling performance related to spoil conditioning agents, tool wear, soil type, head RPM, and thrust loads (Leinala et al., 1999). Among the conclusions developed from the construction data are that tool wear was probably the most significant factor in tunneling advance rate followed by the addition of foam. In cohesive soils, the liquid component of the foam had the most significant influence on mining times. It is presently unclear what effect the use of foams had on reducing the tool wear and further study of this relationship is underway.

5 CONCLUSIONS

A detailed monitoring program was essential to the success of this project. Regular observations of settlement measurement points and the surrounding areas by survey crews quickly detected ground movements. Responses to the observed conditions and repairs, where necessary, were prompt and consequential effects were minimal. Monitoring crews retained by the owner provided high-quality data as they were able to learn from early contracts and transfer knowledge, personnel, and equipment familiarity between contracts. Electronic data transfer, though initially unfamiliar to many construction personnel, allowed rapid sharing of data between many parties. The geotechnical database worked well in producing standardized plots for rapid interpretation during and following construction. The TBM and guidance data systems also worked well. For future projects it would be beneficial to develop links between these systems to permit better tracking of discrete measurements relative to TBM position. Although muck volume measurements are often considered crude and imprecise, the value of visually-judged and manually recorded muck volume data and other construction activities was indispensable for this project. Most importantly, a well-planned, integrated, and well-executed inspection and monitoring program can provide valuable insight into the causes of variability in construction performance for the benefit of the monitored project and future work.

6 ACKNOWLEDGEMENTS

The writers would like to acknowledge the effort of site personnel also involved in the monitoring program including: Morrisson-Hershfield for surveying, Peto MacCallum for data collection, and the Hatch Mott McDonald site staff. The writers also thank the TTC for permission to publish this paper.

REFERENCES

Boone, S.J., McGaghran, S.M., Pennington, B.N. and Marsland, D. (1997). Performance of a Machine Bored Tunnel in Toronto. Proc. 50th Can. Geotech. Conf., Ottawa, 226 - 233.

Boone, S.J., Garrod, B., and Branco, P. (1998). Building and utility damage assessments, risk, and construction settlement control. Tunnels and Metropolises, A. Negro and A. Ferreira (eds.), Balkema, Rotterdam, 243 - 248.

Boone, S.J., Westland, J., and Nusink, R. (1999). Comparative Evaluation of Building Responses to an Adjacent Braced Excavation. Canadian Geotech. Journal, 36(2), (in press).

Boone, S.J., Westland, J., Busbridge, J.R., and Garrod, B. (1998). Prediction of Boulder Obstructions. Tunnels and Metropolises, Proceedings of World Tunnel Congress 1998, Sao Paulo, Brazil, A. Negro and A. Ferreira, editors, Balkema, Rotterdam, pp. 817 - 822.

Busbridge, J.R., Westland, J., Boone, S.J. and Pennington, B.N. (1998). Underground Construction for the Extension of the Toronto Subway. ASCE Geotechnical Special Publication No. 86, 273 - 296.

Chambon, P. and Corte, J-F. (1994). Shallow tunnels in cohesionless soil: stability of tunnel face. Journal of Geotechnical Engineering, ASCE, 120(7), 1148 - 1165.

Clough, G.W. and Leca, E. (1993). EPB shield tunneling in mixed face conditions. Journal of Geotechnical Engineering, ASCE, 119(10), 1640 - 1656.

Garrod, B., Feberwee, J., Wheeler, C. (1996). Sheppard Subway - Design of Twin Tunnels. Canadian Tunnelling, 51 - 60.

Kanayasu, S., Yamamoto, I., and Kitahara, Y. (1995). Stability of excavation face in earth pressure balance shield. Underground Construction in Soft Ground, K. Fujita and O. Kusakabe, eds., A.A. Balkema, Roterdam, 265 - 268.
Leinala, T., Delmar, R., Collins, J.R., Pennington, B., and Grabinsky, M. (1999). Soil conditioning for enhanced TBM performance in differing ground conditions. Proc. 52th Canadian Geotechnical Conference, Regina.
Meidl,B., Herrenknecht, M., and Anheuser, L. (1995). Mechanised Shield Tunnelling. Ernst & Sohn, Berlin.
Peck, R. B. 1969. Deep Excavations and Tunnelling in Soft Ground: State of the Art Report. Proc. 7th Int. Conf. on Soil Mech. and Found. Eng., Mexico City, 225-290.
Peron, J.Y., and Marcheselli, P. (1994). Construction of the 'Passante Ferroviaro' link in Milan, Italy, lots 3P, 5P, and 6P: excavation by large earth pressure balanced shield with chemical foam injection. Proc. of Tunneling '94, British Tunnelling Society, 679 - 707.
Westland, J., Boone, S.J., Branco, P., MacDonald, D., and Meschino, M. (1999). Shoring for Leslie Station: Design Assessment and Construction Performance. Geotechnical Special Publication No. , ASCE 1102 - 1115.

North American Tunneling 2002, Ozdemir (ed.)
© 2002 Swets & Zeitlinger, Lisse, ISBN 90 5809 376 X

Tunneling beneath railway tracks – Criteria for design and construction

R.J.F. Goodfellow, P.E.
URS Corporation, Gaithersburg, Maryland

M.A. Piepenburg, C.E.G.
Hatch Mott MacDonald, Cleveland, Ohio

ABSTRACT: Because urban areas are increasingly crowded, pipeline crossings of infrastructure links using tunneling techniques are becoming more common. Railways are particularly sensitive structures to tunnel beneath and pipeline crossings of railway tracks are being designed and constructed across the United States. This paper presents and examines the American Railway Engineering and Maintenance-of-way Association (AREMA) guidelines for Railway Companies and how these compare to the requirements of five railway companies in the Northeastern United States. This comparison and discussion is illustrated by two case histories from Ohio, one project crossing a heavy freight rail property and the other crossing a rapid transit property. Comparisons of construction permit requirements are made in tabular format for the designer's reference. Conclusions on important design issues, anticipated schedule impacts, and comparisons of design criteria for tunneling under railways are drawn.

1 INTRODUCTION

With urban areas becoming increasingly crowded, the use of trenchless technology for pipeline crossings of major infrastructure links, such as railroads and highways, is becoming more common. This paper considers railroad crossings and focuses in particular on the railway companies of the Northeastern United States. Railway crossings have unique and specific issues to consider associated with public safety and operational continuity. These issues are addressed in this paper with reference to the American Railway Engineering and Maintenance-of-way Association (AREMA) guidelines, railway company guidance manuals, and case histories.

We have researched guidelines of several companies and recorded project experiences to determine the consistency of requirements for the design of tunnels under railways. We have investigated the consistency of approach to help designers better predict time and cost for obtaining construction permits from railway companies.

A conventional tunnel design process is described, followed by a discussion of the primary issues considered by AREMA and by five different railway companies in their design guidelines. Comparisons of construction permit requirements are made in tabular format for the designer's reference. Two case histories, one for a heavy-rail railway company and one for a light-rail transit system with overhead power, are provided to illustrate design issues raised and schedule impacts of a permitting process.

We have drawn conclusions on the issues of consistency of approach between AREMA, heavy freight rail companies, and rapid transit companies, as well as schedule impacts and design issues that should be considered when tunneling under railway tracks.

2 THE TUNNEL DESIGN PROCESS

A successful design process typically requires a high level of communication between the project owner, designer, and all third parties from which permits are required (typically highway administrations, utility companies, railroads, and municipalities). Once the project alignment has been established and a preliminary design developed, the typical design process considering the railroad as the specific third party concern might proceed as follows:

The first task is often a kick-off meeting with the project team, owner, railway company and other third parties to establish the owner's objectives and scope for the project. This meeting establishes communication with the railroad by allowing transfer of information and documents including contract drawings and specifications, relevant subsurface information, written guidelines on contents of submittals, review procedures, and the proposed design and construction schedule. It is generally beneficial to make sure that all parties buy into the design approach. The focus of these meetings is to identify any major impacts to the design, keeping in mind that changing course and making necessary alterations in approach are much more economical at this stage rather than later on in the process.

Design progress meetings between the owner, designer and third parties to present modifications to the design based upon the requests and criteria established by the third parties. More refined and updated project documents may be presented, and a workshop format may be advantageous for particularly complex projects soon after the 30% submittal to the owner. Workshops allow all parties to share their thoughts on the issues at hand (operational, geotechnical, or other issues). The owner and designer should be warned, however, that not all design scenarios are suitable for a productive workshop. The parties must be well briefed in advance to discuss the options presented in the workshop. Also, it is beneficial to have decision-makers present at the workshop so that any design decisions on approach can be made and ratified.

After completion of design and before issuance of Contract Documents for bid, an application is made to the railroad to cross their property. Review periods vary but after this period the railroad will issue a construction permit, usually with detailed conditions attached.

Depending upon the railroad and the project complexity, some railroads like to have a representative present at the owner's pre-construction meeting between the owner and the contractor, where the overall coordination of project and the understanding of the contract documents are confirmed with the contractor. In some cases the Railway Company representative reiterates concerns and requirements to the Contractor at this time on matters such as personnel safety training.

3 DESIGN ISSUES IMPORTANT TO RAILROAD COMPANIES

The primary concerns of all infrastructure owners when a tunnel is proposed underneath their right-of-way or infrastructure link are safety and disturbance of operation. For railway companies, these concerns are specified in criteria that address surface settlement, durability of the structure, instrumentation to verify performance, access to the safety zone around the track, limitations on work hours, and other constraints detailed in Table 1.

AREMA considers all phases of the pipeline's service life in the design guidelines, including construction, service and maintenance, and abandonment. In addition to the published guidelines, the comments received from railway companies fall into three general categories: contractual issues, operational issues, and technical issues.

1 Contractual issues

Contractual issues include comments of a general and administrative nature, such as the need to obtain written approval for all construction work within the railway company right-of-way. While this requirement is clearly stated in all guidelines, other requirements of this type are either implicit or not stated and are often only detailed in initial correspondence or reviews.

These comments include: review and approval of design documents; review of shop drawings, contractor's submittals, and design modifications during construction; and immediate notification of the railway company if any deviation occurs from the previously documented and approved procedure.

Comments of this type could lead to additional reviews during the bid phase or construction. This should be considered in determining both schedule and budget for the project.

2 Operational issues

Many of the operational issues and comments raised by the railway company are repeated as terms and conditions of the construction permit itself. Typical comments include: restriction of construction work to night or weekends; safety procedures for access to the tracks and training of construction personnel; no interruption or speed restriction on rail service during construction; and similar conditions for maintenance and abandonment of the pipeline.

Although there may be some flexibility in some of the initial demands made, the burden is firmly on the designer to convince the railway company that the proposed tunneling will not adversely impact operation of the railroad.

3 Technical issues

A basis of design report or design narrative is often required to outline the designer's thought process in selecting the preferred construction method. This report provides the basis for method selection and subsequent changes during the bidding and selection phase.

The basis of design report is written to answer many of the railway company's technical design questions without the need to search the contract documents. The report includes design details such as: method of tunneling (often demanding continuous face support); method of analysis (numerical analysis often preferred whether or not this is the preferred technical approach); ground modification; instrumentation (often including remote sensing and real time monitoring); and contingency plans (particularly for excessive settlement).

The maximum allowable settlement for railroad tracks is not specified in the AREMA guidelines but this issue is often a focus for designers to specify a method of construction resulting in minimum impact on the railroad. Many railroads ask for zero settlement and these unrealistic expectations need to be managed throughout the design to obtain limits that allow construction to proceed and at the same time do not impact railway operation.

Other important technical issues are durability of the structure, including the need for pipeline secondary casings, and cathodic protection.

The AREMA guidelines give the Chief Engineer of the railway company absolute discretion to approve or disapprove any aspect of the proposed design. Design and review guidelines are split into several sections for different materials – flammable or non-flammable. As both case histories given in this paper are water tunnels, the following comparative study of railway companies considers the requirements for conveying non-flammable fluids.

Table 1 compares the guidelines given by AREMA with particular rail company guidelines. The layout of the table gives a brief description of the design criteria in column 1 and then for each company, a check mark if this information is required, a "No" if it is not required, or "N/A" if it is not applicable. Column 2 of the table describes the AREMA guidelines followed by three heavy freight rail companies in columns 3 to 5 [Consolidated Rail Corporation (Conrail), CSX Transportation (CSX), Norfolk & Southern Corporation (N&S), respectively]. Columns 6 and 7 contain the requirements for two transit systems [the Greater Cleveland Regional Transit Authority (GCRTA) and the Washington Metropolitan Area Transit Authority (WMATA), respectively].

4 DISCUSSION

4.1 *AREMA Guidelines*

As shown in Table 1, the AREMA guidelines request very specific information on geometry of the pipeline in relation to the tracks. Also required is information regarding the contents to be handled by the pipeline, working pressure, type of joint, pipeline coating, method of installation, presence of seals, type size and spacing of insulators, cathodic protection, railway line survey station, location of valves, and right-of-way lines.

Table 1. Comparison of railway company requirements during construction permit process.

Permit Application Requirements	AREMA Manual	Conrail	CSX	N&S	GCRTA	WMATA
Written approval required?	✓	✓	✓	✓	✓	✓
Statutory form provided?	N/A	✓	✓	✓	✓	No
Standard guidelines provided?	N/A	✓	✓	✓	✓	No
Application fee?	N/A	✓	✓	✓	✓	*
Location relative to railway milepost?	✓	✓	✓	✓	✓	✓
Location of railway property line?	✓	✓	✓	✓	✓	✓
General arrangement drawing?	✓	✓	✓	✓	✓	✓
Full set of contract documents required?	No	✓	No	No	✓	✓
Show location of nearby railway facilities?	✓	✓	No	✓	✓	✓
Show wires and poles to be relocated?	✓	✓	✓	✓	✓	✓
Length of tunnel within right-of-way?	No	No	✓	✓	✓	✓
Angle of crossing?	✓	✓	✓	✓	No	✓
Distance and depth of jacking and receiving pits?	✓	✓	No	No	No	*
Contents of pipeline?	✓	✓	✓	✓	✓	*
Casing pipe specification?	✓	✓	✓	✓	✓	No
Geotechnical Data?	No	✓	No	No	✓	*
Method of installation?	✓	✓	✓	✓	✓	✓
Carrier pipe material?	✓	✓	✓	✓	✓	✓
Working pressure?	✓	✓	✓	✓	✓	*
Outside diameter?	✓	✓	✓	✓	✓	✓
Inside diameter?	✓	✓	✓	✓	✓	✓
Wall thickness?	✓	✓	✓	✓	✓	✓
Specification and grade of pipeline material?	✓	✓	✓	✓	✓	✓
Type of joint?	✓	✓	✓	✓	✓	✓
Depth of cover under tracks?	✓	✓	✓	✓	✓	✓
Depth of cover under ditches?	✓	✓	✓	✓	✓	✓
Location of nearest emergency shut-off valves?	✓	✓	✓	✓	✓	No
Cathodic protection for pipeline?	✓	✓	✓	✓	✓	*
Pipeline coating?	✓	✓	✓	✓	✓	No
Contact grout material and mix?	No	No	No	No	✓	✓
Are Contractor's means and methods required for review?	No	✓	No	✓	No	*
Railway attends pre-construction meeting?	No	✓	No	No	No	*
Are settlement limits quantified for Contractor?	No	No	No	No	No	*

* Required during review but not specified in guidelines document

The AREMA guidelines are explicitly valid for pipelines up to 72 inches inside diameter but the principles can be used as a framework to assess larger tunnels. These guidelines prefer the use of a boring or pipe jacking operation to minimize the potential for damage to the railway. If there should be damage, however, the guidelines state that tunneling should continue if the railway would suffer further damage during a work stoppage.

A secondary casing is not always required and can be omitted if the under-crossing is:

- Of a secondary track
- Where leak proof construction and material that can withstand the combination of internal pressure and external load
- At non-pressure wastewater crossings.

The crossing should be as close to perpendicular as possible to the tracks – this prevents stray current effects from affecting structural durability, and minimizes the length of pipe under tracks. The minimum cover is 4.5 feet to the crown and emergency shut-off valves are required within an "effective" distance from the track crossing. The distance in question is not quantified and is therefore left to the discretion of the railway company. Pressure grouting is specified as the preferred method to fill voids caused by tunneling.

4.2 *Comparative Study*

It is clear from Table 1 that the railway companies in this study chose to follow the AREMA guidelines and focus on geometry and structural durability. Each company has extended and adapted these guidelines to suit their own experience and specific local needs. The discussion below will point out some areas where the railway companies did not follow AREMA guidelines and some differences between heavy and rapid transit operations in their requirements for construction permits.

AREMA provides minimum distances (45 feet) to nearby rail facilities and poles, with closer locations requiring special design in coordination with the railroad engineer. The railroad companies require more specific information on the location of these structures. AREMA requests the depth and location, relative to the track, of jacking and receiving pits associated with tunneling. Only Conrail explicitly states this as a requirement, although other companies may expect to see this information as part of an overall general arrangement drawing.

The requirements of transit companies were found to differ in a few respects from those of heavy freight rail companies. Both types of companies in this study required a full set of contract documents, geotechnical data and reports, and details of contact grouting materials. Of the heavy freight railway companies, only Conrail requested geotechnical details and test data for its files.

WMATA required continued involvement throughout the bid phase, and detailed review and assessment of anticipated settlement that no other company requested. This is a reflection of WMATA's institutional experience gained during the relatively recent construction of its underground system.

The permit application process described here applies only to underground work with the jacking, receiving, or work shafts located off of the railroad property. Additional applications, contacts, training, or the need for flagmen or a track foreman may be required by the railroad prior to geotechnical drilling or any construction work within the railroad's right-of-way or property.

Access to the information and the ease with which a designer can find out what is required varies widely between railway companies. Based upon an Internet search of all Class I railroads in the United States in November, 2001, we found that only Burlington Northern Santa Fe (BNSF) and Conrail provided permit application forms in an electronic format. Union Pacific Railroad provided the names of individuals to be contacted regarding the permitting process. Amtrak, CSX, and N&S did not provide contact or permit application information on their web pages. N&S was contacted via telephone for the permit application described below in Case History 1. We expect this situation to change with more information being made available using the Internet.

5 CASE HISTORIES

5.1 *Case History 1: Norfolk & Southern Railroad, Warrensville Heights, Ohio (2001/2002)*

The Northeast Ohio Regional Sewer District (NEORSD) Contract MCI-1 consists of the replacement of an undersized and damaged section of the brick- lined Mill Creek Interceptor beneath the southern Cleveland suburb of Warrensville Heights, Ohio. The replacement sanitary sewer will be constructed as a 600-foot long open cut and a 2,350-foot long 96-inch diameter tunneled excavation through clay, wet sand, and weak Chagrin Shale bedrock. A 175-foot long section of the tunnel will be excavated into shale bedrock beneath a 20-foot high embankment and secondary track belonging to the N&S.

Due to design-related questions regarding the location of the soil-rock interface along the proposed tunnel alignment, the designer contacted the railroad for permission to access the right-of-way to advance a geotechnical boring. Once the design was advanced to about the 75% stage, the railroad was again contacted and the N&S "Application for Pipeline Agreement" (Form 11367) was obtained, completed, and returned to the railroad with the $335 application fee. A copy of the completed permit application is included as Attachment 1.

To provide the railroad with as much background information as possible, the design team included the following information with the permit application package:

- Relevant geotechnical boring logs and interpretive geologic cross sections;
- Relevant specification sections for tunnel excavation and support, instrumentation, carrier pipes, and annular grouting;
- Contract drawings showing the location of the railroad crossing, details of the initial support system, and location of the instrumentation along the railroad.

We assumed that addition of these items helped the railroad with the permit issuance process. Indeed, no delays occurred in the permitting process due to questions or uncertainties regarding the permit information and the application was accepted by the railroad without modification.

The time to obtain this permit was approximately five months. However, this may be atypically long as the railroad was in the process of transitioning all of their permitting to an external contractor. The construction contract was bid in December 2001 for construction in early to mid-2002.

5.2 *Case History 2: Greater Cleveland Rapid Transit Authority, Shaker Heights, Ohio (1996/1997)*

The Greater Cleveland Rapid Transit Authority (RTA) "Blue Line" extends from downtown Cleveland to Shaker Heights, Ohio, with electrical power supplied by overhead catenary lines. In 1996 and early 1997, the Northeast Ohio Regional Sewer District (NEORSD) began subsurface investigation and design for the 8,500-foot long, 60-inch (finished) diameter Heights/Hilltop Interceptor Contract 7C (Contract 7C) sewer interceptor, and a smaller diameter trunk sewer in the city of Shaker Heights. To convey flow from the southern part of the city, the Contract 7C sewers crossed the RTA Blue Line tracks at two locations shown on Figure 1 and consisting of the main sewer crossing at Lee Road and a trunk sewer crossing at either Onaway Road or Ashby Road.

NEORSD began presenting the Contract 7C project details to the RTA prior to completion of the design with a letter of explanation that included:

- Plan and profile drawings of the crossing areas;
- Contract specifications related to tunneling in rock, and microtunneling and pipe jacked tunnels;
- Calculations related to carrier and casing pipe design and pipe jacking;
- Geotechnical boring logs performed adjacent to both sites;
- Data sheet for pipelines crossing under the RTA tracks (the form described above).

Based upon results and discussions from initial meetings with the RTA, and upon completion of the sewer design package, the NEORSD team advanced finalized copies of the project drawings, specifications, and data report to the RTA, along with a completed data sheet for each pipeline crossing. The modifications to the design documents included modifying the pipeline crossing angle to avoid catenary poles, adding RTA survey stations on the data sheet, modifying the casing and carrier pipe cross sections, and increasing the depth of the Onaway crossing from 20 feet to 50 feet to place the excavation in rock instead of soil.

Reiterating reservations about the amount and control of settlement (a safety and operational issue), the RTA granted the NEORSD permission to obtain applicable access easements and proceed with the project with the following provisions to be included with the NEORSD specifications and contract documents:

- The contractor shall monitor the tracks for settlement and immediately repair any track settlement or heave at no cost to the RTA.
- Flagmen operating in compliance with the RTA's Standard Rail Flagging Procedures are to be employed for any work performed within 10 feet of the RTA tracks.
- RTA specification sections for Maintenance of Rail Traffic and Standard Rail Flagging Procedures were to be incorporated into the NEORSD specifications.

In addition, the RTA approval letter noted that the NEORSD would reimburse the RTA for any expenses for labor and materials, including flagmen, in connection with the construction of the Contract 7C sewers.

Approximately 75 calendar days were needed between the initial presentation of the project to the RTA and receipt of their permission to proceed with the NEORSD work. The Contract 7C sewers were successfully completed with negligible recorded settlement or heave of the RTA tracks.

6 CONCLUSIONS

- AREMA has published guidelines for railway companies to follow when issuing construction permits for tunneled pipelines crossing railroad tracks.
- There is a great deal of consistency between heavy freight rail companies that broadly follow the AREMA guidelines. However, each company has extended these guidelines to suit its own experience and local conditions.

- Rapid transit systems studied in this paper required complete sets of contract documents to review and also required geotechnical testing data and reports.
- It is strongly recommended that the designer contact the railway company early in the process to obtain the relevant paperwork and design requirements. It is equally important to keep strong lines of communication open throughout the design process.
- No length of time is included in permit application information for the companies studied. The case histories show a wide range of review times between 75 days and five months.
- It would be prudent to expect a construction permit application for a tunneled pipeline crossing of a railroad to take between 60 and 90 days to complete.
- It is important for a designer to have a flexible approach and be ready to listen to the requirements of the railroad in workshops and review meetings.
- The designer should have a broad knowledge of different methods and technologies regarding tunneling method, ground modification and instrumentation to monitor and mitigate surface settlement, and operational risk to the railway company.

7 ACKNOWLEDGEMENTS

We would like to acknowledge Mr. Richard Switalski, Mr. David Klunzinger, and Ms. Alison Schreiber of the Northeast Ohio Regional Sewer District for their assistance with background information and for providing critical review of this paper. We also appreciate the efforts of Mr. Tom Richardson and Mr. Michael Vitale, who provided internal peer review for our respective companies.

REFERENCES

American Railway and Maintenance of Way Association (AREMA), 1993. Manual for Railway Engineering, Part 5.

Consolidated Rail Corporation, 1995. Specifications for pipeline occupancy of Consolidated Rail Corporation Property.

CSX Transportation, 1998. Form reference CSXT-7455: Application for pipeline crossing under/over properties and track.

Greater Cleveland Rapid Transit Authority, 1997. Standard Specifications and Data Requirements for Crossing under GCRTA Tracks.

Norfolk Southern Corporation, 1991. Form reference 11367: Application for Pipeline Agreement.

Washington Metropolitan Area Transit Authority, 1999. Adjacent Construction Design Manual (Rev. 1).

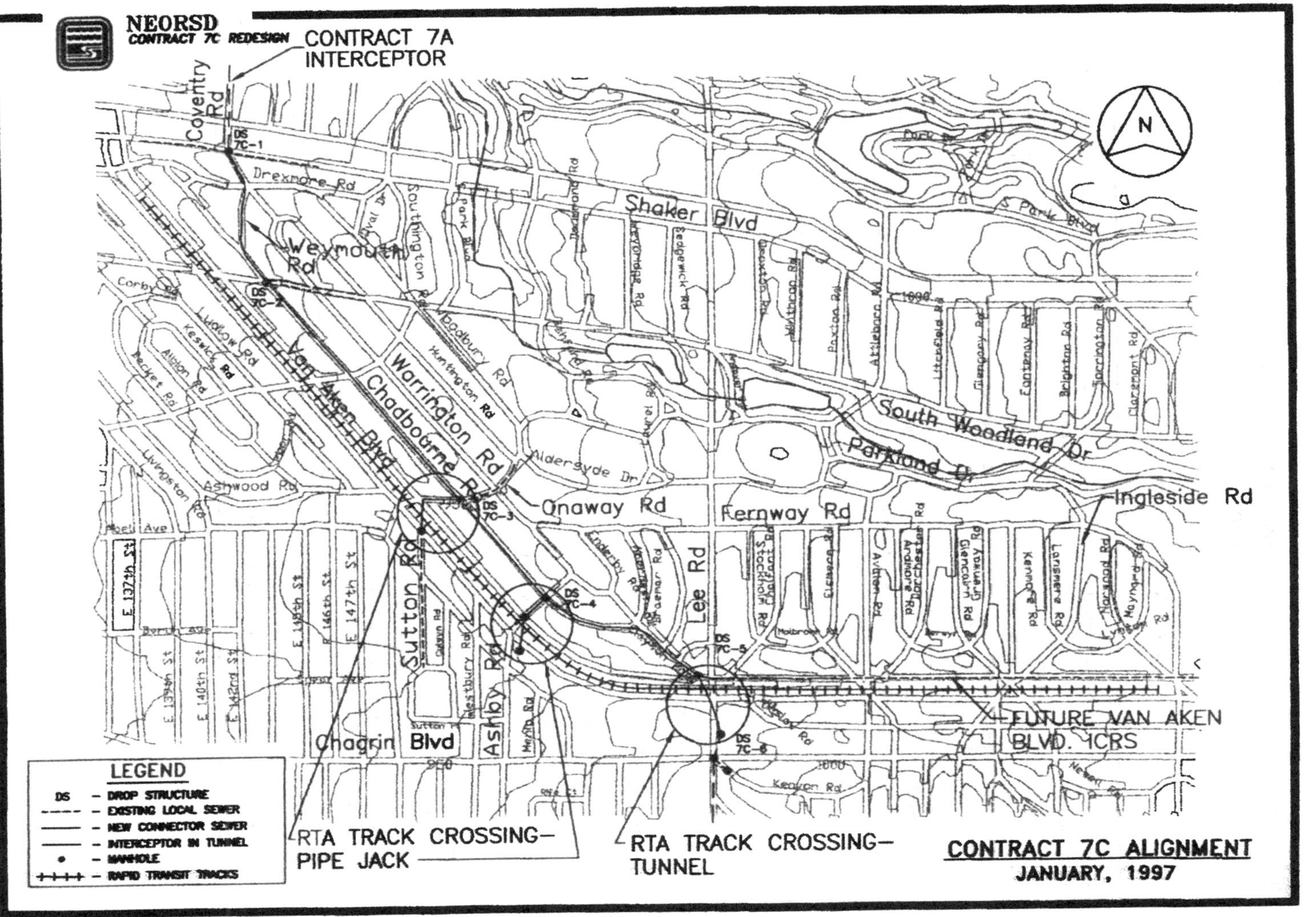
NEORSD
CONTRACT 7C REDESIGN
CONTRACT 7A
INTERCEPTOR
Coventry Rd
DS 7C-1
DS 7C-2
DS 7C-3
DS 7C-4
DS 7C-5
DS 7C-6
Drexmore Rd
Shaker Blvd
Weymouth Rd
Van Aken Blvd
Chadbourne Rd
Warrington Rd
Huntington Rd
Southington Rd
Park Blvd
South Woodland Dr
Parkland Dr
Aldersyde Dr
Onaway Rd
Fernway Rd
Ingleside Rd
Lee Rd
Sutton Rd
Ashby Rd
Chagrin Blvd
Ashwood Rd
Livingston Rd
E 137th St
E 147th St
E 146th St
FUTURE VAN AKEN
BLVD. ICRS
LEGEND
DS - DROP STRUCTURE
- EXISTING LOCAL SEWER
- NEW CONNECTOR SEWER
- INTERCEPTOR IN TUNNEL
- MANHOLE
- RAPID TRANSIT TRACKS
RTA TRACK CROSSING—
PIPE JACK
RTA TRACK CROSSING—
TUNNEL
CONTRACT 7C ALIGNMENT
JANUARY, 1997

NS NORFOLK SOUTHERN

APPLICATION FOR PIPELINE AGREEMENT

FORM 11387 (Rev. 9/91) (Item No. 875589)

I.D. No. ______

Date Issued: ___/___/___

'PLICANT MUST ANSWER ALL QUESTIONS AND RETURN THIS FORM TO:

SUPT.
COMPANY
STREET
CITY, STATE, ZIP
PHONE

NORFOLK SOUTHERN CORPORATION
Superintendent's Office
8111 Nelson Road
Fort Wayne, Ind. 46803

. Legal Name of Applicant (party to agreement): NORTHEAST OHIO REGIONAL SEWER DISTRICT

. Mailing Address:
Street 3826 EUCLID AVENUE
City CLEVELAND State OH Zip 44115

4. Billing Address:
Street 3826 EUCLID AVENUE
City CLEVELAND State OH Zip 44115

. Name of Applicant Representative:
MR. DAVID KLUNZINGER, P.E
Title: PROJECT MANAGER
Telephone Number: (216) 881-6600

5. Name of Contact for Billing Purposes:
MR. DAVID KLUNZINGER, P.E.
Title: PROJECT MANAGER
Telephone Number: (216) 881-6600

. Applicant is a:
[ ] Corporation – Give state of formation: ______
[ ] Limited Partnership – Give state of formation: ______
[ ] General Partnership – Give state of formation: ______
[ ] Sole Proprietorship – Give name of owner: ______
[ ] Individual
[X] Government Entity
[ ] Other: ______

. Location of Pipeline:
A. Nearest Street MILES RD (OHIO RT. 43)
B. Nearest Town WARRENSVILLE HEIGHTS
C. County CUYAHOGA
D. State OHIO
E. Railroad Milepost Reference: MP 11 N E S (W) (Circle one) of Milepost MP 11
F. Will pipeline be located entirely within confines of public street? [ ] Yes [X] No
If yes, provide conclusive evidence for verification.
Street width: ______ ft. Street Right of Way width: ______ ft.

. Pipeline [X] is to be installed [ ] already exists and is being upgraded
Are there any agreements covering the pipeline? [ ] Yes [ ] No [X] Do not know
If yes, identify and attach copies: ______

. The pipeline will: (check all appropriate boxes)
[X] Cross tracks – How many? 1
[ ] Cross communication and/or signal lines – Separation ______ ft.
[ ] Parallel Tracks – Minimum horizontal distance to centerline of adjacent track ______ ft.
[X] Be underground -- Depth below base of rail 63 ft. Minimum depth on right of way 45 ft.
[ ] Be aerial – ______ ft. above top of rail

Identify facilities on Railroad right of way (manholes, pipe bridges, etc.): NONE

(Continued on other side)

Plans for proposed installation are to be submitted to and meet the approval of the Railway Company before construction is begun. Material and installation are to be in strict accordance with specifications of the American Railway Engineering Association and requirements of Norfolk Southern Corporation. Original and six copies of this form shall be submitted, accompanied by six prints of a drawing showing plan, elevation section of crossing from field survey, location in respect to Milepost, width of Railroad's Right of Way, width of Public Right of Way, location of adjacent structures affecting crossing, and all information required in Figures 1 and 2 of A.R.E.A. Specifications, Part 5 Pipelines. If tunneling is necessary details of sheeting and method of supporting tracks or driving tunnel must be shown.

1. Correct name of Applicant: NORTHEAST OHIO REGIONAL SEWER DISTRICT (NEORSD)
2. Post Office address: 3826 EUCLID AVENUE, CLEVELAND OHIO 44125
3. Location 2,250 ft. WEST (direction) from nearest R.R. Milepost MP11
4. Nearest city or town: WARRENSVILLE HEIGHTS County: CUYAHOGA State OHIO
5. Within limits of public highway, name: -N/A- Fed-State-County No.: -N/A-
6. Will temporary track support or rip rapping be required? [ ] Yes [X] No - explain
7. Are there any wires, poles or obstructions to be relocated? [ ] Yes [X] No Temperature
8. Product to be conveyed: SEWAGE Flammable? [ ] Yes [X] No Temperature 55°F ±
9. Maximum working pressure: NONE (psi) Field test pressure: -N/A- (psi) Type of test -N/A-
10. Location of shut-off valves: -N/A- SEWER IS GRAVITY FLOW
11. PIPE SPECIFICATIONS

Carrier pipe will be either RCP pipe or "Hobas" pipe per contractors option →

	CARRIER PIPE: Pre-Cast Reinforced & Conc. Pipe	CARRIER PIPE: Centrifugally Cast Fiberglass reinforced polymer mortar pipe	CASING PIPE
Material			LINER PLATES w/ STEEL RIBS
Material specifications and grade	ASTM C76 CL IV	ASTM D 3262	Plates per ASTM A 569 Ribs per ASTM A 572 Grade 50
Minimum Yield Strenght (psi) of material	4000 psi	10,500 psi	50 KSI (50,000 psi)
Mill test pressure (psi)	-N/A-	-N/A-	50,000 psi (min)
Inside diameter	60"	60"	78" (min)
Outside diameter	72"	63.5" (min)	86" (min)
Wall thickness	6"	1.75" (min)	6.38"
Type of seam	ASTM C443	ASTM D 4161	-N/A-
Laying lengths	8'-0" min	15 to 20 ft.	Ribs @ 4'-0" c/c spacing
Type of joints	ASTM C443	ASTM D 4161	-N/A-
Total length within Railroad right of way	175 ft.		175 ft.

VENTS - number: size: height above ground: -N/A

SEALS - both ends: -N/A- one end: -N/A-

BURY - base of rail to top of casing: 63 ft. — in.

BURY - (not beneath tracks): 45 to 63 ft. — in.

BURY - (roadway ditches): -N/A- ft. -N/A- in.

CATHODIC PROTECTION - [ ] Yes [X] No

PROTECTIVE COATING - [ ] Yes [X] No Give description -N/A-

Specify type, size, and spacing of insulators or supports: Pipe will be installed in a tunnel boring machine (TBM) driven tunnel, blocked against the initial tunnel supports & the annulus backfilled w/ grout

Define any special specifications of the pipeline: see attached specification sections

Method of installation: pipeline installed in a TBM driven tunnel.

If application is approved, applicant agrees to reimburse the Railroad for any cost incurred by the Railroad incident to installation, maintenance, and/or supervision necessitated by this pipeline installation, and further agrees to assume all liability for accidents or injuries which arise as a result of this installation.

(Date) (Signature and title of officer signing application)

(SEE SAMPLE DRAWING (Exhibit A), and SAMPLE SHEET (Exhibit B), attached)

Please refer to the attached contract specifications and drawing details.

The subsurface conditions along the alignment were determined from three geotechnical borings and are shown on Figures 3a and 3b (11 x 17 dwgs.) from the project geotechnical data report. The borings show that the pipeline will be excavated through weak shale bedrock at depths between 45 and 63 feet beneath the railroad right-of-way.

A closed face, tunnel boring machine (TBM) will excavate a 96-inch (minimum) diameter tunnel beneath the railroad right-of-way. The entire tunnel circumference will be immediately supported with steel ribs and steel liner plates blocked against the excavation walls as shown on Detail 1 taken from the Contract Drawings. As indicated in attached Specification Section 02302, Paragraph 3.11A, the tunnel beneath the railroad right-of-way (approximately Stations 16+75 to 18+50) will be excavated on a "round the clock" basis without stopping. Surface settlement markers (SSM) will be installed along the railroad alignment at the locations shown on Detail 2 to monitor the ground reaction to the excavation. Instrumentation details and monitoring schedule are provided in the attached Specification Section 02355.

Upon completion of the tunnel excavation, a 60-inch inside diameter carrier pipe will be installed in the tunnel. The pipe will be either a Class IV reinforced concrete pipe (RCP) or an equivalent strength centrifugally cast fiberglass reinforced polymer mortar pipe manufactured by Hobas (or approved equivalent) as presented in the attached Specification Sections 02560 and 02728, respectively. The annulus between the pipeline and the tunnel liner will be backfilled with a cementitious grout injected through the pipeline as described in attached Specification Section 02330.

Pipe Line and Crossing to be installed and maintained in accordance with latest approved AMERICAN RAILWAY ENGINEERING ASSOCIATION'S "Specifications for Pipelines for Conveying Flammable and Non-flammable Substances.".

	Carrier Pipe	Casing Pipe
Contents to be handled	Sewer	—
Outside diameter	~72"	96" min
Pipe material	RCP or "Hobas"	Steel Ribs & Liner Plates
Specifications and Grade	ASTM C-76 or D-3262	ASTM A572 Gr.50 & A569
Wall thickness	6"	6"
Actual working pressure	NONE	—
Type of joint	ASTM C443 or D4161	—
Coating	NONE	NONE
Method of Installation	TUNNELING	TUNNELING

Vents: No. ______ **Size** ______ **Hgt. above ground** N/A

Seal: Both ends - N/A **One end** N/A - Entire pipe to be grouted

Bury: Base of rail to top of casing 65 **ft.** — **In.**

Bury: (Not beneath tracks) 45-65 **ft.** — **In.**

Bury: (Roadway ditches) -N/A- **ft.** ______ **In.**

Cathodic protection: - N/A -

Type, size and spacing of insulators or supports. N/A

NS NORFOLK SOUTHERN

OPERATING COMPANY

OFFICE OF CHIEF ENGINEER - DESIGN & CONSTRUCTION
ATLANTA, GA

Warrensville Heights, Ohio
Northeast Ohio Regional Sewer District
~2,250' W of MP 11

PX

Session 4a, Track 2

Policy, economics and underground space

North American Tunneling 2002, Ozdemir (ed.)
© 2002 Swets & Zeitlinger, Lisse, ISBN 90 5809 376 X

Denmark – A country with connections

Stephen Slot Odgaard & Ole Peter Jensen
COWI A/S, Kongens Lyngby, Denmark

ABSTRACT: Denmark is located at an important position between Germany and the rest of Scandinavia. In the last five years, two important fixed links have been completed, and a third is presently being studied. This paper will present the completed connections (Great Belt & Øresund), and the missing connection across Fehmarn Belt. The different possible solutions will be presented with focus on technical, environmental and safety aspects. Furthermore, with reference to the two major connections already built, possible ways of financing the third fixed link will also be discussed.

The paper will focus on the tunnel parts of the 3 projects and discuss the limitations of large-length road and rail tunnels. Both immersed and bored tunnels will be discussed.

1 INTRODUCTION

COWI has had major roles in all of the three projects. On the Great Belt COWI was lead in a JV with Mott and responsible for design and technical follow-up during construction and assisted the owner's organization with the design check, supervision and construction management of the East Tunnel. COWI was also responsible for the detailed design of the East Suspension Bridge. On the Øresund project COWI, as lead, performed the detailed design of the cable stayed bridge with SWECO as JV partner. COWI also performed a tender design for the Øresund immersed tunnel. On the Fehmarn Belt Link COWI formed a joint venture with the German company Lahmeyer to carry out the feasibility study.

Through these diverse roles COWI has gained a detailed insight into various parts of these record-braking projects.

This article gives the reader an insight into some of this experience.

The paper is divided into two major parts:

- A general description of the 3 important fixed links in Denmark
- A more detailed description of the tunnel element in the three projects.

This should allow the reader to get an overview of each of the projects, before going into a more thorough description of the tunnel element of the projects.

2 SETTING THE SCENE

Denmark lies in the southern part of Scandinavia, between Germany, Sweden and Norway.

It consists of many islands, and regular ferry traffic has been an important part of the internal transport system the last many years.

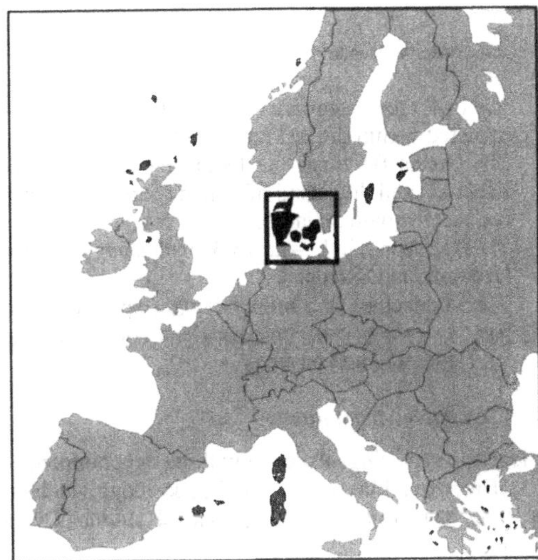

Figure 1. Map of Europe.

The capital Copenhagen is located on the main island Zealand. The western part, the peninsula Jutland, borders Germany. Between Zealand and Jutland is the island of Funen.

Large ferry services have been running on major links carrying both cars and trains.

Connections were built across fjords and bays during the 1960's, -70's, and -80's. After their completion, the major crossings remained to be constructed. There were three of these. One was a domestic link, and the two others would link the eastern part of Denmark to Sweden and Germany. All 3 are between 15 km and 20 km long.

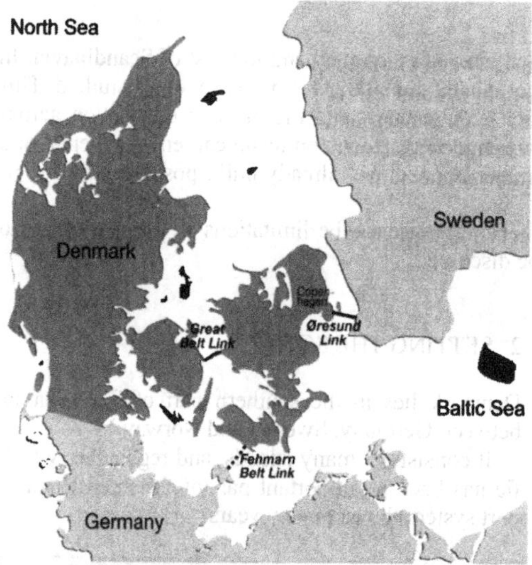

Figure 2. Map of Denmark.

Two of these important fixed links have been completed within the last five years.

- The Great Belt link – a 18 km crossing between Zealand and the island of Funen (Funen is already connected with Jutland).
- The Øresund link – a 16 km crossing between Denmark and Sweden.

One connection of similar length, approximately 20 km, is missing: the Fehmarn Belt connection between Denmark and Germany.

2.1 *Common denominators*

The three crossings have many things in common:

They are all between 15 and 20 km long, which is a long leap from the links previously built in Denmark.

They all provide similar technical challenges, which would have made it almost impossible to construct the structures 20-30 years ago.

They all cross major shipping routes connecting the Baltic Sea to the North Sea and the Atlantic. A guaranteed depth of 17 m allows vessels of up to 250,000 tonnes to pass the Great Belt and the Fehmarn Belt.

The routes through all three straits have navigational turning points within few kilometres of the crossing. These induce high risk of collisions and groundings in the case of malfunction or lack of awareness from the personnel operating the vessels. For both the Great Belt and Øresund the routes have required modification either temporarily during construction or permanently. The modification of these routes required dredging to fulfil the guaranteed depth of the current routes.

The ground consists of limestone and various glacial deposits. Especially the Great Belt and Fehmarn Belt have layers of glacial deposits composed of clay till with subordinate sand till and scattered boulders.

Important sanctuaries and protected natural habitats exist in the close vicinity of all three crossings. This reduces the acceptable disturbance of the sea to a very low level, and only very little spillage of dredged materials can be tolerated.

The three crossings all pass the only waterways connecting the Baltic Sea and the North Sea. During certain periods there is an acute lack of oxygen in the Baltic Sea, and a main source of renewal of the water is exchange through the Great Belt, Øresund, and Fehmarn Belt. A zero blocking effect has therefore been imposed on the fixed links, requiring compensation dredging.

3 THE EXISTING LINKS

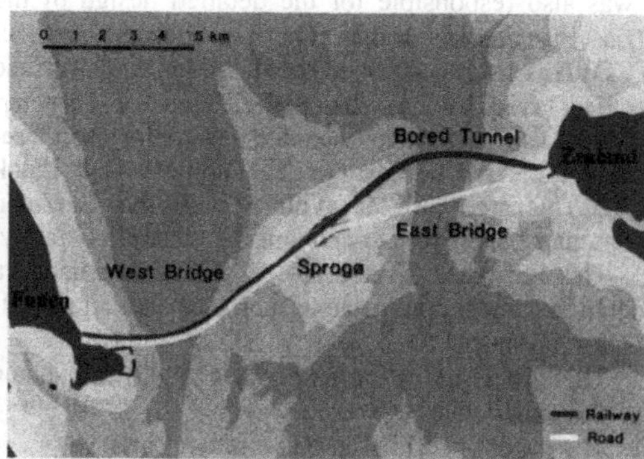

Figure 3. Plan of the Great Belt Link.

3.1 *The Great Belt link*

The Great Belt link is entirely within Denmark, and thus, primarily Danes benefit from this connection. The decision to proceed with the construction was therefore much easier to reach than for the Øresund and Fehmarn, where two governments have to agree. Furthermore, the advantages for the two countries may not be equal. The main advantage of the Great

Belt bridge for domestic users was cutting 2-3 hours off domestic travel between the eastern and the western part of Denmark.

3.1.1 *General description*

The fixed link comprises a motorway and rail connection across the Great Belt between the islands of Funen and Zealand.

The main elements are:

- An exisiting island Sprogø, artificially enlarged to 4 times its original size.
- The West Bridge, a 6.5 km low-level bridge with combined motorway and railway.
- The East Bridge, a 6.8 km combined low-level and suspension bridge for road traffic.
- The East Tunnel, an 8 km double-tube bored tunnel for rail traffic, with a maximum depth of 75 m below sea level.

3.1.2 *Special considerations*

As described earlier in this paper, the main shipping route from the Baltic Sea to the North Sea goes through the Great Belt. The route has a guaranteed depth of 17 m. The existing route passed the Great Belt Link in a zigzag pattern, requiring precise maneuvers to avoid collision and grounding. This traffic highly influenced the chosen road-bridge solution.

To encourage the use of public transport a political decision was taken by the Danish parliament to open the rail connection at least 3 years in advance of the road connection.

The rail and road were led on separate crossings on the eastern part of the connection to facilitate this construction in two stages, Due to considerable problems during the tunnel construction the actual head start was only 12 months.

3.1.3 *The island of Sprogø*

Sprogø lies in the middle of the Great Belt. The only habitants on the island lived in the lighthouse. The important shipping route T lies to the east of the island. The island was well suited as interchange between different structure types on the east and west crossing.

Due to its small initial size of 0.38 km^2, the island was enlarged to 1.56 km^2, leaving the original island untouched.

3.1.4 *The West Bridge*

The 6.5 km waterway on the western side of Sprogø reaches a maximum depth of more than 20 m. A low girder bridge solution was selected, with a clearance of 18 m and a span of 110 m. The structure consists of 5 concrete elements:

A bridge girder for the motorway, a bridge girder for the railway, each with its own pier shaft, and one caisson providing support for both bridges. The caissons were based on compacted stone at a maximum depth of 30 m.

3.1.5 *The East Bridge*

The 6.8 km waterway on the eastern side of Sprogø has gentle slopes except the central 1-1.5 km where the bottom drops from 20 m depth to 55 m depth. Simulations of the vessels passing the bridge were performed by pilots assisting large tankers passing the waters to determine the required span of the bridge. Tests were run in bad weather and with different bridge designs. Based on these results, a design with a suspension bridge with a main span of 1624 m was chosen. A vertical clearance of 65 m was internationally accepted. Architectural considerations played an important part in the development of the bridge design due to its monumental size.

Several protective measures were adopted to reduce the risk of ships colliding with the bridge or other ships: The Route T was straightened to avoid zigzag maneuvers in the vicinity of the bridge. Protective islands were created at the sides of the anchor blocks, a Vessel Traffic Service was established warning vessels of risks of collision, and in the worst case closing the bridges for traffic.

3.1.6 *The East Tunnel*

A bored tunnel solution was the preferred solution due to the profile of the seabed and the required free span. The main problem with this solution, though, was the complicated tunneling conditions in the dominating soft glacial deposits. The alignment was taken further north than the East Bridge, to fulfil the gradient requirements. This allowed a more shallow descent and avoided deeper areas along the straight alignment.

The tunnel and its construction is described in more detail later in this paper.

3.1.7 *Organization and financing*

The Danish Parliament agreed to establish a fully state-owned limited liability company to be responsible for the planning, design, construction and operation of the link. The link was financed by domestic and foreign loans guaranteed by the Danish State. The cost was almost divided equally between the railway and motorway.

The Danish State Railway will repay its share over 30 years, and the motorists will repay their share through toll payments. It was decided that the price for the toll be equivalent to the price of the ferry tickets.

The cost of the link is divided as follows:

- 28% for the East Tunnel
- 23% for the West Bridge
- 34% for the East Bridge
- 15% for the railway installations, landworks and construction works at Sprogø.

3.1.8 *Predicted effect*

The public opinion about the new project was mixed as many saw the one-hour ferry journey as a pleasant break on the trip. However, the ferry trip was also a source of irritation, since the busy car ferries required long waiting in peak periods, especially if users had not reserved a ticket well in advance. On peak travel days the wait could be many hours.

The traffic prognosis available in 1988, when it was agreed to proceed with the link, predicted 12,200 vehicles / day.

A revised prognosis made in 1997, six months prior to the opening, predicted a substantial increase: The ferry service transported 7100 vehicles / day prior to the opening. The predicted traffic was these 7100 plus 3800 from other crossings, plus an additional traffic generated solely by the existence of the fixed link of 5500, totaling 16,350 vehicles / day.

3.1.9 *Actual effect*

The actual situation has proven this drastic effect, with the road traffic totalling:

- 17,400 veh / day in 1998
- 19,000 veh / day in 1999
- 20,600 veh / day in 2000
- 21,600 veh / day in 2001.

One effect of this high amount of traffic was that the demand exceeded the toll booths' resulting in fairly long queues at peak hours initially. This problem was solved during the first year.

3.2 *The Øresund Fixed Link*

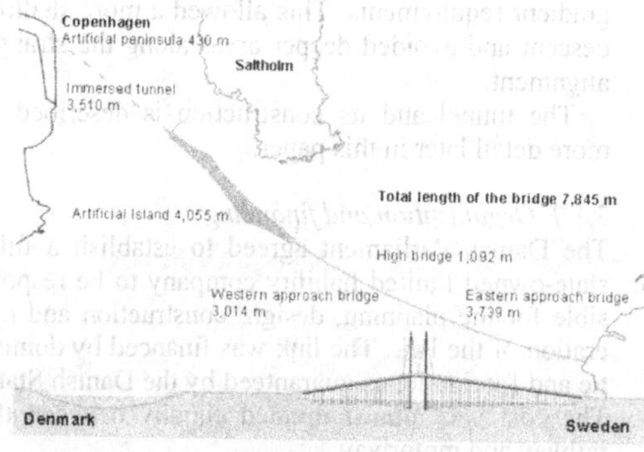

Figure 4. Plan of the Øresund Link.

The Øresund Strait is a strait between the island of Zealand (Denmark) and the southern province of Scania (Sweden). The capital of Denmark, Copenhagen, and the provincial capital of Scania, Malmö, are located on either side of the strait. In an area of 14,000 km^2 a total of 3 million people (2000) reside. 60% of these live on the Danish side and 40% live on the Swedish side.

Several types of ferry services existed between the two countries. Hydrofoil service made passenger travel between the centres of the two cities in 45 minutes. Slower car ferries made the journey in twice the time. Rail traffic had to cross the strait some 50 km further north of Copenhagen. It was believed that these services were important obstacles in the integration of the two regions. The decision was therefore made to construct a combined rail and motorway. The construction was combined with a new rail connection from Copenhagen Central Station to Copenhagen International Airport.

3.2.1 *General description*

The fixed link comprises a 16 km motorway and rail connection across the Øresund between Copenhagen in Denmark and Malmö in Sweden. The main elements are:

- The western peninsula: an artificial peninsula off the Danish coast.
- The tunnel: a 4 km immersed tube tunnel under the Drogden navigational channel.
- The artificial island: a 4 km artificial island named Peberholm.
- The bridge: an 8 km combined low-level and cable stayed bridge.

3.2.2 *Special considerations*

The fact that this structure was to be built between two countries caused a significant increase in the initial work for approval of the project. Although Denmark and Sweden are very closely connected, governmental, cultural, and managerial styles are significantly different.

The fact that the connection is between two countries has required special attention to the organization of emergency response. All alarms send for a response team from both countries, regardless of the location of the accident. As the couplers on the fire hoses differ in the two countries, two outlets are provided at each fire hydrant.

3.2.3 *The western peninsula*

The west-end of the tunnel surfaces just outside the territory of the Copenhagen International Airport, Kastrup. The approaches were built just of the end of one of the main runways. An area of 0.9 km^2 of land was reclaimed. The western approach for the tunnel was constructed in this area.

3.2.4 *The tunnel*

A 4 km combined motorway and dual railway immersed tunnel was chosen. Each road tube had two lanes but no emergency lanes. Between the two road tubes, a central gallery exists along the whole length of the tunnel. This is divided into an upper service gallery and a lower escape gallery.

The tunnel is described in more detail later in this paper.

3.2.5 *The artificial island*

An artificial island was constructed for the transition between the western tunnel and the eastern bridge. The island was 4.4 km long and consisted of stone, sand and dredged material from the tunnel trench. Stones were placed along the perimeter of the island, and geo-textiles installed on the inside of the embankments to avoid seepage.

3.2.6 *The bridge*

The Øresund Bridge is composed of a high bridge and two approach bridges. The high bridge has the longest cable-stayed main span in the world for a combined road and rail traffic bridge.

The two-level superstructure is made of steel and concrete. The steel girder supports the upper deck, which accommodates the motorway and the lower deck where the railway is located.

The high bridge is 1092 m long, with a main span of 490 m and a navigation clearance of 55 m.

The pylon and anchor pier foundations are made up of concrete caissons some 13-28 m below sea level. Protective islands, designed to prevent ship collisions, surround each caisson.

The approach bridges are made up of two-level composite girders with 120/140 m span.

3.2.7 *Organization and financing*

The decision to construct the Øresund Fixed link was taken by the Danish and Swedish government in 1991. An owner's organization was established in 1992, to be responsible for the design, construction, and operation of the link. The link was financed by loans guaranteed by the Danish and Swedish state through a 'joint and several liability'. The loans are repaid by tolls from motorists and a fixed nominal fee to be paid by the railway operators. As with Great Belt, the prices were fixed by law to be comparable to the price of a ferry ticket.

3.2.8 *Owner-Contractor relationship*

The project was tendered according to the Design and Construct philosophy. This has proven advantageous in many cases, as it leads to good planning and effective construction giving the Contractor the possibility to adapt his particular methods.

The owner's organization was quite unique and proved very successful. From the start of the project the owner made it clear that they wanted to enter a relationship based on trust, giving the Contractor as much freedom as possible. The relationship is best described as a partnership with clear goals laid out from the start. The owner took responsibility of the process and urged open up-front discussions about potential problems. It required a high level of technical expertise on the owner side to be able to enter into these discussions. The result of this unique partnership was that the construction was completed within the agreed timeframe and budget, and funds set aside to cover claims were intelligently used on various bonus schemes.

3.2.9 *Predicted effect*

The vision of the Øresund Fixed Link was to join the regions of the two countries and create a synergy effect.

The link should also facilitate transport between Sweden and the eastern parts of Norway towards Germany. The construction of the Fehmarn Link will merely help this transport. Although Danish and Swedish are different languages, Danes understand Swedish and vice-versa, and a great potential for cooperation was seen.

3.2.10 *Actual effect*

The expected integration has proven more difficult than expected. There are significant differences in culture and company culture. Both countries have high income taxes, but structured in different ways. In order to avoid tax speculation an agreement between the two states has resulted in complex tax rules for cross-border employees. These rules are almost impossible to live and work under. Several international companies tried to encourage their employees to work in offices on the other side of the link but have abandoned these trials due to the rules. The use of the road bridge for leisure purposes has also been very limited due to the high toll price.

The connection has now been open for 17 months, and there is no clear positive trend in the development of the road traffic. This is in clear contrast with the user success of the Great Belt Link. Less than 10,000 vehicles / day pass the Øresund. Contrary to the road, the railway has experienced a significant number of passengers. A fair number of these are Swedish passengers traveling via Copenhagen International Airport, the first station on the Danish side.

The integration between the two regions is off to a slow start, and it may take some years before the current obstacles are overcome, and the traffic rises to the levels this magnificent structure deserves.

4 THE MISSING LINK - FEHMARN BELT

For centuries the Fehmarn Belt has been an important route between Scandinavia and Europe. Since 1963 and until today a combined train and car ferry connection as existed between Rødbyhavn in Denmark and Puttgarden in Germany. Currently this service runs with half-hour service.

After the opening of the Great Belt and the Øresund Fixed Link the 19 km Fehmarn Belt crossing remains "The Missing Fixed Link".

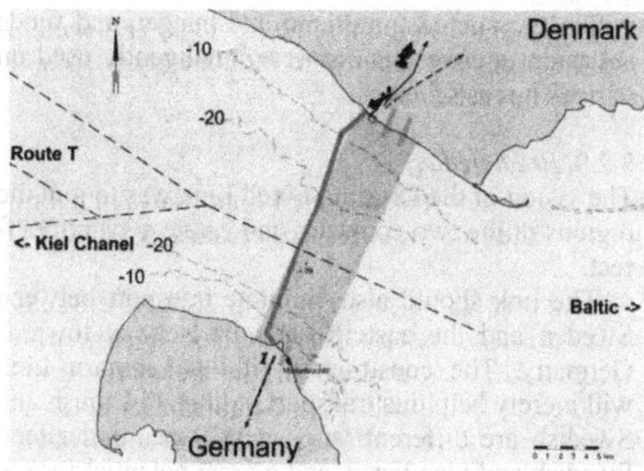

Figure 5. Plan of the Fehmarn Link.

4.1 *The fixed link study*

In 1992 the Danish and German Governments agreed to start preliminary studies. As a part of these studies a feasibility study was completed in 1999. In this study four different levels of capacity were examined and compared:

1 Continued ferry operations, the reference level
2 0+2: No road connection, but dual track railway with shuttle service for the road traffic.
3 3+1: Two-lane road-traffic (one in each direction) and a third lane used for service/emergency operations, plus single-track railway.
4 4+2: Four-lane motorway and dual track railway.

For each of these configurations, several technical solutions exist. Eight different solution models were defined and examined. These are briefly described below.

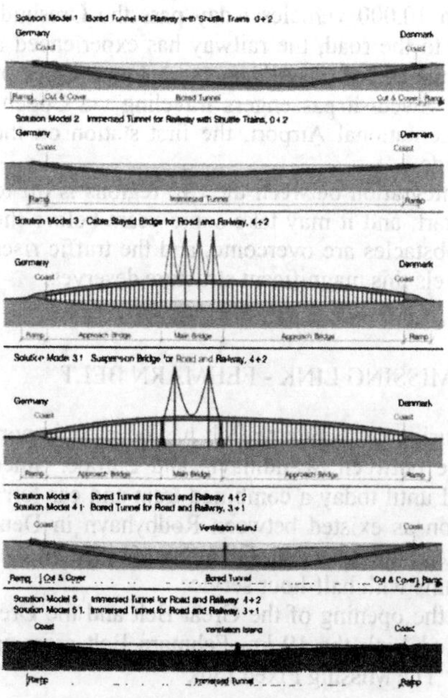

Figure 6. Fehmarn solution models.

4.2 *Description of solution models*

1 Solution Model 1: 0+2 Bored, consists of a dual tube bored railway tunnel with an internal diameter of 8.9 m. Road traffic is transported on a railway shuttle-service.
2 Solution Model 2: 0+2 Immersed, consists of a dual tube immersed steel tunnel with a central escape corridor. The width of the elements is 19 m.
3 Solution Model 3: 4+2 Cable stayed, consists of a combined motorway and dual track railway bridge. The bridge is a three-span cable stayed bridge with main spans of 724 m and approach bridges.
4 Solution Model 3.1: 4+2 Suspension, consists of a combined motorway and dual track railway bridge. The bridge is a one-span suspension bridge with a main span of 1.8 km and its approach bridges.
5 Solution Model 4: 4+2 Bored, consists of four bored tunnels. Two for the motorway with an internal diameter of 10.8 m. The railway tunnels have an internal diameter of 8.0 m.
6 Solution Model 4.1: 3+1 Bored, consists of two bored tunnels. One bore for a three lane road with an internal diameter of 14.2 m, and one bore for one rail track with an internal diameter of 8.0 m.
7 Solution Model 5: 4+2 Immersed, consists of a motorway and dual track railway. The element width is 43 m.
8 Solution Model 5.1: 3+1 Immersed, consists of 3 road lanes and one railway track. The element width is 27.4 m.

These solution models are evaluated according to the following criteria:

- Economy
- Construction Period
- Capacity
- Ventilation
- Risk aspects
- User comfort and safety
- Ventilation

4.3 *Risk policy and management*

The policy adopted to evaluate the risk in the feasibility study is formulated as:

- The safety level for the users of the link shall be reasonable and comparable to other traffic installations.
- Risk shall be systematically identified, evaluated and managed.

To assure that the risk management has a central position it shall be regarded as an integral part of the project management.

The upper limit of individual risk to railway passengers and road users of the fixed link has been chosen slightly higher than the risk to railway passengers and road users on land. The upper limits were selected at:

- Road users: $100*10^{-9}$ fatalities / passenger
- Railway passengers: $5*10^{-9}$ fatalities / passenger.

4.4 *Special investigations*

4.4.1 *Risk aspects*

Large infrastructure projects as the Fehmarn Belt Fixed Link are complex systems. Safety in the operation as well as in the construction phase is of major importance to the planning of the structures. The risks, therefore, have to be identified, analyzed and controlled.

The risk add-on, the consequences in terms of fatalities as well as disruption, economic loss, and environmental impact were assessed. The different types of consequences were weighted with the defined preferences, and an integration of all identified hazards was made.

4.4.2 *Ship collision study and ship traffic observation*

The purpose of the ship collision study was to analyze and quantify the risks to the different solution models associated with navigation activity in the Fehmarn Belt.

Two traffic scenarios have been developed based on historic information from the Great Belt and the Kiel Channel. This resulted in two estimates: a low with 40,006 annual passages and a high with 44,678 annual passages. The high estimate was used for the study.

To obtain a more reliable basis for the ship traffic projections, a 12-month ship traffic observation was carried out in the Fehmarn Belt. The German Navy made observations from their station at Marienleuchte on the island of Fehmarn. The observations indicated 45,477 ships per year.

4.5 *Project cost and construction period*

From the financial evaluation it appears that the solution models 4.1, 5.1 and 3 are the solutions, which show an acceptable rate of return as a basis for an implementation either following the Øresund model or the private financing using a Build Operate and Transfer (BOT) model.

The construction periods for the eight solution models vary between 6 and 8 years.

4.6 *Status and future*

The time plan for the future project steps after the Feasibility Study has been investigated and defined as follows:

The conclusion of the Feasibility Study is the starting point for the planning, which is limited to the coast-to-coast connection across Fehmarn Belt.

All results available from the feasibility study are regarded as a basis for the planning and no solution model has been selected at present. The decision to be made by the two Traffic Ministries is still open.

Neither in Germany nor in Denmark is a funding of a future Fehmarn Belt Link from governmental funds likely. Thus, focus is on project development processes with alternative financing schemes.

The Øresund Link offers answers to several open questions, and some basic arrangements are considered in the planning too:

- Set-up of an owner organisation owned by Germany and Denmark.
- Use of state guaranteed loans to finance planning and construction.
- Prepare functional tender designs and request contractors to perform their tender design within their own responsibility.
- Refinance the accumulated loans through user tolls from rail and road users.

Three phases are defined for the time period under consideration. These phases are:

- Decision Phase - until a national legal agreement is achieved in Denmark and Germany.
- Approval Phase - until the project based on the selected solution has been approved in both countries.
- Realisation Phase - from preparation of tender design to final start of operation.

In December 2000, the German and Danish Traffic Ministries agreed to continue developing the Fehmarn Project. The agreement included a market survey to examine the private interest in planning, constructing, and operating the fixed link. This process requires dialogue with possible investors and contractors and was named *Enquiry of Commercial Interest*.

To lead the dialogue the two governments have created a joint Danish-German project organization, *The Fehmarnbelt Development Joint Venture*. The operators of the Great Belt and the Øresund Fixed Link have been appointed partners by the Danish Government.

The market survey will continue until the summer of 2002. A political debate will follow, and the crossing is expected to open 10 years after the go-ahead has been given by the two governments.

5 THE TUNNELS

5.1 *The Great Belt Link - East Tunnel*

5.1.1 *The requirements*

The Civil Works were tendered as both an immersed tunnel and bored tunnel.

The requirements to the tunnel included inter alia:

- Design speed of 160 km/h
- Design life of 100 years
- Gradients not exceeding 1.56 %
- Main tunnel c-c distance sufficient to assure undisturbed ground between the two
- Emergency walkways on either side of the tracks

Figure 7. The Great Belt Tunnel.

- Escape routes to the other tunnel tube at maximum 250 m interval.

5.1.2 *Geology*

The bored tunnel passes two ground types:
- Quartenary glacial till
- Palaeocene marl

The quartenary glacial till contains boulders of granite and gneiss up to 2 m diameter or perhaps even more. Sand and gravel deposits are frequent (up to 20% of total mass) and their thickness may exceed 15 m. The layers dip at angles up to 40°.

The marl is a weak to moderately weak rock according to British Standard Classification. It is fissured and jointed, especially in the nadir zone where very fissured and crushed zones exist.

The interface between the layers represent special tunneling problems as they are inhomogeneous with high permeability.

5.1.3 *Main tunnels*

A dual tube bored tunnel was selected for economic and environmental reasons. The design was generally made in accordance with BS8110: *Structural use of Concrete*.

An alignment with a northern diversion was chosen to pass at a location with shallower water depth, to fulfil the gradient requirements and to pass an area with better ground conditions. The chosen alignment is 8 km long.

The main tunnel lining was designed as a 0.4 m thick pre-cast concrete lining with eight segments plus a key. The segment length was 1.65 m and the inside radius was 3.85 m.

The shallow parts of the tunnels were built as cut and cover tunnels due to insufficient cover for the TBM operations. They averaged a length of 250 m each.

5.1.4 *Cross passages*

The passage lining was designed as a cast iron (SGI) lining fully bolted and gasketted, with a mass concrete junction to the main tunnel.

The passages are used to store all equipment that is not essential in the main tunnels. They are also required as escape routes for passenger evacuation. All equipment was installed along the sides of the passages behind screen doors, leaving a central walkway. At each end of the passages there are emergency doors to each main tunnel tube. The doors are designed to the varying pressures generated by passing trains and are fire rated to 60 minutes. The dimensions of the passages were dictated by the escape route requirements of 1.85 m width and 2.1 m height. The internal diameter was fixed at 4.5 m.

Due to the risk of ground failure, a special construction sequence was introduced:
- Ground investigations
- Ground treatment
- Installation of safety doors
- Excavation of a 1.8 m diameter pilot tunnel
- Excavation of a 4.5 m diameter SGI lining using head and face boards.
- Staged excavation of the cross-passage/main tunnel junction structure, supported by a primary lining of shotcrete and steel ribs.

This ensured the maximum information on ground conditions and the safety of the operation.

5.1.5 *Sumps and drainage*

Great effort has been made to ensure the underground works are sealed against ingress of groundwater.

Tunnel leakage was expected to be very low:
- 0.1 liter/m²/day average
- 0.3 liter/m²/day maximum.

Leakage between the segments is channeled via the tunnel circumferential joints to a central drain channel in the tunnel invert leading to sumps at the low point of each tunnel. The drainage system is also designed for extreme situations such as fire fighting, and spillage of hazardous fluids. The sumps are placed at the low point of each main tunnel and consist of storage space below a track slab 50 m to each side of the low point. Two pump sumps with two pumps each are located in each tube. The full pumping capacity is reached with three pumps. The sumps and two similar but smaller sandtraps were excavated by hand using similar techniques as the cross-passages.

5.1.6 *Construction challenges*

The bored tunnel was constructed with four Earth Pressure Balanced (EPB) Tunnel Boring Machines, two from each side. The use of 4 machines turned out to be well justified. With the geological difficulties experienced, carrying out the construction with only two machines would have resulted in catastrophic delays.

The complex ground conditions with pressures exceeding 3 bar, made the tunneling extremely difficult. The high pressures especially impaired two operations:

- Entry in the TBM working chamber for repairs.
- Construction of the cross passages.

Any measures that could reduce the working pressure would be very beneficial. To facilitate these operations an innovative technique named Project MOSES was developed. The technique was to reduce the pore water pressure to manageable levels by de-watering wells from the seabed down into the marl. The water in the marl tended to flow in horizontal bands, and the de-watering had an effect over considerable distances. One example showed the effect 3 km away for a reduction of pore water pressure of only 3 m.

During a weekend, while maintenance was being carried out on the screw conveyer of a TBM, a water path was established between the working chamber and the Great Belt through some 12 m of clay till cover. Up until this time the earth surface seen had been stable. On the weekend, in contradiction with procedures, both bulkhead door and manlock had been left open. The result was rapid inundation of both this tunnel, part of the western side and the parallel tunnel. Clay was quickly excavated and dumped on the seabed above the location of the TBM, plugging the hole. After the situation was considered stable divers entered the tunnel, closed the openings in the bulkhead and the screw, after which the floodwater could be pumped out. This resulted in the replacement of a great amount of electrical components, and a delay of eight months.

Special precautions were also taken on the eastern side, due to the unstable ground conditions experienced on the western side. Water ingress from sand bodies around the TBMs often resulted in extensive ground failure and voids around the machines, which, in turn, led to water paths being established through the till cover. Large voids around the cutterhead and the obvious air leakage paths made manned interventions at the cutterhead unsafe. Scheduled maintenance was therefore deferred for too long. The lack of maintenance, together with the high thrust pressures employed in various attempts to build a plug of soil in the chamber, led to excessive wear on the cutterhead to the extent that the structural integrity of the head was at risk. The solution was to create a jet-grouted ground treatment zone in front, into which they could be advanced for repair. A shotcrete-lined repair chamber was formed around the TBM with access from the surface of the Great Belt through 2 m diameter bored shafts. From within these two chambers strengthening and replacement operations could be executed on the heads of both TBMs. This delayed operations for nine months.

Later maintenance operations were performed via a combination of dewatering and freezing in front of the cutterheads. These operations were time consuming but successful.

A month after one of these operations, the North-Eastern TBM was struck by a very serious fire. The most likely cause was a burst hydraulic hose. The fire had been fed by approximately 2000 l of hydraulic oil escaping from one of the tanks. The fire activated a gas alarm, set to trip the main power supply, and, unfortunately, the sprinkler and foam systems were not connected to the explosion-proof emergency power supply. The crew was rapidly forced to withdraw. Smoke divers were unable to reach the site due to the intensive heat, and an attempt to put out the fire by pumping in carbon dioxide also failed. It was decided to turn off ventilation and let the fire burn out itself. After 36 hours, firemen reached the front of the TBM and found the concrete lining severely damaged. Heavy spalling of the front five concrete rings was observed, and only 130 mm of the original 400 mm concrete thickness remained. Emergency repairs by shotcreting were performed immediately. However, a continued risk of collapse existed, because a connection to the seabed had developed while excavating the damaged rings. The tunnel was abandoned and closed by temporary bulkheads, while final repair solutions were developed. The chosen scheme was protection of the damaged section by a new inner lining of SGI segments. This reduced the inner diameter by 0.5 m, and necessitated some modifications to the railway installations. The repairs were performed under compressed air to increase safety against collapse. The TBM was not used further, and the rest of the northern tunnel was bored from the west side. The fire delayed the whole tunnel project by nine months.

5.1.7 *Conclusion*

The construction of the Great Belt Tunnel is undoubtedly one of the most difficult tunneling projects carried out this century. It required such advanced techniques that it could not have been constructed 20 years ago. The extremely difficult ground conditions required innovative techniques, like the subsea de-watering, considerable use of ground improvement and the construction of the cross passages through emergency doors.

In total, the construction was delayed more than three years.

The tunnel was completed with the full commitment of the Contractor and the Tunnel Boring Machine Manufacturer until the end.

5.2 *The Øresund fixed link*

The project was tendered as a Design and Build project based on an illustrative design.

5.2.1 *The requirements*

The link was designed in accordance with the Eurocode together with a Project Application Document (PAD) according to the following main requirements:

- The design life is 100 years.
- Road maximum gradient 2.5 %.
- Railway maximum gradient 1.56%.
- Two-way motorway, two lanes in each direction, and dual track railway.
- Minimum 40 MPa concrete mix.

5.2.2 *Geology*

The upper levels of the Øresund seabed are mostly limestone.

5.2.3 *The tunnel*

The immersed tunnel is 3.5 km long, 40 m wide, and 9.5 m high. It consists of 20 elements each 175.5 m long, made up of 8 segments.

The tunnel is ventilated longitudinally. The fans installed are 710 mm diameter reversible fans. Each road tube has 80 fans in four zones. Each zone has five rows of four fans. Each rail tube has 20 fans in four zones. Each zone has five fans installed in-line.

5.2.4 *Casting area*

Figure 8. The Øresund casting area.

The casting area was located at the northern end of the Copenhagen harbour, some 10 km away from the tunnel location.

The main elements were:

- Batching plant placed right next to the pre-casting factory.
- Pre-casting factory allowing for the assembly of reinforcement cages, casting and curing in a protected environment.
- The launch basin.

5.2.5 *Two level Launch Basin*

The Launch Basin was approximately 150 m wide and 400 m long. It was protected by 10 m high earth dams on all sides with a 100 m long sliding gate in front of the factory and a floating gate towards the sea. The basin was divided in two levels, with the upper level being at level with the factory, and the lower level being more than 10 m below the sea level.

5.2.6 *Production plant*

There were two parallel production lines allowing for the continuation of one production line in case of problems.

Reinforcement cages were prefabricated inside the factory. The segments were cast in one operation of 22 m length or 2700 m^3 concrete.

After curing the segments were jacked forward on installed skidding beams, each segment being supported on 36 hydraulic flat jacks.

Eight segments were assembled to a 175.5 m element on the skidding beams. Bulkheads were installed at the ends of the segment and main outfitting was performed in the dry.

The launch basin was then closed and flooded with water to 10 m above sea level. The elements were towed to the deep part of the basin for final outfitting and prepared for tow-out. The water level of the basin was returned to the level of the sea and gates reopened allowing for progress on the next elements at the high level.

5.2.7 *Preparing the seabed*

In total 2,000,000 m^3 of limestone and flint had to be removed. The dredging was restricted to a maximum allowed spillage of 5% of the dredged weight, for environmental reasons. A 56 m wide trench was dredged by cutter-suction of the seabed.

Strict requirements were laid out for the tunnel foundation. A gravel-bed had to be placed at depths of up to 22 m in a thickness of 1 m and with an accuracy of ±25 mm.

5.2.8 *Construction challenges*

There were many complex obstacles to be overcome in connection with the construction of the Øresund tunnel:

As mentioned above, the required accuracy of the gravel bed required new innovative techniques.

The Øresund is notorious for strong currents (more than 4 knots) ice and fog. The direction of the currents changes with changing weather systems. The salinity changes with the current resulting in a density fluctuation between 1.008 and 1.025 kg/m^3.

The Øresund is one of the most trafficked channels in the world. Approximately 40.000 commercial vessels pass per year (more than 100 per day). Øresund is one of two routes between the Baltic Sea and the Atlantic. A Vessel Traffic Service was established to ensure the safe construction. Russian translators were employed to assist the Russian crews. Four guard boats operated simultaneously. During the construction 143 violations and 600 near misses were reported.

During immersion of element number 13 one of the temporary bulkheads failed, and the element dropped to the bottom of the trench. At the time of the bulkhead failure, the element was approximately 1.3 m above its theoretical foundation level. The element sank almost in line on the gravel bed, approximately 3 m from the previous element. The bulkhead failure happened at the outer railway tube, and caused the element to tilt and sink to the bottom in seconds, taking the pontoons with it. Air escaped through the access shaft at a speed of several hundreds m^3 per second! After the element had been filled, the shaft functioned as a large waterhose with a plume reaching some 30 m above sea level. Immediate action to identify the cause of the failure was undertaken. The main concern was the already installed elements number 11 and 12 together with the floating element 14, as all three had bulkheads currently under water pressure. Diver inspections showed that a temporary concrete beam supporting the base of the bulkhead had failed at the connection to the base-slab. Later investigations revealed missing reinforcement in the connection of the temporary beam to the base-slab.

As the element had come to rest in the tunnel trench, the next tasks were to remove the element from its position and to allow for inspection and clear the location for repositioning of the element or its replacement. After initial repairs the element was taken to a temporary gravel bed and prepared in the trench some 300 m away. A number of criteria were set out regarding structural integrity and durability. Special considerations were given to shear keys and pre-stressing tendons. It was considered probable that some of the tendons (especially at the base) had been broken. In order to reduce the load on the tendons at the base, the element was ballasted in a hogging condition, and the element was only lifted 3 m above the trench bottom giving the benefit of the hydrostatic pressure on the ends. The element was examined and damages evaluated. The element was found reusable, and after a second preparation of the gravel bed the element was towed to its final position. Two months were lost. Operations on the remaining elements were accelerated with the result that the delay was not only recovered but, in fact, the last element was installed one month prior to the timing of the original schedule.

5.3 *The Fehmarn Belt link*

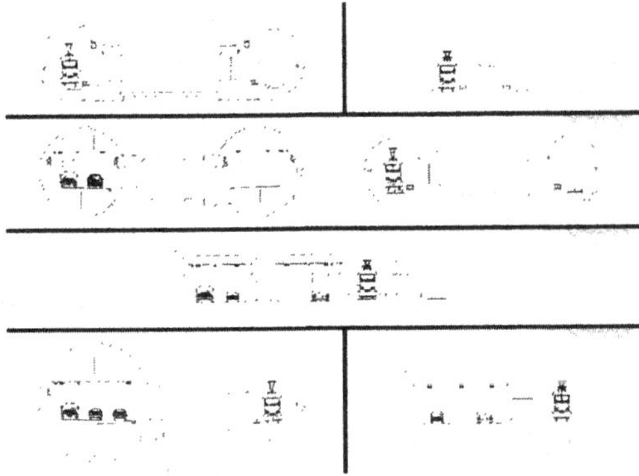

Figure 9. The Fehmarn tunnel solutions.

5.3.1 *Project basis*

The coast-to-coast investigations were limited to a corridor of about 5 km width at about the same location as the current ferry route.

5.3.2 *Site conditions*

The international sea-route (Route T) with guaranteed water depths of 17 m or more passes the corridor under investigation. Traffic from both the Great Belt and the Kiel Channel to the Baltic Sea passes the corridor. About 10 km west of the corridor the Kiel-Ostsee route joins Route T.

A substantial amount of large vessel traffic, which is not able to pass the Øresund due to lower water depths, use this route.

Geological and geo-technical investigations have given the following information:

- Post-glacial deposits dominated by sand with local occurrences of gravel, silt, gyttja and peat and late glacial deposits composed of clay, silt and sand.
- Glacial deposits composed of clay till with subordinate sand till, gravel, sand, silt and scattered boulders.
- Tertiary unit composed of plastic clay and clay silt.
- Limestone unit composed of Cretaceous chalk.

The evaluation of the geology and profile with respect to the suitability of the different solution models has lead to the following alignments within the corridor:

- Bored Tunnel - westerly
- Bridges - central
- Immersed Tunnel – easterly.

5.3.3 *Design basis*
As mentioned previously the Eurocodes have been adopted for the design of the Øresund link, with the addition of a PAD. On the Metro in Copenhagen, currently under construction, the Eurocodes have also been used. For a bi-national project like the Fehmarn Link, classified as a major European infrastructure project, it is considered appropriate to adopt the Eurocodes.

A number of functional requirements have been established. These are similar to the ones used for the Øresund project.

5.3.4 *Tunnel ventilation*
The railway tunnels of all the Solution Models do not need artificial ventilation due to the piston effect in single-track tunnels. During the study it has been discussed if the use of piston relief ducts would save traction power, but the forced ventilation will be disturbed and an artificial ventilation system could be necessary. At this stage of the project it is therefore recommended to avoid pressure relief ducts.

With an enclosed length of around 20 km special solutions are required in order to ventilate the road tunnels.

The solution is highly dependent on emission data and concentration limits, especially CO, diesel fumes, and NO.

Simple longitudinal ventilation is not effective for distances longer than 4 km, which rules out this solution for the Fehmarn Belt Crossing.

The options are semi-transverse and transverse ventilation. To minimize the overall tunnel cross-section and thereby reducing the cost, semi-transverse ventilation was selected. The ventilation of the tunnel will be broken down in four sections longitudinally, two being ventilated from shore and the two central being ventilated via a ventilation island in the middle of the Fehmarn Belt.

This configuration will reduce the road capacity from the theoretical limit of the tunnel cross-section. To attain the theoretical capacity two to three ventilation islands would be required. However, the capacity with one island will be largely sufficient according to the traffic prognosis currently available.

5.3.5 *Ventilation island*
As the alignments of the bored and the immersed tunnel differ, so does the optimal location of the ventilation island.

For the immersed tunnel it is located above a lime stone dome east of the straight line connecting Rødbyhavn and Puttgarden. The tunnel alignment passes this dome about 10 km and 8 km from the Fehmarn and the Lolland shorelines, respectively. Placing the island above the dome reduces harmful settlements of the immersed tunnel. The limestone top is in elevation -42.5 m, and the bottom of the immersed tunnel is foreseen to be at level -37.0 m.

The alignment for the bored tunnel is chosen in order to have the boring operation made in uniform and favorable soil. The alignment of the bored tunnel is more westerly compared to the alignment for the immersed tunnel. The bored tunnel is deep below the sea bottom, elevation -72 m, and is, therefore, less vulnerable to uneven settlements of the island. The ventilation island for the bored tunnel solution is located about midway between Fehmarn and Lolland.

To protect the ventilation shaft against ship impact, the island is designed with two different mechanical principles with different kinematic characteristics, namely:

The vessel slides up on the island surface with limited penetration (friction island concept).

The vessel penetrates deeply into the island core (penetration island concept).

In Fehmarn Belt super tankers (up to 250,000 DWT) with large pollution potential cruise in the East-West direction, parallel to the longitudinal axis of the tunnel. Damage to the ship hull should be avoided. The main water transport by current is East-West; thus, the blocking effect of a long island will not be significant. A friction island is the proper choice in this direction.

Ships cruising in the North-South direction are smaller, and they will normally not carry significant quantities of polluting goods. Damage to the ship is no major issue, but it is of imperative importance to achieve the necessary protection of the ventilation shaft and a minimum blocking of current. A penetration island is the proper choice in this direction.

Based on the above requirements the island is shaped as a truncated cone with an elliptic base. The cross section of the island perpendicular to the current in Fehmarn Belt has been kept at a minimum, while, in the longitudinal direction, the island has gentle slopes.

6 LONG TUNNELS

Figure 10. Fehmarn solution model 4.1.

In Europe, the public uneasiness in connection with tunnels has increased significantly the last years due to severe accidents in road tunnels, especially in the Alps. Long tunnels do pose special problems, which have to be addressed to ensure that tunnels will be adequately safe and people will continue to use them. Some of the problems are mentioned below.

6.1 *Ventilation*

Ventilation schemes are very important as the supply of fresh air must be available to the users of the tunnels.

Solutions to ventilation on both links already built have been to use conventional ventilation methods. The Øresund with its 4 km tunnel is, however, close to the limits for what can be achieved with longitudinal ventilation. On the Fehmarn Belt crossing, the road tunnel solutions are in the vicinity of 20 km, and ventilation is an important issue. Finding the right concept can save enormous amounts of money and may makes the difference between a project being Financially sound or not.

6.2 *Road layout*

The recent accidents have mainly occurred in tunnel tubes with road traffic going in both directions. Future tunnels will most probably focus even more on safety against head-on collisions with alternate layouts.

6.3 *Evacuation routes*

In many tunnels evacuation is mostly based on self-help. In the long tunnels, evacuation becomes increasingly complex, and special measures must be applied to assure the right behavior of the users.

6.4 *Tunnel control & emergency team access*

It has shown to be of paramount importance to have precise procedures and commando lines, so all involved personnel knows who does what. The lack of coordination between different operators of the tunnel can lead to wrong decisions being taken. These problems are of particular importance for cross border links.

The long tunnels must not lead to excessive response times to emergencies, and the whole safety concept must be thoroughly investigated.

7 CONCLUSION

A number of important and challenging infrastructure projects have been under development in Denmark over the last 15 years. Several records were broken during the construction of these, and several very complex construction challenges were overcome. Two projects are already in operation, and a third is currently being investigated, with a go-ahead being given within the next few years.

REFERENCES

COWI. 1999. *Femer Bælt-forbindelsen. Forundersogelser - resumerapport.* Copenhagen: Trafikminsteriet.

Gimsing, N.J. et al. 1997. *East Tunnel.* Copenhagen: A/S Storebæltsforbindelsen.

Gotfredsen, H.H. 1993. *International symposium on technology of bored tunnels under deep waterways.* Copenhagen: Teknisk Forlag.

Lundhus, P. et al. 2000. *Immersed Tunnel Conference, proceedings 5-7 April 2000.* Copenhagen: Øresundskonsortiet, Øresund Tunnel Contractors I/S and International Tunnelling Association.

North American Tunneling 2002, Ozdemir (ed.)
© 2002 Swets & Zeitlinger, Lisse, ISBN 90 5809 376 X

Why is policy important for successful management of complex underground and infrastructure projects?

J.J. Reilly, P.E., C.P.Eng.
John Reilly Associates International, Ltd.

ABSTRACT: The successful planning, management, design, construction and operation of major, complex, underground projects requires careful, long-term, strategic, sometimes Machiavellian, thinking. Many issues that affect the success of projects are related to public policy and public processes – therefore, the strategic use of such policies and processes is a key to meeting essential project goals and objectives – as well as satisfying the many diverse "stakeholder interests".

1 INTRODUCTION

"It shall be the Policy of the United States that....". We know these words and most of us can probably fill in the next few phrases. Therefore we understand what policy is. However, we generally do not participate in policy formulation or implementation, even though policy is important in terms of our projects – including their funding, direction, effectiveness and return-on-investment.

2 WHAT IS POLICY?

The process of defining, establishing, communicating and implementing public policy, for public works and underground construction, is neither well understood nor well discussed in engineering literature. Less well understood is the process for changing established policy.

Policy is closely related to politics - one cannot be considered without the other. Policy making involves all factors which effect the current and future environment of a project.

2.1 *Policy as process*

Policy is not static - it is primarily a process, that involves the whole population, conceived, shaped, communicated and implemented by interested and motivated citizens. These include governmental and managerial leaders as well as citizen groups, concerned with the project and its impact.

The policy-making process also includes elected representatives who will subsequently codify policy in laws and regulations. The fluidity of policy determination and acceptance stands in contrast with the more static and rigid environment which follows codification and legislation.

Policy formulation is fluid, dynamic, evolutionary and opportunistic - driven by factors such as societal needs, economics, competition, profit and recognition.

Codification produces a static, stabilizing, definitive and inhibiting environment, that is difficult to change and revise.

This means that there is a tension built into the process of development, codification and implementation of policy. A new paradigm may be required to allow us to be able to grasp the significance of events at the immediate, day-to-day level, and then to put the issues in context and perspective in order to satisfy long-term societal, generational and heritage needs.

3 POLICY FOR INFRASTRUCTURE

As planners, engineers and contractors we find it difficult to approach policy and the "soft" questions required by today's infrastructure and underground construction - which must increasingly take place in dense, complex, restrictive urban environments.

However, we are concerned about those issues which affect the quality of our lives, those of our children and the heritage which we leave to future generations.

3.1 *Demands and publicity*

Demands on our professional skills continue to increase and projects today must meet ever more stringent environmental and other regulatory standards – based on policy determinations.

We know that media attention can be selective, transitory, and biased. While the media increases public awareness, it can also prejudice a cause, derail an initiative, delay or stop a necessary project and unnecessarily raise unreasonable expectations in the general public (Dawson 1995, Tunnel 1995). Articles such as these mean that we need to respond to media initiatives clearly and early.

3.2 *Participation in policy determination*

As a profession therefore, we need to be involved in policy and the policy process itself. We should identify those policies which will be of help to our industry that need to be formulated, debated and communicated by key policy makers and leaders. We need to answer new challenges facing our urban environment, which include conflicting development alternatives, public security, public-private ownership, privatization, social equity and, environmental justice.

4 ENGINEERS AND POLICY

As engineers we believe that problems can be solved by the application of scientific principles, practical considerations, hard work and commitment based on our training and experience. This is our profession, our training and our aptitude.

Our training is a good basis for technical decisions. However, with respect to policy, it may be necessary to broaden the base from which we understand the complexity of project issues in a political and policy environment. With our projects subject to increased publicity and heightened public awareness, a more coherent and integrated approach to project planning and management may be necessary.

4.1 *Effects on management of projects*

As the complexity of our urban environment increases and environments become interdependent on a national and perhaps an international scale, it becomes more difficult to integrate day to day decisions into a coherent long term vision - which satisfies all policy, environmental and stakeholder requirements, demands and conditions.

This is a challenge for engineers, and affects their relationships with other 'stakeholders' - in the context of complex, urban underground projects. It argues for a more comprehensive view of the project development and delivery process, as discussed in the following sections.

An example of such thinking is the ITA 1990 Report "Cost - Benefit Methods for Underground Urban Public Transportation Systems" which implies that major policy-related decisions can be made based on cost considerations. Perhaps this approach has passed its time.

5 CONCLUSION

If we, as engineers and contractors, do not appreciate the influence of policy and related disciplines, we will continue to struggle against legitimate needs of key project "stakeholders" with resulting delays and cost increases to our projects. By "stakeholders", we mean all individuals, groups or organizations that are affected in any way, positively or negatively, by the project.

REFERENCES

R.J. Dawson 1995, 'Communication Strategies for Major Metro Rail Projects - the Los Angeles Metro Rail Program Under Siege' *Proc Project Management Institute*, Annual Conference, October 16-18.

Tunnel 1995, 'What is going on in Los Angeles? – Metro Los Angeles: Inkompetenz auf ganzer Linie?" Vol. 6/95 pp 8-18.

Session 4b, Track 2

Security and special uses of the underground

North American Tunneling 2002, Ozdemir (ed.)
© 2002 Swets & Zeitlinger, Lisse, ISBN 90 5809 376 X

Applications of underground structures for the physical protection of critical infrastructure

Don A. Linger
Deputy for Technical Programs, Defense Threat Reduction Agency, 6801 Telegraph Road, Alexandria, VA 22310

George H. Baker
Associate Professor, James Madison University, College of Integrated Science and Technology, 701 Carrier Drive, Harrisonburg, VA 22807

Richard G. Little
Director, Board on Infrastructure and the Constructed Environment, National Research Council, 2101 Constitution Avenue, NW, Washington, DC 20418

ABSTRACT: The U.S. President's Commission on Critical Infrastructure Protection (PCCIP), convened in the wake of the bombing of the Murrah Federal Building in Oklahoma City, concluded that the nation's physical security and economic security depend on our critical energy, communications, and computer infrastructures.[1] While a primary motivating event for the establishment of the commission was the catastrophic physical attack of the Murrah Building, it is ironic that the commission focused its attention primarily on cyber threats. Their rationale was that cyber vulnerabilities posed a new, unaddressed challenge to infrastructure security. This approach was further questioned by the events of September 11, 2001 and the subsequent biothreat events in America.

During and shortly after the convention of the President's Commission on Critical Infrastructure Protection in 1997-99, a working group met to look into the physical protection of critical U.S. infrastructure using underground structures. This "Underground Structures Infrastructure Applications" (USIA) group provided a timely balancing discussion of issues surrounding the physical vulnerability aspect of the infrastructure protection problem. The group convened a workshop on the subject under the auspices of the National Research Council's Board on Infrastructure and the Constructed Environment.[2] The present paper draws on the working group deliberations and NRC Workshop proceedings to address infrastructure assurance in the context of maximizing the physical protection of high value infrastructure by the use of underground construction. Although the PCCIP did not directly address a role for underground facilities (UGFs), its final report recommended a program of government and industry cooperation and information sharing to improve the physical security of critical United States infrastructure.

We develop specific recommendations for the use of underground facilities to support implementation of the general PCCIP challenge. It is our contention that underground facilities reduce the risk of infrastructure disruption from both physical and cyber attacks. In the latter case, undergrounds provide secure reserve operations and back-up data storage locations to ensure the reconstitution of critical electronic information systems. In addition, noting that the Norwegians have placed much of their critical infrastructure underground, we provide information on the Norway's experience gathered on a fact-finding trip to that country organized by one of the authors. The events of September 11th, 2001 and the use of underground locations for terrorist security by the Taliban, provide a strong impetus for the use of underground construction to protect vital infrastructure.

[1] Critical Foundations: Protecting America's Infrastructures, Report of the President's Commission on Critical Infrastructure Protection, October 1997
[2] Use of Underground Facilities to Protect Critical Infrastructures, National Academy Press, Washington, D.C., 1998

1 INTRODUCTION

Civil infrastructures are vital public artifacts that support a nation's economy and quality of life. They present a massive capital investment, and, at the same time, are an economic engine of enormous power. Modern economies rely on the ability move goods, people, and information safely and reliably. Consequently, it is of the utmost importance to government, business, and the public at large that the flow of services provided by a nation's infrastructure continues unimpeded in the face of a broad range of natural and manmade hazards. This linkage between systems and services is critical to any discussion of infrastructure. Although it may be the hardware (i.e., the highways, pipes, transmission lines, communication satellites, and network servers) that initially focuses discussions of infrastructure, it is actually the services that these systems provide that is of real value to the public. Therefore, high among the concerns in protecting these systems from harm is ensuring the continuity (or at least the rapid restoration) of service.

In light of the importance of these systems, there is growing concern over the national dependence on "soft" facilities (aboveground, physically exposed, unhardened) for housing high value civilian infrastructure systems. The terrorist attacks of September 11, 2001 mandate renewed attention and efforts to develop better solutions to this problem. Underground facilities provide the most effective physical protection available and can be designed to provide immunity to aircraft impact, truck bombs and even nuclear weapons. For many reasons, underground construction has not been widely used in the past. However, heightened concern about terrorist capabilities and threats within the United States, the recognized importance of infrastructure to our physical and economic well-being, and significant improvements in underground construction technology and cost-effectiveness argue for a fresh look at underground structures. Most critical above ground functions could be safely housed in underground structures with attendant benefits related to protection, energy savings, and environment preservation. Officials looking to broaden their options for protecting citizens and workers from future terrorist attacks and other emergency situations should seriously consider underground structures as a major resource for high value infrastructure elements.

2 PAST UGF PARADIGMS

During the Cold War, high value defense-related facilities were designed and built underground in a large part to counter the threat of strategic nuclear attacks. Underground facilities were designed to help ensure the continuity of government, assure survivable military command and control of the U.S. strategic forces, and provided the necessary safe havens from which to respond to attacks against the United States.

Within the over-arching national security framework of the Cold War, the general public came to view the federal government's UGFs as necessary tools for military and civil defense purposes, but little else. Under the current national defense paradigm, many of these UGFs have been or are being closed or phased down because no compelling case has been made for other domestic security benefits. Furthermore, stigmas associated with the Cold War in general, and civil defense (e.g., fallout shelter) programs in particular have fostered a negative impression of UGFs in the minds of many Americans.

Over the past several years, growing national concerns have emerged regarding the increased vulnerability of our domestic civil infrastructures to attack or disruption, and the evolution of threats against high value civilian targets. These concerns were tragically validated by the events of September 11th, 2001. The continued reshaping of our national defense strategy and the related transition of many national security functions to civil and commercial enterprises have increased the risks and consequences of infrastructure attack. This growing dependence of national defense on civilian infrastructure was noted with concern in the report of the President's Commission on Critical Infrastructure Protection:

> National defense is no longer the exclusive preserve of government, and economic security is no longer just about business… we are convinced that our vulnerabilities are increasing steadily, that the means to exploit those weaknesses are readily available, and that the costs associated with an effective attack continue to drop.[3]

The current emphasis on Homeland Defense is now further reshaping the national security debate over such topics as infrastructure protection, information assurance, and preparedness for catastrophic terrorist attacks. Physical and cyber attacks on America's critical infrastructures – its telecommunications, electrical power, gas and oil, banking and finance, transportation, water supply, government services, and emergency service systems – are viewed as an increasingly attractive avenue by which terrorists, rogue nation states, and non-state actors can attack the U.S. directly and seek to coerce national public opinion and decision-makers into responding to their political demand. The potential multiplicity of threats to our critical infrastructures (e.g., physical, cyber, biological, chemical, and elec-

[3] President's Commission on Critical Infrastructure Protection, *Critical Foundations: Protecting America's Infrastructures*, Report, October 1997, pp. ix-x (hereafter cited as *Critical Foundations*).

tromagnetic) provides an opportunity for a fresh look at UGFs – their advantages and new, more widespread applications in the 21st century.

3 ATTRIBUTES OF UNDERGROUND FACILITIES

Underground structures possess a number of inherent attributes that make them attractive options for physical protection. They provide increased security from direct physical attack, lower energy costs and maintenance costs, improved environmental control, and better land use efficiency through conservation of the earth's surface environment.

The physical protection provided by UGFs is superlative. They can be built to withstand effects from essentially any explosive device including nuclear weapons. Their physical security benefits make them particularly well suited to ensuring the continuity and reconstitution of critical infrastructure functions. Dollar for dollar, underground construction provides higher levels of physical protection than similarly sized hardened above-ground structures since specially designed facade treatments, interior wall reinforcement and blast-resistant window glazing are not needed (more on cost benefits in Section 5). Although UGFs do not provide direct protection against cyber attacks, their physical strength makes them a safe haven for critical backup media crucial for recovery following a cyber attack.

Despite the unprecedented scale of the recent attacks in New York City and Washington, DC, experience over the past several decades has shown that the preferred weapon of choice for attacks against buildings and other infrastructure is the vehicle bomb. Although the magnitude of the threat and the likelihood of an attack against a specific facility will vary considerably, there are four basic concerns of protective design that can be satisfied through the use of underground construction. These protective design principles are: (1) the establishment of a secure perimeter; (2) the prevention of progressive structural collapse; (3) the isolation of internal threats from occupied spaces; and (4) the mitigation of debris resulting from the damaged façade and window glazing. Other considerations, such as the tethering of nonstructural components and the protection of emergency services, are also key design objectives that require special attention.

Secure underground facilities can readily address all of these concerns. For example, with a limited number of access points, the facility can be more easily secured against unauthorized access and interior security maintained with a combination of technology and human assets. By providing full blast protection from conventional weapons, the threat of progressive collapse and debris-related injuries is eliminated. Depending on their geographic location, critical infrastructures and facilities will also be faced with a wide range of natural hazards such as earthquakes, extreme wind events, landslides, and floods. Underground facilities can provide true multi-hazard protection against these events as well.

Underground facilities exhibit some disadvantages when compared to surface facilities. These include limited ingress and egress points and potential safety hazards, particularly from fire, in some underground locations. Additionally, a small fraction of the population is averse to working underground although modern design approaches can make these facilities virtually indistinguishable from above-ground spaces. First costs for small underground structures can be higher than for surface facilities. However, for large facilities, costs per square foot for underground structures are comparable to conventional above ground construction and may be lower for comparably hardened above ground structures.

In the United States, underground facilities have become associated predominantly with defense applications. Their applicability for civilian and commercial purposes is less well known and appreciated. However, most critical infrastructure functions and missions, both military and civilian, are amenable to underground locations. Potential infrastructure applications for underground structures are primarily seen to apply in five infrastructure sectors: (1) Information and Communications, (2) Banking and finance, (3) Physical distribution systems (pipelines), (4) Energy, (5) Vital human services.

In many applications, underground facilities can be designed as "dual-use" facilities – serving as locations for normal day-to-day community functions and as well as emergency/ crisis centers.

It is recognized that underground facilities are only one of many potential solutions to the Nation's emerging infrastructure protection problem. Nonetheless, their increased use would provide a major line of defense against physical attack of our most critical domestic systems.

4 UNDERGROUND INFRASTRUCTURE EXAMPLES AND EXPERIENCE

There are many well known defense applications of underground facilities in the U.S. These include the Cheyenne Mountain Underground Complex, hardened underground switching centers throughout the AUTOVON communications network, ICBM silos, and the former Congressional relocation facility at Greenbrier, West Virginia.

In the United States, applications of underground technology to industrial infrastructure are limited and less well known. Examples include buried telecommunications and cable routes, back-up data media storage, underground warehouse and storage

Figure 1. Norwegian Underground Electric Power Generation Station.

Figure 2. Norwegian Underground Oil Storage Facility.

network in Kansas City, petroleum storage, buried electric utilities, and subsurface petroleum and natural gas pipelines. UGFs have been adapted for use as subterranean occupancies for commercial, residential, recreational, and educational use.

In contrast, European countries such as Norway, Sweden, and Switzerland have aggressively exploited underground technology to make full use of underground space. These countries have demonstrated the feasibility, affordability and efficiency of using underground facilities, not only for military applications, but critical infrastructure protection as well. The experience of Norway is particularly instructive. Norway integrates underground space into all aspects of its national emergency preparedness planning. For example, all electric power generation is housed in underground structures, as are the national archives, water supply and treatment facilities, civil defense, war headquarters, financial centers, and an air traffic control facility. Many of Norway's facilities serve dual-uses for civilian and national security purposes.

The authors participated in a fact-finding tour of Norway's underground infrastructure in October 1998. Norway uses underground technology in every infrastructure sector. Underground facilities visited included a water treatment plant, a munitions factory, a waste water treatment plant, an oil storage complex, a banking center, an air traffic control center, a dairy product processing and storage plant, a dual-use Olympic ice arena – emergency shelter facility, a telecommunications center, and an electric power generation station.

Common site features included multi-layered security (i.e., card-key entryways, CCTV, guards). Many sites had electromagnetic protection against EMP and RF weapon environments. Supporting utilities and personnel were protected by blast doors, air filters, back-up power (including large fuel reserves), alternate cooling services, uninterruptible power supplies and personnel shelters. The Norwegians also provided alternate operating locations for all sites. Plans for relocation were in place and regularly exercised.

In addition to superior physical protection of critical assets, major motivating factors for underground construction in Norway include increased security, lower life cycle costs, conservation of above ground space and environmental protection.

The geology and topography of Norway are particularly favorable to underground construction and the integration of surface and subsurface facilities. In the Norwegian experience, building underground is equal to or less costly than building comparable surface facilities and provides far greater security.

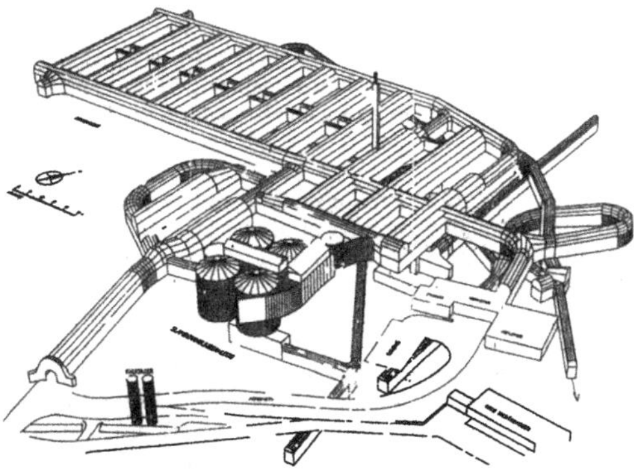

Figure 3. Norwegian Underground Waste Treatment Plant.

5 COST BENEFITS

Although there is very little published cost data on commercial underground facilities in the U.S., available evidence and life-cycle cost analysis indicates that UGF costs are competitive with standard above ground construction, and over the life of a facility, savings may be realized due to decreased energy and maintenance costs.

Two separate case studies are available. The first is based on data graciously provided by the Norwegian Defence Construction Service on the cost of in-rock versus above-ground facilities[4]. The second is a preliminary analysis of U.S. construction costs comparing the up-front and life cycle costs of hardened above ground versus underground facilities.[5]

The Norwegian Defence Construction Service has compiled comparative data on the investment, operations, and maintenance costs associated with both above-ground and in-rock underground defense facilities. The Norwegian study indicates that capital costs associated with constructing in-rock, inherently hardened, underground facilities are 25% higher than those of standard, unhardened above-ground facilities. The data confirm that up-front costs are higher for underground construction. However, contrary to conventional wisdom, the Norwegian experience shows that <u>life cycle costs for underground facilities</u> larger than 5,000 square meters (~50,000 square feet) <u>are lower</u> than those for above ground facilities. In fact, the Norwegian data shows that large underground facilities are 40% less expensive to operate and maintain on a unit area basis.

Figure 4 is a comparison curve depicting life cycle costs in Norwegian Krone (presently 1 dollar ≈ 9 Kroner) per spare meter. The costs are for facilities with internal areas of up to 30,000 square meters and a life cycle of 20-60 years. Factors included in calculations were investment cost, O&M costs, facility area, operational lifetime, and a 7% discount rate.

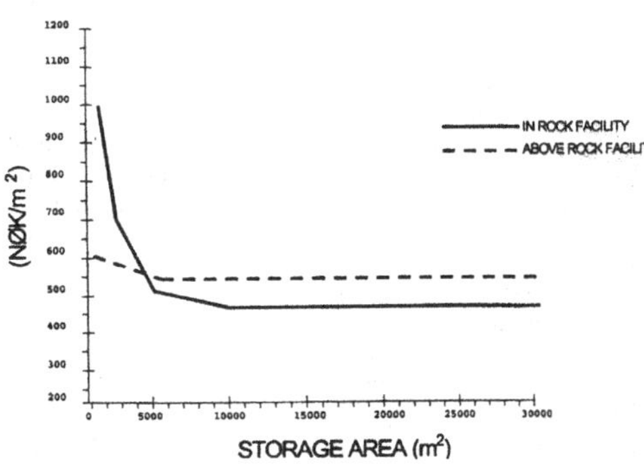

Figure 4. Lifetime cost comparison of facilities in and above rock (Costs in Norwegian Kroner).

It should be noted that the cost of hardening above ground facilities to a level approaching that of underground facilities, significantly reduces any up-front cost differential. This is apparent from the second case study of U.S. facilities.

---

[4] Norwegian Defence Construction Service, Defence Facilities in Rock, Oslo, Norway, 1998

[5] Frank Gertcher, Benefits and Costs of Protecting Infrastructure Systems Against Terrorist and Related Threats: Cost Analysis, Defense Threat Reduction Agency, Report No: RT-0103-99.

The U.S. facility cost study indicates that underground, blast-protected, cut and cover buildings can reduce the present value costs compared to semi-buried or above ground buildings, even at lower levels of hardening.

Table 1, from reference 5, presents new construction costs for buildings of 10,000; 20,000; 50,000; and 100,000 square feet of protected interior floor space. Costs per square foot are presented for protection levels 1, 2, 3, 4, and 5. Details on the nature of the cost estimates presented are covered in the notes that follow the table.

As can be seen in Table 1 – new, smaller semi-buried and underground buildings with lower levels

Table 1. Blast-Hardened New Building Construction Costs for Data Processing Center.[7]

Interior Floor Area (Ft^2)	Stories		Construction Cost (per square foot[2] FY98 Dollars)[1]				
	Above-ground	Underground[8)]	Level 1[2)]	Level 2[3)]	Level 3[4)]	Level 4[5)]	Level 5[6)]
10,000	1	0	178	202	204	269	---
	1	1	199	204	216	288	394
	0	1	191	191	199	260	352
	0	2	195	196	208	270	356
20,000	1	0	146	153	171	225	---
	1	1	157	165	178	224	267
	0	1	157	163	173	213	250
	0	2	154	160	177	217	256
50,000	1	0	127	134	149	189	---
	1	1	134	142	143	183	227
	0	1	136	144	149	186	219
	0	2	133	139	140	178	212
100,000	2	0	114	119	133	172	---
	3	0	123	128	130	171	---
	2	1	122	127	126	164	---
	1	1	117	129	133	163	205
	0	2	122	127	133	163	197

NOTES

1) Estimates of construction costs per square foot of interior floor area were obtained as an output from the Construction Cost Management Analysis (CCMAS) model. This is a patented, validated, Government-owned model. The model equations are based on actual cost and design data for a large number of Government-owned data processing centers in the continental United States. The model has proven to be a reasonably accurate predictor of actual construction costs. Errors above and below the expected value are stochastic (normally distributed).

2) Level 1 buildings are not hardened.

3) Level 2 buildings blast-hardening features include: steel or reinforced concrete roof, steel or concrete frames, non-load bearing reinforced masonry or concrete walls, a seismic zone 3 foundation and floor, a minimal number of thickened, protected glass windows, and steel doors. Level 2 buildings are designed to withstand the blast pressure of 50 pounds of TNT at a distance of 40 feet. Underground portions of such buildings have substantially greater blast survivability.

4) Level 3 buildings blast-hardening features: steel or reinforced concrete roof, a structural ceiling slab 10 feet below the roof level, steel or concrete frames, non-load bearing concrete or reinforced concrete walls, a seismic zone 3 foundation and floor, a minimal number of thickened, protected glass windows, and steel doors. Level 3 buildings are designed to withstand the blast pressure of 220 pounds of TNT at a distance of 40 feet. Underground portions of such buildings have substantially greater blast survivability.

5) Level 4 buildings features: concrete, heavy steel or heavy steel joists for the roof, a structural ceiling slab 10 feet below the roof level, concrete or heavy steel frames with lateral load resisting elements designed for seismic zone 4, cast-in-place 18-inch thick reinforced concrete walls with 14-foot floor-to-floor heights, concrete, heavy steel, or heavy steel joist floors strengthened to seismic zone 4, a seismic zone 4 foundation, and a minimal number of thickened, protected glass windows. Doors are blast-hardened at a cost of $25K per door (see Fairchild back-up satellite operations center emergency exit door estimates, updated to 1998 dollars). Level 4 buildings are designed to withstand the blast pressure of 500 pounds of TNT at a distance of 40 feet. Underground portions of such buildings have substantially greater blast survivability.

6) Level 5 buildings blast-hardening features: use only semi-buried or underground options, add two feet of earthen cover and a concrete burster slab, a 24-inch thick reinforced concrete slab roof (no ceiling slab), 24-inch thick cast-in-place concrete exterior load-bearing walls, 12-inch thick cast-in-place concrete interior load-bearing walls, and a continuous, mat-type foundation, strengthened to seismic zone 4. Note that a frame is not required, since walls are load-bearing. The building has no windows. Doors are blast-hardened (4 entries) and add $50K per door to the cost of the building (see Fairchild back-up satellite operations center main entry door estimates, updated to 1998 dollars). Level 5 buildings are designed to withstand the blast pressure of 1000 pounds of TNT at a distance of 40 feet. Underground portions of such buildings have substantially greater blast survivability.

7) Costs presented do not include cyber, chemical, or biological threat protection costs. Such costs are assumed to be approximately the same for above-ground, semi-buried, and underground buildings (see discussions in main text). Such costs will increase the per-square-foot costs by the same amount for each entry in the table, but will not change the relative rank order by cost of the alternatives considered (above-ground, semi-buried, and underground).

8) It is assumed that excavation and backfill for semi-buried and underground structures occur in moderate soil conditions (alluvial, clay, soil/gravel mix, etc.). Blasting and other heroic measures are not required. Also, it is assumed that ground water conditions do not require pumping, special barriers, etc., for excavation, backfill, or construction. If extensive rock or groundwater problems exist, they could increase the cost of the building substantially from the amounts presented.

of hardening have higher when compared with above-ground buildings. Semi-buried and underground buildings are cost-competitive for higher levels of hardening and larger buildings. However, in almost all cases, the cost differences per square foot are not large. Table 1 also shows that economies of scale are clearly present for larger buildings. Finally, initial construction costs are slightly lower for multi-stories underground compared to single stories underground, assuming that excavation and backfill occur in moderate soils (see note **8**).

The U.S. cost study supports three important conclusions:

(1) Economies of scale work in favor of larger underground hardened buildings over hardened above ground buildings
(2) Underground or semi-buried buildings can be expected to have significantly reduced annual exterior maintenance and HVAC costs compared to above ground buildings, which lends itself to favorable life cycle cost considerations
(3) For higher levels of hardening (higher threat and risk profiles), the inherent hardness of UGFs often makes them less expensive than above ground facilities.

Although no data are available, the authors believe that these figures suggest that the retrofit of critical infrastructure functions into existing underground space can also provide substantial up front cost savings for infrastructure owners and operators.

In conclusion, the initial construction costs of UGFs can be considerably higher than for above ground facilities, but may be less for higher levels of hardening. Over the entire life cycle, operations and maintenance cost savings greatly improve UGF cost benefits. The life cycle costs of newly constructed UGFs can be very competitive with above ground facilities, i.e., in many cases a life cycle cost advantage can be anticipated.

6 SUMMARY

Domestic infrastructure services and national security functions are increasingly reliant on soft infrastructure and services provided by non-defense organizations. The attacks of Based on the September 11, 2001 greatly increased the national perception of the threats against these infrastructures and facilities. Future attacks may be physical, cyber, or electronic in nature. Underground facilities offer physical protection to critical infrastructure elements, can be hardened against electronic attacks, and can provide secure alternate locations for backup data and equipment. The costs of underground facilities costs are competitive with hardened above ground structures and life cycle costs are actually lower for large hardened structures. The Norwegians have shown that critical infrastructure functions (including electric power generation, water purification and treatment, telecommunications centers, banking and finance functions, and air traffic control) are <u>all</u> amenable to underground construction.

7 RECOMMENDATIONS[6]

Core competencies in underground technology need to be nurtured and revitalized as necessary and redirected to infrastructure assurance applications. In the post 9-11 security environment, the U.S. Department of Defense can play a critical role in providing advice and assistance to private and public sector partners on the construction and operation of secure underground facilities.

Education and consensus building in the infrastructure community regarding the capabilities and utilization of UGFs will be essential. The underground technical community must develop and present a clear to the larger corporate world and the public. The establishment of an academic center for underground studies would enhance the visibility and encourage the acceptance and use of underground construction. A repository for UGF cost data and capabilities is needed to enable informed decisions on building and using underground space. This center could also serve as a designated clearinghouse organization to hold and distribute information and serve as a "matchmaker" for users in search of suitable underground sites.

Efforts are needed to encourage research and innovation and promote federal government-industry partnerships. Community coordination and partnerships with smaller organizations at the state and local level are also vitally important. A pilot, or demonstration project would do much to make the case for underground applications for critical infrastructure protection. In this regard, it would be beneficial to select a prime infrastructure application as a point of focus for discussions between government and industry on costs, benefits and implementation issues.

[6] See also <u>Use of Underground Facilities to Protect Critical Infrastructures</u>, National Academy Press, Washington, D.C., 1998

Session 4, Track 3

Site conditions and investigations

North American Tunneling 2002, Ozdemir (ed.)
 ISBN 90 5809 376 X

Tomographic imaging using seismic reflection from a 6.5-m TBM in Japan

D.M. Neil
NSA Engineering, Inc., Golden, Colorado

K. Nishioka
Kajima Corporation, Tokyo, Japan

ABSTRACT: Efficient, economic, and safe excavation of tunnels depends on a detailed understanding of rock conditions to be encountered at the working face. Unanticipated joints, faults, or shear zones could lead to potentially hazardous conditions that may result in work stoppages and resultant claims and disputes. New methods of risk identification during site investigation and active tunnel excavation can have a significant impact on contractors' competitive advantage or on overall project costs for owners. Kajima has applied a new method of seismic tomography in Japan pioneered by Kajima and NSA Engineering called True Reflection Tomography "TRT™." Using seismic energy from multiple sources including the TBM cutting action, TRT™ creates a 3-D isometric map of the geologic structure some 100 m out and up to 30 m around the tunnel alignment. Kajima has applied TRT™ on seven tunneling projects since its introduction in February 1999 to predict rock conditions ahead of the tunnel face. The images produced by TRT™ have been used satisfactorily to manage the risk caused by encountering unforeseen geologic conditions.

1 INTRODUCTION

Reflection tomography TRT™ methods developed by NSA process reflections produced by seismic waves generated from a number of source types and tunneling machines commonly found in mining and tunneling. An isometric plot of detected reflection anomalies along the tunnel alignment give a 3-D image of the geologic structure for some distance along the tunnel alignment. This information allows the site engineer and geologist to more accurately assess expected conditions for risk management and abatement. Kajima Corporation has used TRT™ extensively on the Fujikawa and Kanaya Tunnels in often very difficult and varied geologic conditions to manage risk for improved safety and operational management resulting in a major economic benefit to the owner.

TRT™ data analysis is complex and requires special skills to extract accurate and complete geologic information. Data collection typically takes about four hours, including hardware installation, data collection, and teardown. The data from a particular section are then transferred to NSA via the Internet for processing. Upon receipt at NSA offices, the data are input, filtered, picked, and processed for analysis by NSA geophysicists and engineers. This process usually requires between 8 to 20 hours depending on the complexity of the geologic situation. The resultant analysis is then returned to the site ready for the next morning's work.

2 FUJIKAWA TUNNEL

The 4,520-m Fujikawa Highway Tunnel in Shizuoka Prefecture, Japan, is being driven through fractured and sheared andesites with tuff breccia lenses. TRT™ has been applied in two sections. The objective of the TRT™ survey for the first section was to delineate the boundary of the gravel deposit overlying an andesite and tuff breccia formation above the alignment of the TBM-excavated tunnel. The objective of the second survey was to image ahead of the TBM where ground conditions had deteriorated due to frequent faulting and shear zones where the tunnel passes under the Umuse River flowing into the Fuji River and continues in a 470-m section between Sta. 1041+40 to Sta. 1036+70.

2.1 *Gravel Boundary*

Figure 1 shows the location of TRT™ arrays and the extent of generated images combining data from two locations between Sta. 1050+00 and 1048+00 in the Fujikawa tunnels. The first site is in the existing 3.5-m-diam pilot tunnel; the second is from the face of the 5-m-diam main TBM tunnel. Blue sections along each tunnel mark arrays of accelerometers installed at each site. Red sections mark the range of sledgehammer strikes used as seismic impact sources for each site.

Figure 2 shows an isometric projection north-northeast of the same vertical tomogram. Horizontal anomalies appear natural for an undisturbed gravel

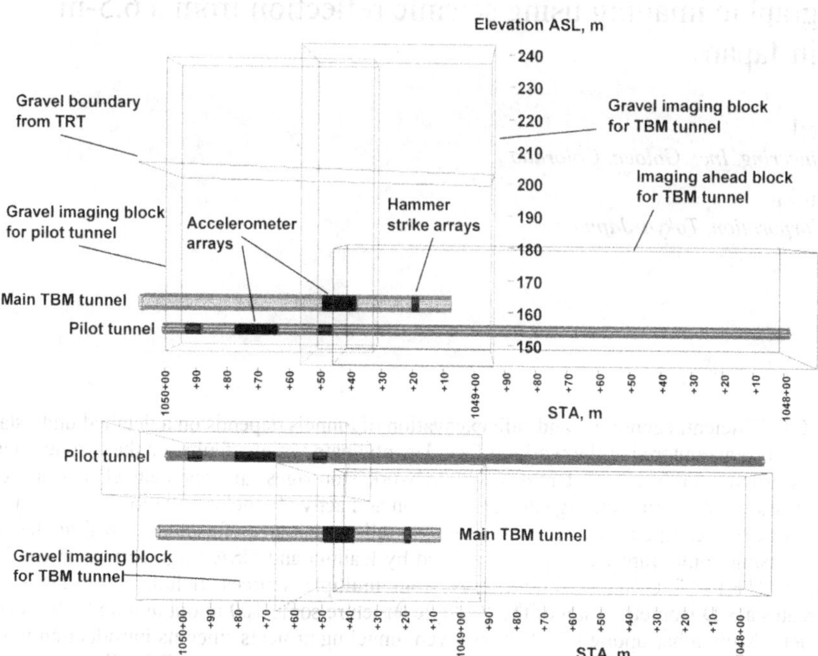

Figure 1. Side and vertical projection of ground imaging blocks around Fujikawa Pilot and TBM Main tunnels for imaging gravel bottom and ahead of TBM.

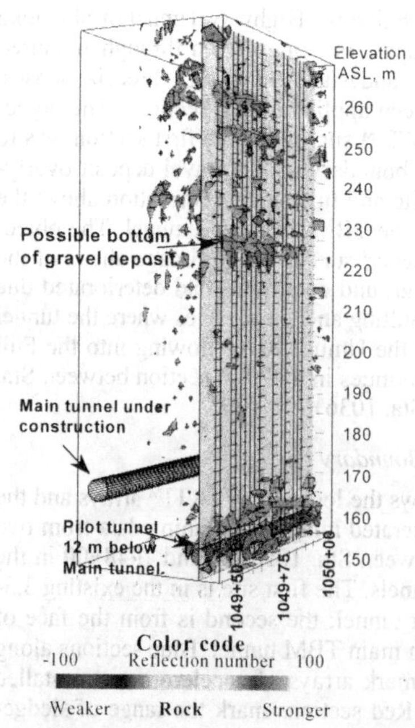

Figure 2. Isometric projection due north-north-east of preliminary TRT reflection tomogram through Pilot tunnel, and contours of negative reflective anomalies showing three-dimensional profile of possible boundary of gravel deposit.

and similar sedimentary deposit. The original assessment of the gravel bottom was at elevation approximately 220 m (722 ft) ASL, or 55 m (180 ft) above the centerline of the TBM tunnel. After final adjustment for attenuation of seismic waves, another horizontal reflective boundary was detected at elevation 200 m (656 ft) ASL, or 35 m (115 ft) above the centerline of the TBM tunnel. A nearly identical sequence of reflective anomalies was reconstructed for the TRT™ survey in the TBM tunnel. However, the elevation of the lowest boundary was at 202 m (663 ft) ASL, or 37 m (121 ft) above the TBM tunnel centerline. Combined images for the pilot tunnel and for the TBM tunnel shown in Figure 3 allow three-dimensional assessment of the shape of the gravel bottom boundary. The horizontal alignment of this boundary appears to correspond well with surface reflection data, but its elevation is higher by approximately 5 to 7 m (16 to 23 ft). Also, the boundary appears to dip slightly from the TBM tunnel toward the pilot tunnel.

No other explicit horizontal boundaries were identified in the elevation range between 200 m (656 ft) ASL and the TBM tunnel.

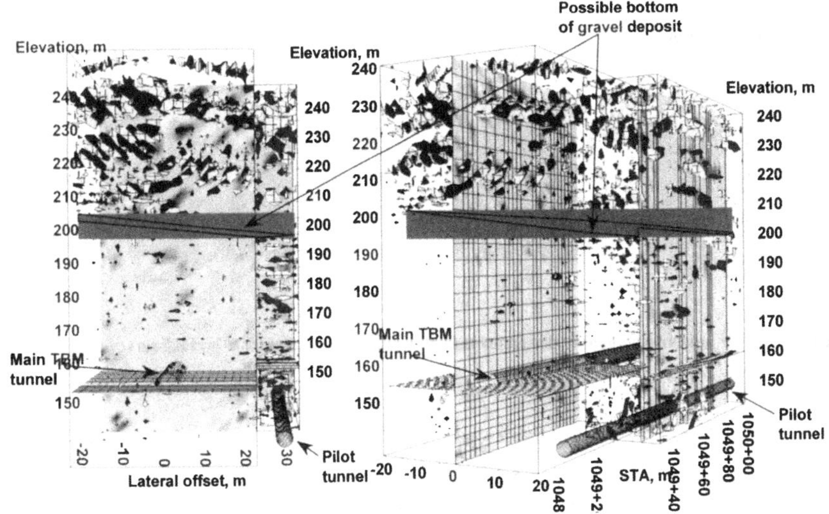

Figure 3. Isometric projection due east (left), and due northeast of vertical and horizontal tomograms and contour reflective anomalies above Fujikawa tunnels.

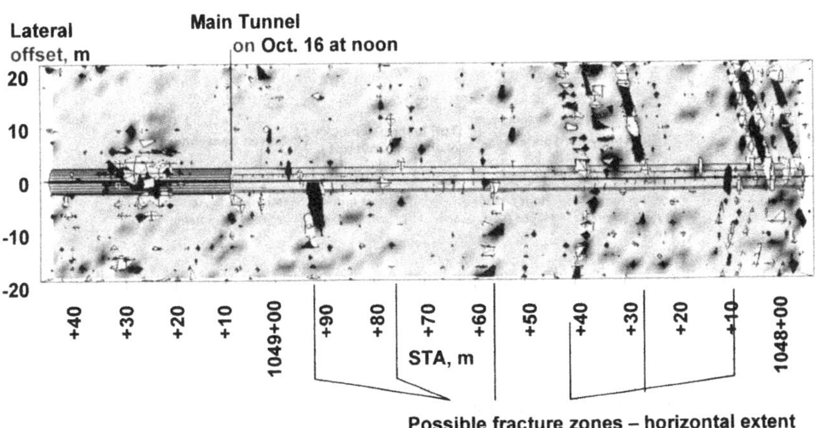

Figure 4. Plan view of reflective anomalies detected by TRT ahead of the main Fujikawa tunnel.

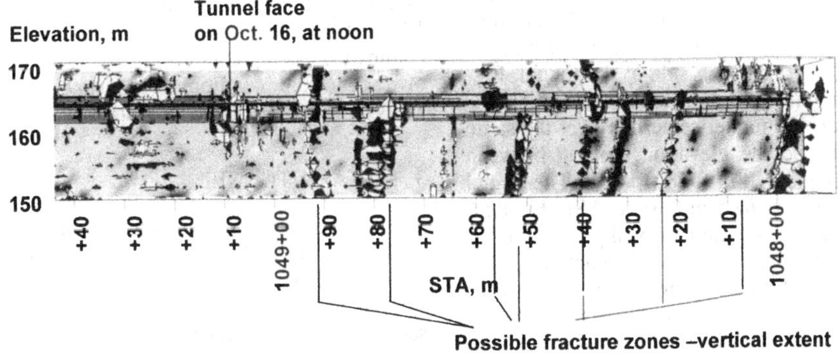

Figure 5. Side view due south of reflective anomalies detected by TRT ahead of the main TBM Fujikawa tunnel.

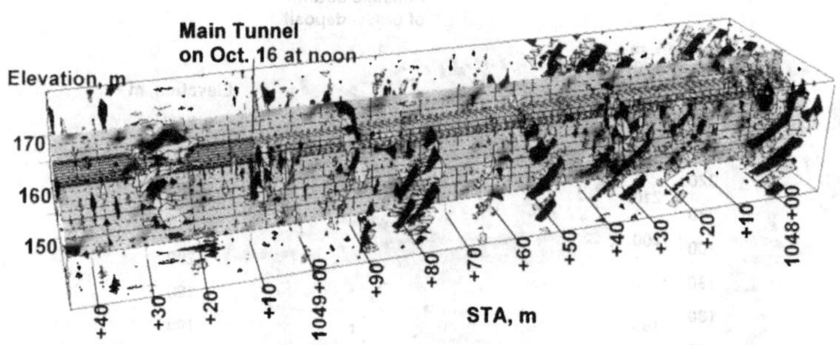

Figure 6. Isometric projection of reflective anomalies detected ahead of the main Fujikawa tunnel by TRT survey.

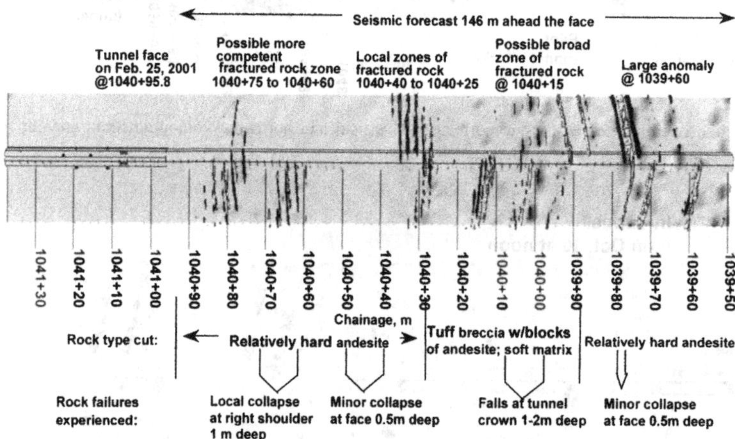

Figure 7. Horizontal section of TRT seismic reflective image and correlation between forecasted and detected rock features in Fujikawa tunnel.

2.2 *Imaging Ahead of TBM*

Figures 4 to 6 show a number of anomalies in front of the TBM, possibly associated with interchanging zones of andesite and tuff breccia. Figure 4 shows a plan view of a horizontal tomogram through the tunnel centerline and a number of contour reflective anomalies over the range of 125 m (410 ft) ahead of the TBM on October 16, at noon. The image was extended 20 m (66 ft) on each side of the tunnel. Figure 5 shows a side view due south of a vertical tomogram through the tunnel centerline over the same distance range and the elevation ranging between 145 m (476 ft) and 170 m (558 ft) ASL. The tunnel centerline was about 162.5 m (531 ft) ASL. Figure 6 shows an isometric three-dimensional projection due southwest of both tomograms and the contour reflective anomalies. All reconstructed anomalies appear nearly vertical, or slightly tilted due west. Some of the anomalies appear terminated from the top. The anomalies are most likely boundaries between softer tuff and stronger andesite. The more massive anomalies are possibly associated with more unstable ground conditions.

During the period February 28 to March 14, 2001, three separate but continuous sections up to 150 m each were imaged. The images were then correlated with the machine forces and observations. Figure 7 shows the original image between Sta. 1041+38 and Sta. 1039+50 along with forecast comments, geologic information, and observations of encountered conditions. Figure 8 overlays the analysis with TBM force measurements.

3 KANAYA TUNNEL

A TRT™ survey was conducted at three sites in the 4,454-m Kanaya Highway Tunnel in Shizuoka Prefecture, Japan. The objective for the Kanaya tunnel was to image the lithological boundary between sandstone and mudstone at two sites, to image features associated with a faulting at the third site, and to allow prediction of ground conditions along the alignment of the westbound tunnel. To date, some

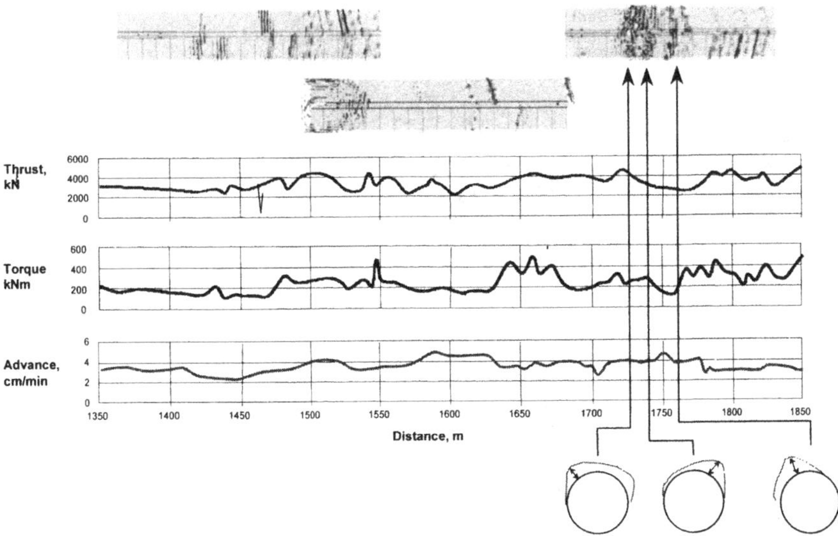

Figure 8. Reflective anomalies reconstructed by TRT compared with rock cutting parameters measured on TBM and with observed instability along Fujikawa tunnel.

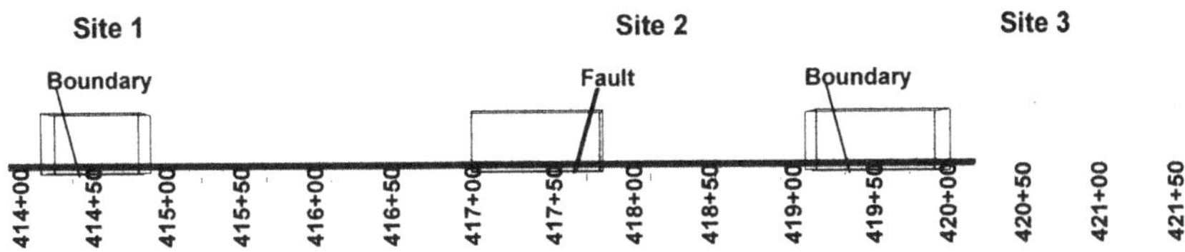

Figure 9. TRT survey sites in Kanaya tunnel showing accelerometer arrays and hammer strikes, image blocks, reconstructed fault zone, and alignment of strata boundaries between weaker and stronger rock formations.

1,240 m between Sta. 413+70 and Sta. 426+10 have been imaged. Figure 9 shows the location of accelerometer arrays and hammer strikes and the alignment of strata boundaries between weaker and stronger rock formations.

3.1 *Survey Results – Site 1 – Sandstone-Mudstone Boundary*

Figure 10 presents a tomogram (slice) and contour reflection anomalies through the image block at Site 1. This figure combines a plan view of a horizontal tomogram through the tunnel centerline at elevation 138.3 m (453.6 ft) ASL and a side view due north of vertical tomogram along the tunnel alignment with an isometric projection due northwest of contour reflective anomalies above horizontal tomogram. This projection emphasizes three-dimensional characteristics of TRT™-generated images. The images indicate the presence and alignment of the strata boundary by producing a pattern of larger size colored spots along and on the side of the boundary opposite the sources and accelerometers. The boundary appears to cross the tunnel at approximately STA 414+53, and at an angle about 105° counterclockwise from the tunnel alignment. Due to rather small differentiation in rock properties across the strata boundary, the efforts to determine which side of the boundary represented stronger rock were inconclusive at this time.

The images also show a zone parallel to the boundary and crossing the tunnel adjacent to the array of accelerometers. The nature of this zone is presently unknown and may be caused by guided, possibly refracted waves traveling along directions parallel to the detected boundary.

3.2 *Survey Results – Site 2 –Fault Zone*

Figure 11 shows a tomogram and contour reflective anomalies through the image block at Site 2.

This figure shows a plan view of a horizontal tomogram through the tunnel centerline at elevation 137.5 m (451 ft) ASL combined with a side view due north of vertical tomogram along the tunnel along the tunnel alignment and an isometric projection due north of contour reflective anomalies above the horizontal tomogram through the tunnel

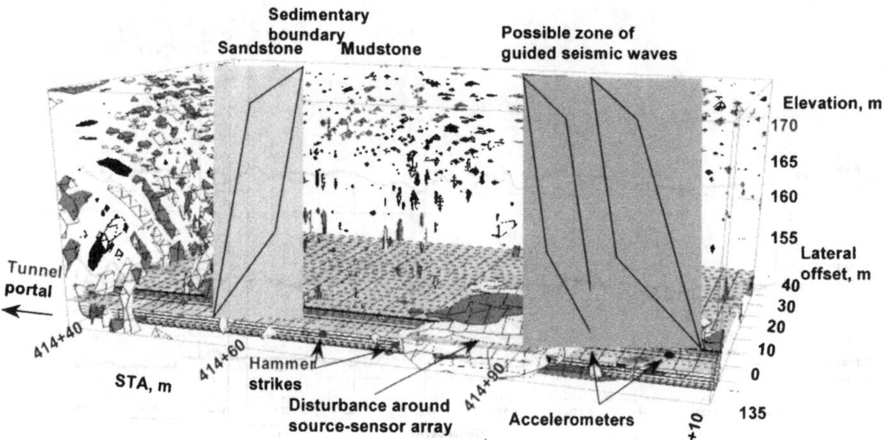

Figure 10. Isometric projection of TRT-generated horizontal tomogram and contours of reflective anomalies for Site 1 of Kanaya tunnel.

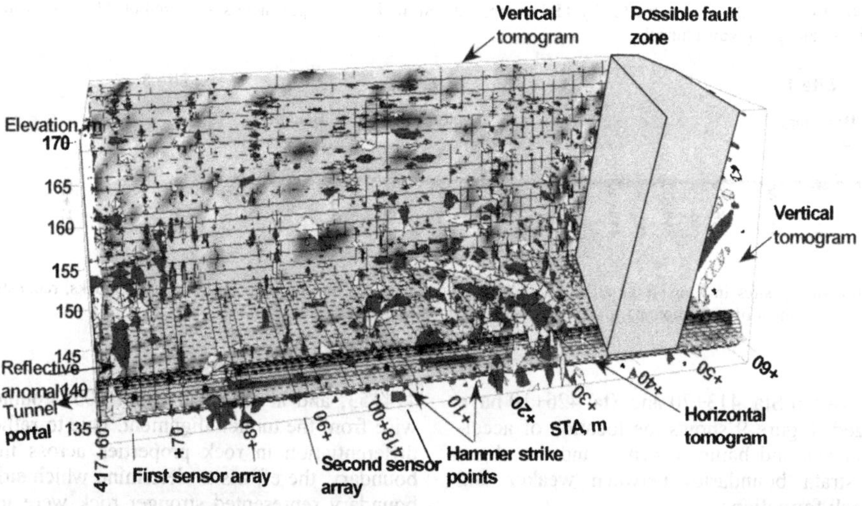

Figure 11. Isometric projection of TRT-generated tomograms and contours of reflective anomalies for Site *2* of Kanaya tunnel.

and in front of vertical tomogram 43 m (141 ft) north from the tunnel. The shaded area in this figure indicates general extent and orientation of the fault zone. This zone crosses the tunnel approximately at Sta. 418+40. It appears oriented approximately 75° counterclockwise from the tunnel alignment and is tilted approximately 10° of vertical to the east. Similar to Site 1, small differentiation in rock properties combined with strong attenuation of seismic waves did not allow a reliable distinction between stronger and weaker rock mass associated with the fault. No other explicit structural features were detected at Site 2.

3.3 *Survey Results – Site 3 – Mudstone-Sandstone Boundary*

Figure 12 presents a tomogram and contour reflective anomalies through the image block at Site 3.

This figure shows a plan view of a horizontal tomogram through the tunnel centerline at elevation 136.5 m (448 ft) ASL combined with a side view due north of a vertical tomogram along the tunnel alignment and an isometric projection due northwest of contour reflective anomalies above the horizontal tomogram through the tunnel and in front of a vertical tomogram 43 m (141 ft) north from the tunnel.

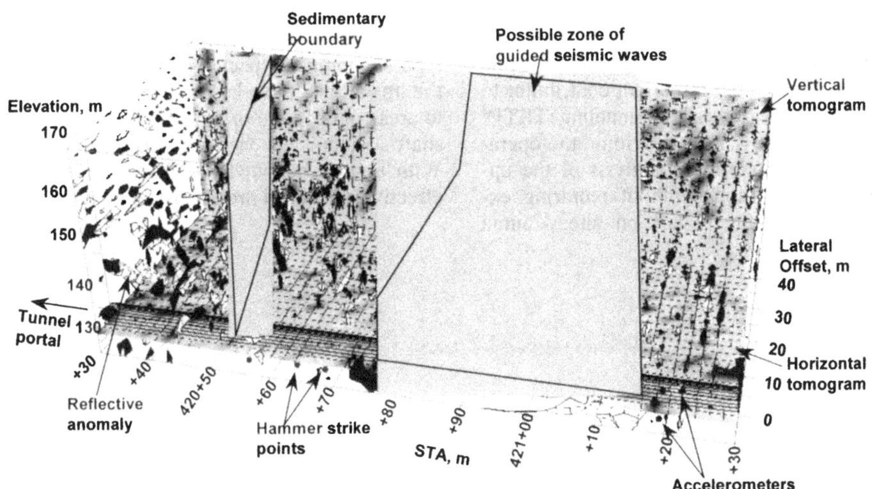

Figure 12. Isometric projection of TRT-generated tomograms and contours of reflective anomalies for Site 3 of Kanaya tunnel.

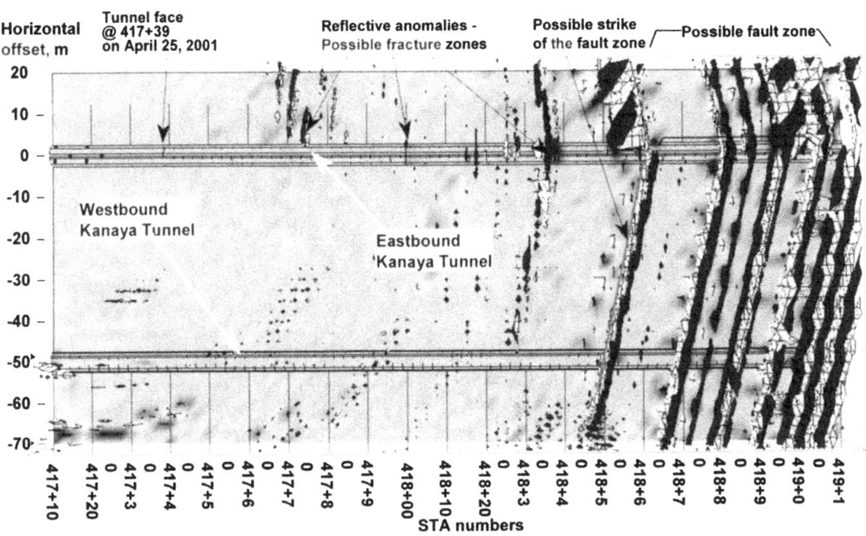

Figure 13. Plan view of horizontal reflective tomogram extended south at tunnel elevation 138.4 m for TRT survey in the eastbound Kanaya tunnel on April 25, 2001.

According to the change in pattern of contour anomalies/color spots, the sandstone-mudstone boundary crosses the tunnel at Sta. 420+51. Its orientation is at an angle about 105° counterclockwise from the tunnel alignment, and its tilt appears to match the orientation of the boundary at Site 1. A wide zone crossing the tunnel near the array of accelerometers coincides with a zone of poor rock conditions identified in the tunnel while selecting locations for accelerometers.

3.4 *Survey Results – Fault Locations*

A major fault zone was detected in front of the TBM in the eastbound Kanaya tunnel between Sta.

417+40 and Sta. 419+6. Figure 13 illustrates an explicit fault plane crossing at the tunnel alignment at Sta. 480+60. This fault plane cuts the westbound tunnel at Sta. 418+50. This result corresponds with the TRT™ survey conducted earlier in the westbound tunnel.

4 CONCLUSIONS

Encountering unforeseen geologic conditions while tunneling is the single most important contributor to cost overruns and increased risk in tunneling. TRT™ effectively reduces that risk by providing the operator with a timely and complex analysis of the upcoming geologic conditions without requiring expensive and highly trained staff on site. Kajima engineers have effectively utilized TRT™ images in widely varying rock units to predict the location and type of anomaly expected. In a number of situations, the image data have been reevaluated and projected to areas outside the original alignment to study vent shaft locations as well as peripheral excavations. With improved knowledge, Kajima engineers have effectively reduced project risk and improved safety.

North American Tunneling 2002, Ozdemir (ed.)
© 2002 Swets & Zeitlinger, Lisse, ISBN 90 5809 376 X

Settlement prediction on an operational immersed tube tunnel

Hongwei Huang
Dept. Geotech. Engrg., Tongji University, Shanghai China

Pierre-Yves Hicher
Laboratory of Civil Engineering, Ecole Centrale de Nantes, Nantes France

Dongmei Zhang
Dept. Geotech. Engrg., Tongji University, Shanghai China
Laboratory of Civil Engineering, Ecole Centrale de Nantes, Nantes France

ABSTRACT: Immersed tube tunnel has a trend application with increasing recent years in China. The earliest immersed tunnel has operated about 6 years. The present measured results show that there is a trend with no stop for settlement of some elements. People who administer this tunnel want to know the long-term settlement. In this paper, a settlement prediction method was proposed based on in-situ measurement and rheology model. Then, the models to stimulate time-dependent deformation were discussed and optimized based on the measurement values. After considering their different character that was closely related with the markedly different settlement, the foundation is divided into several parts. The visco-elastic back-analysis method was employed to calculate their visco-elastic parameters of soil in a certain tube using the measurement settlement during recent years. Lastly, the subsequent settlements can be predicted with the back-analyzed parameters. It was turned out that the method was suitable to analysis the long-term settlement of immersed tube tunnel.

1 INTRODUCTION

The procedure of immersed tube tunnel was first developed by an American engineer, Wilgus,W.J. and this idea of immersing a prefabricated tube tunnel was first put into practice in American in 1896, for a water line crossing Boston Harbor (Gursoy 1995). Before 1997, there are 106 existing traffic tunnels worldwide (Rasmussen 1997). In the mainland of China, only three tunnels for road, rail and sewage emission respectively were constructed in the past years (Wang 2000). Now immersed tube tunnel has a trend application with increasing recent years in China. At present there are two immersed tube road tunnels being constructed for city traffic in Shanghai and Ningbo, respectively. The immersed tube tunnel presented here was completed six years ago for a roadway crossing a river. There are 5 prefabricated concrete tube elements, which are called as E1, E2, E3, E4 and E5, three of them are 80m length and the rest are 85m length, all these elements have a 7.6m width. These tube elements were reposed on the mucky and silty clay, which is saturated flow plasticity sea deposit with 10^{-7} coefficient of permeability. The measured records show that there is a trend with no stop for settlement of some elements, even though over muck on the tube was removed two times one year after several years operation.

It is known that the compressed Gina is crucial for water tightness of all the joints between tunnel elements. From the inside a second Omega profile rubber sealing was installed to provide a second seal against leakage of water. Though the total tunnel is flexible in its longitudinal direction and is able to absorb some settlement of its foundation due to flexible joints, the capacity of Gina and Omega seal to bear the shear force are limited. Excess deformation of joints will pull out and further damage Omega and Gina seals, and that will result in seepage of water into tunnel. The handler of this tunnel wants to know in more urgent whether the settlement of tunnel will be smaller or go on in the coming year.

Settlement research or prediction on soft soil under the strip load is a traditional issue. Manifold methods were proposed based on the theory of consolidation or rheology, that can be classified as empirical, semiempirical, in-situ monitoring, and analytical methods, for instance Taylor (1946), Scott and Ko (1969), Taylor (1983), Ladayi et al., (1991) and Ranjan (1994), Briaud and Gibbens (1999). Whereas normally, a slight net load on the subsoil is considered during the design and construction of an immersed tube tunnel. Simple method for immersed tube tunnel, for instance the theory of elastic foundation beam, is used in more generals (Mainwaring et.al 2000). Few care about settlement and even long-term settlement should be required for immersed tube tunnel on subsoil. It is encouraging that Janbu et.al. back-calculated creep rates using case records, and it is possible to predict coming deformation through measured records (Janbu et. al. 1989). In this paper, a settlement prediction method was proposed based on in-situ measurement and rheology model. Due to the operation of several years, the behavior and mechanics parameters of soil are changed; the tunnel foundation is not the same one as before construction. Therefore, firstly, the models to stimulate time-dependent deformation were discussed and optimized based on the measurement values. Then considering their different character that was closely related with the markedly different settlement, the foundation is divided into several parts. The visco-elastic back-analysis method was employed to calculate their visco-elastic parameters of soil in a certain tube using the measurement settlement during recent years. And later, the subsequent settlements can be predicted with the back-analyzed parameters. It was turned out that the method was suitable to analysis the long-term settlement of immersed tube tunnel.

2 SOME STIPILATIONS AND ASSUMPTIONS

For the convenience of prediction, some necessary stipulations and assumptions should be outlined bellow:

1. Settlement during immersed construction is not considered for long-term settlement prediction, and the settlement arising from foundation layer has finished after construction;
2. The measurement begins to be implemented before operation, and the star time in this paper is from post-construction, not from operation; and the time relating to measurement and prediction is based on days;
3. The concrete tube element is regarded as a rigid structure, then the settlement of element structure can be deemed as the one of subsoil;
4. The subsoil is homogeneous and continuous with undrained creep, in this way, the volumetric deformation of subsoil can be neglected; and also the effect of river temperature on settlement is ignored.

3 THE PRINCIPLE OF SETTLEMENT PREDICTION

In-situ records show that it is an urgent issue to know the coming settlement of tunnel. From a phenomenological point of view in measured records, it is possible to use visco-elastic rheology models to describe the time-dependent behavior of subsoil (Scott and Ko 1969). Therefore, the visco-elastic model is employed for predicting the long-term settlement considering the forenamed assumptions.

3.1 *Rheology model and settlement calculation*

It is known that several rheology models used to predict settlement of soil are composed of three basic elements, namely spring, dash-pot and slider (Schiffman et. al. 1964, Ranjan and Sharma 1994). The various models were conceived just to adapt to different state of soil (Komamura and Huang 1974). Since 1964 three-element model has been employed successfully to predict settlement of clay (Schiffman et. al. 1964). Whereas, it was the very model to be chosen and used to predict the long-term settlement of immersed tube tunnel. The three-element model is illustrated in Fig.1, which is used to simulate the shear deformation of soil, whereas the volumetric deformation was assumed as no contribution to settlement according to the former hypothesize.

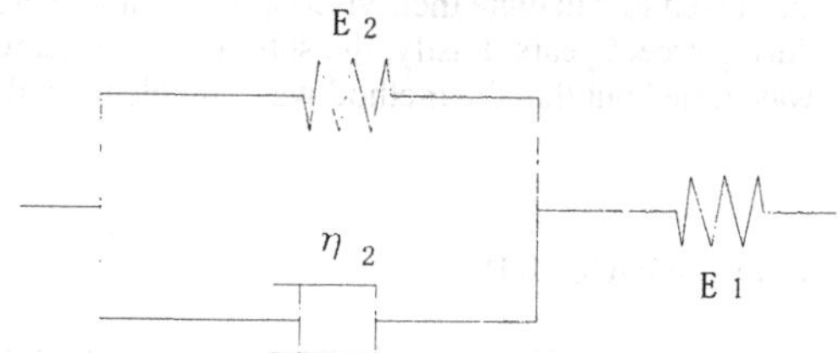

Figure 1. The sketch of three elements model.

It is distinctly to suppose the subsoil of the immersed tube tunnel as a semi-infinite space issue under the strip load. Using the correspondence principle of elastic and visco-elastic, it is potential to deduce the visco-elastic settlement from the elastic one through visco-elastic constitutive relation and Laplace transform. The elastic solution of semi-infinite space was showed as the formula (1) (Jumikis 1969).

$$w(t) = \frac{1-\mu^2}{E} pa\omega \tag{1}$$

where $w(t)$= surface settlement; μ =Poission's ratio; E = elastic modulus of soil; p = uniform pressure on the subsoil; a = width of strip load; ω=settlement calculation coefficient.

Using the correspondence principle combined with Laplace transform as well as counter-transform, the settlement of visco-elastic soil can be deduced finally as formula (2):

$$w(t)=\frac{pa\omega}{4}\left[\frac{A}{B}+\frac{3A}{3KA+B}-\frac{1}{E_2}e^{-\frac{E_2}{\eta_2}t}-\frac{3E_1^2}{(3K+E_1)[3KA+B]}e^{-\frac{3KA+B}{(3K+E_1)\eta_2}t}\right] \quad (2)$$

where $A=E_1+E_2$; $B=E_1E_2$; E_1, E_2 and η_2 are illustrated as in Fig.6, which denote elastic modulus and viscosity coefficient respectively.

Substituting $J(t)$ for the item inside bracket in formula (2), we have

$$J(t)=\left[\frac{A}{B}+\frac{3A}{3KA+B}-\frac{1}{E_2}e^{-\frac{E_2}{\eta_2}t}-\frac{3E_1^2}{(3K+E_1)[3KA+B]}e^{-\frac{3KA+B}{(3K+E_1)\eta_2}t}\right] \quad (3)$$

Then the formula (2) can be written as follows:

$$w(t)=\frac{pa\omega}{4}J(t) \quad (4)$$

Here $J(t)$ is called equivalent creep flexible modulus of subsoil.

3.2 *Optimization and back-analysis of parameters*

In formula (3), there are 4 unknown parameters that are E_1, E_2, η_2 and μ. Generally, Poisson's ratio μ can be considered as a time-independent known parameter relative to other parameters. Therefore, the unknown parameters are E_1, E_2, η_2 in formula (3). Evidently, $J(t)$ is a nonlinear power exponential function, in which the back-analysis could not be lightly implemented using optimization method. Owing to not considering volumetric strain of soil, equivalent creep flexible modulus $J(t)$ can be simplified as:

$$J(t)=\frac{E_1+E_2}{E_1E_2}-\frac{1}{E_2}e^{-\frac{E_2}{\eta 2}t} \quad (5)$$

For the sake of back-analysis, substituting x_1, x_2, and x_3 for $\frac{E_1+E_2}{E_1E_2}$, $\frac{1}{E_2}$ and $\frac{E_2}{\eta_2}$ respectively, then the objective function in optimization is as follows:

$$f(x)=\min\sum_i^m (J(t)-J_i)^2 \quad (6)$$

where $J(t)$= equivalent creep flexible modulus at t through formula (3); J_i= creep flexible modulus through measurement; m= measurement number for each point. In this way, a developed small program can obtain visco-elastic parameters quickly.

4 VISCO-ELASTIC PREDICTION

From measurement results, tunnel elements could be separated into three sections, that is E2-E3, E3-E4 and E4-E5, and it should be reasonable to back-analyze parameters and predict settlement using the principle as above. Table 1 is the back-analyzed parameters of different sections, that show E2-E3 section has a harder subsoil and E3-E4 slightly harder and E4-E5 softer subsoil. And then we can use these back-analyzed parameters to predict settlement of joints between two elements. Due to limit pages, a typical comparison sketch between measured and predicted settlement for E2-E3 is presented in Fig.2. And also Table 2 gives out the correlation coefficient of predictions with measurements.

Fig.2 and Table 2 show that there are good concords between visco-elastic prediction and measurement. They have relatively high correlation coefficient with more than 0.9583. Fig.2 shows there is a slightly agreement trend between measurement and visco-elastic prediction. The biggest absolute error is 4.493mm and the relative error is 28.2% to measurement value. This is mainly because of different behavior of subsoil of different elements. Besides, Fig.1 indicates that rheology model is more suitable for softer soil, and it comes into view that back-analysis in part is more important and reasonable.

Considering gradual compression on foundation layer finished quickly during construction, settle-

Table 1. Back-analyzed parameters of subsoil.

Sections / Parameters	E2-E3	E3-E4	E4-E5
E_1 (Mpa)	69.11	12.4	11.6
E_2 (Mpa)	20.317	19.4	9.824
η_2 (Mpa.d)	8930.6	4853.3	3434

Table 2. Correlation coefficient.

Different joints	Visco-elastic	Hyperbola
E2/E3	0.9583	0.9443
E3/E4	0.9768	0.9618
E4/E5	0.9951	0.9941

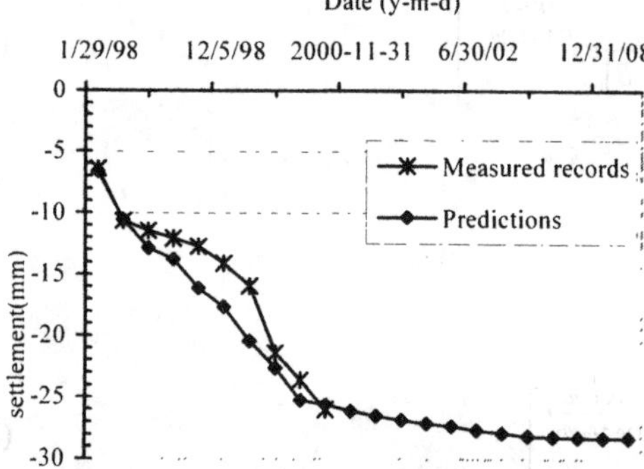

Figure 2. The comparison between measured and predicted settlements for E2-E3.

ment arising from construction was neglected in this paper. In fact this part of settlement may similarly induce the tension and shear of Gina and Omega seals, and makes Gina and Omega finish their partial function before operation. As not estimating this already capacity well and truly, appraisal on residual capacity of seals is more difficult. In another way, we can not but to say settlement arising from construction has somewhat contribution on subsoil behavior. However, from the point of long-term settlement, not considering this part of settlement may cause little impact on ultimate settlement after using the back-analysis method.

Usually, there is a small additional load on the bottom of tunnel according to the workmanship of immersed tunnel. As to the cyclic action of tide, measured records show it submits to a normal distribution with almost zero means, there is inattentive contribution from the long-term view. In this way, we prefer to think the settlement during operation is induced mainly from the rheology behavior of subsoil.

5 CONCLUSIONS

In this paper, settlement prediction of an immersed tube tunnel was carried through. The visco-elastic model was employed to predict long-term or ultimate settlement, and together with comparison between prediction and measurement. Two main conclusions were attained.

1. Rheology model has good agreements with measured records. The comparison between them shows that it is slightly better to choose rheology model and back-analysis from the view of deformation mechanism and real engineering appraisal.
2. Due to inevitable longitudinal differential settlement, reasonable parameters of soil behavior should be back-analyzed from measurements in term of the different elements.

REFERENCES

Briaud, J. L., and Garland, E. (1985). "Loading rate method for pile response in clay." *J. Geotech. Engrg.*, ASCE, 111(3), 319-335.

Briaud, J. L., and Gibbens, R. (1999). "Behavior of large spread footings in sand." *J. Geotech. and Geoenvir. Engrg.*, ASCE, 125(9), 787-796.

Gravesen, L. and Rasmussen, N. S. (1993). "Milestone in tunneling: Rotterdam's Maas Tunnel celebrates its fiftieth anniversary." *Tunnelling and Underground Space Technology*, 8(4), 413-423.

Gursoy, A. (1995). "Immersed steel tube tunnels: an American experience." *Tunnelling and Underground Space Technology*, 10(4), 439-453.

Janbu, N., Sintef, G. S., and Sintef, S. C. (1989). "Back-calculated creep rates from case records." *Proc. 12th Int. Conf. on Soil Mech. and Found. Engrg.*, Rio de Janeiro, A.A.Balkema, 1809-1812.

Jumikis, A. R. (1969). "Theoretical soil mechanics." *Van Nostrand Reinhold Company*, New York.

Kazuya, Y., Masayuki, H., Adrian, H. And Kazutoshi, H. (1991). " Cylic-induced settlement in soft clays." *Proc. of the 10th European Conference on Soil Mechanics and Foundation Engineering*, Florence, A.A.Balkema, 887-890.

Komamura, F., and Huang R. J. (1974). "New rheological model for behavior." *J. Geotech. Engrg.*, Div., 100(7), 807-824.

Ladayi, B., Lune, T., Pierre, V. And Bernard, L.(1991). "Predicting creep settlement of foundations in permafrost from the results of cone penetration tests." *Can. Geotech. J.*, 32(5), 835-847.

Lin, H. D. , and Wang, C. C. (1998). "Stress-strain-time function on clay." *J. Geotech. and Geoenvir. Engrg.*, ASCE, 124(4), 289-296.

Loganathan, N., Balasubramaniam, A. S., and Bergado, D.T. (1993). "Deformation analysis of embankments." *J. Geotech. Engrg.*, ASCE, 119(8), 1185-1206.

Mainwaring, G.D., Weeks, C.R., and Brandsen, C. (2000). "Detailed design of the Medway tunnel project." *Proc. of the Institution of Civil Engineers, Transport*, 141(1), Thomas Telford Services Ltd, Engl, 9-24.

Ranjan, G. and Sharma, R. P. (1994). "Nonlinear visco-elastic constitutive model for time dependent behavior of clays." *Proc. 13th Int. Conf. on Soil Mech. and Found. Engrg.*, New Delhi, A.A.Balkema, 421-424.

Rasmussen, N. S. (1997). "Concrete immersed tunnels- forty years of experience." *Tunnelling and Underground Space Technology.*, 12(1), 33-46.

Scott, R. F., and Ko, H. Y. (1969). "Stress-deformation and strength characteristics." *State-of-Art reports, Proc. 7th Int. Conf. on Soil Mech. and Found. Engrg.*, Mexico, 1-47.

Schiffman, R. L., Ladd, C. C. et.al. (1964). "The secondary consolidation of clay." *IUTAM Symp. Rheology and Soil mech.*, Grenoble, 273-304.

Taylor, D. W. (1946). "Fundamentals of soil mechanics." *John Wiley & Sons, Inc.*, New York.

Taylor, B. B. and Matys, E. L. (1983). "Settlement of a strip footing on a confined clay layer." *Can. Geotech. J.*, 20(3), 535-542.

Vaid, Y. P., and Campanella, R.G. (1977). "Time-dependent behavior of undisturbed clay." *J. Geotech. Engrg.*, Div., 103(7), 693-709.

Vaid, Y. P., and Eliadorani, A. (2000). "Undrained and drained (?) stress-strain response." *Can. Geotech. J.*, 37(5), 1126-1130.

Wang, J. Y. (2000). "Tunnelling and technological progress in tunnelling in China." *Proc. of the Inter. Conference on Tunnels and Underground Structures.*, Singapore, Zhao, J., and Shirlaw, J.N., and Krishnan, R. eds., A. A. Balkema, 97-106.

North American Tunneling 2002, Ozdemir (ed.)
 ISBN 90 5809 376 X

Underground utility exploration, depiction, and planning at Seattle-Tacoma International Airport

K. Oja & M. Spaur
Kennedy/Jenks Consultants, Federal Way, Washington, USA

S. Rao
Parsons Infrastructure, SeaTac, Washington, USA

B. White
Port of Seattle, Seattle, Washington, USA

ABSTRACT: Over the next 15 years, the Port of Seattle plans to complete modernization of Seattle-Tacoma International Airport (Sea-Tac), construct a third runway, and add a new terminal, rental car facility, roadways, and other improvements to expand the airport's capacity from 28 million to 60 million air passengers per year. Since opening in 1942 the airport has gone through a series of expansions and modifications: tenants have moved; concourses and satellites were added; runways were expanded; and underground utilities were installed, expanded, used and abandoned as needs changed. The utility records were filed by year on a project-by-project basis, without the benefit of comprehensive utility system mapping. To gain better control of its utility records, the Port of Seattle recently initiated a project that collated available underground utility information into a single mapping model that can be used for future planning and development of the airport. This paper describes the process applied to create the model and concepts from subsurface utility engineering that were introduced.

1 INTRODUCTION

1.1 *Description of Sea-Tac*

Sea-Tac, owned and operated by the Port of Seattle (Port), began as Bow Lake Field on March 30, 1942. After World War II, a series of expansions transformed Bow Lake Field into an International Airport in the 1950s and eventually into the major air transportation hub that it is today. The process of converting a small, private, wartime airfield into an international transportation hub was gradual. The airport underwent a series of transformations and expansions as tenants moved, concourses and satellites were added, and runways were expanded.

In the mid 1990s the Port completed a master plan identifying a capital improvement program to modernize the existing airside and landside operations areas at Sea-Tac and add a third runway (Port of Seattle 2000a). In addition, plans were developed to add a new north end terminal, rental car facility, roadways, and other improvements to expand the airport's capacity from 28 million air passengers per year (MAP) to 60 MAP (Landrum & Brown 1998).

Sea-Tac is the primary air transportation hub serving Washington State and the Northwestern United States. Located 12 miles south of downtown Seattle and 20 miles north of Tacoma, Sea-Tac is the only airport in the Seattle/Tacoma metropolitan area that offers scheduled commercial airline service. The airport's primary service market is the central Puget Sound region, comprised of King, Pierce, Snohomish, and Kitsap counties. The region has a population base of approximately 3.5 million with total employment of approximately 1.7 million. In year 2000 Sea-Tac ranked 17th on the list of the top 50 US commercial service airports, serving over 28 MAP (Port of Seattle 2000b).

1.2 *Underground utilities*

As the airport developed, underground utilities were also installed, expanded, used, and abandoned as needs changed. Initially, the Port kept track of utility information on a project-by-project basis, filed by the year the project was initiated. This system made the design of new underground facilities difficult since it required the designer to know the year in which the existing facility was originally built and perhaps a project title or sponsor in order to retrieve the information. As the airport continued its transformation, the level of underground utility congestion continued to increase. Added complexities were created in tenant lease areas where improvements were frequently made without providing record drawings to the Port.

To support the planning and development of the capital improvement program, a project was initiated to collect and collate available underground utility information into a single mapping model that could be used for future planning and development of the airport. Key elements of the project included:

- Applying industry best practices for depicting utility information
- Demonstrating the cost benefit of creating an underground utility mapping model
- Creating a mapping model that would support utility operations and capital project planners.

2 UNDERGROUND UTILITY RISK MANAGEMENT

2.1 *Traditional utility record recovery*

Accurately locating underground utilities is an important element of Sea-Tac's risk management program. The traditional approach for obtaining utility information during planning and design often included requesting information from individual utilities or contacting one-call services. In some cases, the utility owner or one-call service may mark the location of the facility in the field, but this step is frequently left until just prior to start of construction. In areas where potential conflicts are identified during design, a limited number of potholes may be dug to collect more accurate information on depth, size, and location. Typically the construction documents came with disclaimers by the designer regarding the accuracy or completeness of the utility information. Utility conflicts encountered during construction have often resulted in increased costs to the contractor and the Port, and often disrupted service to the airport's customers. Continuing to apply the traditional approach for locating utilities on Sea-Tac's proposed capital improvement program would have resulted in an increased risk of service disruptions and potential for contractor claims.

2.2 *Airport security*

By knowing where each underground utility is located, Sea-Tac can reduce the potential for damage and disruption of service during construction, and also allow for more timely location during periods of emergency response. The events of September 11, 2001 have increased the importance of security at Sea-Tac and all airports across the country. Understanding where utilities are located, which facilities they serve, and imposing access restrictions have become more critical components of airport security planning. Federal Aviation Regulation (FAR) Part 107 outlines the criteria that airports are required to meet in establishing a security program (FAA 2001). Advisory Circular (AC) 107 provides guidance and recommendations on how airports can meet the requirements of FAR 107 (FAA 1972).

The Federal Aviation Administration (FAA) has identified a security survey as the first step toward achieving adequate protection at airports. The purpose of the security survey is to complete a detailed physical inspection of all areas, facilities, and operations. Knowing the location, size, control attributes, and the service area for each utility is understood as a vital component of the security plan. In addition, utility barriers are required where traverse culverts (allowing access between airside and landside), troughs, or other openings larger than 96 square inches (approximate 12-inch diameter) are present. In those instances, the openings need to be protected by fencing, iron grills, or other suitable barriers to preclude unauthorized access (FAA 1972).

2.3 *Underground utility quality guidelines*

The concept of Subsurface Utility Engineering (SUE) was developed, refined and put into professional practice in the 1980s. The Federal Highway Administration (FHWA) has served as a lead agency in promoting and using this concept. Other organizations such as the American Society of Civil Engineers (ASCE), FAA and various state departments of transportation (DOTs) have applied this process (Lew 2000).

ASCE is currently developing guidelines for establishing a standard quality level approach for depicting underground utilities using the SUE approach (ASCE 2001). The following is a description of the quality levels in ASCE's standard guidelines:

- Quality Level D (QL D): Information derived solely from existing records or verbal recollection.
- Quality Level C (QL C): Information obtained by surveying and plotting visible above ground features and by using professional judgment in correlating this information with "Quality Level D" information.
- Quality Level B (QL B): Information obtained through the application of appropriate surface geophysical methods to identify the existence and approximate horizontal position of subsurface utilities. QL B data can be reproduced by surface geophysics at any point of their depiction. This information is surveyed to applicable tolerances and reduced onto plan documents.
- Quality Level A (QL A): Information obtained by the actual exposure (or verification of previously exposed and surveyed utilities) of subsurface utilities, using (typically) minimally intrusive excavation equipment to determine their precise horizontal and vertical positions, as well as their other utility attributes. This information is surveyed and reduced onto plan documents. Accuracy is typically set at 15mm vertical, and to applicable horizontal survey and mapping standards.

Applying the appropriate quality level designation of QL D (lowest) through QL A (highest) to each utility as it is placed on drawings allows the reader to clearly understand the level of confidence concerning the reliability and accuracy of the information. A representative sample is shown on Figure 1.

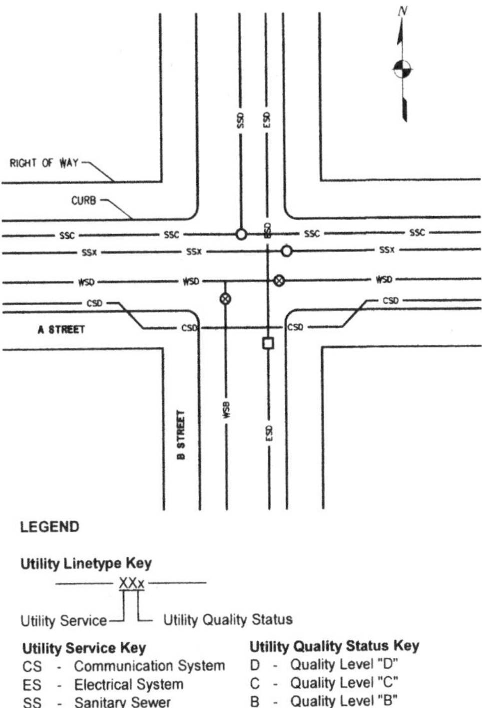

Figure 1. Examples of applying underground utility quality levels.

Presenting the utility data in this manner allows the user to make more informed decisions during project development. QL D information may be sufficient for general planning, while acquiring QL A may be both cost and schedule effective in the areas where proposed construction activity is to occur, or where response to security incidents may be critical. Determining the appropriate quality level should be addressed as part of the project planning process with the owner during the early stages of the project.

3 COST BENEFIT OF APPLYING THE QUALITY LEVEL CONCEPT

3.1 *Purdue University study*

In 1996 the FHWA commissioned Purdue University to study the cost savings from four state DOTs that routinely used the quality concept for locating and mapping underground utilities. A total of 71 projects from Virginia, North Carolina, Texas, and Ohio were studied. The total construction costs of these projects exceeded $1 billion. These projects included a mix of interstate, arterial, and collector roads in urban, suburban, and rural settings (Lew 2000). Their major findings included:

- A savings of $4.62 was achieved for every $1.00 spent on collecting quality level B and A data on the 71 projects.
- The study determined that the cost of collecting and mapping quality level B and A data amounted to less than 0.5% of the total $1 billion construction cost.
- Only three of the 71 projects had a negative return on investment.

3.2 *Benefits*

The Purdue study identified several benefits from applying the quality level concept (Lew 2000). Some of the more significant benefits include:

- Reduction in unforeseen utility conflicts and relocations
- Reduction in project delays due to utility relocations
- Reduction in claims and change orders
- Reduction in delays due to utility cuts
- Lower project bids
- Reduction in cost of project design
- Improvement in contractor productivity and quality
- Reduction in utility company costs to repair damaged facilities
- Minimization of customer loss of service
- Facilitation of electronic mapping accuracy
- Inducement of savings in risk management and insurance costs.

4 MAPPING MODEL

4.1 *Description*

A master drawing was created of the 2,400 acre Sea-Tac project area located within the airport boundary and properties adjoining the airport in the north end area. The master drawing was created in AutoCAD model space format, referenced to Sea-Tac's survey coordinate system. AutoCAD's external reference (XREF) process was used to display all utility data (Autodesk 1998). Utilizing a special file naming system in the XREF utility drawings preserved the flexibility for selecting composite or individual utilities. This method can be used to highlight an entire utility system, such as the Storm Drain System, or to highlight the utilities associated with a specific project. This approach is accomplished by unloading extraneous XREFs or by utilizing the layer filtering in AutoCAD to alter the display of utilities as desired.

Before it is inserted into the master drawing, the utility data associated with a project or survey is first saved as a combined drawing of all utilities, then subsequently broken down into separate utility layers by system. A given project or survey may have any number of utilities as part of its drawings. The separated drawings are attached to the master drawing as individual XREFs. The combined drawing, consisting of the same utilities, is also attached to the master drawing but left "unloaded" to serve as a reference to the utilities as they are related to their project. The individual XREF drawings are compared to current survey data to determine if there is any overlapping information. Where overlaps are found, the duplicated data is removed in favor of current survey data.

4.2 *Quality level of data*

The initial model included a combination of QL D and QL C utility data. QL D data was collected from past airport project records, tenants, airlines, public utility records, and inspection records from airport maintenance personnel. Records that were available electronically were merged into the AutoCAD model. Paper drawings were first scanned and then merged into the model. Where data gaps were subsequently identified, follow up contacts were scheduled with the data providers to search for additional records. Over 3400 drawings were reviewed during follow up contacts. Of these, 800 included utility data, and 250 included utility data that were not in the model.

Gravity systems recently surveyed by the Port were shown as QL C in the model. These systems were limited to storm drain, sanitary and industrial waste sewers where the surface attributes had been tied to the Port coordinate system.

Pressure lines including water, natural gas and fueling; and communications, security, and electrical systems were all shown as QL D, where records were available. With the exception of water line records, the quantity of information available for these systems was very limited.

After additional review of the mapping model and discussion with Port operations personnel, the study team concluded that there were additional pressure lines, communications, security, and electrical conduits, but records identifying their location were not available. This resulted in a recommendation to complete QL B depiction to confirm the presence of the additional systems.

4.3 *Pilot studies*

During development of a work plan to complete QL B depiction services, the study team identified several complexities. There were several airside issues related to working in the vicinity of aircraft and vehicle traffic from support operations. Landside issues were related to working in the vicinity of high volume traffic corridors, tenant operations areas, and airline customer pathways. In addition, approval to perform the work at the airport required separate work permits from both airside and landside business units.

A pilot study approach was selected to address the complexities. A representative section of each of the business unit areas was identified for QL B depiction. The objective of the pilot study was to confirm the accuracy of the geophysical methods applied and to collect information on production rates that could be achieved in each of the business unit areas. Accuracy of the geophysical methods will be confirmed by vacuum extraction sampling at random locations. The intent is to apply the results of the pilot study to create a work plan and budget for completion of QL B depiction of the entire project area. At the time this paper was submitted the pilot study plan was approved and implementation is anticipated in 2002.

5 APPLICATION OF THE MAPPING MODEL

5.1 *Drawing grids*

AutoCAD paper space format drawings were created on a grid system encompassing the project coverage area. The grid system was referenced to the Port's coordinate system on each drawing grid. The drawings were created on a scale of 1 in. = 50 ft. and set up for plotting one grid per standard "D" size drawing sheet. The grid drawings included all utility systems in the AutoCAD model. CD-ROMs of the grid drawings were provided to airport planners, including airport operations, capital program project managers, and consultants assigned to the capital program.

5.2 *Planning synergy*

Information meetings were held periodically during development of the model to present the drawings and to describe the quality level concept. As the quantity of utility information in the model increased, and project planners achieved an understanding of the quality level concept, use of the model also increased. The model became a valuable resource for responding to requests for information (RFIs) from planners and for addressing utility conflicts encountered during construction. Planning for future projects such as the terminal, roadways, and rental car facility in the north end area applied the utility model to planning maps to identify utility conflicts and for planning new utility corridors. As the quality level of the utility information increases to QL B, the model will be of even greater value to the planners.

The utility model also supports airport security planning by identifying locations where barriers need to be maintained to restrict access to utilities to airside operations from landside. Some utilities cross beneath the perimeter fence and are larger than the 96 square inch limit identified by FAA.

5.3 *Link to attribute database*

The next phase, currently in development, includes linking a database table to the utility model. Utility attribute information is currently being collected to include in the database. This information will initially include attributes such as manhole and vault coordinate locations and elevations, utility depths, and sizes. Future additions under consideration include utility capacities and maintenance history. A component could also be included to address the frequency of utility security barrier inspections.

5.4 *Addition of quality level data to design documents*

All construction documents prepared for projects at the airport must be referenced to the port coordinate system. This allows the airport to maintain a master site plan that identifies all proposed project footprints. The utility model can also be integrated into the construction document process. Information from the utility model can be extracted and inserted into site drawings for new projects. This will allow the planners to work with site drawings that have current utility information that is identified at the appropriate quality level. As updated utility models are issued, the utility information in the construction document can also be updated.

6 CONCLUSIONS

In the process of generating the AutoCAD mapping model and applying it to the design of future expansion of the airport, the Port of Seattle has gained some important insights:

- The value of the American Society of Civil Engineers' Collection and Depiction of Existing Subsurface Utility Data. Use of ASCE's definitions of levels of quality of subsurface utility data enhances the depiction of the utilities and clarifies planning efforts.
- The need for pilot studies before undertaking large underground utility location projects in congested areas. The Sea-Tac Airport has numerous areas where congested underground utility corridors are located under and around active gates and taxiways. These airport operations will have an important impact on utility location.
- Having all the underground data in one place on a common coordinate system creates opportunities for synergy. The ability to overlay planning sketches for roadways and buildings on top of the utility information has brought people together to talk about the need for a coordinated utility strategy for the airport expansion.
- This approach has enabled the Port to plan strategies for maintaining utility systems and to distribute this information to stakeholders for update and use.
- The model will allow the Port to realize cost and time savings by planning underground utility systems in advance of detailed design and construction.
- The potential risk of utility service disruptions during construction will be reduced with more reliable information about the location of utilities.

By gathering available utility information into a single model, the Port of Seattle will be able to anticipate and plan for underground utility impacts before undertaking major construction projects at the airport.

REFERENCES

American Society of Civil Engineers (ASCE) (Collection and Depiction of Existing Subsurface Utility Data Committee of Management Group F, Codes and Standards) 2001. *Standard Guidelines for the Collection and Depiction of Existing Subsurface Utility Data, Public Ballot Revision #4.* New York: American Society of Civil Engineers.

Autodesk 1998. *AutoCAD Map.* San Rafael: Autodesk.

Federal Aviation Administration (FAA) 2001. *Federal Aviation Regulation Part 107 – Airport Security.* Washington, DC: Federal Register.

Federal Aviation Administration (FAA) 1972. *Advisory Circular 107-1 Aviation Security at Airports.* Washington DC: Federal Aviation Administration.

Landrum & Brown, Inc. & Arai/Jackson 1998. *North End Airport Terminal Strategic Planning Study Report Summary.* Seattle: Landrum & Brown.

Lew, J. 2000. *Cost Savings on Highway Projects Utilizing Subsurface Utility Engineering (FHWA-HIF-00-014).* Washington, DC: Federal Highway Administration.

Port of Seattle 2000a. *Port of Seattle 2000 Annual Report.* Seattle: Port of Seattle.

Port of Seattle 2000b. *Seattle-Tacoma International Airport 2000 Activity Report.* Seattle: Port of Seattle.

North American Tunneling 2002, Ozdemir (ed.)

Tomographic ground imaging for the Henderson CSO Treated Tunnel Alignment, King County, Washington

David M. Neil
NSA Engineering, Inc., Golden, Colorado

Roberto J. Guardia
Shannon & Wilson, Inc., Seattle, Washington

ABSTRACT: A seismic tomographic survey was conducted to image the subsurface ground conditions along an inaccessible portion of the alignment for the planned Henderson CSO tunnel project in King County, Washington. The survey produced two- and three-dimensional tomographic images of the ground conditions under seven BNSF and UP railroad tracks and under Interstate 5 to detect logs, stumps, construction rubble, and boulders that would impact tunneling operations. Subsequent to the imaging, areas on each tunnel section were trenched at different locations to confirm the position and nature of anomalies indicated in the tomographic images prior to tunneling.

1 INTRODUCTION

The Henderson Combined Sewer Overflow (CSO) will prevent wastewater from entering Lake Washington during periods of high rainfall. An approximately 1,000-ft segment of the project will consist of a 72-inch-diameter concrete pipe that will be jacked under the heavily used eight-lane Interstate-5 and the Burlington Northern and Union Pacific railroad corridor into Seattle.

Due to the inaccessibility of this portion of the alignment to perform characterization borings, three small-diameter pipes were installed using Horizontal Directional Drilling (HDD) methods in a triangular pattern around the proposed 72-inch-diameter pipe for conducting a seismic tomographic survey.

The objective of the survey was to produce two- and three-dimensional tomographic images of the ground conditions under the railroad tracks and I-5 highway to…

1 locate hard zones and any structural anomalies within the survey area and/or volume;
2 detect man-made obstructions, boulders, wood stumps, and logs that may impact tunneling operations.

This segment of the Henderson CSO is located in the floodplain of the Duwamish River. The alluvial and estuarine deposits generally consist of very loose to medium dense saturated sand, and very soft to medium stiff silt, clayey silt, and organic silt. The alluvial and estuarine deposits extend to depths greater than 80 feet. The natural in-place soils are covered by 10 to 22 feet of mixed soil fill. The deepest fills (22 feet) are present at I-5. The shallowest fills (10 feet) are present at the railroad. The groundwater table is generally located between the springline and the top of the proposed 6-ft-diameter pipe, at a depth of 17 to 20 feet below ground surface.

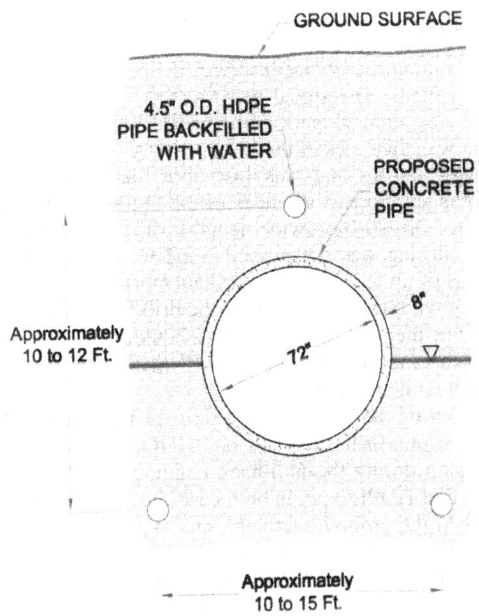

Figure 1. HDD access holes.

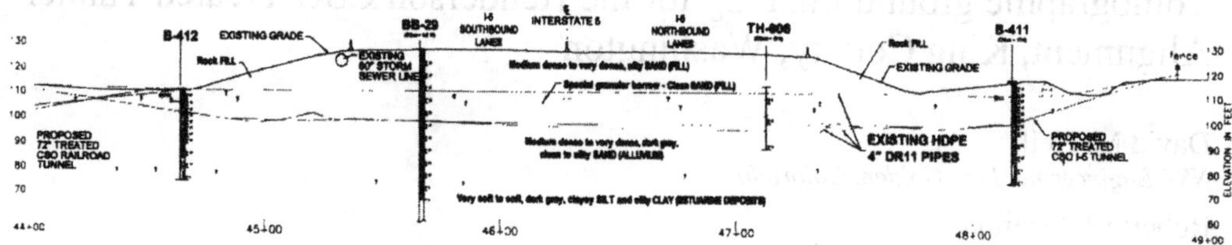

Figure 2. Plan and elevation position of HDPE pipes for I-5 segment.

1.1 *Horizontal Directional Drilling*

Three 4.5-inch-outside diameter, Standard Diameter Ratio (SDR) 11 High Density Polyethylene (HDPE) pipes (wall thickness of 0.409 inch) were installed by HDD methods at approximately the 12, 4, and 8 o'clock positions around the proposed pipe alignment as shown in Figure 1. The approximate location of the HDPE pipes in elevation for the I-5 segment is shown in Figure 2. The slope of the proposed pipe is 0.06%.

The staging area for HDD was located between the railroad tracks and I-5. The length of the boreholes in the railroad section (including Airport Way) was between 360 to 440 ft and in the I-5 section varied between 492 to 505 ft.

The HDPE pipes were installed utilizing a Vermeer D30x40 and a Ditch Witch directional drilling machine working concurrently. Directional drilling beneath the I-5 tunnel portion of the project was problematic since the tracking signal from the DigiTrack™ sonde located at the drill bit could not be monitored while drilling under the highway pavements. The signal was tracked from the toe of the west embankment to the west shoulder. The signal was then picked up in the vicinity of the median jersey barrier and was lost until the drill bit approached the east shoulder of the highway. This resulted in 80-foot wide lengths of drilling where monitoring was not possible, and in drilling deviations of up to 30 feet. Several attempts were made to improve the alignment once the drill bit was located at the median or at the east shoulder of I-5. The depth range of the DigiTrack™ tracking system is 50-ft with an accuracy of 5 percent.

During the drilling of the railroad segment HDPE pipes, one drill steel and one HDPE pipe failed in tension during the pull-back reaming/installation of the first HDPE pipe. In both cases, the drill steel was left in the ground overnight, and apparently the soil collapsed around the drill steel after being subjected to the vibrational loading of repeated railroad traffic during the night. The railroad crossing goes under four mainlines and three sidings.

Subsequently, the HDPE pipe was successfully pulled immediately after completing the bore, rather than waiting overnight, with no apparent collapse of the hole.

1.2 *Seismic Tomography*

A cross-borehole method using NSA's RockVision 3D™ ground imaging technology was applied between directional horizontal boreholes to provide information on the subsurface conditions without the need for surface borings along the proposed tunnel alignment. The imaged zone extended from the top borehole to the two boreholes at the base of the triangle. The boreholes were located approximately 2 feet above and below the proposed tunnel, respectively. A series of 3-D images of high-velocity units within the imaged area and 2-D tomography cross-sections on 5-foot centers along each alignment were prepared to show the 3-D spatial relationship of high-velocity materials impinging on the tunnel horizon.

2 SURVEY PROCEDURE

A three-dimensional seismic survey using the RockVision3D™ cross-borehole and surface velocity tomography methods was conducted at the test site on August 2 and 3, 2000. Table 1 shows major components of the seismic system used for the survey.

The HDPE pipes were backfilled with water prior to the survey in order to conduct seismic energy from the seismic source placed in one hole to the hydrophone strings placed in the other two holes.

Table 1. RockVision 3D™ equipment and instrumentation.

Recording Device	Geometrics R-24, 24-bit, 24-channel seismograph with sampling rate from 0.25 to 16 kHz
Downhole Receivers	Sercel DH5 hydrophones with 20 dB preamplifier with a frequency response from 7 Hz - 4.7 kHz
Downhole Source	Etrema – a magnetostrictive controlled waveform source

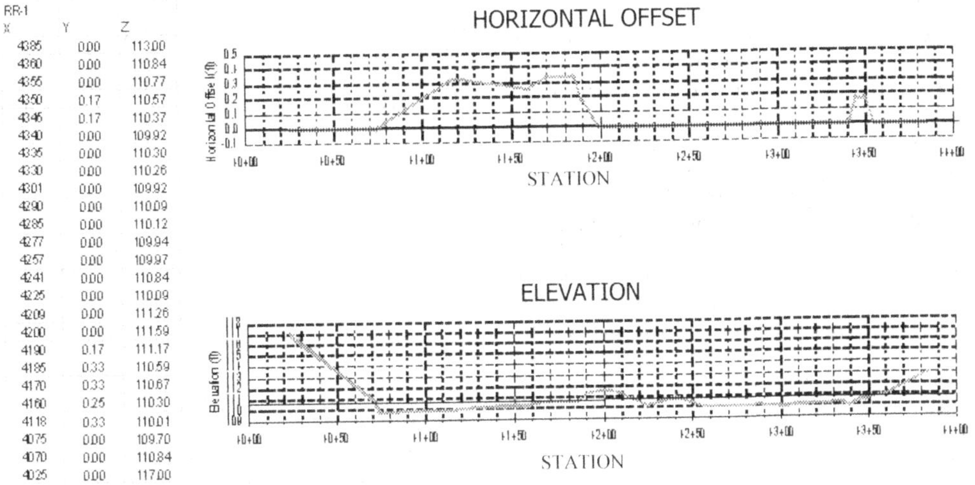

RR-1

X	Y	Z
4385	0.00	113.00
4360	0.00	110.84
4355	0.00	110.77
4350	0.17	110.57
4346	0.17	110.37
4340	0.00	109.92
4335	0.00	110.30
4330	0.00	110.26
4301	0.00	109.92
4290	0.00	110.09
4285	0.00	110.12
4277	0.00	109.94
4257	0.00	109.97
4241	0.00	110.84
4225	0.00	110.09
4209	0.00	111.26
4200	0.00	111.59
4190	0.17	111.17
4185	0.33	110.59
4170	0.33	110.67
4160	0.25	110.30
4118	0.33	110.01
4075	0.00	109.70
4070	0.00	110.84
4025	0.00	117.00

Figure 3. Borehole locations and survey coordinates used for the BNSF Railway site.

Figure 3 shows typical borehole locations and survey coordinates used for the tomographic survey at the BNSF Railway location. The holes were used to place and activate the seismic source at different depths in the ground and to install two strings of 12 hydrophones each that recorded seismic waves from seismic source locations.

An Etrema magnetostrictive electronic source was used as the seismic source in selected drill holes. Each pulse of the Etrema produced a seismic impulse against the walls of the hole. The source was activated in the holes at 5-foot intervals over the length of the boreholes.

Two strings, each consisting of 12 hydrophones (ITI DH-5) at 6.6-foot centers were placed in holes #2 and #3 to detect seismic waves from each activation of the seismic source in hole #1. The source and the receiver locations were advanced until all holes were surveyed. In each hole, the hydrophones were pulled approximately 75 feet after completing one series of source impulses, and the process was repeated to develop the appropriate pixel size for the required resolution. Resolution is approximately ±3 feet with a spacing of the seismic source at 5-foot intervals. Electric signals generated by the hydrophones in response to passing seismic waves were transmitted via cables to the Geometrics R24 Seismograph – a digital data acquisition system. The range of Etrema source and hydrophone locations in the boreholes assured horizontal coverage along the tunnel alignment.

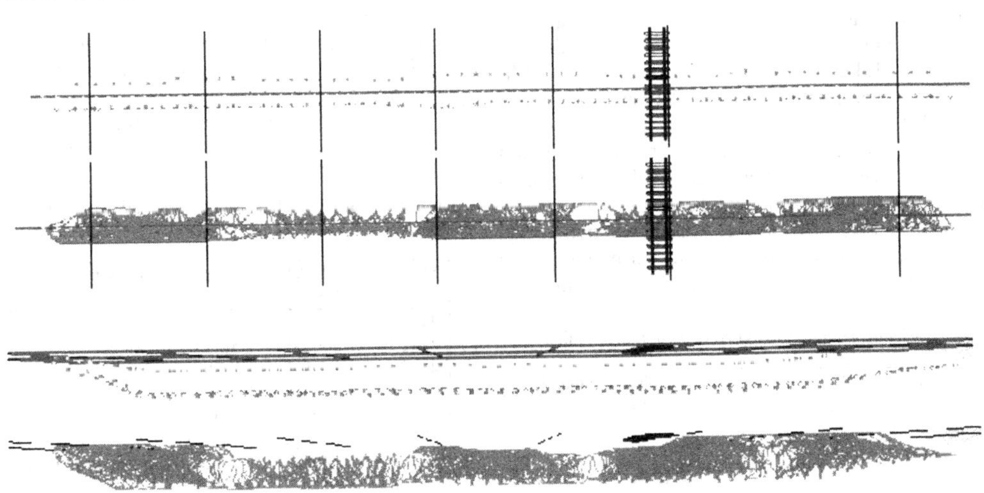

Figure 4. Ray path coverage and source and receiver locations at the BNSF site.

2.1 *Signal Quality*

Seismic waves detected by the system were saved in the seismograph as digital data files. After the survey, the data files were transferred to a computer for processing and reconstruction of images using velocity of seismic P-waves. The common frequency range used for data processing was selected between 100 and 800 Hz. In general, 90% of the signals detected by the hydrophones were of good quality. The heavy traffic on the freeway and railroad created a high noise-to-signal ratio that required a number of steps to secure adequate data. Stacking 10 times at each source location, multiple filtering, and extensive automated picking were used. Figure 4 shows the BNSF ray path coverage and source and receiver locations.

2.2 *Data Processing*

The proprietary RockVision3D™ software was used to process the acquired seismic data and to produce three-dimensional images of the ground within the perimeter of the array of all source and receiver locations. Seismic records for each pair of source and receiver locations were used to measure the travel time for P-waves between these locations. The travel times, combined with coordinates of sources and receivers, were used to construct a seismic velocity volume of the subsurface between the sources and receivers. This volume was then interpreted to produce an image of the subsurface geology.

2.3 *Data Analysis*

The velocity data from the borehole surveys were used to identify predominant soil types in the survey area. For each type, a range of representative seismic velocity was assigned. The color scale allows both velocity and soil types to be identified at any point within a tomogram. For example, blue indicates velocity of approximately 750 ft/sec corresponding to local soils. Likewise, green and yellow colors indicate velocities greater than 1,000 ft/sec, or various mixed fill.

The soil velocities are consistent between 700 and 750 ft/sec, which are analogous to dry and mixed soils with fairly uniform compaction.

An initial velocity model was developed to reduce image distortion outside the perimeter of the source and receiver locations. To generate this model, generalized geologic columns, using the soil types from Table 2, were created for each borehole in the survey area. Velocities of each grid point in the velocity model were interpolated from velocity ranges assigned to the generalized geologic column from nearby boreholes.

Table 2. Seismic velocities assigned to predominate units.

Soil type	Velocity range, ft/s
Alluvial and Estuarine Soil: Medium dense silty sand to clayey and gravelly sand	700 – 750
Mixed fill: Lenses of silts, compacted sand, gravels, and anomalous fill material	1,100 – 1,500

3 DATA INTERPRETATION

The three-dimensional representation of the survey area was constructed to produce cross-sections at selected locations every 5 feet along the alignment to display 3-D contours of various velocities.

The seismic tomography survey detected a number of anomalies (potential obstructions) in the railroad and I-5 alignments. Along the railroad alignment, the anomalies reflect the probable presence of wood as either logs or pile tops and possibly cement blocks, as well as weak rock debris or more compact soil. The anomalies do not appear to be boulders. The possible obstructions identified in the railroad alignment are tabulated in Table 3 by station. A similar table was prepared for the I-5 tunnel. The size of the obstruction is classified as small or large. "Small" is less than about 3 feet in the largest observed dimension, and "large" is greater than 3 feet in the smallest observed dimension. This classification is related to the resolution of the tomographic images.

Anomalies observed from Station 41+20 to 41+80, on the east side of Airport Way and the west side of the railroad appear to protrude 2 to 5 feet above the invert. If they are indeed wood, one can visualize these being the tops of timber piles. A possible pile stub is shown in the tomography section shown in Figure 5, at approximately Station 41+35. These tomography sections are shown looking east. The top elevation of the piles is approximately elevation 104 ft, which is within the organic, clayey silt. It would seem that the tops of piles should have been at elevation 109 ft to correspond to the top of the original ground surface before the fill was placed. However, the tops of the timber piles may

Table 3. Location of high-velocity anomalies at railroad tunnel horizon.

Station	Relative Size	Location in section
41+05	Small	Upper 2 ft of crown
41+25	Small	South rib at invert
41+20 to 41+80	Small	Protruding 2 to 5 ft from center of invert
41+95	Small	Protruding from center invert
42+05	Large	South side
42+80 to 42+85	Small	7 ft above invert on south side
43+25 to 43+30	Small	2 ft above invert
43+85	Small	Upper 2 ft of the crown

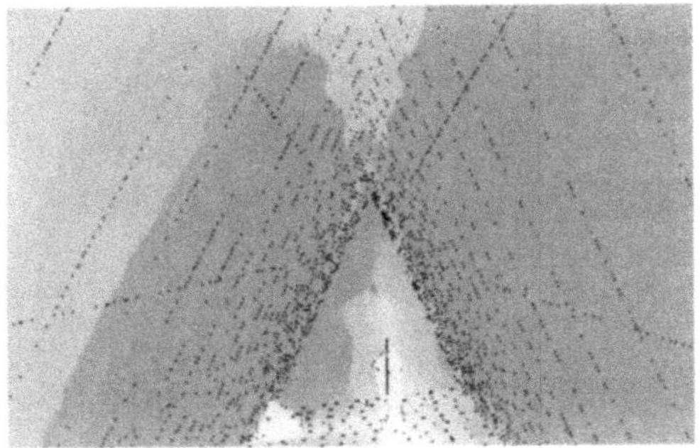

Figure 5. Station 41+35 looking east.

correspond approximately to the groundwater elevation, indicating that the timber piles in the zone above the groundwater surface may have disintegrated.

Figure 6 is a cross section of the I-5 alignment. Tomographic images in this area indicate that most of the anomalies are located near the top of the proposed tunnel. Most of the high-velocity anomalies are protruding 2 feet to 4 feet below the crown. These anomalies are mostly in the special granular borrow placed at the base of the fill in this portion of I-5. There is a concentration of anomalies on the east edge of the embankment. The anomalies probably consist of crushed rock and/or concrete debris.

4 CONFIRMATION TEST PITS

Three test pits were excavated at the east embankment of I-5 and in the accessible areas between the railroad tracks and I-5 to find obstructions identified in the seismic tomography survey.

Figure 7 shows longitudinal tomography cross-sections and highlights the area investigated in the vicinity of Test Pit TP-3 to verify the location and type of anomaly seen in the group of images in Figure 8.

Test pit TP-3 was located on the east side of the railroad tracks at Station 43+85. The test pit was excavated to a depth of 13 feet below ground surface and supported with a trench box. Abundant wood fragments were encountered in a gray-to-black, silty sand from 7 feet to the bottom of the excavation. A large tree log was observed in the northwest corner of the excavation at Station 43+80 approximately 7 feet below ground surface. The groundwater table during the excavation was also observed at a depth of about 7 feet.

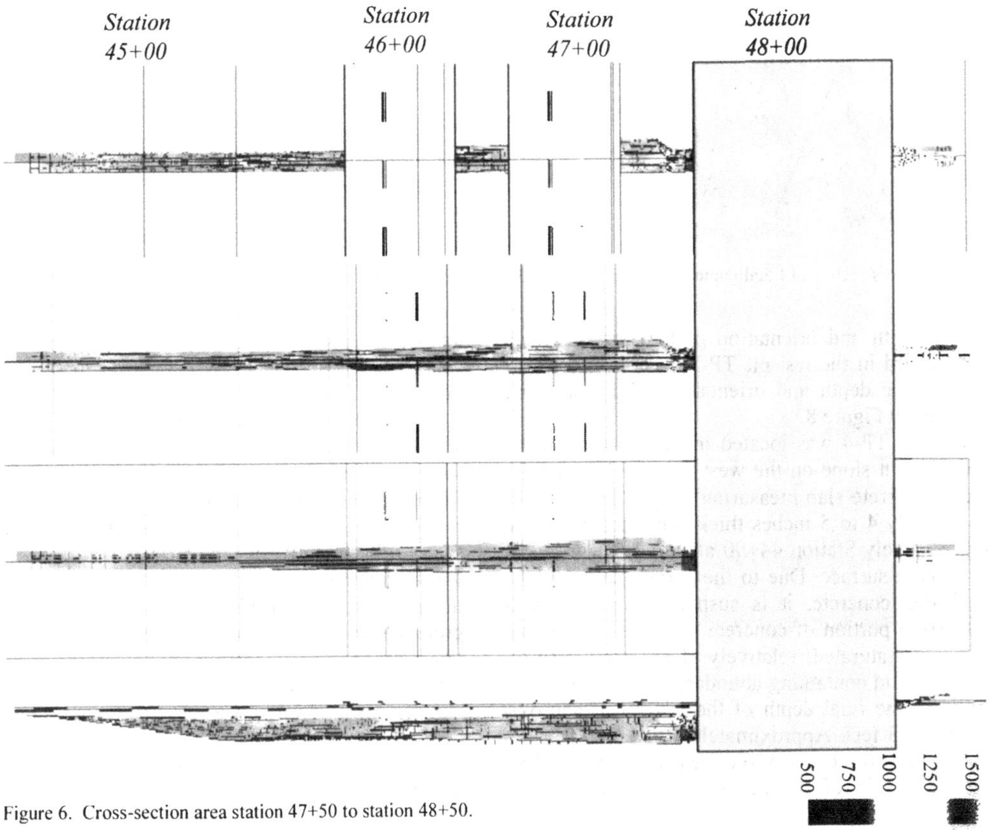

Figure 6. Cross-section area station 47+50 to station 48+50.

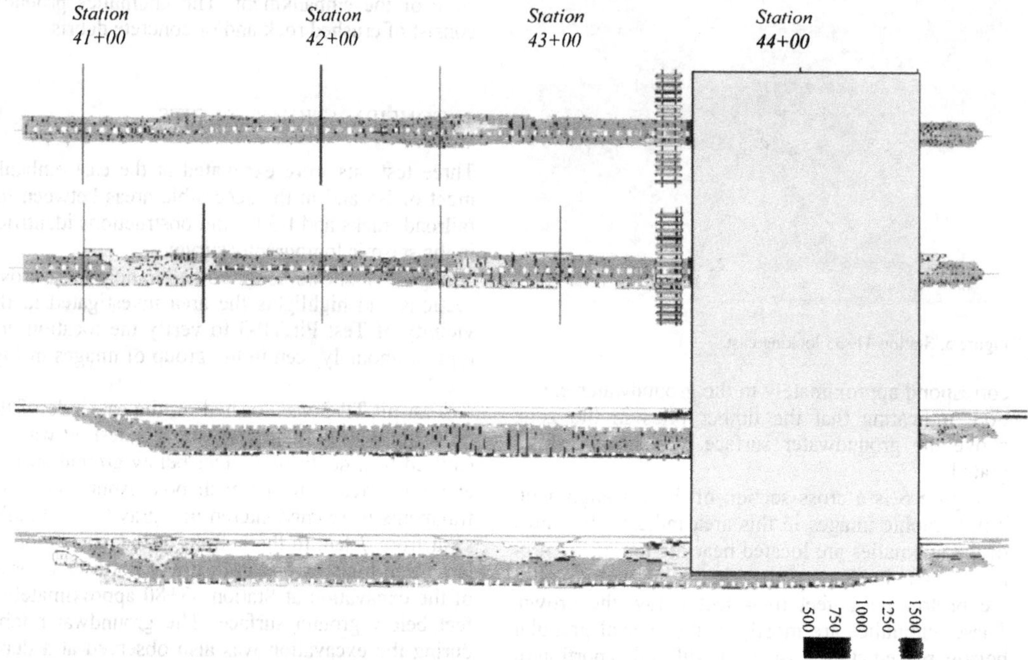

Figure7. Cross-section area Station 43+50 to 44+50.

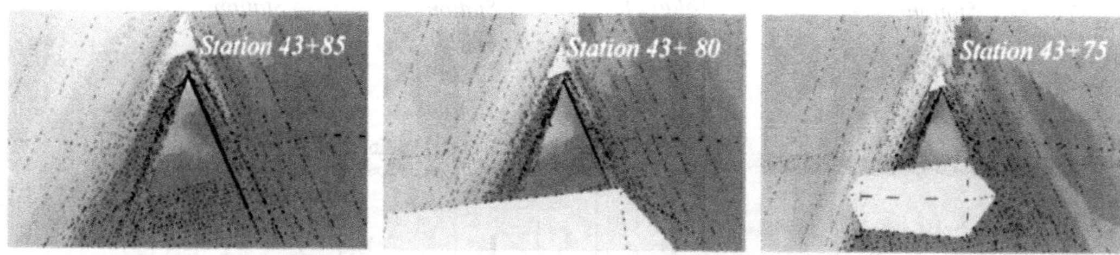

Figure 8. Cross-section of I-5 alignment showing Stations 43+85, 43+80, and 43+75.

The depth and orientation of the large tree log encountered in the test pit TP-3 corresponds to the approximate depth and orientation of the anomaly indicated in Figure 8.

Test pit TP-4 was located in the toe area of the embankment slope on the west edge of I-5. A reinforced concrete slab measuring 4 by 4 feet and approximately 4 to 5 inches thick was encountered at approximately Station 44+90 at 1 to 1.5 feet below the ground surface. Due to the curved shape of the reinforced concrete, it is suspected that this was formerly a portion of concrete pipe. Below a depth of 9 feet, saturated, relatively clean, gray, fine-to-medium sand containing abundant wood debris continued to the final depth of the test pit at approximately 15.5 feet. Approximately 10 and 11 feet below ground surface, two tree logs were observed on the south side of the pit at Stations 44+80 and 44+95, as shown in Figure 9. The groundwater table was observed at a depth of 12 feet during the excavation.

The tomographic images in the area of test pit TP-4 were not conclusive in showing a distinct anomaly in this area.

Test pit TP-5 was located on the east side of I-5 at Station 47+95. Below 7 feet, gray, fine-to-medium sand with variable silt content and abundant wood was encountered. This saturated alluvial sand continued to the bottom of the test pit at 14 feet below ground surface. Two large tree logs were encountered during excavations between 9 and 13 ft below ground surface, lying nearly diagonally across the test pit. Groundwater was observed approximately below 11 ft.

Figure 10 is a group of images representative of the area investigated with Test pit TP-5 on the east

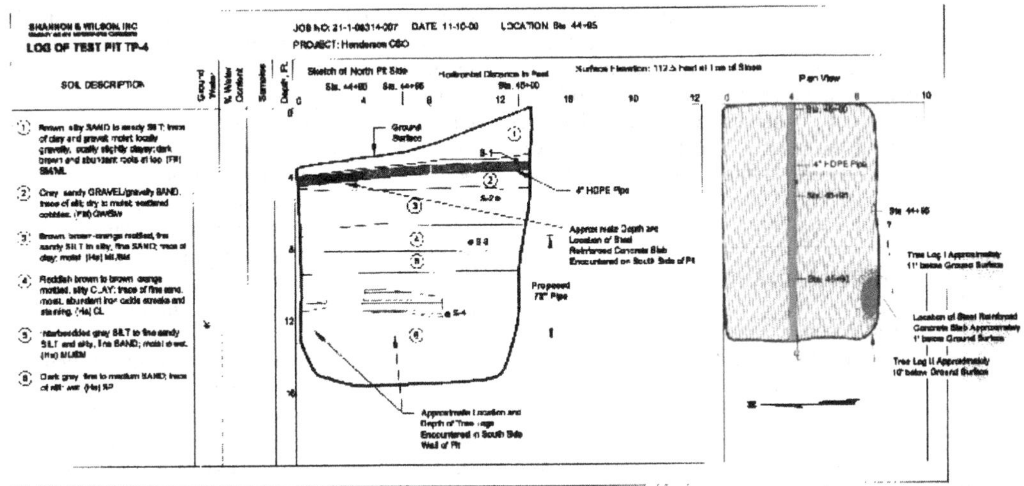

Figure 9. Log of Test Pit TP-4 showing concrete slab at Station 44+90 and tree logs at Stations 44+80 and 44+95.

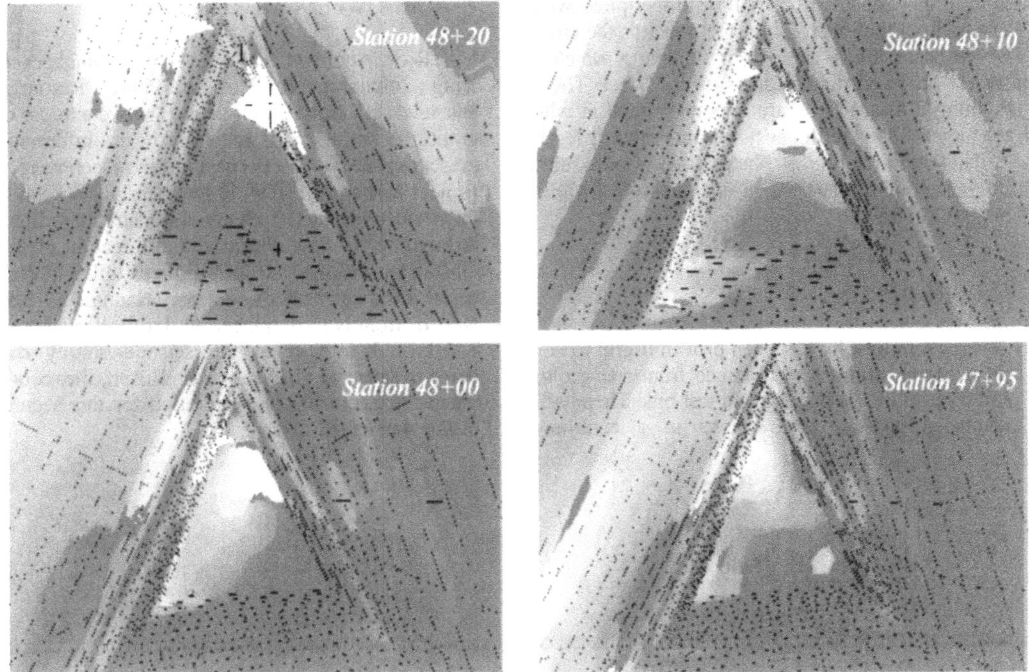

Figure 10. Group of images representative of survey area.

side of the I-5 survey zone. The image of frame 47+95 indicates an anomaly in the lower right side that may represent the logs encountered in TP-5.

During HDD, the mud return carried wood fibers. This happened mostly during the reaming/pulling of the pipe. Samples of the wood fiber were collected, and the pieces had both weathered and unweathered surfaces, and may indicate that the wood fibers were scraped from larger pieces of wood, possibly stumps or logs.

While installing the southern-most bottom horizontal drill hole beneath I-5, the drilling crew reported an obstruction at Station 47+45, which is under the embankment slope on the east side of I-5. The tomography survey shows some anomalies in the vicinity of this station. The irregular drill path in this area as shown in Figure 2 evidences the drilling difficulties. After pulling back the northern-most bottom horizontal drill pipe under the highway, a piece of reinforcing bar approximately ½-inch in diameter and about 4 inches long was observed wrapped around the reamer. Possibly the hole was drilled near or through a piece of reinforced concrete.

5 GEOTECHNICAL BASELINE REPORT

A Geotechnical Baseline Report (GBR) was prepared for the Henderson CSO project that included the I-5 and RR tunnels. The baseline report identified the number and types of obstructions that can be reasonably anticipated in the tunnel alignment. For example, the GBR stated that, for baseline purposes, the Railroad Tunnel will encounter 13 obstructions between Station 40+73 and 43+50. All 13 obstructions will consist of wood, piling, logs, or stumps. Four of these obstructions will be timber piles, which extend from the invert 2 to 6 feet towards the crown. Seven obstructions are logs or stumps in which the largest observed dimension is up to 3 feet. The remaining 2 obstructions are logs or stumps in which the largest observed dimension is more than 3 feet.

This interpretation of obstructions was based on inferences of the anomalies encountered in the tomography survey.

6 CONCLUSIONS

Seismic tomography has been proven useful to complement the information obtained from site geology and a series of test borings and test pits. The method identified possible obstructions in a continuous manner along the proposed tunnel alignment where standard investigation methods were inconclusive or impractical.

The 3-D images produced from velocity and attenuation are a complement to the basic geotechnical information presented in the Geotechnical data reports. The identification of anomalies in areas that were inaccessible because of surface access restrictions was verified by subsequent excavation of several test pits. This allowed gaps in the baseline data to be filled with some confidence by extrapolating the baseline boring geologic information with the image results.

The use of horizontal directionally drilled cased holes to provide access under the highway and railroad was cost effective because it helped to quantify the presence of obstructions and anomalies at this site. The seismic survey was approximately 25% of the horizontal directional drilling and casing cost.

7 ACKNOWLEDGMENTS

Mr. Rick Andrews was the Project Manager for King County Department of Natural Resources, Wastewater Treatment Division, and we express our gratitude for allowing us to utilize the information presented above. Dan Pecha and John Koch were the Project Managers for HDR Engineering Inc. who is the project designer. Trenchless Construction, Inc. of Arlington, Washington, under the direction of John Gustafson, installed the HDPE pipes utilizing directional drilling methods. Kanaan Hanna and Dave Neil Jr. from NSA Engineering, Inc. collected, processed, and interpreted the seismic survey data. Roberto Guardia of Shannon & Wilson, Inc. coordinated the geotechnical exploration for this segment of the Henderson CSO project.

North American Tunneling 2002, Ozdemir (ed.)
© 2002 Swets & Zeitlinger, Lisse, ISBN 90 5809 376 X

FE analysis of combined effects for adjoining braced excavations

B. Altabba
HNTB Corporation, Boston, MA, USA

A.J. Whittle
Massachusetts Institute of Technology, Cambridge, MA, USA

ABSTRACT: The paper summarizes a program of 2-D and 3-D non-linear numerical analyses performed to evaluate the effects of changes in the excavation sequencing of two adjacent ninty foot deep contracts. The purpose was to substantiate the structural design of lateral earth support systems and to estimate ground movements beneath the historic Russia Wharf buildings. The analyses simulate the sequential excavation of Vent Building #3, braced principally by corner braces, followed by the Central Artery NorthBound (CANB) tunnel. The difficulties in the analyses relate to the combined effects of two excavations joined by a common diaphragm wall and the corner effects in boxed excavations. Two-dimensional analyses confirmed that simplified models of soil behavior and uncoupled analyses could be applied to this problem. Three-dimensional analyses demonstrated that corner bracing is effective in minimizing ground deformations and hence, revisions in the excavation sequence are unlikely to cause damage to the Russia Wharf buildings.

1 INTRODUCTION

Major sections of Boston's new underground Central Artery have been constructed through densely developed downtown areas using conventional cut-and-cover techniques. During this process, contractors have played a leading role in refining the designs of the temporary lateral earth support systems (perimeter walls and cross-lot bracing), generating significant cost savings for the project. The key to more efficient structural design has been the widespread use of non-linear finite element analyses that simulate the response of the surrounding soil continuum through successive stages of the excavation process (e.g., SEI, 2000).

Although these analyses capabilities are now becoming more accessible to structural engineers, through the availability of specialty commercial software packages, reliable predictions of ground deformations (and their effects on adjacent structures) remains a difficult task. Settlement predictions are strongly linked to the modeling of soil behavior, and to the representation of in-situ soil stresses, soil properties and groundwater conditions (e.g., Whittle et al., 1993,) in addition to properties of the structural elements and support systems.

This paper summarizes some experiences in the use of finite element analyses for a complex structural and geotechnical problem caused by the interaction effects of two separate excavation contracts adjacent to the historic Russia Wharf buildings in Boston.

2 PROBLEM DESCRIPTION

Russia Wharf comprises a group of three seven story high, unreinforced masonry buildings that were constructed in the early 1900's and are designated as historic structures by the Massachusetts State Historic Preservation Officer. The buildings have a one story basement and are founded on massive granite block footings bearing on ten inch diameter timber piles whose length is unknown (more details of this type of foundation can be found in Parkhill, 1998).

The northbound tunnel for the new I-93 Central Artery (CANB; contract C17A1) lies immediately beneath Atlantic Avenue and within 15ft of the Russia Wharf structure; while a large ventilation building no. 3 (VB3; contract C17A3) is located within 20ft along the eastern side, Figures 1 & 2. Both contracts required temporary lateral earth support for excavations that are more than 90ft deep and share a common wall marked in Figure 1. The effects of these excavations on the adjacent Russia Wharf buildings were further complicated by plans for construction of twin tunnels for the MBTA Transitway project immediately beneath these structures (Fig. 1). The MBTA tunnels are much shallower than either of the two Central Artery excavation contracts dis-

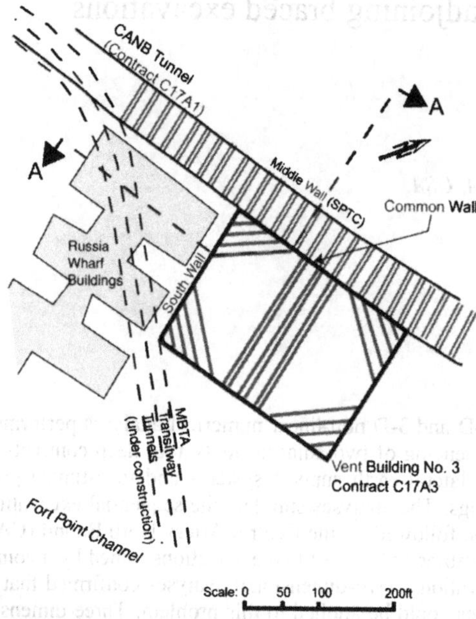

Figure 1. Site layout.

Figure 2. Site Photo Looking East. Russia Wharf on the right.

cussed in this paper and are being constructed concurrently using NATM tunneling techniques (in combination with ground freezing and structural underpinning).

In order to protect Russia Wharf buildings from damage, the project engineering team had to demonstrate that the proposed designs of temporary lateral earth support systems for the two Central Artery contracts would generate angular distortions, $\beta < 1/2000$ beneath these buildings. This represents a conservative interpretation of damage criteria proposed by Boscardin and Cording (1989) and provides a similar allowance in separate calculations for the MBTA tunnels. The project team also had to show that consolidation settlements would not exceed 1 inch due to effects of dewatering during construction.

2.1 *Original Design*

According to the original schedule, subsurface construction of the ventilation building VB3 was to be completed prior to excavations for the CANB tunnels in the vicinity of Russia Wharf. Indeed, the design of the wall and bracing system was already finalized (November 1996) using the following project wide analysis criteria:

1. Conventional beam-spring analyses were used to design the wall and bracing elements. Excavation was modeled using a sequential beam on elasto-plastic soil springs approach using the finite element program GTSTRUDL.
2. Separate geotechnical analyses of wall deflections and ground deformations were made using the displacement-based finite element code SoilStruct (incorporating a hyperbolic model of soil behavior). Separate 1-D consolidation calculations were used to assess potential effects of dewatering.

Each of these analyses simulated conditions corresponding to bracing of a single wall (combined settlements of the two contracts were estimated by superposition). Further structural analyses were subsequently carried out in conjunction with a value engineering proposal (VECP made by the contractor) to reduce the size of the support wall for VB3 and the number of bracing levels. These calculations were carried out using the finite element program ANSYS. This analysis, which models the soil continuum (and used soil properties very similar to those in SoilStruct), confirmed that the beam-spring models were unduly conservative (i.e., overestimate both the strut forces and wall bending moments).

2.2 *Revised Excavation Sequence*

One major proposed change in the schedule was to advance the construction of the Central Artery NorthBound (CANB) tunnel such that both the CANB and VB3 excavations could take place almost concurrently in the vicinity of Russia Wharf. This sequence of events raised further questions regarding the design of the lateral earth support system (most notably the design of the common wall between the two contracts) and the control of ground deformations beneath the Russia Wharf buildings.

The three-dimensional aspects of the problem are obvious and presented a major challenge compared to prior 2-D analyses. The proposed excavation support system for VB3 used corner bracing to transfer load to the common wall (i.e., between VB3 and CANB excavations, Fig. 1). This load is then carried across the CANB excavation to the 'middle wall' between the north and southbound tunnels of the Central Artery (Fig. 1). Two-dimensional models of this three wall system must inevitably focus on a critical skew-cross-section involving the largest width (and lowest bracing stiffness) of the combined excavations of VB3 and CANB as shown by A-A in Figure 1. The 2-D models assume plane strain conditions (i.e., all soil deformations and flow occur in the plane of the analysis). In all likelihood, this type of calculation will provide a conservative estimate of ground movements for a selected set of soil properties (initial stresses and groundwater flow conditions).

There are then two basic outcomes that could be envisioned from 2-D analyses of the skew-section A-A:

1. If the predicted movements were within the allowable tolerances for Russia Wharf, then construction could proceed based on the original design.
2. If the predicted movements exceeded the prescribed limits, then the design team would either have to modify the support system design or constrain the construction schedule.

This second outcome can scarcely be justified by highly simplified 2-D analyses that completely ignore the bracing effects at the southwest corner of vent building 3. More realistic 3-D finite element analyses are capable of modeling the support systems for the two adjacent contracts. However, there is a massive increase in complexity associated with the preparation of a 3-D finite element model, minimal prior experience in the use of such analyses for design of lateral earth support systems, and practical limitations on parametric analyses that can be carried out.

As a result of these considerations, the project team adopted a strategy involving the parallel use of 2-D and 3-D finite element analyses. The 2-D analyses provide a framework for evaluating effects of individual parameters (e.g., excavation and support sequence, groundwater control) and design assumptions, while the 3-D calculations provide refined estimate of critical parameters such as the settlements (and angular distortions) beneath the Russia Wharf buildings.

3 FINITE ELEMENT ANALYSES

3.1 *General Considerations*

There are several commercially available finite element (or finite difference) programs that are capable of performing non-linear analyses of soil-structure interaction and simulating prescribed construction sequences (e.g., Whittle, 1999). Prior to this study, two FE programs, Soilstruct and ANSYS, had already been used to analyze single-wall sections for the two separate construction contracts (for design of the bracing system and ground settlement predictions, respectively). After evaluating the available software, the design team decided to use two other FE packages for analyzing the combined effects of the C17A1 and C17A3 contracts. The introduction of these new packages was agreed after careful checking to establish that the analyses could reproduce single wall analyses comparable to the earlier design calculations.

The 2-D calculations were carried out using the PLAXIS program (Brinkgreve & Vermeer, 1998). This particular software has a user friendly graphic interface for pre- and post-processing of data (including an efficient and flexible 2-D mesh generator) and is equipped with a relatively robust procedure for automatic load stepping. The program solves coupled flow and deformation behavior of the soil mass, using mixed elements with pore water and deformation degrees of freedom. Hence, it can model the generation and (partial) dissipation of excess pore pressures (and their effects on structural forces) that can occur during the excavation process and associated groundwater pumping activities. The program has a closed architecture with a limited range of built-in soil constitutive models. Baseline analyses of excavations can be carried out using linearly elastic-perfectly plastic (EPP) models (with conventional Mohr-Coulomb yield criterion), while a Hardening model (Schanz et al., 2000) is included to simulate more realistically the non-linear shear-stiffness properties of typical soils (modeled using a hyperbolic function). The Hardening soil model represents an upgraded version of the well known Duncan-Chang formulation (Duncan et al., 1980). The program includes a separate module for computing steady, unconfined groundwater flow, with automatic computation of the phreatic surface, and can introduce hydraulic conductivity parameters that vary with void ratio. The PLAXIS program also includes a limited library of structural elements that enable modeling of the wall and associated bracing systems (cross-lot struts, prestressed tiebacks etc.), with controlled interface shear strength and transmissivity characteristics.

The principal difficulties in performing 3-D finite element analyses of braced excavations are practical considerations relating to model complexity and computational resources. Three-dimensional analyses of the combined VB3 and CANB excavations were performed using ABAQUS (v5.8, HKS, 1998). ABAQUS was one of the first general purpose finite element codes that included capabilities for handling coupled flow and deformation, and included effective stress models of soil behavior. Whittle et al. (1993) and Whittle (1998) has previously used ABAQUS as a platform for evaluating the role of advanced soil models in predictions of excavation performance. The current 3-D analyses adopt much simpler models of soil behavior and resort to simplified representation of groundwater conditions in order to reduce the computational complexity.

3.2 *Features of 2-D FE Models*

Figure 3 shows a typical example of the three-wall, 2-D FE model at a critical stage in the construction sequence, where the Vent Building excavation is completed (and the concrete foundation mat already in place), and the CANB tunnel is excavated to its final grade (El. +10ft).

The analysis includes many assumptions which can be summarized as follows:

1. Dimensions of the mesh are set in order to minimize far field boundary effects on the predicted ground movements and structural forces. The ground surface is level (El.+110ft,) and three slurry walls each extending more than 100ft deep into the underlying intact rock (layer B2, Fig. 3). Although minimal movements are expected at the toes of the walls, the FE mesh is extended a further 100ft into the rock to enable effects of local dewatering to be modeled accurately. The lateral boundaries are set more than 200ft from the excavation where horizontal soil displacements are minimal and pore water pressures are unaffected by the excavation process.
2. The analyses assume that the soil profile can be represented by a series of horizontal layers. This is a reasonable approximation of the available borehole data, and assumes maximum thickness of the weaker strata (fill, organics and clay layers). It should be noted that the marine clay (Boston Blue Clay) is only 30ft thick at this site and that the interface to the underlying till layer occurs 50ft below the ground surface.
3. The groundwater table in the fill is perched close to the ground surface and is closely linked to the tidal conditions in the adjacent Fort Point Channel, seen in the background in Fig. 2. The pressures in the underlying till and rock layers, however, may be related to regional bedrock topography and pumping activities. The FE analyses considered two scenarios: i) worst case scenario, '100 year flood', where the water table is at the ground surface and pore pressures are hydrostatic (u_0); and ii) best estimate conditions with constant piezometric head in the fill and organics (El. 106ft), a reduced head in the till and rock (El. 96ft) and linear head loss through the clay.

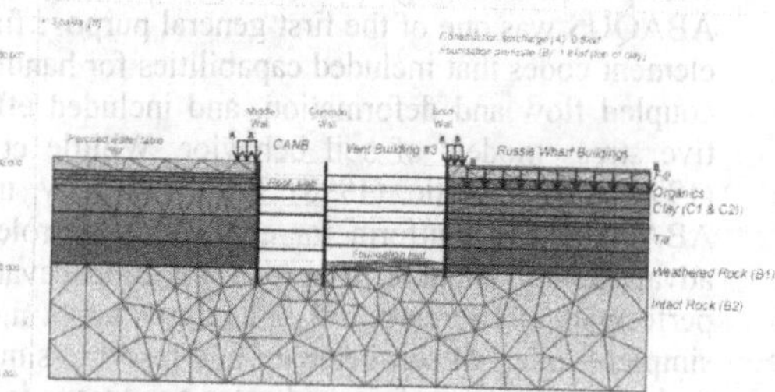

Figure 3. Typical three-wall model of skew-section.

4. The in-situ vertical stresses in the soil profile are derived from estimates of the unit weights of the layers and the assumed pore pressures (i.e., the initial effective vertical stress, $\sigma'_{v0} = \sigma_{v0} - u_0$), while lateral stresses are based on expected K_0 values ($K_0 = \sigma'_{h0}/\sigma'_{v0}$) listed in Table 1. The controlled selection of K_0 provides a more realistic estimate of in situ soil stresses in the horizontally layered profile than prior FE analyses where initial stresses were generated by 'gravity turn-on'.
5. The Russia Wharf column loads are supported on granite block, spread footings (extending to approximately 10ft depth) and groups of timber piles whose length is unknown but may extend through the clay to the underlying till. The FE analyses model represent these foundations by a surcharge load of 1.8 kips/ft^2 applied at the top of the clay. The analyses assume that the underlying soils are in fully drained equilibrium with these applied loads prior to wall installation.
6. A temporary construction surcharge of 0.6kips/ft was applied on a strip 15ft wide either side of the excavations. This load was applied after the installation of the first level of struts.
7. The analysis models the reinforced concrete slurry walls using elastic beam elements with (axial and bending) stiffness parameters corresponding to the cracked section, Table 1. The Plaxis program uses Mindlin beam theory, hence the wall deflections occur due to shearing and bending. Cast in-place concrete walls have a rough interface with the surrounding soil, such that the shear strength of the interface is equal to that of the adjacent soil. The corner and cross-lot bracing struts are represented by 2-node elastic spring elements, which can be preloaded when the elements are activated. The current analyses do not consider the effects of thermal expansion and contraction on strut loads or earth pressures. Figure 3 shows the elevations of these struts for both VB3 and CANB excavations. The upper tier of braces for the CANB tunnel corresponds to the final roof beam.
8. The soil layers are modeled using 6-noded triangular elements (quadratic interpolation of displacements). Information on soil properties for each of the layers was derived from local site investigation reports (GEI, 1992) and prior experience in the local area (e.g., Whittle et al., 1993). The till, rock and fill layers have relatively high hydraulic conductivity (estimated in the range $k \approx 0.3 - 3.0$ft/day; Table 1) and hence, can be expected to respond as free draining materials within the timeframe of the excavation events. In contrast, there will be very limited migration of pore water within the low conductivity organic and clay layers ($k \approx 0.02 - 0.0003$ft/day). The 2-D FE analyses assume that these layers undergo undrained shearing during each stage of excava-

Table 1. Input parameters for base case analyses.
a) Soil properties.

Layer	Total Unit Weight γ_t	Earth pressure at-rest K_0	Hydraulic Conductivity k	Shear Strength c' [s_u]	ϕ'	Deformation Properties G/σ'_{v0}	ν'
	pcf		ft/day	ksf	°		
Fill Dry	125	0.5		0	30	25	0.33
Wet	130		3				
Organics	105	0.5	0.014	[0.46]	--	25	0.3
Clay C1	125	1.0	0.00028	[2.00]	--	100	
Clay C2	118	0.8	0.00028	[1.30]	--	100	0.3
Till T2	140	1.0	0.14	0	43	110	0.3
Weathered Rock B1	140	1.0	0.28	3.6	32	1100	0.3
Intact Rock B2	150	1.0	2.8	36	32	4400	0.3

Notes: σ'_{v0}(ksf) ≈ 8.9840 - 0.0823El.(ft)
G is the elastic shear modulus

b) Properties of structural elements (all elastic.)

Component	EA	EI	ν
	kips/ft	kips-ft^2/ft	
Slurry Wall	1.56×10^6	5.19×10^5	0.2
Strut	2.24×10^5	--	--

tion. The effects of partial drainage (on earth pressures, ground deformations etc) have also been considered by allowing dissipation of excess pore pressures (within the organic and clay layers) over prescribed time periods corresponding to the expected rates of construction (typically 10 days for each 10ft of soil removed).

9. Baseline 2-D analyses assume linearly elastic (with effective stress stiffness parameters; E', ν') perfectly plastic properties for the soil and rock layers with conventional Mohr-Coulomb strength parameters (c', ϕ') and zero dilation (i.e., no volume change, ψ = 0°, Fig. 4a) at yield, Table 1. Tensile strengths of the rock layers are specified as 10% of the unconfined compressive strengths. The clay and organic layers are assumed to have constant undrained shear strengths, s_u, throughout the excavation. This approximation has minimal impact on the predictions due to the proximity of these layers to the ground surface.

   More realistic modeling of shear strength properties is needed elsewhere in the CA/T project where deep clay layers underlie the excavation (and can soften with partial drainage).

10. There is very limited data available on the engineering properties of the soil layers to enable more refined modeling of their stress-strain-strength behavior. Further analyses have been carried out using the hardening soil model in PLAXIS. This model includes a hyperbolic shear stress-strain relation in loading and a linear elastic response in unloading, Fig. 5. Calculations were carried out using input parameters reported previously for single walls (as part of the validation of

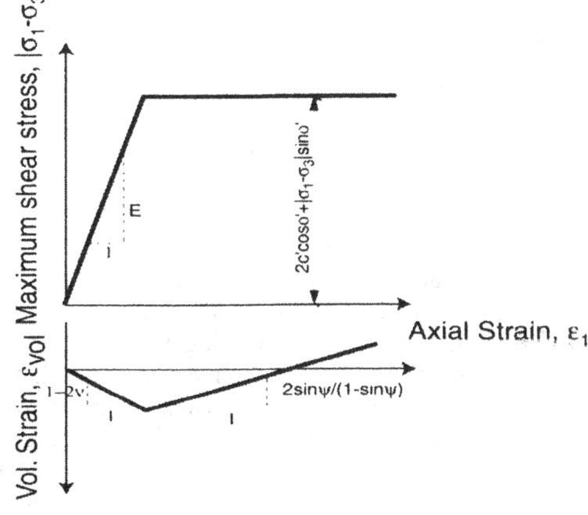

a) Shear stress-strain

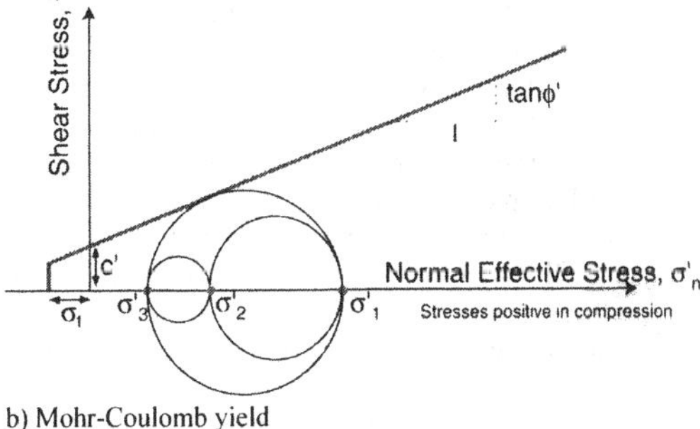

b) Mohr-Coulomb yield

Figure 4. Key features of EPP model.

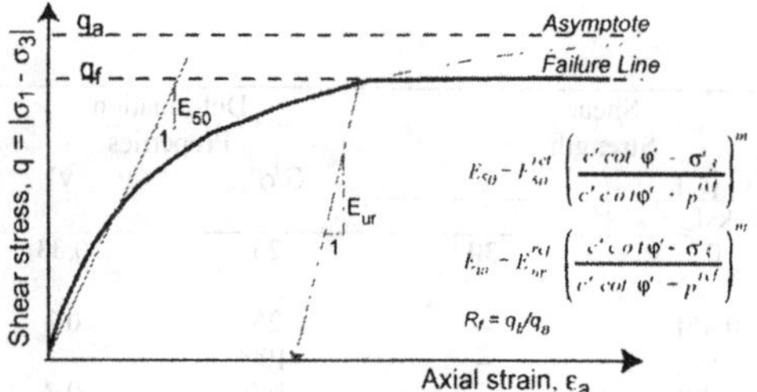

a) Shear stress strain behavior, hardening soil model.

Layer	E_{50}^{ref}	m	E_{ur}^{ref}	ν_{ur}
	ksf		ksf	
Organics	180	0.0	540	0.3
Clay C1	940	0.0	1880	0.3
Clay C2	630	0.0	1260	0.3

where p^{ref} = 2.05 kips/ft^2 (atmospheric pressure)

b) Stiffness parameters for low permeability layers

Figure 5. Hardening soil model.

the PLAXIS software). However, the lack of site specific data for estimating the non-linear stiffness parameters (except in the clay layer) discouraged more widespread use of the hardening model at this site.

Apart from these general considerations, there are two other aspects of the simulations which proved very important for this project:

1. Pre-loading of struts. According to the design, struts are to be loaded to 50% of their maximum expected loads. Great caution should be exercised in selecting the appropriate amount of preloading as the higher the preload the higher the effective support stiffness. In an indeterminate system this will lead to the attraction of lager amounts of load to the strut. The 2-D FE analyses assume that the struts are installed at the excavated grade, while the predicted maximum strut loads occur after the excavation has proceeded down to the next level of bracing. By simple iteration, it is then possible to reduce the magnitude of the pre-load imposed at strut installation.
2. De-pressuring base of excavation. The design specifications for the project place an upper limit on the pore pressures at the base of the till (relative to the excavation depth) such that there are no instability problems associated with high upward seepage gradients below the base of the excavation. De-pressuring must be simulated in the 2-D PLAXIS analyses to avoid numerical instabilities (convergence problems) that can occur when effective stresses are very small in the till, but have minimal effect on predictions in the retained soil (wall deflections or surface settlements) as the walls are securely embedded in the intact rock.

3.3 *Features of 3-D FE Model*

Although there is no conceptual difficulty in performing non-linear FE analyses of braced excavations in 3-D, there are major practical difficulties associated with the preparation of the model and significant constraints due to the large computational requirements. This may explain why there are only a few examples of 3-D non-linear FE analyses for braced excavations published in the literature (e.g., Ou et al., 1996; Lee et al., 1998).

The first key decision is to identify the (plan and elevation) dimensions of the 3-D model. Figure 6a shows that the current 3-D analyses included the full footprint area of the Russia Wharf Buildings, but considers only half the plan area of the Vent Building. This decision implicitly assumes that the central cross-lot bracing (Fig. 1) will ensure planar deformations. The FE mesh is extended 250ft laterally behind the perimeter slurry walls of both VB3 and CANB and assumes that the ground surface and soil layers are horizontal (with K_0 stresses). This is clearly a simplifying assumption (when compared to available borehole data) and also ignores the presence of the Fort Point Channel (Fig. 1). The 3-D FE model only extends to El.-20ft (much shallower than used in 2-D calculations). This is sufficient to model the embedment of the slurry walls within the intact rock (B2 Argillite). It is not adequate for simulating flowfields within the underlying rock and hence, the current 3-D analyses do not attempt to simulate coupled flow and deformation. The analyses make the conventional assumption that low permeability layers (organics and clay) remain undrained throughout the excavation, while the other layers are fully drained. The calculations apply differential pore pressures on the support walls based on assumptions of steady flow conditions within the till and rock layers (these are updated at each stage of excavation to reflect depressuring within the base of the excavation). Other assumptions regarding the foundation pressures from the Russia Wharf buildings and construction surcharge conditions are identical to those used in 2-D analyses.

The 3-D FE mesh uses 20-node quadratic 'brick' elements to model the soil mass (reduced integration methods were used to reduce computational times and appear to have little influence on the accuracy of the solutions). The walls are simulated using (8-noded) doubly-curved, thick shell elements, and the walers by (3-noded) quadratic beam elements, all struts were pre-loaded to 50% of their design loads. Figure 6b gives a clear illustration of the detail incorporated in modeling the entire Vent Building and CANB support systems. The 3-D model also assumed simplified sequences for the CANB excavation in order to expedite the calculations.

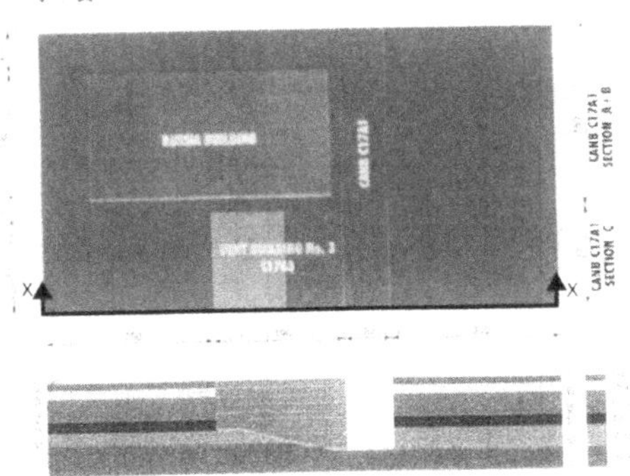

Figure 6a. Extent of 3-D FE model.

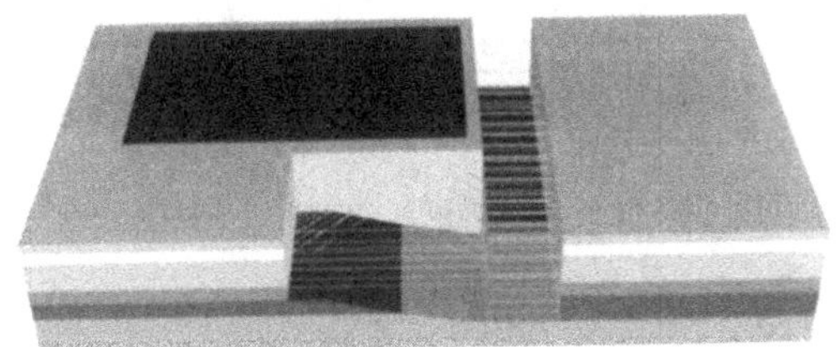

Figure 6b. Isometric view of 3-D model at end of CANB excavation.

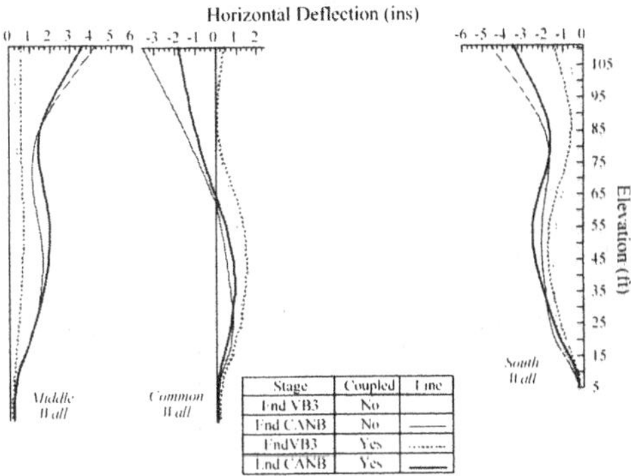

Figure 7. Effects of flow-deformation coupling on wall displacements.

4 RESULTS OF 2-D ANALYSES

4.1 *Effects of Coupled Flow and Deformation*

One of the key advantages of FE programs such as PLAXIS is the ability to represent coupled effects of flow and deformation within the soil mass. In principle, this enables the analyses to simulate soil deformations (partial consolidation) and associated changes in structural forces due to transient groundwater flow, as well as changes in soil stiffness and shear strength properties (due to changes in effective stresses within the soil skeleton). These coupled effects are most significant for excavations in deep layers of soft clay or in situations where the support wall is not embedded within a competent bearing layer. Changes in soil properties require relatively sophisticated constitutive models of soil behavior (e.g., Hashash & Whittle, 1996). In the current study, the slurry walls are embedded within the intact argillite, while the soil profile includes approximately 30ft of overconsolidated clay (and less than 10ft of compressible organics).

Figure 7 compares predictions of wall deflections from two PLAXIS analyses of section A-A assuming 1) uncoupled behavior, with undrained shearing in the clay and organics, and fully drained behavior (and steady flow conditions) in the fill, till and rock layers; and 2) coupled behavior, assuming that each excavation stage is represented by an undrained initial response (all soil layers), followed by a 10 day consolidation period (representative of typical field excavation rates). The figure compares horizontal deflections of the South, Common and Middle walls (cf. Fig. 1) at the end of a) the Vent Building excavation and b) the CANB excavation.

At the end of Vent Building excavation, maximum inward deflection of the common and south walls is in the range 1.4 – 1.6 inch and occurs at approximately El. 50ft. Slightly higher deflections occur in the coupled analysis, while differences in deflections of the two walls reflect the foundation pressures of the Russia Wharf buildings and the assumed surface construction surcharge loads (adjacent to the south wall). Excavation of CANB causes significant racking (up to 4 inch) of the Vent Building (south and common) walls towards the northbound tunnel and similar inward deflection of the Middle wall. Large deflections at the top of all three walls reflects the layout of bracing used for the CANB excavation (in this example, the first level of bracing is 25ft below the ground surface). However, the racking of the walls is an unrealistic feature of the 2-D analysis where the stiffening effects of corner bracing are not represented.

Although there are clearly some differences in predictions between the coupled and uncoupled analyses, the overall impression is that it is a secondary factor in predictions of wall deflections (as well as wall bending moments and strut forces).

Figure 8 summarizes predictions of surface settlements from the same analyses at both the ground surface and at the top of the till layer. Differences between the uncoupled and coupled analyses are clearly seen at the end of the Vent Building excavation. In this case, partial drainage within the compressible organic and clay layers causes an increase in the predicted settlement (at both the surface and top of till). Large increments in settlements occur beneath the surcharged areas at the end of CANB excavation. These are principally due to the inward

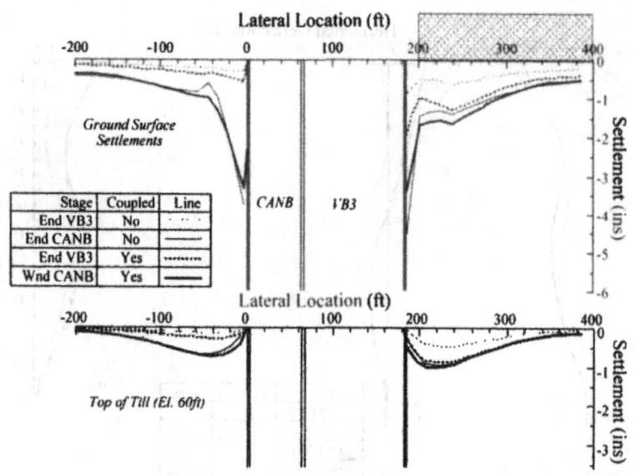

Figure 8. Effects of flow-deformation coupling on vertical settlements.

cantilever movements of the walls during the initial unsupported phase of CANB excavation. Beneath the Russia Wharf buildings, the maximum surface settlements increase from 0.5 inch to 0.75 inch at the end of VB3 to 1.5 inch after complete excavation of CANB (much smaller changes occur at the top of the till). There is minimal difference in settlements predicted by the uncoupled and coupled analyses at the end of the excavation.

Figures 9a-d provide further insight to explain the role of coupling between flow and deformation for section A-A. Figures 9a and 9b compare contours of the displacements (magnitudes of the displacement vector) immediately upon excavation of VB3 from El. +75ft to El. +60ft (top of till) and after a further 10 days of consolidation, respectively. The results show that this period of partial drainage is associated with significant vertical heave below the base of the excavation and much smaller changes in displacements within the retained soil (centered around the grade elevation). Figure 9c and d show piezometric head contours at the same two states. It is clear that the undrained excavation produces negative excess pore pressures beneath the excavated grade (in both the till and rock layers; Fig. 9c). However, due to the relatively high permeability of the underlying layers, the flowfield in Fig. 9d closely resembles steady state conditions, with reduced piezometric head (H = 83ft, specified at El. 33ft) to prevent excess upward hydraulic gradient within the till. It is clear that there is little connection between the response in the retained soil and below the excavated grade and hence, approximations of uncoupled conditions can be justified in the more complex 3-D analyses.

4.2 *Effects of Soil Model*

Although advanced soil models do have an important role to play in understanding the performance of lateral earth support systems and in predicting ground movements (e.g., Hashash & Whittle, 1996, 2001), their application can only be considered if

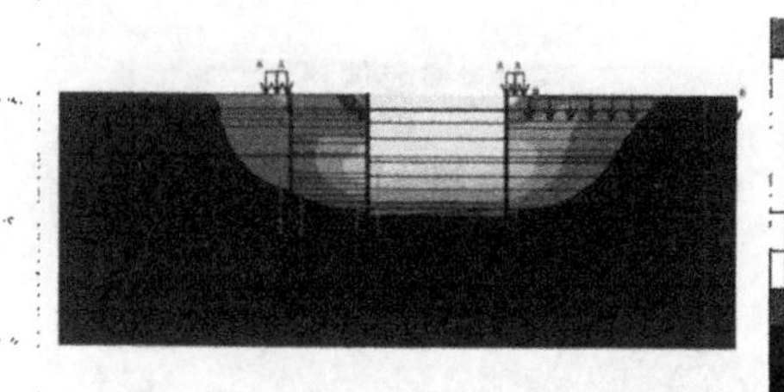

a) Displacements after undrained excavation.

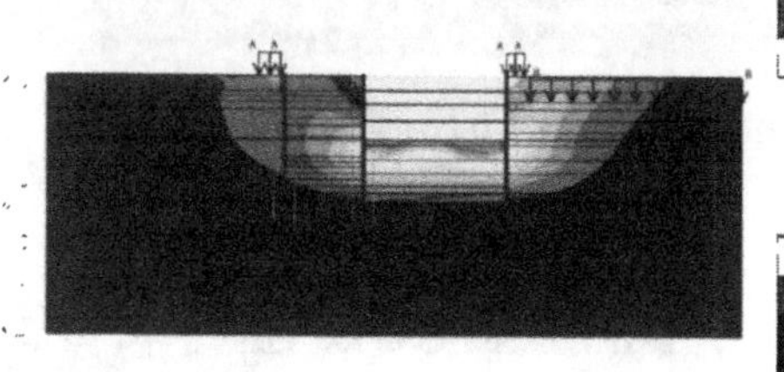

b) Displacements, 10 days after excavation stage.

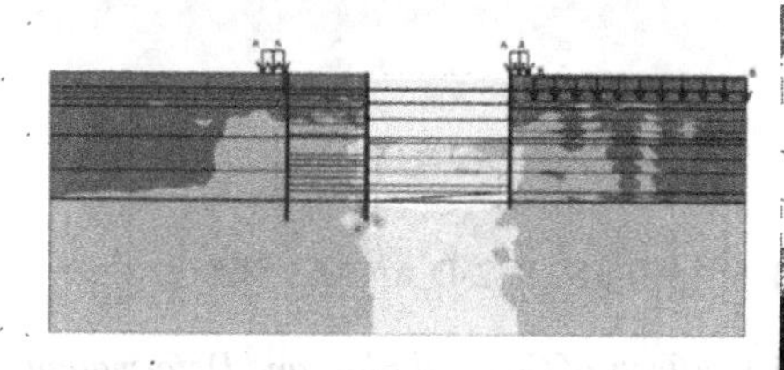

c) Piezometric head after undrained excavation.

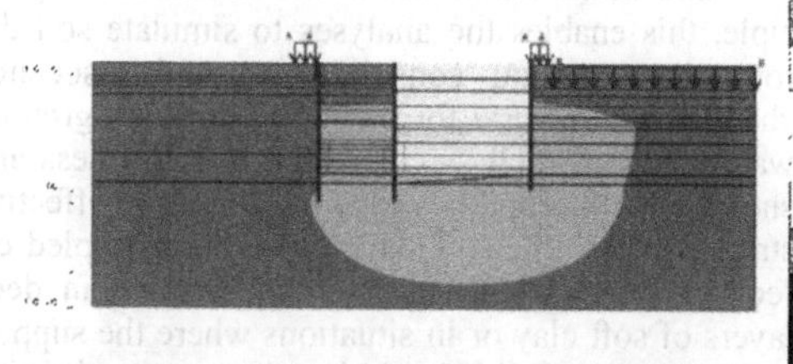

d) Piezometric head, 10 days after excavation stage.

Figure 9. Illustration of coupled flow and deformation.

there is sufficient experimental data to enable reliable parameter selection. For this project, calculations have been performed using a combination of linearly-elastic, perfectly plastic (EPP) and Hardening Soil (hyperbolic, perfectly plastic; Fig. 3) models. Figure 10 compares predictions of wall deflections at the end of the Vent Building and CANB excavation stages for the base case analysis using the EPP model, and for the case where the organic and clay layers are modeled using Hardening Soil

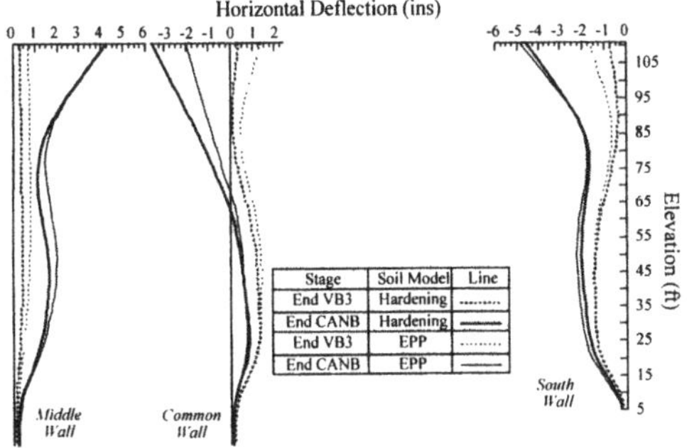
Figure 10. Effects of soil model on wall deflections.

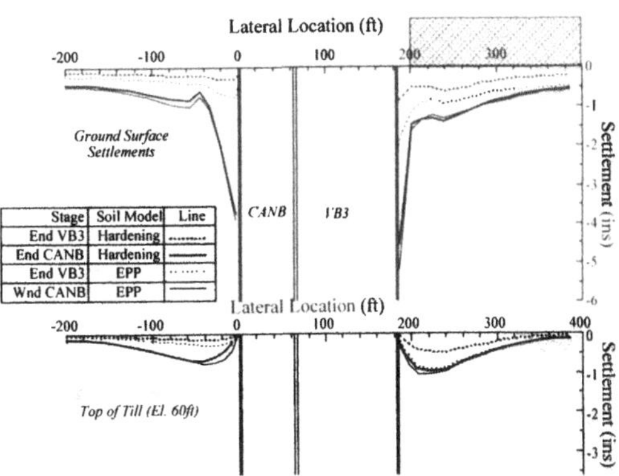
Figure 11. Effects of soil model on vertical settlements.

(HS) in PLAXIS. Both sets of calculations are for uncoupled analyses.

The characteristic deflection mode shapes are not influenced significantly by the selection of soil model. Indeed, the deformed shapes of the middle and south walls are practically identical at the end of CANB excavation. The base case analysis using the EPP model shows slightly higher wall deformations at the end of VB3 excavation.

Figure 11 summarizes the predicted settlement at the ground surface and top of till for the same two analyses. The soil model has a noted influence on the magnitude of the settlements at the end of VB3 excavation (base case, EPP calculations again producing larger settlements), but results at the end of CANB are practically identical. It is important to note that there is minimal difference in the predicted shapes of the settlement troughs below Russia Wharf for the EPP and HS models. This is a significant result as many previous authors have advocated for the use of hyperbolic models in settlement predictions. Indeed, Hashash and Whittle (1996) have shown that non-linear stiffness parameters have a profound influence on the settlement distribution for deep excavations in soft clay. These effects may be of secondary importance in this case study, where there is only a limited thickness of overconsolidated clay. The results in Figure 11 justify using the EPP model in 3-D analyses (in order to achieve significant reductions in computational times).

5 RESULTS OF 3-D ANALYSES

Figure 12 shows predictions of wall deflections along section A-A (Fig. 1) from the 3-D finite element analysis at the end of VB3 and CANB excavation stages. The 3-D analyses assume the same excavation steps, bracing elevations and pre-load forces as in the prior 2-D calculations. At the end of the Vent Building excavation the common and south walls move inward by approximately 0.75 inch and 1.25 inch, respectively. Although the mode shape differs from the base case analyses (EPP, Fig. 10), the magnitudes of the wall deflections are quite similar. Subsequent excavation for the CANB produces almost negligible additional movements of the Vent Building walls (indeed there is a small racking movement towards Russia Wharf), while maximum movement of the Middle wall is approximately 1.25 inch. These results differ very substantially from the 2-D calculations and provide a very clear indication of the importance of the corner bracing in stabilizing the Vent Building excavation.

Isometric views of the surface settlement distribution in Figure 13 provide a clear understanding of the effectiveness of corner bracing for the Vent Building excavation. At the end of the Vent Building excavation, the largest settlement occurs close to the south wall (adjacent to the Russia Wharf buildings). After the CANB tunnel is excavated, the largest surface settlements (approx. 0.48 inch, occur adjacent to the Middle wall (at section A-A), and there is only minimal increase in settlements predicted beneath Russia Wharf. Previous assumptions that the combined effect of the two excavations could be estimated by superposition are unnecessarily conservative.

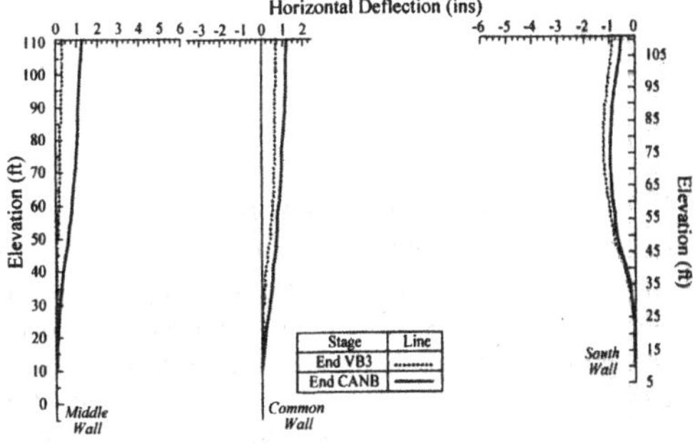
Figure 12. Wall deflections from 3-D model.

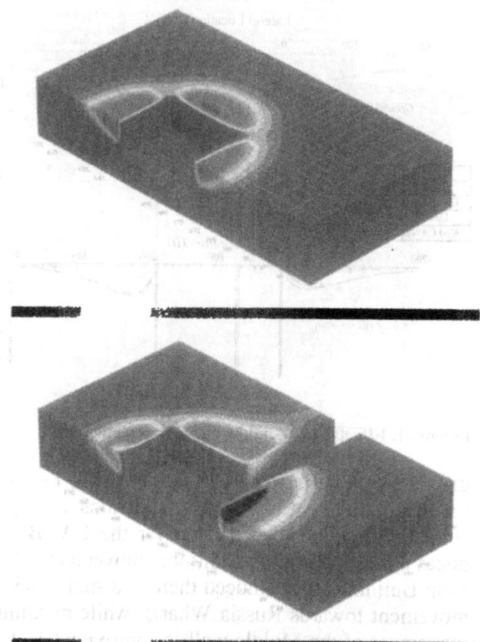

Scale [ft]

Figure 13. Predictions of surface settlements, 3-D FE analysis. a) after Vent Building excavation, b) after CANB excavation.

Predictions of relative distortion were interpreted from deformations computed in the 3-D FE analyses. Figure 14 shows discrete ranges relative distortion predicted at the top of the till layer (i.e., expected base of pile foundations for Russia Wharf buildings) after completion of the CANB excavation. The largest relative distortions (range 1/2000 – 1/5000) occur adjacent to the south wall where surface surcharge loads were assumed along an access roadway. Much smaller relative distortions occur elsewhere beneath the buildings and hence, the analyses suggest that sequential excavations for the Vent Building and CANB tunnels can be carried out within tolerances imposed for protection of these historic buildings.

6 CONCLUSIONS

It is now possible to perform reliable 2-D analyses of soil-structure interaction for braced excavations routinely using non-linear finite analyses that can account for effects of coupled deformation and flow in the soil. These calculations represent a significant advance over pre-existing beam-spring models (used in the original structural design of the bracing systems for this project), but require careful assessment of ground conditions (in situ stresses and groundwater conditions) and soil properties. This information

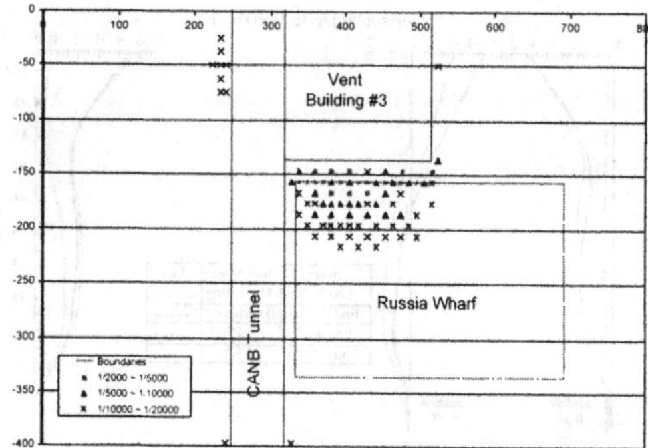

Figure 14. Evaluation of vertical distortions at top of till (after CANB excavation).

must be obtained from a well coordinated site investigation program.

Two-dimensional FE analyses for this project confirmed previous results indicating surplus capacity in the design of the structural elements (walls and bracing system). The analyses also demonstrated that a simple modeling approach using linearly elastic, perfectly plastic soil models and uncoupled flow, would be sufficient for this project (i.e., for the given soil profile and lateral earth support design) to obtain realistic predictions of wall deflections and ground deformations.

The combined effects of the two excavations occurring sequentially could not be realistically modeled by two-dimensional analyses due to the essential role of the corner bracing in the Vent Building excavation. Non-linear three-dimensional FE analyses of braced excavations represent a significant challenge, both for input data preparation and in computational efforts, compared to 2-D calculations. The analyses for this project were accomplished with a number of judicious simplifications and provided compelling evidence of the stiffening effects of the corner bracing. Sequential excavation of the CANB tunnel generated minimal additional settlements beneath the historic Russia Wharf buildings. The outcome of these calculations demonstrated that the revised construction schedule could be accomplished with minimal additional risk of damage to the existing buildings.

7 ACKNOWLEDGMENTS

A joint venture team of HNTB Corporation and Fay Spofford and Thorndike, Inc. carried out the design of the Central Artery contracts C17A1 and C17A3. The project management team, Bechtel-Parsons Brinckerhoff, commissioned this study. The Authors would especially like to thank their colleague Jim Branch (FST) for his contributions to this work, to Bruno Mattle (ILF Consulting Engineers, Innsbruck) who supervised the 3-D analyses and to Yo-Ming Hsieh (graduate student, MIT) who assisted in validating these calculations.

REFERENCES

Boscardin, M.D., and Cording, E.J. 1989. Buildings Response to excavation induced settlement. *ASCE Journal of Geotechnical Engineering.* 115(1), 1-21.

Brinkgreve R.B.J & Vermeer P.A. 1998. PLAXIS Version 7. Rotterdam: Balkema.

Duncan J.M., Byrne P., Wong, K.S., and Mabry P. 1980. Strength, Stress-Strain and Bulk Modulus Parameters for Finite Element Analysis of Stresses and Movements in Soil Masses. *University of California*, Berkeley.

GEI Consultants, Inc 1992. Final Geotechnical Engineering Report, Design Section D017A, I-93/Central Artery-Congress Street to North Street.

GI 1997. Guidelines of Engineering Practice for Braced and Tied-Back Excavations, Geo-Institute, *ASCE.*

GT STRUDL. User Reference Manual. Georgia Institute of Technology. Altanta. Georgia.

Hashash Y.M.A. & Whittle A.J. 1996. Ground Movement Prediction for Deep Excavation in Soft Clay. *ASCE Journal of Geotechnical Engineering.* 122(6), 474-486.

Hashash, Y.M.A. & Whittle, A.J. 2001. Load transfer mechanisms and arching in braced excavations in soft clay. To appear *ASCE Journal of Geotechncial & Geoenvironmental Engineering.*

HKS 1998. ABAQUS Version 5.8 User's Manual. Hibbitt, Karlsson & Sorensen, Inc., Providence, RI.

Lee, F.H., Yong, K.Y., Quan, K.C.N., Chee, K.T. 1998. Effects of corners in strutted excavations. *ASCE Journal of Geotechncial & Geoenvironmental Engineering,* 124(4), 339-349.

Ou C.Y., Chiou D.C. & Wu T.S. 1996. Three-Dimensional Finite Element Analysis of Deep Excavations. *ASCE Journal of Geotechnical Engineering,* 122(5), 337-345.

Parkhill, S.T. 1998. Geotechnical Design and Construction from 1848 to 1998. *Civil Engineering Practice*, BSCE, Fall, 7-30.

Schanz, T., Vermeer, P.A. & Bonnier, P. 2000. Formulation and verification of the hardening soil model. To appear *International Journal for Numerical and Analytical Methods in Geomechanics.*

SEI. 2000. Effective Analysis of Diaphragm Walls. Structural Engineering Institute, *ASCE.*

Whittle, A.J. 1996. Prediction of excavation performance in clays. *BSCE Journal of Civil Engineering Practice*, 12(2), 65-88.

Whittle, A.J. 1999. Role of finite element methods in geotechnical engineering *BSCE Journal of Civil Engineering Practice*, 14(2), 81-88.

Whittle A.J., Hashash Y.M.A. & Whitman R.V. 1993. Analysis of Deep Excavation in Boston. *ASCE Journal of Geotechnical Engineering.* 119(1), 69-91.

North American Tunneling 2002, Ozdemir (ed.)
© 2002 Swets & Zeitlinger, Lisse, ISBN 90 5809 376 X

Author index

9789058093769